Human Virology in Latin America

Juan Ernesto Ludert • Flor H. Pujol
Juan Arbiza
Editors

Human Virology in Latin America

From Biology to Control

Springer

Editors
Juan Ernesto Ludert
Instituto Politécnico Nacional
Center for Research and Advances Studies
Mexico City, Mexico

Juan Arbiza
Facultad de Ciencias
Universidad de la Republica
Montevideo, Uruguay

Flor H. Pujol
Centro de Microbiologia y Biologia Celular
Instituto Venezolano de Investigaciones Científicas (IVIC)
Caracas, Venezuela

ISBN 978-3-319-54566-0 ISBN 978-3-319-54567-7 (eBook)
DOI 10.1007/978-3-319-54567-7

Library of Congress Control Number: 2017937370

Printed on acid-free paper

This Springer imprint is published by Springer Nature
The registered company is Springer International Publishing AG
The registered company address is: Gewerbestrasse 11, 6330 Cham, Switzerland

Foreword

The publication of this book, which includes the main aspects of human, animal and environmental virology in the Latin American and Caribbean region (LAC) in 2016, coincides with important achievements and challenges in public health in the region.

According to the Pan American Health Organization/World Health Organization, in September 27, 2016, "The Region of the Americas is the first to have eliminated measles in the world, a viral disease that can cause severe health problems, including pneumonia, brain swelling and even death. This achievement culminates a 22-year effort involving mass vaccination against measles, mumps and rubella throughout the Americas. Measles is the fifth vaccine-preventable disease to be eliminated from the Americas, after the regional eradication of smallpox in 1971, poliomyelitis in 1994, and rubella and congenital rubella syndrome in 2015". Another important achievement and innovative action in LAC occurred from March 2006, when rotavirus A vaccines were introduced in national immunization programmes, considerably reducing hospitalizations and mortality related to rotavirus A diarrhoea.

Despite these important achievements, new challenges have arisen in the region. Considering only arboviruses, the four dengue virus serotypes circulate here, and recently the introduction and the pandemic dispersion of the Chikungunya and Zika viruses were observed. In this context, both the original and scientifically robust contributions of virologists and the regional scientific community who generated knowledge of these viruses should be highlighted. Thus, considering the achievements and challenges, it is possible to state that virologists and virology in LAC followed an innovative course and contributed in expressive ways in the generation of scientific knowledge and to the understanding of different phenomena related to the existing "virosphere" in LAC.

This book, *Human Virology in Latin America: From Biology to Control*, edited by J. E. Ludert, F. H. Pujol, J. Arbiza, contains 22 chapters that have been contributed by expert virologists on topics of great relevance to Public Health in the LAC. Starting with the history of virology in the region that occurred between the sixteenth and twenty-first centuries, Chap. 1 describes the scientific knowledge generated over time about the main viruses, and their impact on public health and prevention actions relevant to the region. This volume covers five fields of virology: viral gastrointestinal diseases, including the unprecedented environmental virology (Chaps. 2, 3, 4, 5, 6 and 7); mosquito-borne viral diseases (Chaps. 8 and 9); hemorrhagic, skin and respiratory viral diseases (Chaps. 10, 11, 12, 13, 14 and 15); sexually and blood-borne transmitted diseases (Chaps. 16, 17, 18, 19 and 20); prevention and treatment (Chaps. 21 and 22). This book undoubtedly reflects the resizing of virology in LAC. We hope that this initiative will be the catalyst for the creation of a Latin American and Caribbean Society for Virology (LACSVI), similarly to the European Society for Virology.

The LAC countries bear a rich socio-biodiversity that concentrates the greatest extension of forests dedicated to biodiversity conservation (26%). In addition, out of the ten countries with the largest freshwater reserve on the planet, three are in LAC (Brazil, Colombia and Peru). The concept of One Health aims to reduce the risks of emergence and spread of infectious diseases resulting from the interface between animals, humans and ecosystems. In this context, viruses are important agents of emerging and re-emerging diseases, since they do not occupy any specific and permanent ecological niche in a conservative way. On the contrary, due to their intrinsic capacity for evolutionary mechanisms, they present potential to parasitize alternative host species.

Considering the infectious diseases of viral aetiology presented in this text, integration between human health, animal health and environment propels new challenges for LAC, mainly in qualified human resources that can generate new knowledge of new multidisciplinary approaches and translational researches with respect to socio-biodiversity and maintenance of healthy ecosystems for a most just, responsible and sustainable economic and social development.

José Paulo Gagliardi Leite, PhD
Senior Researcher Laboratory of Comparative and Environmental Virology
Oswaldo Cruz Institute/Fiocruz/Ministry of Health, Brazil
Senior Researcher 1A – National Council for Scientific and Technological Development (CNPq)
Designated Member of Scientific Commission of Parasitology and Microbiology, CNPq (10/2016–06/2019)
Scientist of Our State, Carlos Chagas Filho Foundation for Research Support of Rio de Janeiro State (FAPERJ)
Temporary Advisor, Pan American Health Organization
Member, Rotavirus Technical Working Group, World Health Organization
President of Brazilian Society for Virology (1997–1998)

Preface

Viruses were discovered at the very end of the ninetieth century, and although many known viruses to date cause no harm to humans, some of them, like Ebola virus, influenza or the human immunodeficiency virus (HIV) for example, are among the most fearful pathogens affecting humankind. Moreover, viruses have the ability to affect large areas or populations in short periods of time causing epidemic outbreaks that in occasions may become pandemic. Out of the 11 most important epidemics affecting the world in the last 14 years, 9 have been caused by viruses.

Latin America is an extensive region populated by more than 600 million inhabitants, containing an exuberant biological richness constituting one of the most diverse ecological regions on Earth. Since the 1980s the Latin America region has been affected, like the rest of the world, by the HIV, but also by the hemorrhagic manifestation of dengue. More recently the region was affected by a large outbreak of chikungunya and currently the region is battling to control the Zika epidemic. All these emerging viral diseases add to the more "classical" endemic viruses such as papilloma, viral hepatitis and those causing respiratory and gastrointestinal infections. Moreover, the region is under the constant threat of the emergence or reemergence of highly pathogenic human viruses such as yellow fever or Mayaro, some of which are currently silent under well-established Amazonia sylvatic cycles. Another threat is the introduction of viruses from elsewhere such as the Middle East Respiratory Syndrome (MERS).

Through the 22 chapters of this book, some of the most respected virologists working in Latin America provide their views of the state-of-the-art of virology in the region. They address issues that range from history to biology, pathogenesis, epidemiology, prevention and treatment of the most important human viral diseases in the region. Almost in every case, the answer to an emerging disease in the region has been reactive, even though lessons from past epidemic experiences, in combination with the current epidemiological, medical and scientific knowledge, should allow for a more proactive, early and, therefore, more efficient reaction. It is the hope of the Latin American community of virologists to generate original and valuable scientific knowledge that will not only impact the universal knowledge, but that will also provide effective tools to alleviate the current burden caused by viral diseases in the region and to be better prepared for future contingencies.

We like to express our gratitude to all the authors who so generously and enthusiastically contributed their chapters. We hope that their work reaches and informs graduate students, scientists and public health authorities with updated, authoritative and useful information about the virology endeavor in Latin America, and no less important, that they inspire a new generation of scientists to become virologists!

Mexico City, Mexico — Juan Ernesto Ludert
Caracas, Venezuela — Flor H. Pujol
Montevideo, Uruguay — Juan Arbiza

Contents

Part I
The Early Days

Chapter 1
Viral Epidemics in Latin America from the Sixteenth to the Nineteenth Centuries and the Early Days of Virology in the Region

José Esparza

1 Introduction

The known epidemiological history of Latin America started with the encounter between American and European populations after the voyages of Christopher Columbus [24, 30]. In the sixteenth and seventeenth centuries, newly introduced viral epidemic diseases rapidly spread among the American aborigines, causing "virgin soil" epidemics and affecting populations without previous exposure to the new pathogens, causing high mortality rates and resulting in the decimation of the American aboriginal populations [9, 10, 42]. To some extent, those early American epidemics biologically resembled the more recent "virgin soil" epidemics of chikungunya and Zika in the Americas [16].

By the second half of the nineteenth century, the concepts of epidemics and contagion began to be understood more scientifically, especially after the germ theory of disease was formalized thanks to the work of Louis Pasteur (1822–1895) and Robert Koch (1843–1910). This new understanding of disease led to the discovery of many microorganisms that were found to be specifically associated with different diseases. Although a vaccine against smallpox was developed in 1796 by Edward Jenner (1749–1823) [26] and one against rabies was developed by Louis Pasteur in 1885 [3], it was not until 1892 that the concept of filterable viruses was proposed by the Russian botanist Dmitri Ivanovsky (1864–1920) for the agent of the tobacco mosaic disease [25]. This discovery led to the identification of other filterable agents of disease (now known as viruses), which also resisted cultivation in standard bacteriological media. The first observation that an animal disease is caused by a filterable

J. Esparza (✉)
Institute of Human Virology, University of Maryland School of Medicine,
Baltimore, MD, USA
e-mail: Jose.Esparza5@live.com

J.E. Ludert et al. (eds.), *Human Virology in Latin America*,
DOI 10.1007/978-3-319-54567-7_1

virus was that in Germany by Friedrich Loeffler (1852–1915) and Paul Frosch (1860–1928) in 1898 of food-and-mouth disease [28]. In 1901, Walter Reed (1851–1902) and collaborators were the first to identify, in Cuba, a human filterable virus as the cause of yellow fever [34]. It was during this exciting time of the birth of virology that the first steps in this science were taken in Latin America.

2 Viral Epidemics in Latin America (Sixteenth to Nineteenth Centuries)

2.1 Influenza

The Spanish medical historian Francisco Guerra [23] proposed that the first viral epidemic introduced by the Spanish explorers in the Americas may have been swine influenza, which broke out in Santo Domingo (Hispaniola) during the second voyage of Columbus in 1493. The epidemic might have been originated from infected pigs brought from the Canary Islands. Guerra suggested that the influenza epidemic spread to a number of islands in the Caribbean, causing large mortality among the aboriginal population.

Since 1510 and until the end of the nineteenth century, there appeared to have been ten or more pandemics of influenza [32], with evidence that most of them reached Latin American countries where they received different popular names: zamparina (Brazil, 1771), pasa diez (Colombia, 1808), susto de la pinacata (Mexico, 1826), la corcunda (Brazil, 1826), la jardinera (Peru, 1875), emisión (Peru, 1878), etc.

During the last influenza pandemic of the nineteenth century in 1889–1890, researchers used the new methods of medical microbiology in an effort to identify the microbial cause of influenza. An apparent breakthrough came in 1892 when Richard Pfeiffer in Germany, a protégé of Robert Koch, announced that he had found a bacillus as the cause of influenza, known as Pfeiffer's bacillus or *Bacillus influenzae*, which was probably *Haemophilus influenzae.* Moreover, bacterial vaccines were prepared and used at that time in an attempt to prevent influenza. Another 30 years later, the British investigators Andrewes, Smith, and Laidlaw [1] isolated, in 1933, the virus that causes influenza in humans.

2.2 Smallpox

However, the major epidemic killer in the Americas was smallpox, first introduced from Africa to Santo Domingo in 1518. From there, the smallpox epidemic spread in 1519 to Puerto Rico and Cuba, reaching Mexico in 1520 [14]. It has been repeatedly proposed that this first smallpox epidemic in the Americas contributed to the fall of the Aztecs and, a few years later, to that of the Inca Empire, where the epidemic arrived in 1525–1527.

That first American smallpox epidemic was followed by other epidemics in different countries in Central and South America, caused by multiple reintroductions of the disease from slaves of African origin. One of the worst smallpox epidemics of the sixteenth century took place between 1585 and 1591, affecting the Andean region, from Colombia to Argentina, causing great mortality among aborigines and Spaniards. The Amazonian forest may have protected Brazil from these initial epidemics that ravaged the Spanish territories; however, an epidemic introduced directly from Portugal affected all the Brazilian coast in 1562–1563. It is believed that epidemic diseases, especially smallpox, were the major cause of the demographic catastrophe of the American Indian population in the sixteenth and seventeenth centuries. By 1650, when censuses became more accurate, the American indigenous population was estimated as 6 million, down from an estimated of 50 to 100 million in 1492.

Multiple epidemics of smallpox occurred in the Americas during the seventeenth to the twentieth century. The standard societal response to those early epidemics was based on the isolation of cases to prevent further spread of the disease. However, from the 1760s, variolation (the inoculation of smallpox) began to be used in Latin America, some 40 years after the procedure was first introduced in England by Lady Mary Wortley Montagu [44]. The beginning of the end of the smallpox epidemics in Latin America came in 1804 with the introduction of vaccination (inoculation of cowpox) by the Royal Philanthropic Expedition of the Vaccine [15, 33]; the last case of smallpox in the Americas occurred in Brazil in 1971.

2.3 *Yellow Fever*

Yellow fever was one of the most feared epidemic diseases in the Americas from the seventeenth to the end of the nineteenth century [7]. The origins of the disease most likely lie in Africa, and the virus and its mosquito vector were probably brought to the Caribbean and South America by ship after 1492 [4]. Although epidemics that resemble yellow fever had been described in the Caribbean and Mexico since 1527, one of the first well-described outbreaks, mentioned in the *Chilam Balam* de Chumayel, was in 1648 in Yucatan, Mexico. During the sixteenth and seventeenth centuries, yellow fever became endemo-epidemic in the Americas, causing numerous outbreaks mostly in the Caribbean, Mexico, Venezuela, and Brazil. By the late seventeenth century, yellow fever had also reached ports in the United States, and in the nineteenth century, the epidemics also made inroads in European ports with commercial contacts with America. In 1871, a severe epidemic of yellow fever occurred in Buenos Aires, killing about 8% of the inhabitants of the city [37].

Today we know that yellow fever is caused by a virus and transmitted by mosquitoes, but that knowledge was only acquired at the very beginning of the twentieth century. In the urban cycle of yellow fever, the virus is transmitted from human to human by the *Aedes aegypti* mosquito. In the sylvatic or forest cycle, different species of *Aedes* serve as a vector between nonhuman primates and humans. The sylvatic cycle of yellow fever was described in Brazil in 1932 by the Rockefeller Foundation epidemiologist Fred Soper (1893–1977) [40].

2.4 *Other Viral Epidemics*

In addition to influenza, smallpox, and yellow fever, many other infectious diseases, bacterial and viral, affected colonial Latin America [24]. Their precise identification remains difficult because of the limited clinical information available and also because very often two or more epidemic diseases occurred at the same time.

What seems to be clear is that the third emerging viral infection in colonial Latin America was measles, which, similar to smallpox, was first introduced in the Caribbean around 1529 [9]. Measles arrived in Mexico in 1531, soon extending to all Mesoamerica and the Andes. Epidemics of a disease compatible with mumps were described in Mexico in 1550–1560, also extending to Mesoamerica. Both measles and mumps remained in Latin America as endemo-epidemic diseases. Dengue may have been introduced in the Caribbean as early as in 1635 [21]. However, cases compatible with poliomyelitis were only reported for the first time at the end of the nineteenth century, in Argentina (1860) and in Cuba (1879).

3 The Early Days of Virology in the Region

The Latin American society responded to those epidemic diseases with a number of activities that were pioneers in the field.

3.1 *The Royal Philanthropic Expedition of the Vaccine*

On May 14, 1796, Edward Jenner performed his best known experiment when he showed that the inoculation of an 8-year-old boy, James Phipps, with cowpox "material" protected him from developing disease after a smallpox challenge. This observation and others were privately published by Jenner in 1798 [26]. Information about Jenner's experiments was initially known in Spain in 1799, and the vaccine itself (cowpox material) reached Spain in December 1800; by the end of 1801, several thousand vaccinations had been performed in that country. At that time, dozens of booklets about vaccination were published in Spain and widely circulated in Spanish America. Moreover, several attempts were made to bring the vaccine to the New World using cowpox material in impregnated silk threads or sealed between small glass plates, but these methods proved unreliable on lengthy journeys and in warm climates.

In 1802, the Viceroy of New Granada (now Colombia) asked His Majesty the King of Spain for help after informing him of an epidemic of smallpox that had caused thousands of deaths. Consequently, on March 1803, King Charles IV instructed the Council of Indies to evaluate the means to introduce the vaccine to his American and Asian possessions. The process resulted in the selection of Francisco

Fig. 1.1 Francisco Xavier de Balmis, the director of the expedition that in 1804 brought the smallpox vaccine from Spain to Latin America

Xavier de Balmis (1753–1819) (Fig. 1.1) as director of an expedition that would take the vaccine to all corners of the Spanish colonies and around the world [12, 29, 33, 41]. The expedition was carefully planned and executed with three objectives: (1) to provide cost-free vaccinations to the general population, (2) to train local physicians in correct vaccine administration, and (3) to establish central and regional vaccination boards to ensure the preservation and distribution of the vaccine. The expedition left the Spanish port of La Coruna on November 30, 1803, aboard the corvette *María Pita*. It was directed by Balmis with the assistance of Joseph Salvany as Vice Director, two physicians, two surgeons, and four male nurses. To transport the vaccine on the 1-month voyage to the Americas, 22 nonimmune orphan boys, aged 3–9, would be sequentially vaccinated during the crossing by serial arm-to-arm inoculation. The expedition also included the rectoress of the La Coruna foundling house, Isabel Zendal Gómez, who took care of the children. The expedition lasted for 3 years, and new children were recruited along the way to maintain the arm-to-arm transfer of the vaccine.

The corvette *María Pita* made a first stop in the Canary Islands, and from there the expedition continued westbound to reach the island of Puerto Rico in February 1804. In March, the expedition sailed to Caracas, Venezuela, where it was received with open arms and enthusiastic collaboration. In less than 1 month, more than 12,000 people were vaccinated in Venezuela, where the first Central Vaccination Board was established on which other Spanish American boards were modeled. In Caracas, the expedition was divided into two groups. One group, led by Salvany, proceeded south to the Viceroyalty of Nueva Granada (Colombia) and Peru, with the final goal of reaching Rio de la Plata (Argentina). The other sub-expedition, directed by Balmis himself, sailed from Caracas to La Havana, Cuba, and from there to the Viceroyalty of New Spain (Mexico), where it arrived in June 1804. From there, the vaccine was taken to different places in Mexico and neighboring regions. In February 1805, Balmis sailed from Acapulco to the Philippines accompanied by

26 Mexican children to serve as vaccine carriers. For health reasons, Balmis sailed to the Asian mainland in September 1805, reestablishing vaccination in the Portuguese colony of Macau. Balmis returned to Europe by sailing around the African continent, finally landing in Cadiz on September 7, 1806.

An estimated 250,000 people were directly vaccinated by the expedition, with many more reached by the programs they established in different countries. When Jenner learned of the expedition, he said that he does not *imagine the annals of history furnish an example of philanthropy so noble, so extensive as this.*

3.2 *Discovering the Etiology and Mechanism of Transmission of Yellow Fever*

The full emergence of the germ theory of disease during the 1880s gave impetus to the search for the "germ of yellow fever," and many putative microorganisms were proposed at that time [11]. In Brazil, Domingos José Freire (1843–1899) reported that the pathogenic agent of yellow fever was *Cryptococcus xantogenicus* [6], whereas in the same country, João Baptista de Lacerda (1846–1915) proposed that the cause was the fungus *Cogumelo*. In Mexico, Manuel Carmona y Valle (1827–1902) believed that the cause was the mold *Peronospora lutea*. And even Carlos Finlay (1833–1915) (Fig. 1.2) in Havana had a candidate that he called *Micrococcus tetragenus febris flavae*.

In view of those multiple claims, in 1887, the U.S. government commissioned George Miller Sternberg (1838–1915) to clarify the situation. Sternberg was an army physician who is considered to be the first U.S. bacteriologist. After traveling to two endemic countries, Brazil and Cuba, in 1890, he emitted the verdict stating that none of the proposed microorganisms was the cause of yellow fever. However, Sternberg himself suggested that perhaps a new germ that he identified, known as *Bacillus x*, could be the cause. After these very definite conclusions by Sternberg, it came rather as a surprise when in 1897 Dr. Giuseppe Sanarelli (1864–1940)

Fig. 1.2 Carlos Finlay, the Cuban physician who in 1881 formally proposed the mosquito transmission of yellow fever, information that was used in 1900–1901 by Walter Reed and collaborators to advance our knowledge regarding the etiology and transmission of the disease

Fig. 1.3 Giuseppe Sanarelli, a distinguished Italian bacteriologist working in Uruguay, conducted pioneering work on the etiology of yellow fever and in 1898 discovered the viral etiology of rabbit myxomatosis

(Fig. 1.3), a distinguished Italian bacteriologist and professor of experimental hygiene in Montevideo, Uruguay, formerly from the Institut Pasteur in Paris where he was a disciple of Élie Metchnikoff (1845–1916), announced that he has discovered the true cause of yellow fever, *Bacillus icteroides* [38]. This time some American experts believed Sanarelli to be right, and even Sternberg considered the possibility that *Bacillus icteroides* and *Bacillus x* were one and the same.

Consequently, in 1900, Sternberg, who by then was the U.S. Army Surgeon General, sent to Cuba a new Yellow Fever Commission (the fourth one) with the main goal of confirming or refuting the claim that yellow fever was caused by *Bacillus icteroides*. It is important to remember that at that time Cuba was occupied by the United States as a consequence of the conflict known as the Spanish-American War that was initiated in 1898. The commission was led by Major Walter Reed; acting Assistant Surgeon General, James Carroll (1854–1907); Aristides Agramonte (1868–1931); and Jesse Lazear (1866–1900). After conducting the necessary research, the conclusion of the Commission was that *Bacillus icteroides* was simply a contaminant. Walter Reed concluded that *At this stage of our investigation… the time had arrived when the plan of our work should be radically changed.* So, the Commission went through a major paradigm change and decided that rather than studying the causes of the disease, they would focus on what transmits it. At this point, in August 1900, the Commission decided to consult with Carlos Finlay and to test his mosquito theory in human volunteers [17].

Carlos Finlay, a Cuban physician trained in Europe and the United States, had sent in 1865 a paper to the Academy of Sciences of Havana outlining his theory on weather conditions and its relationship to yellow fever. However, after a more careful study of the epidemics, he finally made the correct observation that the appearance of epidemics in the hot and wet summer months was not caused by a worsening of some of the miasmatic conditions but rather by an increase in the population of a mosquito that Finlay referred to as the *Culex* mosquito, known at that time as *Stegomyia fasciata* and today as *Aedes aegypti*.

On August 14, 1881, Finlay read a paper before the Academy of Sciences of Havana, proposing that yellow fever was propagated by mosquitoes. He began by admitting that the cause of the disease was a *material transportable cause which may be either an amorphous virus, an animal or vegetable germ, bacterium, etc., but which consists in all cases of a tangible something which has to be communicated from the sick to the healthy in order that the disease may be propagated.* By the 1880s, the germ theory of disease was gaining followers within the medical community, and it was clear that Finlay accepted that concept rather than the popular miasmatic theory ("something in the air"). Finlay continued his presentation to the Academy, proposing that *it seems natural* that the agent that transmits the pathogenic material of yellow fever *could be found in that class of insects which, by penetrating in the interior of the blood vessel, could suck up the blood together with any infecting particles contained therein, and carry the same from the diseased to the healthy.* Through acute observation, Finlay correctly identified *Aedes aegypti* as the vector of the yellow fever virus [18, 19].

His presentation was received with total indifference, and for many years, Finlay was ridiculed because of his ideas. Nevertheless, that same year of 1881, Finlay commenced experimental inoculations on a series of 20 nonimmune Spanish soldiers, completing in 1900 a total of 102 cases. It has been argued that Finlay could have not convincingly transmitted the infection because his volunteers were not adequately isolated and the observed infections could have occurred by natural exposure to the infection rather than by the experimental mosquito transmission. Moreover, it has been argued that the experiments were not optimized for the "extrinsic" and "intrinsic" incubation periods of the virus. Transmission of yellow fever from human to human requires a competent mosquito to feed on an infected human and survive an extrinsic incubation period in which the virus replicates in the mosquito and disseminates to its salivary glands, whereupon it finally feeds on a susceptible human. Similarly, disease in humans occurs after an intrinsic incubation period in which the virus replicates and disseminates within the person. The extrinsic incubation period in *Aedes aegypti* has a median of 10 days at 25°C, and the intrinsic incubation period had a median of 4.3 days. Those concepts were unknown to Finlay, who did not consider the mosquitoes as biological vectors of the yellow fever virus, but rather as only "flying pins" that mechanically carried the yellow fever agent from human to human.

In 1900, Carlos Finlay had an opportunity to have his mosquito theory tested by the American Commission, and he personally provided the eggs of the mosquitoes to be used in the experiments [8, 19]. This time, the Commission was careful in isolating the volunteers to ensure that any infection observed in the volunteers was indeed experimentally transmitted by the mosquitoes. After some initial failures, the Commission achieved positive results when the experiments were designed to allow for the appropriate extrinsic and intrinsic incubation periods. Significantly, the mosquitoes had fed on cases within the initial 3 days of the disease and had been allowed to ripen for at least 12 days before the transmission experiments were conducted. One of the volunteers was a member of the Commission, Jesse Lazear, who developed a severe infection and died of the disease.

With these initial results, Walter Reed rapidly prepared a preliminary note that he presented on October 23, 1900, at the annual meeting of the American Public Health Association in Indianapolis, Indiana (USA) [35]. Reed returned to Cuba where he conducted additional experiments in "Camp Lazear," west of Havana near the adjacent suburban towns of Quemados and Marianao. Most of the volunteers were recent Spanish immigrants who were not immune to yellow fever (were not "acclimatized") and who signed what are probably the first examples of informed consent. More than 30 volunteers participated in the experiments, and 22 developed yellow fever.

On August 1, 1901, less than 1 year after the transmission experiments in Camp Lazear, James Carroll, from the Walter Reed Commission, who returned to Cuba and was aware of the foot-and-mouth experiments of Loeffler and Frosch, proceeded to demonstrate that filtered serum from one yellow fever patient can induce yellow fever when injected in a healthy volunteer. The subsequent filtration and passage of serum from the second subject to a third, with the same results, provided the evidence of a replicating infection agent rather than a toxin, and this was the demonstration that the yellow fever agent was a filterable virus [34].

Soon after the yellow fever work of Walter Reed and collaborators was publicized, there was a search in the literature for visionary individuals who have previously suggested a possible role of insects in the transmission of disease. Several such "predecessors" were identified, but among those one remains apart, Louis Daniel Beauperthuy (1808–1871) [5, 13, 20, 27] (Fig. 1.4). Beauperthuy was born in the island of Guadalupe in the French Antilles and studied in Paris, graduating as physician and surgeon in 1837 with a thesis entitled "De la climatologie" in which he analyzed the environment and its relationship to diseases. In 1839, he moved to Venezuela where he worked for a brief period as a naturalist for the Musée d'Histoire Naturelle de Paris. During a yellow fever epidemic that occurred in 1853 in Cumaná, Venezuela, he made the connection between mosquitoes and the spread of the disease, pointing out that among the many species of mosquitoes he studied, *the agents of this infection come in many varieties that are not all harmful to the same extent. The silly mosquito*

Fig. 1.4 Louis Daniel Beauperthuy, based on observations made in Venezuela, suggested in 1853 that yellow fever is transmitted by mosquitoes

variety, with white streaks on its legs, a somehow domestic variety, is the most common. Entomologists accept that Beauperthuy correctly identified the *Aedes aegypti* vector. In 1853, he published his observations locally in Cumaná, Venezuela, but also in the *Comptes Rendus des Séances* and in the *L'Abeille Médicalle* of Paris.

A major difference between the 1853 theory of Beauperthuy and the 1881 suggestion of Finlay was that Beauperthuy believed that the mosquitoes mechanically carried the yellow fever agent from unhealthy swamps to humans, whereas Finlay correctly proposed the human-to-human transmission of the yellow fever virus by mosquitoes, although it took several years for virologists to fully understand the biological role of mosquitoes in the epidemiology of arboviruses.

3.3 Early Identification of Rabbit Myxomatosis as a Viral Disease by Sanarelli in Uruguay

Myxomatosis was first recognized when it killed European rabbits *(Oryctolagus cuniculus)* in Giuseppe Sanarelli's laboratory in Montevideo, Uruguay, in 1896. After failing to detect bacteria in the lymph from the vesicles of diseased rabbits, Sanarelli described the myxomatosis agent as invisible. In 1898, Sanarelli reported his findings at the Ninth International Congress for Hygiene and Demography in Madrid, classifying the myxomatosis agent as a virus on the basis of its submicroscopic size, even though its filterability was not reported for some years [39]. However, Sanarelli noted in 1898 that centrifugation produced an infectious serum that did not contain microorganisms. Sanarelli's difficulties in demonstrating a filterable virus were probably the result of using a fine filter and the relatively large size of the myxomatosis virus (a poxvirus). It is noteworthy that the initial identification of the myxomatosis was made the same year that the German scientists Loeffler and Frosch reported the agent of foot-and-mouth disease as the first filterable animal virus [25, 43].

In 1911, workers in the Oswaldo Cruz Institute in Rio de Janeiro, Brazil, correctly further classified the agent of myxomatosis as a large virus, and Henrique de Beaurepaire Rohan Aragão (1879–1956) showed that it could be transmitted mechanically by insect bite.

3.4 Rabies Vaccination Arrives in Latin America as a Sign of Modernity

In 1880, Louis Pasteur began to work on rabies, a disease that plagued Europe in the nineteenth century. Starting in 1884, he presented the successful results of preventive rabies vaccines in dogs, establishing the principle of vaccination before exposure to rabies in animals. Pasteur then sought to improve his method and developed a means of attenuating the virulence of the rabies microorganism, which consisted in exposing the spinal cords of rabies-infected rabbits to the air in specially designed flasks. Pasteur then had the idea of using this vaccine to create immunity after a dog bite and to give it to humans [3].

Pasteur took the next step in 1885 with the vaccination of a 9-year-old child, Joseph Meister, who was presented to Pasteur in his laboratory at the Ecole Normale in rue d'Ulm in Paris. The young boy arrived from Alsace presenting multiple deep dog bites, received 13 injections of rabbit medulla homogenate, 1 per day, and survived. Three months later, Pasteur repeated the experiment on a young shepherd, Jean-Baptiste Jupille, severely bitten by a dog. On October 26, 1885, Pasteur showed the promising results of his treatment against rabies in humans to the French Academy of Sciences. From then on, patients bitten by rabid animals flocked to Pasteur's laboratory. On March 1, 1886, Pasteur presented a paper to the French Academy of Sciences with the results from the inoculation of 350 people, concluding that *rabies prophylaxis after a bite is justified. There is a cause to create a rabies vaccine establishment.* He immediately launched an international fund, and as a result, in November 1888, the Institut Pasteur was created, dedicated not only to rabies treatment but to Pasteur's study of science.

News about Pasteur's vaccination spread rapidly, and people from all over the world began to arrive in Paris to receive the rabies vaccine. Pasteur also opened several vaccination centers in Russia. In 1887, Dr. Valentine Mott (1852–1918) opened a center to administer the rabies vaccine in New York. That year, Dr. Mott went to Paris as the representative of the "American Pasteur Institute," and when he returned to the United States, Pasteur permitted him to bring back a rabies-inoculated rabbit. In 1891, Albert Calmette (1863–1933) was sent to Saigon to administer the rabies vaccine, leading to the creation of the first overseas Pasteur Institute [22].

It appears, however, that the first vaccination against rabies in Latin America was done in Argentina on September of 1886, only 1 year after the vaccination of Joseph Meister in Paris. The vaccination was done by Dr. Desiderio Davel (1857–1943) using a strain provided by Pasteur *which was maintained by repeated passages in rabbits during the steamship travel from Paris to Buenos Aires.* Some claim that the Pasteur Laboratory of Buenos Aires, now the Institute of Zoonosis, was the first institution outside Europe that conducted rabies vaccination in humans [2].

Rabies vaccination was introduced in Mexico in 1888 by the eminent physician Eduardo Liceaga (1839–1922), the most distinguished hygienist of late nineteenth-century Mexico, who brought the vaccine in the brain of an inoculated rabbit maintained in glycerin during the travel from Paris [36].

The accounts presented here are interesting examples of the early transfer and adoption of vaccine technologies in Latin America.

4 Conclusions

The Latin American colonial society was exposed to numerous emerging viral infections that were imported from Europe or Africa during the process of conquest and colonization, resulting in significant cultural and demographic impact. Society today is confronting some of the same diseases with vaccines that were mostly developed during the twentieth century. Smallpox was declared eradicated in 1980 thanks to a vaccine first developed in 1796 by Edward Jenner. Poliomyelitis may be eradicated by 2018 using vaccines developed by Jonas Salk (1914–1995) in 1955 and Albert Sabin (1906–1993) in 1961. Although a highly effective yellow fever vaccine was developed in 1937 by Max Theiler (1899–1972) [31], we are still experiencing severe epidemics of yellow fever, especially in Africa. Similarly, rabies causes tens of thousands of deaths every year, mostly in Asia and Africa, and 15 million people every year receive post-bite vaccination with new versions of the vaccine first developed by Louis Pasteur in 1885, thus preventing hundreds of thousands of deaths.

The growth of the global population from 2 to 7.4 billion during the past century and the increased mobility of the population will surely allow new pathogens to emerge and spread locally, regionally, and globally. Vaccines will continue to have a role in the control of current and future epidemics and pandemics.

We reviewed how, at the beginning of the nineteenth century, it took only 7 years from the discovery of smallpox vaccination to the launching of the Balmis expedition, the first-ever global health campaign. We also reviewed how, soon after the germ theory of disease was formulated in Europe, scientists working in Latin America reported their experiences in identifying new viruses, specifically those causing yellow fever and rabbit myxomatosis. The rapid transfer of rabies vaccination from Paris to Argentina and Mexico was interpreted as a sign of progress and modernity and of the commitment of the new nations in Latin America to become an integral part of the civilized world.

The twentieth century continued producing examples of excellent virological work in Latin America. The younger generations of Latin American virologists, those who are working in the twenty-first century, should find encouragement in the successes of the past to address the challenges of the future.

References

1. Andrewes CH, Laidlaw PP, Smith W (1934) The susceptibility of mice to the viruses of human and swine influenza. Lancet ii:859–862
2. Anonymous. Desiderio Davel y la vacunación antirrábica. Todo es Historia. http://www.todoeshistoria.com.ar/nota_detalle.php?nota=24. Accessed on 11 June 2016
3. Bazin H (2011) Vaccination: a history. John Libbey Eurotext, Montrouge
4. Bryant JE, Holmes EC, ADT B (2007) Out of Africa: a molecular perspective of yellow fever virus into the Americas. PLoS Pathog 3:e75
5. Beauperthuy de Benedetti R (1971) Beauperthuy. Gráfica Americana, Caracas
6. Benchimol JL (1995) Domingos José Freire e os primordios da bacteriologia no Brasil. Hist Cienc Saude Manguinhos 2:67–98
7. Carter HR (1931) Yellow fever: an epidemiology and historical study of its place of origin. Williams & Wilkins, Baltimore
8. Chavez-Carballo E (2005) Carlos Finlay and yellow fever: triumph over adversity. Mil Med 170:881–885

9. Cook ND (1988) Born to die: disease and New World conquest, 1492–1650. Cambridge University Press, Cambridge
10. Cook ND, Lowell WG (1992) Secret judgment of God: Old World disease in colonial Spanish America. University of Oklahoma Press, Norman and London
11. Delaporte F (1991) The history of yellow fever. An essay on the birth of tropical medicine. MIT Press, Cambridge and London
12. Diaz de Yraola G (2003) La vuelta al mundo de la expedición de la vacuna. Consejo Superior de Investigaciones Científicas, Madrid
13. Esparza J (1973) Yellow fever and stamps. ASM News 39:800
14. Esparza J (2000) La introducción de la viruela en América: epidemias en América Latina durante el siglo XVI. Tierra Firme (Caracas) 72:527–561
15. Esparza J, Yépez Colmenares G (2004) Viruela en la Venezuela Colonial: epidemias, variolización y vacunación. In: Ramírez S, Valenciano L, Nájera R, Enjuanes L (eds) La Real Expedición Filantrópica de la Vacuna. Consejo Superior de Investigaciones Científicas, Madrid, pp 89–118
16. Fauci AS, Morens DM (2016) Zika virus in the Americas—yet another arbovirus threat. N Engl J Med 374:601–603
17. Ferreira Moreno VG (2016) Evocation to the Dr. Carlos J. Finlay Barres on the centennial of his death. Colomb Med 47:63–66
18. Finlay C (1881) El mosquito hipotéticamente considerado como agente transmisor de la fiebre amarilla. An Real Acad Cien Med Fis Nat Habana 1881:147–169
19. Finlay CE (1942) Carlos Finlay y la fiebre amarilla. Minerva, La Habana
20. Godoy GA, Tarradath E (2010) Short biography of Louis Daniel Beauperthuy (1807–71); pioneer of microbiology and medical science in Venezuela. J Med Biogr 18:38–40
21. Gratz NG, Knudsen AB (1996) The rise and spread of dengue, dengue hemorrhagic fever and its vectors. A historical review (up to 1995). World Health Organization, Geneva
22. Guénel A (1999) The creation of the first overseas Pasteur Institute, or the beginning of Albert Calmette's Pastorian career. Med Hist 43:1–25
23. Guerra F (1988) The earliest American epidemic: the influenza of 1493. Soc Sci Hist 12:305–325
24. Guerra F (1999) Epidemiología Americana y Filipina, 1492–1898. Ministerio de Sanidad y Consumo, Madrid
25. Hughes SS (1977) The virus: a history of the concept. Heineman, London
26. Jenner E (1978) An inquiry into the causes and effects of the variolae vaccinae. Sampson Low, London
27. Lemoine W, Suárez MM (1984) Beauperthuy, De Cumaná a la Academia de Ciencias de París. Fundación para la Ciencia, Caracas
28. Loeffler F (1898) Berichte der Kommission zur Erforschung der Maul- und klauenseuche be idem Institut für Infektionskrankheiten in Berlin. Zentbl Bakt Parasitkd 23:371–391
29. Mark C, Rigau-Perez JG (2009) The world's first immunization campaign: the Spanish smallpox vaccine expedition, 1803–1813. Bull Hist Med 83:63–94
30. Moll AA (1944) Aesculapius in Latin America. Saunders, Philadelphia and London
31. Norrby E (2007) Yellow fever and Max Theiler: the only Nobel Prize for a virus vaccine. J Exp Med 204:2779–2784
32. Patterson KD (1986) Pandemic influenza 1700–1900: a study in historical epidemiology. Rowman & Littlefield, Totowa
33. Ramirez S, Valenciano L, Nájera R, Enjuanes L (2004) La Real Expedición Filantrópica de la Vacuna: Doscientos años de lucha contra la viruela. Consejo Superior de Investigaciones Científicas, Madrid
34. Reed W, Carroll J, Agramonte A (1983) Landmark article. Feb 16, 1901: the etiology of yellow fever. An additional note. By Walter Reed, Jas. Carroll and Aristides Agramonte. JAMA 250:649–658

35. Reed W, Carroll J, Agramonte A, Lazear JW (1983) Classics in infectious diseases. The etiology of yellow fever: a preliminary note. Walter Reed, James Carroll, A. Agramonte, and Jesse W. Lazear. Surgeons, U.S. Army. The Philadelphia Medical Journal 1900. Rev Infect Dis 5:1103–1111
36. Rodriguez de Romo AC (1996) La "ciencia pasteuriana" a través de la vacuna antirrábica: el caso mexicano. Dynamis 16:291–316
37. Ruiz Moreno L (1949) La peste histórica de 1871: Fiebre amarilla en Buenos Aires y Corrientes. Nueva Impresora, Parana
38. Sanarelli G (1897) A lecture on yellow fever. With a description of the *Bacillus icteroides*. Delivered before the University of Montevideo on June 10th, 1987. Br Med J 2:7–11
39. Sanarelli G (1898) Das myxomatogene virus. Beitrag zum studium der krankheitserreger ausserhalb des sichtbaren. Zentbl Bakt Parasitkd 23:865–873
40. Soper FL (1936) Jungle yellow fever; new epidemiological entity in South America. Rev Hyg Saude Publ Rio de Janeiro 10:107–144
41. Tuells J, Duro-Torrijos JL (2015) El viaje de la vacuna contra la viruela: una expedición, tres continentes y miles de niños. Gac Med Mex 151:416–425
42. Verano JW, Ubelaker D (1999) Disease and demography in the Americas. Smithsonian Institution Press, Washington and London
43. Waterson AP, Wilkinson L (1978) An introduction to the history of virology. Cambridge University Press, Cambridge
44. Weiss R, Esparza J (2015) The prevention and eradication of smallpox: a commentary on Sloane (1755) "An account on inoculation". Philos Trans R Soc Lond B Biol Sci 370(1666):20140378. doi:10.1098/rstb.2014.0378)

Part II
Viral Gastrointestinal Diseases

Chapter 2
Rotavirus Biology

Susana López and Carlos F. Arias

1 Introduction

Acute, infectious diarrhea is one of the most common causes of morbidity and mortality among children living in developing countries. In 2010, 1.7 billion cases of diarrheal events were estimated to have occurred worldwide in children under 5 years of age [123]. Diarrheal diseases are the third cause of death in this age range, after perinatal problems and respiratory infections [1]; however, it is complicated to calculate the number of deaths associated with a particular enteric pathogen [39, 115]. Updated global estimates of rotavirus mortality in children less than 5 years of age indicate a decline from 296,000 deaths in 2008 to 215,000 in 2013, with a slight decrease in the proportion of diarrheal deaths caused by rotavirus, from 39% to 37% in this same period [115]. Developing countries bear the major burden of mortality from rotavirus, with about 85% of these cases occurring in six countries in Africa and Asia and very few in industrialized nations [39, 115].

Rotaviruses continue to be the leading etiological agent of severe diarrheal disease, even though two live attenuated vaccines have been licensed in more than 100 countries since 2006,[115]. These live oral vaccines have shown a lower efficacy in countries with a high burden of diarrheal disease [17, 39, 114], and the majority of those currently using rotavirus vaccines are low-mortality countries, so the impact of vaccine use on global estimates of rotavirus mortality has been limited [115]. Furthermore, the recent Global Enteric Multicenter Study showed that rotavirus was the leading cause of infant diarrhea among more than 20,000 children studied in seven sites across Asia and Africa [60]; this study also reported that each episode of severe diarrhea in children increased the risk of delayed physical

S. López (✉) • C.F. Arias
Instituto de Biotecnología, Universidad Nacional Autónoma de México,
Av. Universidad 2001, Colonia Chamilpa, Cuernavaca, Morelos, Mexico
e-mail: susana@ibt.unam.mx; arias@ibt.unam.mx

J.E. Ludert et al. (eds.), *Human Virology in Latin America*,
DOI 10.1007/978-3-319-54567-7_2

and intellectual development as well as increased mortality by 8.5 fold [60, 114]. Thus, the development of improved vaccines and therapeutic strategies is needed to efficiently control rotavirus infection, and in fact new rotavirus vaccines are under investigation or have recently been licensed in various parts of the world [121]. Fundamental to these developments is a basic understanding of the molecular mechanisms by which rotaviruses interact with their host cell.

Although rotavirus can infect older children and adults, severe diarrheal disease is primarily observed in children less than 2 years of age [37]. Rotavirus infection is primarily restricted to mature enterocytes located at the tip of intestinal villi. However, additional extraintestinal spread of rotavirus during infection of animals indicates a wider host tissue range than previously appreciated [88, 99]. In vitro, rotaviruses bind to a wide variety of cell lines, although only a subset of these, including cells of renal or intestinal origin and transformed cell lines derived from breast, stomach, bone, and lung, are productively infected [23]. The initial stages of rotavirus interactions with the host cell are complex and are the focus of intense current research. Most of these studies have been performed using model cell culture lines, the monkey kidney epithelial cell line MA104 and the human colon carcinoma cell line Caco-2, both of which are highly permissive to rotavirus infection and are the most commonly employed.

The mature rotavirus infectious particles are formed by a triple-layered protein capsid that encloses the genome, composed of 11 segments of double-stranded RNA (dsRNA). The innermost layer, formed by 120 dimers of VP2, contains the viral genome and 12 copies each of VP1, the virus RNA-dependent RNA polymerase (RdRP), and VP3, a protein with guanylyltransferase, methylase, and phosphodiesterase enzymatic activities; these viral elements constitute the core of the virus. The addition of 260 trimers of VP6 on top of the VP2 layer produces double-layered particles (DLPs). The outermost layer is made by 780 copies of the glycoprotein VP7 arranged in trimers, which form a smooth surface layer from which 60 spikes composed of trimers of VP4 protrude to form the characteristic, infectious, triple-layered particles (TLPs) [37].

During or shortly after cell entry, the infecting TLP loses the external protein layer and is converted to a DLP. Once in the cytoplasm, the DLP, which is transcriptionally active, begins the synthesis of viral mRNAs that direct the synthesis of six structural proteins (VP1 to VP4, VP6, VP7) and six nonstructural proteins (NSP1 to NSP6). In addition to their function as mRNAs, the viral transcripts also serve as RNA templates for the synthesis of negative-strand RNAs to form the dsRNA genomic segments. The newly synthesized viral proteins are recruited to viroplasms, electrodense cytoplasmic structures, where the viral genome replicates and double-layered replication intermediate (RI) particles assemble. The DLPs newly formed in the viroplasms mature by budding into the lumen of the endoplasmic reticulum (ER) through the ER membrane, which is modified by the viral glycoproteins VP7 and NSP4. During this process, mediated by the interaction of VP6 with NSP4, the DLPs acquire a transient lipid envelope that is subsequently lost to yield mature infectious TLPs. Finally, in MA104 cells, the virus is released into the medium by cell lysis, whereas in Caco-2 cells, the virus exits through a non-lytic mechanism that is not well characterized [37] (Fig. 2.1).

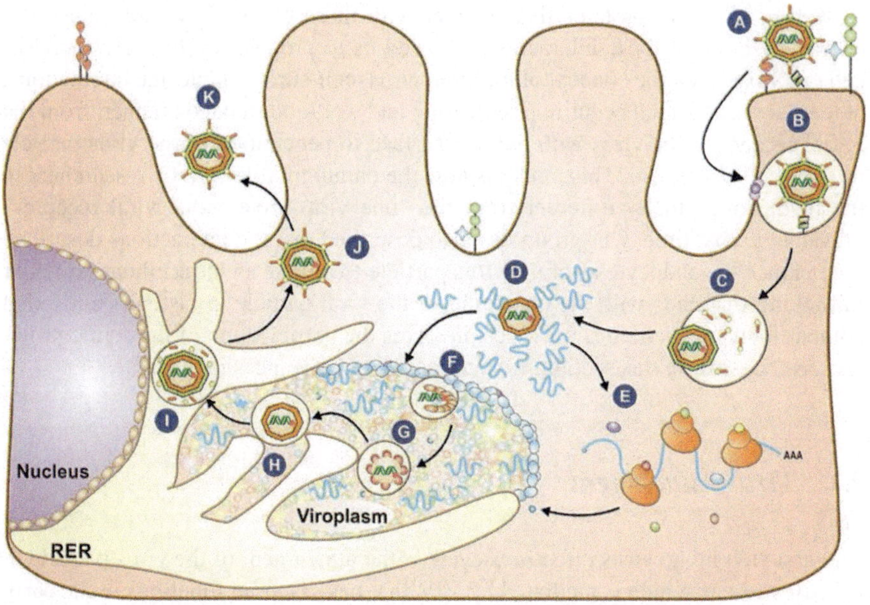

Fig. 2.1 Rotavirus replicative cycle. The virus replication cycle starts with the binding of the virus to the cell surface **A** and its internalization by endocytosis **B**. Inside the cell, the outer protein layer is shed **C**, and the double-layered particle becomes transcriptionally active **D**, giving rise to 11 RNA transcripts that encode 12 viral proteins. **E** Once a critical mass of viral protein is accumulated, the mRNA transcripts also serve as templates for the synthesis of the genomic double-stranded RNA (dsRNA), which occurs in replication intermediate particles within electrodense structures called viroplasms that are composed of viral proteins, viral RNA, and some cellular proteins **F–H**. Newly synthesized single- and double-layered particles assemble concurrently with genome replication, and **I** the double-layered particles then bud through a NSP4- and VP7-modified endoplasmic reticulum membrane into the lumen of the Rough Endoplasmic Reticulum (RER), where the final maturation of the virus particle takes place. **J**, **K**, Triple-layered, infectious particles exit the cell either by lysis or through a non-lytic process, depending on the cell line

Considering the purpose of this book, we describe in this chapter aspects of rotavirus biology where significant contributions by researchers working in Latin America have been made, with emphasis in our own work. This manuscript does not pretend to be a comprehensive review of the area, and we apologize to the colleagues we do not cite because of length restrictions.

2 Rotavirus Cell Entry

Among our principal contributions to the field of rotavirus is the characterization of the early events of virus–cell infection. Our research group has described the existence of at least four distinct interactions between the virus and host cell-surface molecules that mediate the attachment of the virus particle to the cell membrane and its subsequent entry into the cell. We have identified cell receptors and co-receptors, as well as the viral proteins that interact with these cell-surface molecules. The characterization of these interactions allowed us to propose a model for rotavirus cell entry based on the concept of multiple virus–cell-surface molecule interactions, at least some of which occur in a sequential and well-coordinated manner, from the initial contact of the virus with the cell surface to penetration of the virus particle into the cell cytoplasm. This model is now the paradigm followed by researchers in the field, conceptually different from the "one viral protein–one viral receptor" prevalent at that time. Our group has also described that the interactions described here induce the endocytosis of the virus particle to initiate an intracellular vesicular trafficking that ends with the uncoating of the viral particle in distinct endosomal compartments, which, in some cases, involves the participation of the acidic proteases, cathepsins. In this section, we summarize our advances in this area.

2.1 Virus Attachment

The first step in the virus infectious cycle is the attachment of the virus particle to the cell surface, which is mediated by VP4 that has essential functions in the early interactions of the virus with the cell, including receptor binding and cell penetration [27, 64, 65, 79, 80, 84, 133]. The properties of this protein are therefore important determinants of host range, virulence, and induction of protective immunity. To be infectious, the virus depends on the specific trypsin cleavage of VP4, of 776 amino acids, to yield polypeptides VP8 (aa 1–231) and VP5 (aa 248–776), both of which remain associated to the virion [11, 25, 35, 36, 64]. The cleavage of VP4 does not affect cell binding, but rather it seems to be required for virus entry. The VP8 domain of VP4 mediates the attachment of the virus to the cell, whereas VP5 and the surface glycoprotein VP7 interact with downstream post-attachment molecules [72].

Rotavirus strains were initially classified as neuraminidase (NA) sensitive or NA resistant, depending on their ability to infect cells that had been previously treated with NA. Most human rotaviruses are NA resistant, whereas animal rotaviruses can be either NA sensitive or NA resistant [24, 53, 72]. Rotaviruses whose infectivity is decreased by NA treatment bind to the cell surface through terminal sialic acids (SAs), which are susceptible to NA cleavage. On the other hand, some NA-resistant viruses bind to internal SAs, which are not cleaved by NA [49], while yet others bind to human blood group antigens (HBGAs) [51, 52].

In the case of NA-sensitive rotavirus strains, gangliosides have been associated with rotavirus cell attachment for some time [14, 31, 49, 55]. However, knocking down the expression of two key enzymes involved in ganglioside synthesis decreased ganglioside levels as well as the infectivity of both NA-resistant and NA-sensitive rotavirus strains, but did not affect their binding to the cells, suggesting that gangliosides are not essential for cell-surface binding but rather they are needed during a later step of the entry process, regardless of the NA sensitivity of the virus [81].

2.2 Post-attachment Interactions

After the initial attachment to glycans on the cell surface, rotaviruses interact with additional surface molecules to gain access into the cell. Among these molecules are some integrins (α2β1, αXβ2, αVβ3) and the heat shock cognate protein 70 (hsc70) [71, 72, 83, 85, 132, 134, 135]. Whether all these molecules are used by all rotavirus strains and whether the interactions of the virus with them are sequential or alternative is not known; however, in the particular case of the rhesus rotavirus strain (RRV), we showed that some of these interactions occur sequentially [71, 72, 83, 85, 132, 134, 135]. Interestingly, not all rotavirus strains interact with integrins, although all the strains tested require hsc70 for efficient cell infection [42, 45, 46].

The interaction of rotavirus with integrin α2β1 is mediated by a DGE motif located toward the amino-terminal end of the VP5 domain of VP4 and the domain I of the integrin subunit α2 [42, 134]. On the other hand, integrin αVβ3 interacts with rotavirus through a linear sequence in VP7 [135]. The interaction between the viral particle and hsc70 is mediated by VP5 (amino acids 642 and 659) and the peptide-binding domain of hsc70, and it has been suggested that the ATPase domain of hsc70 could be involved in promoting conformational changes in the viral particle to facilitate virus entry or uncoating [96, 134]. Furthermore, it has been shown that gangliosides, as well as integrins α2β1, αVβ3, and hsc70, are associated with detergent-resistant membrane microdomains, where infectious viral particles are also present during cell entry [54], and we showed that the integrity of these microdomains is fundamental for viral infection [44, 46].

Integrins have a polarized distribution in epithelial cells, localizing primarily at the basolateral face of the plasma membrane. Therefore, rotaviruses reaching the intestinal epithelium would find the integrin receptors hidden beneath the tight junctions (TJs). How might then rotavirus, with putative basolateral ligands, infect polarized epithelia? A possible explanation was offered when it was shown that a recombinant VP8 protein was able to decrease the trans-epithelial electrical resistance of polarized Madin–Darby canine kidney (MDCK) cells [89]. The ability of VP8 to generate a leaky TJ could allow integrins to diffuse to the apical surface, so that the virus could bind and infect from the apical side. The ability of virus particles to disrupt TJs during their early interaction with polarized epithelia, however, remains to be shown. Furthermore, we have shown that rotavirus infects polarized cells more efficiently through the basolateral face in comparison to the apical surface [22, 100]. In addition, we recently reported that the TJ protein JAM-A is important for the entry of some rotavirus strains at a post-attachment step, and we also found that occludin and ZO-1 are relevant for virus entry [116, 117].

It is of note that the assays used to block the interaction of rotaviruses with each of these proposed receptors and co-receptors using different approaches, such as proteases, antibodies, peptides, sugar analogues, or siRNAs, only decrease viral infectivity by less than tenfold, suggesting that either a more relevant entry factor for rotavirus has yet to be found, the virus can use more than one route of entry, or the cellular factors that allow the entry of rotavirus are redundant.

2.3 Virus Internalization

The cell entry of rotavirus by endocytosis is supported by several experimental approaches, including pharmacological inhibitors, overexpression of dominant-negative mutant proteins, and knocking down the expression of proteins implicated in different endocytic routes. In addition, actinin 4 and the activation of the small GTPase RhoA and Cdc42, as well as its activator CDGAP, which are involved in different types of endocytic processes, have been implicated in the entry of rotavirus [32, 46, 126]. Of interest, all tested rotavirus strains, with the exception of the RRV strain, enter cells through clathrin-mediated endocytosis [32, 46], whereas RRV uses an atypical endocytic pathway that is clathrin- and caveolin independent but depends on dynamin 2 and on the presence of cholesterol [107, 111]. The requirement for cholesterol and dynamin is also shared by those rotaviruses that are internalized by clathrin-dependent endocytosis [46], although contradictory results were recently reported in MDCK cells [126].

It is interesting to note that the interactions of the virus with the putative receptor and co-receptor molecules characterized so far do not seem to determine the endocytic pathway used, because both NA-resistant and NA-sensitive strains, as well as rotaviruses that interact with HBGAs, can enter cells using a clathrin-dependent mechanism [32]. In addition, using reassortant viruses, our group recently reported that the outer layer protein VP4 determines the endocytic pathway used, and a single amino acid substitution in the VP8 domain of RRV can change its entry pathway from a clathrin-independent to a clathrin-dependent mechanism [32]. We also showed that the infectivity of rotavirus is enhanced by calcium and that internalization of the virus induces an early permeabilization of cells [28, 92].

2.4 Intracellular Vesicular Traffic and the ESCRT Machinery

After internalization, rotavirus travels along the intracellular vesicular traffic moving from the cell periphery to the perinuclear space. During this traffic, the virus is transported by endocytic primary vesicles to early endosomes (EEs), then to maturing endosomes (MEs) that contain intraluminal vesicles (ILVs), and finally to late endosomes (LEs) [33, 111] (Fig. 2.2). The formation of the characteristic ILVs present in the ME is generated by the endosomal sorting complex required for transport (ESCRT) machinery [128]. Independent of the nature of the cell-surface receptor and the endocytic pathway used for cell internalization [32, 33, 46, 107, 111], all rotavirus strains tested converge in EEs during entry [32, 33, 46, 127] and depend on a functional ESCRT machinery, as knocking down the expression of components of the ESCRT complex by RNAi reduces virus infectivity [13, 111]. Why the entry of rotaviruses depends on the ESCRT machinery and what is the role of ILVs in this process has not been elucidated (discussed in [111]).

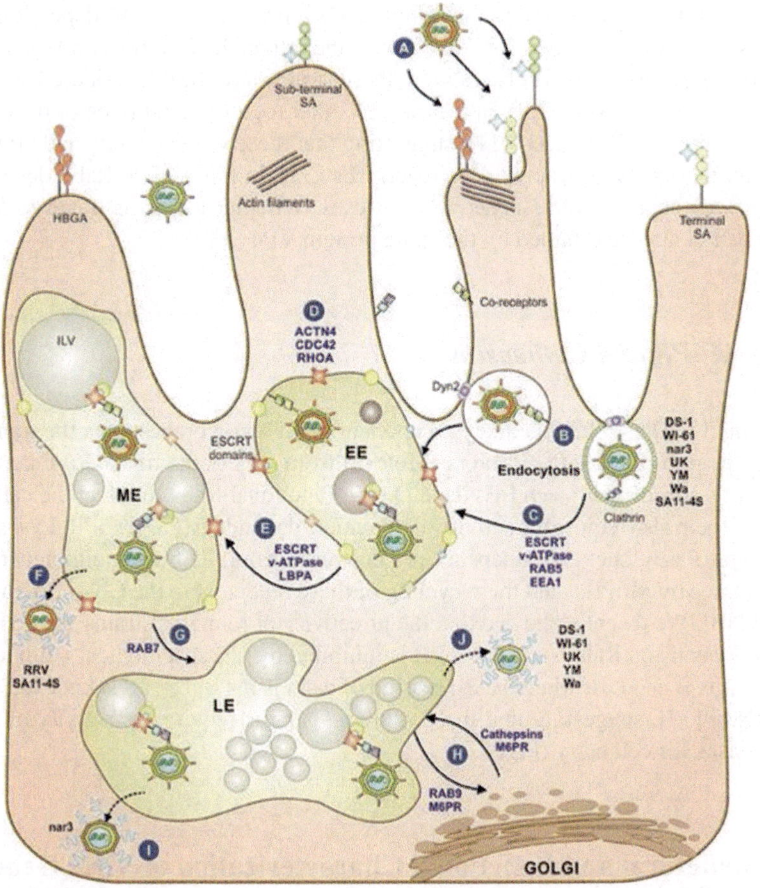

Fig. 2.2 Working model for rotavirus (RV) cell entry pathway in MA104 cells. **A** RVs attach to the cell surface through different glycans, depending on the virus strain. After initial binding, the virus interacts with several co-receptors concentrated at lipid rafts. **B** RVs are internalized into cells by clathrin-dependent or clathrin-independent endocytic pathways, depending on the virus strain. **C** Regardless of the endocytic pathway used, all RV strains reach early endosomes (EEs) in a process that depends on RAB5 and EEA1 and probably on HRS and the v-ATPase. **D** At the EE, the virus probably begins to be internalized into the endosomal lumen through the action of VPS4A. **E** EEs progress to mature endosomes (MEs), with a progressive decrease in pH and intraendosomal calcium concentration through the function of the v-ATPase; during this process, the formation of intraluminal vesicles (ILVs) increases. **F** E-P rotaviruses RRV and SA11-4S reach the cytoplasm from MEs. **G** GTPase Rab7 participates in the formation of late endosome (LE) compartments; ILVs increase in number. **H** The stability and function of LEs depend on the arrival of cellular factors (e.g., cathepsins) from the trans-Golgi network, traffic that is mediated by mannose-6-phosphate receptors (M6PRs) and the GTPase Rab9, among other factors. **I** L-P RV strains reach late endosomes; RV nar3 exit from LEs requires the function of Rab9. **J** RV strains UK, Wa, WI61, DS-1, and YM require, in addition to Rab9, the function of the CD-M6PR and the activity of cathepsins to productively infect cells. **F, I, J** The cytosolic double-layered particles begin transcribing the RV genome to continue the replication cycle of the virus

How far rotaviruses go into the different vesicular compartments depends on the virus strain. For rotaviruses RRV and SA11, the intracellular traffic comes to an end at MEs, and, based on this observation, these strains have been considered as early-penetrating viruses (Fig. 2.2). In contrast, all other rotavirus strains tested depend on the expression of Rab7 [33, 111], suggesting that these viruses continue their travel through the endosomal network to reach LEs [33]. In this regard, Rab7-dependent rotaviruses behave as late-penetrating viruses. Whether the virus travel to EE or reach LE is also determined by the spike protein VP4 [33].

2.5 *M6PR and Cathepsins*

The small GTPase Rab9 is a key component of LEs and orchestrates the transport of mannose-6-phosphate receptors (M6PRs) from LEs to the trans-Golgi network. Rotavirus strains that reach LEs depend on a functional Rab9 to infect the cell, and most of them also require the activity of the cation-dependent (CD) M6PR. Lysosomal acid hydrolases, such as cathepsins, are delivered from the trans-Golgi network to endosomes by M6PRs, and the recycling of these receptors to the Golgi depends on Rab9 [18]. We recently showed that the infectivity of rotavirus strains whose infectivity depends on Rab9 and CDM6PR is inhibited by pharmacological inhibitors of cathepsins B and L or when the expression of cathepsins B, L, or S is knocked down by RNAi [33], suggesting that these rotavirus strains require the activity of these hydrolases for cell entry (Fig. 2.2).

3 Structural and Functional Characterization of Viral Genes and Cellular Proteins Required for Rotavirus Genome Replication and Virus Morphogenesis

Our group was involved from the dawn of rotavirus research in the characterization of the proteins coded by each of the 11 segments of the viral genome [5, 6] and in determining the primary structure of the genes and their encoded protein products [2, 7, 64–69, 104]. This involvement allowed us to identify structural domains and predict potential antigenic and functional regions of the viral polypeptides [2, 7–10, 29, 38, 63–65, 67, 68, 70, 90]; however, the characterization of the role of the different proteins in the rotavirus life cycle was more difficult as it was limited by the technological tools available in the late 1990s. At the beginning of the past decade, a breakthrough for the analysis of gene function of mammalian cells occurred with the adaptation of the RNA interference (RNAi) system to efficiently and specifically knock down the expression of cellular genes [34]. In 2002, we reported that it was possible to inhibit the expression of rotaviral genes using this system: this represented one of the first reports in virology and the first in the rotavirus field that demonstrated the feasibility of inhibiting the expression of animal virus genes by RNAi [12, 30]. Using this technology, we knocked down the expression of all rotavirus genes and characterized the function of the encoded proteins during different stages of the life cycle of the virus, including the replication of the viral genome [16, 76], the formation of viroplasms [75, 76], the assembly of double-layered RI particles [16, 75, 76], and the morphogenesis of mature, infectious viruses [30, 75, 82]. We also characterized the role of viral proteins in the control of the unfolded protein response [119], the mechanism of inhibition of cellular protein synthesis [87, 102, 103, 108], and the control of formation of stress granules [87]. Others have used this technology to prove the role of NSP4 in altering the Ca^{2+} homeostasis in rotavirus-infected cells [130]. Some of these contributions are briefly described following and in the next section.

Regarding the replication of the viral genome, it has been proposed that the synthesis of the negative strand of each genome segment occurs in viroplasms, concurrently with packaging of the positive-stranded RNAs (equivalent to the mRNAs) into core RI particles [93]. The analysis of the kinetics of transcription and replication of the viral genome throughout the replication cycle of the virus allowed us to provide evidence for the existence of a second round of transcription originated from newly assembled, transcriptionally active, double-layered RI particles, resulting in a second wave of assembly of DLPs [16]. In agreement with earlier studies in rotavirus genome transcription and replication by Eugenio Spencer and colleagues [94], this analysis also showed that all the proteins that form the DLPs (VP1, VP2, VP3, VP6) are essential for replication of the dsRNA genome, because in their absence there was little synthesis of viral mRNA and dsRNA [16]. In a parallel study, we also showed that the efficient replication of the viral genome depends on the ubiquitin-proteasome system (see following).

Once DLPs assemble in viroplasms, they mature by budding into the adjacent ER membrane, which is modified by the viral glycoproteins VP7 and NSP4. During this process, mediated by the interaction of DLPs with NSP4, the particles acquire a transient membrane envelope that contains VP4, NSP4, and VP7, which is later removed to yield the mature TLPs [37]. The mechanism of removal of the transient lipid envelope is largely unknown, although we demonstrated that VP4 is not involved and VP7 is important for this step [30], suggesting that rather than the membrane-piercing activity of VP4, as had been previously suggested, the assembly of the VP7 trimers into DLPs is responsible to exclude the lipid membrane from the viral particles. We also showed that the correct assembly of mature, infectious rotavirus particles is influenced by the two folding systems involved in the ER quality control. Grp78, protein disulfide isomerase (PDI), calnexin, and calreticulin were found to promote the timely trimming of the carbohydrate chains of VP7 and NSP4, the correct formation of VP7 disulfide bonds, and the incorporation of properly folded VP7 into TLPs to yield infectious virus, indicating that these chaperones are involved in the quality control of rotavirus morphogenesis [82]. On the other hand, Grp94 and Erp57 do not seem to be required for rotavirus morphogenesis [82]. PDI has also been suggested to be involved in rotavirus cell entry [20].

To better understand the replication of the viral RNA and virus morphogenesis, we have also analyzed the composition, dynamics, and gene function requirements for viroplasm formation, underscoring the highly organized nature and complex regulation of this structure [21, 40, 41, 75–77, 118]. We are currently studying the highly organized nature of viroplasms by super-resolution confocal microscopy.

The approach of knocking down the expression of one-by-one cellular proteins to search for those relevant for virus replication was not practical, but the scenario changed when the RNAi system became amenable for genome-wide screening of cellular functions. Using this system, we identified more than 500 cellular genes involved in rotavirus replication [78, 111]. These genes clustered functionally into several biological processes potentially involved in various steps of the rotavirus life cycle. Among these functional clusters were endocytic processes, the tight junction protein network, and the ubiquitin-proteasome protein degradation system.

The endocytic process of the virus and the role of tight junction proteins in virus infection were described in the previous section. With regard to the proteasome-ubiquitin components, in silico proteomics showed a strong cluster of positive hits in our data set that included E3 ligases regulated by deubiquitinase PAN2, heat shock proteins, and components of the 26S proteasome subunits [111]. These findings were supported by our demonstration, and that of a different group, of the requirement of the proteasome-ubiquitin pathway for rotavirus replication [26, 77]. We showed that both the proteolytic and ubiquitination activities of the ubiquitin-proteasome system were needed for the correct incorporation into viroplasms of the viral polymerase VP1 and the capsid proteins VP2 and VP6, as well as for the efficient replication of the viral genome [77], suggesting that this system has a very complex interaction with the rotavirus life cycle.

4 Rotavirus Strategies to Control the Antiviral Response of the Host Cell

Stress and innate immune cell responses are closely linked and overlap at many levels. The outcomes of these responses serve to reprogram host expression patterns to prevent viral invasions. In turn, viruses fight back against these responses to ensure their replication through various mechanisms, depending on the virus. Interestingly, the first step to control the antiviral response of the cell, and a solution seen in several virus families, is to take over the translation machinery of the host, such that the translation of viral proteins is ensured while the expression of the stress and antiviral responses of the cell is blocked. In addition, immediately upon infection, the cellular RNA decay pathways and the innate immune responses are triggered. To guarantee their successful replication, viruses have evolved different tools to subvert these pathways. Our group has been interested in characterizing the interactions between rotavirus and its host cell to understand the mechanisms by which this virus is able to establish a productive infection based on controlling the antiviral response of the cell [73, 74] (Fig. 2.3).

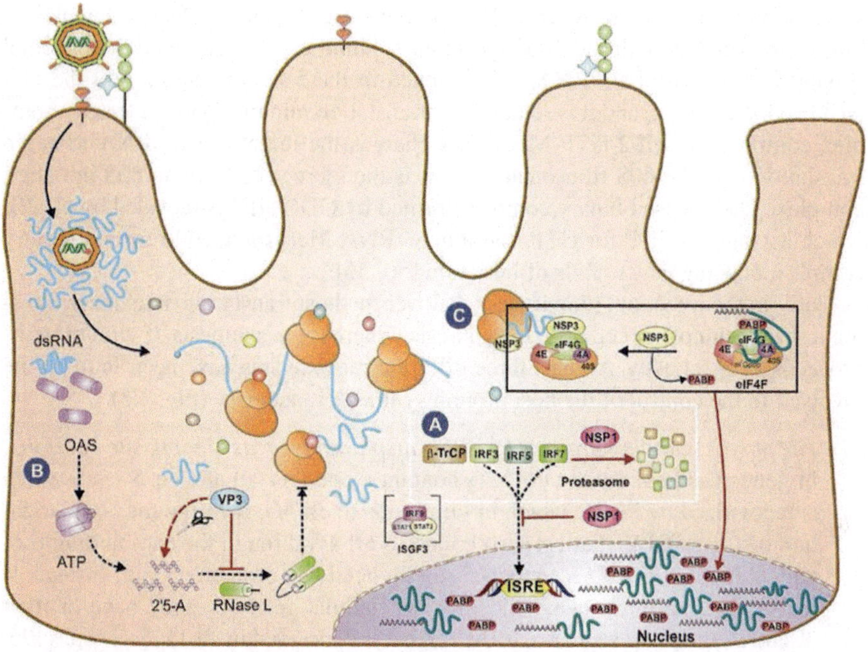

Fig. 2.3 Rotavirus measures to counteract the host response. **A** During the infection, NSP1 interacts with and causes the proteasome-dependent degradation of IRF3, IRF5, IRF7, and beta-TrCP, and also prevents the activation and translocation of ISGF3 in a proteasome-independent manner. **B** In the presence of dsRNA, 2′-5′-oligoadenylate synthetase (*OAS*) oligomerizes and synthesizes 2′-5′-oligoadenylates (*2-5A*), which in turn interact with RNase L, causing its dimerization and activation. The phosphodiesterase activity of VP3 degrades the 2-5As, preventing RNase L activation. **C** NSP3 interacts specifically with eIF4G, displacing poly(A)-binding protein (PABP) and preventing the translation of poly(A)-containing mRNAs. Also in the infection, PABP and poly(A)-containing mRNAs accumulate in the cell nucleus. *2-5A* 2′-5′-Oligoadenylate, *IRF* interferon regulatory factor, *ISRE* interferon-stimulated response element, *OAS* 2′-5′-oligoadenylate synthetase, *dsRNA* double-stranded RNA

4.1 Protein Synthesis in Rotavirus-Infected Cells

As obligate intracellular parasites, viruses depend on the cell translation machinery for the production of their proteins. Although every step of the translation process is amenable to regulation, in general, mRNA translation is regulated mainly at the level of initiation [113], a process mediated by the eukaryotic initiation factors (eIFs). The main checkpoints for the control of polypeptide chain initiation are the formation of the eIF4F complex and the activity of eIF2, both of which are targets of control by viruses (reviewed in [124]). In eukaryotic cells, mRNA translation initiation begins with the recruitment of mRNAs by the eIF4F complex and the subsequent assembly of the 40S and 60S ribosomal subunits. The eIF4F complex is formed by several canonical eIFs; the cap-binding protein eIF4E recognizes the cap structure present at the 5′-end of mRNAs; eIF4A is an ATP-dependent RNA helicase that unfolds secondary structures of mRNAs and eIF4G, which functions as a scaffolding protein where several eIFs bind; and eIF4F complex favors the interaction of the mRNA with the 40S ribosomal subunit [56]. Once the 40S ribosomal subunit is bound to the mRNA, it is scanned in the 5′–3′ direction, until the first AUG codon is found, and it is selected for translation initiation [50]. A ternary complex composed of eIF2-GTP-Met-tRNA charges the initiator Met-tRNA to begin translation, and the 60S ribosomal subunit is then joined to form an 80S initiation complex. The released binary complex formed by GDP-eIF2 is recycled by eIF2B, which exchanges GDP for GTP, and a new tRNA-Met is loaded to form a ternary complex, ensuing new rounds of initiation [50, 56].

Early in the infection, rotaviruses take over the host translation machinery, causing a severe shutoff of cell protein synthesis, whereas the synthesis of viral proteins proceeds very robustly. At least three different mechanisms have been found to be involved in the control of the host protein synthesis machinery (Fig. 2.3).

(i) *The poly(A)-binding protein (PABP) is displaced from its binding site in eIF4G.* In general, all eukaryotic mRNAs contain a poly(A) tail at their 5′-end, which is recognized by PABP, which in turn binds to eIF4G, favoring the circularization of the mRNAs that are also bound to eIF4G through the cap-binding protein. Rotavirus mRNAs contain 5′-methylated cap structures, and, instead of the poly(A) tails characteristic of most cellular mRNAs, they have at their 3′-end a consensus sequence (GACC) that is conserved in all 11 viral genes [95, 98]. The nonstructural protein NSP3 binds through its amino-terminal domain to this consensus sequence, and it also binds through its carboxy-terminal domain to eIF4G, at the same site where PABP binds. Thus, it was proposed that during infection, NSP3 evicts PABP from eIF4GI, impairing the translation of cellular mRNAs while leading to an enhanced translation of rotaviral mRNAs [97, 98]. However, despite the essential role proposed for NSP3 in infected cells, we found that silencing the expression of this protein by RNAi indeed blocks the translation of cellular mRNAs but the viral mRNAs were still efficiently translated. We also found that the knockdown of NSP3 results in an increased production of viral progeny [86]. These findings were questioned with the argument that even small undetected amounts of NSP3 could be able to initiate the synthesis of viral proteins, at a time in the infection where there is little viral mRNA to compete with the cellular mRNAs [43]. Differences in the viral strains used and on the cell lines or experimental paradigms used may also account for these discrepancies.

(ii) *PABP is accumulated in the nucleus of the cell, and there is a block in the nucleocytoplasmic transport of polyadenylated cellular mRNAs.* Interestingly, we and others have recently found that NSP3 has an additional mechanism to prevent the translation of cellular mRNAs: PABP is a protein that assists the transport of mRNAs from the nucleus to the cytoplasm, where they are available to the translation machinery; during rotavirus infection, PABP becomes accumulated in the nucleus of infected cells [15, 47, 87, 103], and it was shown that the eIF4G-binding domain of NSP3 is important for the nuclear localization of PABP [47, 87], although the precise mechanism through which this occurs has not yet been determined [103].

Furthermore, we found that the accumulation of PABP in the nucleus of rotavirus-infected cells also resulted in the accumulation and hyper-polyadenylation of poly(A)-containing mRNAs [97], suggesting that the shutoff of cell protein synthesis during the infection might be caused by a blocking of the nucleocytoplasmic transport of polyadenylated mRNAs [103].

(iii) *Phosphorylation of the translation initiation factor eIF2.* In rotavirus-infected cells, the inhibition of cell protein synthesis is also regulated by a third mechanism because the alpha-subunit of eIF2 becomes phosphorylated early in the infection and it is maintained in this state throughout the virus replication cycle [87]. When eIF2α is phosphorylated, the eIF2-GDP complex binds with higher affinity to eIF2B, preventing the exchange of GDP to GTP catalyzed by eIF2B, which reduces the formation of pre-initiation translation complexes and causes a severe reduction in global translation [56]. The phosphorylated status of eIF2α is beneficial for the virus, because under these conditions the viral mRNAs are efficiently translated but the synthesis of most cellular proteins is prevented. We and others have found that the dsRNA-dependent protein kinase, PKR, is the enzyme responsible for the phosphorylation of this translation initiation factor in rotavirus-infected MA104 [102] and intestinal epithelial cells [122].

The precise mechanism involved in viral protein synthesis has not been identified. However, we have found that during the infection, the amount of viral transcripts produced is in the range of tens of thousands of molecules per cell [103]. The huge number of viral mRNAs in a cell where the translation of poly(A)-containing mRNAs is inhibited by at least three different mechanisms [eIF2α, poly(A)-containing mRNAs sequestered in the nucleus, and eviction of PABP from eIF4G] leaves the translation of viral mRNAs with little competition for the protein synthesis machinery and explains the severe shutoff host translation caused by rotaviruses.

4.2 Stress Response of the Cell

Two of the most common stress responses of the cell are the formation of stress granules [3] and an integrated stress response known as the unfolded protein response [129]. These responses have been characterized in rotavirus-infected cells (Fig. 2.3).

(i) *Stress granules* (SGs) are cytoplasmic aggregates of stalled translational pre-initiation complexes that accumulate during stress [59]. In addition to its direct effect on protein synthesis, the phosphorylation of eIF2α is one of the signals that induces the formation of SGs. It has been proposed that SGs are sites in which the integrity and composition of mRNAs are triaged and then mRNAs are sent either to translation, degradation, or storage (reviewed in [91]). Because the main function of SGs is to arrest protein synthesis until the stressful conditions are resolved, viruses have to interact with these structures to ensure the translation of their mRNAs, and several different viral strategies have been developed to cope with their deleterious effect (reviewed in [125]).

Interestingly, we have found that even though eIF2α is phosphorylated in rotavirus-infected cells, SGs are not formed [87]. Furthermore, we found that rotaviruses prevent the formation of SGs, because these structures are not formed in cells infected with rotavirus when induced to form SGs by treatment with arsenite, a well-characterized SG inducer. The mechanism by which the formation of these structures is prevented during the infection has not been determined.

(ii) *Unfolded protein response (UPR)*. The accumulation of misfolded proteins in the ER causes stress and leads to activation of a coordinated adaptive program called UPR (reviewed in [48, 57, 129]). The function of the UPR is to handle unfolded proteins by upregulating the expression of chaperone proteins and degradation factors to refold or eliminate misfolded proteins and to reduce the incoming protein traffic into the ER by attenuation of translation [106] (Fig. 2.4). Failure to alleviate ER stress leads to activation of apoptotic pathways and cell death [58]. Rotavirus infection induces the UPR; however, this response is modulated by the virus [119, 131]. At least two of the three arms of the UPR appear to be activated in rotavirus-infected cells; the mRNA of Xbp1 was spliced by IRE1, and the transcription of GRP78 and CHOP is induced, indicating that the ATF6 pathway was activated. The UPR, however, is suppressed at the translational level by NSP3 [119]. The consequences of suppressing this response during rotavirus infection have not been addressed.

4.3 Antiviral Response of the Cell

Double-Stranded RNA Double-stranded RNA (dsRNA) is considered a key mediator of interferon (IFN) induction in response to virus infection. When the cell sensors detect dsRNA, a cascade of events is activated that promote the shutoff of cell protein synthesis, the induction of transcription of genes encoding IFN and other cytokines, and finally cell death [101]. Several findings indicate that rotaviral dsRNA is exposed to cell sensors at some point during virus replication: (a) the kinase that phosphorylates eIF2α in rotavirus-infected cells is PKR, which is activated by dsRNA [25]; (b) RIG-I and MDA5 are active and mediate the IFN response in rotavirus-infected cells [19, 110]; and (c) viral dsRNA can be detected in the cytoplasm (outside viroplasms) of rotavirus-infected cells [102]. These observations suggest that during rotavirus infection, either naked viral dsRNA, or highly structured viral mRNA [62], or both are present in the cytoplasm where they are detected by RIG-I and MDA5 with the consequent activation of the IFN response and the PKR activity that leads to the phosphorylation of eIF2α and the modification of the cellular translation machinery. Another pathway that is activated by dsRNA is the 2′–5′-oligoadenylate synthetase (OAS)/RNase L pathway. OAS is activated by

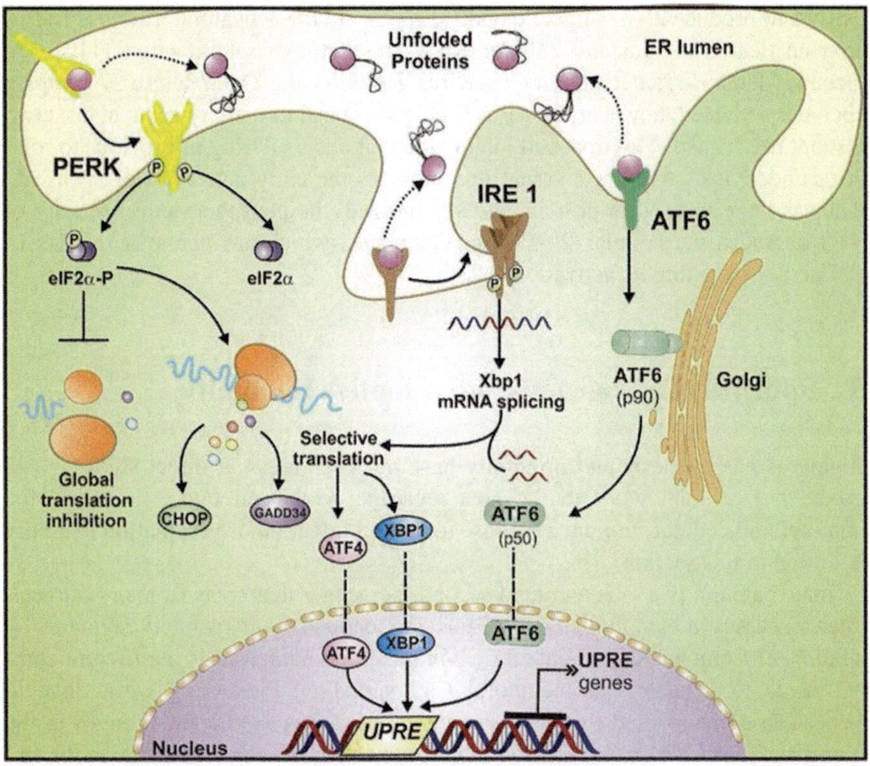

Fig. 2.4 The unfolded protein response. Accumulation of misfolded proteins in the endoplasmic reticulum (ER) causes stress and leads to activation of a coordinated adaptive program called the unfolded protein response (*UPR*). Three ER-resident transmembrane proteins are activated in response to ER stress: the PKR-like ER kinase (*PERK*), the activating transcription factor 6 (*ATF6*), and the inositol-requiring enzyme 1 (*IRE1*). Under normal conditions, the ER chaperone GRP78 is bound to the luminal domain of each sensor. When misfolded proteins accumulate in the ER, GRP78 binds these proteins and releases the sensors. Upon release, PERK and IRE1 homodimerize, causing autophosphorylation and activation, and released ATF6 relocalizes to the Golgi apparatus where it is cleaved and activated. Once activated, PERK phosphorylates the alpha-subunit of IF2 at Ser51. Phosphorylated eIF2a inhibits global translation and stimulates the translation of ATF4, which in turn transcriptionally activates UPR-responsive genes encoding proteins that ameliorate the ER stress. The gene encoding the CCAT/enhancer-binding protein (*CHOP*) is a target of ATF4, and this protein can function as a proapoptotic or prosurvival transcription factor, depending on the strength or duration of the stress. Both transcription factors, ATF4 and CHOP, can induce the transcription of the *GADD34* gene, encoding a protein that interacts with protein phosphatase 1 (PP1) to dephosphorylate eIF2a, resulting in a negative feedback loop that recovers protein synthesis and allows the translation of stress-induced transcripts. When ATF6 is cleaved in the Golgi apparatus, one of its cleavage products becomes an active transcription factor that promotes the transcription of chaperone genes. Finally, upon dimerization of IRE1, it autophosphorylates and mediates the removal of an intron from X-box-binding protein 1 (*XBP1*) mRNA. The spliced form of XBP1 encodes a transcription factor that activates the transcription of genes encoding chaperones and proteins involved in the ER stress-associated protein degradation (ERAD) system. (Figure reproduced from *Current Opinion in Virology* 2012;2:1–10, with permission)

dsRNA to produce 2′–5′-oligoadenylates, which are the activators of RNase L; this enzyme degrades viral and cellular RNAs restricting viral infection [112]. We recently demonstrated that after rotavirus infection the OAS/RNase L complex becomes activated; however, the virus is able to control its activity using at least two distinct mechanisms: a virus–cell interaction that occurs during or previous to rotavirus endocytosis triggers a signal that prevents the early activation of RNase L, whereas later, once viral proteins are synthesized, the phosphoesterase activity of VP3 degrades the cellular 2′–5′-oligoadenylates, which are potent activators of RNase L, preventing its activation [108].

5 Rotavirus Pathogenesis and Adaptive Immunity

Rotavirus pathogenesis and immunity have not been areas of direct study by our group, but relevant work in the area regarding virus–cell interactions and the humoral and cellular immune response to natural infection or vaccination is briefly reviewed in this section.

Ionic calcium (Ca^{2+}) is a crucial second messenger that controls many intracellular processes in mammalian cells. Thus, intracellular [Ca^{2+}] is finely regulated by a number of proteins that maintain Ca^{2+} intracellular homeostasis in different compartments to regulate spatiotemporal Ca^{2+} signaling. Pioneering work done in Venezuela demonstrated that rotavirus infection causes significant changes in the homeostasis of Ca^{2+} of the infected cell. These changes bring alterations in the cell cytoskeleton that may be related to pathogenesis but also help to create favorable intracellular conditions for virus maturation [105]. In addition, work from Venezuela also helped to firmly establish NSP4 as a key function in the Ca^{2+} alterations observed in infected cells.

Understanding the adaptive immune response to rotavirus infection is necessary if efficient preventive measurements are to be developed. Work carried out in Mexico and Colombia has helped greatly in understanding rotavirus immunity. A pioneering work by the Mexican Institute of Nutrition, where a cohort of more than 200 rotavirus-infected children was followed from birth to 2 years of age, answered several of the key questions necessary to launch the development of an effective rotavirus vaccine: a primordial finding of that study was that a rotavirus infection, either symptomatic or asymptomatic, would protect against subsequent infections [120]. Also, work from Mexico has helped in the identification of T-cell epitopes on the main rotavirus structural protein VP6 and its use as a potential recombinant vaccine for veterinary use [61]. Finally, work developed in Colombia has helped in the understanding of the B- and T-cell response to rotavirus infection and in the identification of correlates of protection for rotavirus vaccines [4].

6 Future Challenges

The rhythm of research on fundamental aspects of rotavirus biology has slowed down in the past decade, probably because of the successful incorporation of two rotavirus vaccines in national immunization programs for children around the world. However, it is important to keep in mind that, so far, the impact of vaccine use on global estimates of rotavirus mortality has been limited [115], and in some regions of the world rotavirus infections still place an enormous burden on societies from both health and economic perspectives. There is need for a renovated effort to better understand the life cycle of rotavirus and to improve our knowledge about its epidemiology, evolution, ecology, and pathogenesis, as well as the immune response it elicits, all of which should facilitate the development of improved vaccines and therapeutic approaches.

Our knowledge about virus biology has advanced greatly during the past years; however, most stages of virus replication are incompletely understood, such as rotavirus entry and vesicular traffic, translation of the viral polypeptides, replication of the virus genome, morphogenesis of the newly assembled viral particles, and the egress of the mature, infectious virus from cells. Most of what we know has been learned from studies using nonpolarized MA104 cells or differentiated cultures of intestinal cell lines, such as Caco-2. However, to better understand virus–cell interactions in detail, it is important to incorporate methodological advances that make possible the analysis of the host cell response at a single-cell level instead of characterizing the response of pooled and usually heterogeneous cell cultures.

Furthermore, it is of utmost importance to study the virus replication cycle and the virus–host interactions in the cells that the virus targets in a natural infection. Animal models have been very useful to characterize virus restriction factors that participate in defining host range, and virus pathogenesis, as well as the innate and acquired immune responses induced by rotavirus infection. However, these models represent a complicated system to characterize the different steps of virus replication. In this regard, the recent development of enteroids from human intestinal origin, which have been reported to mimic the complex cellular lineages and tissue architecture of the gut and to efficiently support the replication of rotavirus [109], represents an appealing alternative for these studies. This system, together with the possibility of using the CRISPR/Cas9 technology in these cells and the possibility of characterizing the interactions of the virus with the cell-surface cellular receptors/co-receptors by live cell imaging systems, and the use of novel super-resolution microscopy techniques, are important tools for advances in this field.

Acknowledgments The work in our laboratory relevant to this chapter was supported by grant # 221019 from CONACYT, Mexico. and grants # IG200114, and #IG200317 from DGAPA-UNAM, Mexico. The authors thank Dr. Liliana Sanchez Tacuba for the elaboration of the figures.

Competing Interests We declare no competing interest.

References

1. (2014) World Health Statistics 2014, part III. In: http://www.who.int/gho/publications/world_health_statistics/2014/en/ (ed)
2. Almanza L, Arias CF, Lopez S (1994) Amino acid sequence of the porcine rotavirus YM VP1 protein. Res Virol 145:313–317
3. Anderson P, Kedersha N (2002) Stressful initiations. J Cell Sci 115:3227–3234
4. Angel J, Steele AD, Franco MA (2014) Correlates of protection for rotavirus vaccines: possible alternative trial endpoints, opportunities, and challenges. Hum Vaccin Immunother 10:3659–3671
5. Arias CF, Lopez S, Espejo RT (1982) Gene protein products of SA11 simian rotavirus genome. J Virol 41:42–50
6. Arias CF, López S, Espejo RT (1982) Identification of the RNA segments encoding some structural polypeptides of Nebraska calf diarrhea virus. Rev Lat Microbiol 24:47–54
7. Arias CF, Lopez S, Bell JR, Strauss JH (1984) Primary structure of the neutralization antigen of simian rotavirus SA11 as deduced from cDNA sequence. J Virol 50:657–661
8. Arias CF, Ballado T, Plebanski M (1986) Synthesis of the outer-capsid glycoprotein of the simian rotavirus SA11 in *Escherichia coli*. Gene (Amst) 47:211–219
9. Arias CF, Lizano M, Lopez S (1987) Synthesis in *Escherichia coli* and immunological characterization of a polypeptide containing the cleavage sites associated with trypsin enhancement of rotavirus SA11 infectivity. J Gen Virol 68(pt 3):633–642
10. Arias CF, Ruiz AM, Lopez S (1989) Further antigenic characterization of porcine rotavirus YM. J Clin Microbiol 27:2871–2873
11. Arias CF, Romero P, Alvarez V, Lopez S (1996) Trypsin activation pathway of rotavirus infectivity. J Virol 70:5832–5839
12. Arias CF, Dector MA, Segovia L, Lopez T, Camacho M, Isa P, Espinosa R, Lopez S (2004) RNA silencing of rotavirus gene expression. Virus Res 102:43–51
13. Arias CF, Silva-Ayala D, Lopez S (2015) Rotavirus entry: a deep journey into the cell with several exits. J Virol 89:890–893
14. Arias CF, Silva-Ayala D, Isa P, Díaz-Salinas MA, López S (2016) Rotavirus attachment, internalization, and vesicular traffic. In: Svensson L, Desselberger U, Greenberg HB, Estes MK (eds) Viral gastroenteritis: molecular epidemiology and pathogenesis. Elsevier, London, pp 103–119
15. Arnold MM, Brownback CS, Taraporewala ZF, Patton JT (2012) Rotavirus variant replicates efficiently although encoding an aberrant NSP3 that fails to induce nuclear localization of poly(A)-binding protein. J Gen Virol 93:1483–1494
16. Ayala-Breton C, Arias M, Espinosa R, Romero P, Arias CF, Lopez S (2009) Analysis of the kinetics of transcription and replication of the rotavirus genome by RNA interference. J Virol 83:8819–8831
17. Babji S, Kang G (2012) Rotavirus vaccination in developing countries. Curr Opin Virol 2:443–448
18. Braulke T, Bonifacino JS (2009) Sorting of lysosomal proteins. Biochim Biophys Acta 1793:605–614
19. Broquet AH, Hirata Y, McAllister CS, Kagnoff MF (2011) RIG-I/MDA5/MAVS are required to signal a protective IFN response in rotavirus-infected intestinal epithelium. J Immunol 186:1618–1626
20. Calderon MN, Guerrero CA, Acosta O, Lopez S, Arias CF (2012) Inhibiting rotavirus infection by membrane-impermeant thiol/disulfide exchange blockers and antibodies against protein disulfide isomerase. Intervirology 55:451–464
21. Carreno-Torres JJ, Gutierrez M, Arias CF, Lopez S, Isa P (2010) Characterization of viroplasm formation during the early stages of rotavirus infection. Virol J 7:350
22. Cevallos Porta D, Lopez S, Arias CF, Isa P (2016) Polarized rotavirus entry and release from differentiated small intestinal cells. Virology 499:65–71

23. Ciarlet M, Crawford SE, Cheng E, Blutt SE, Rice DA, Bergelson JM, Estes MK (2002) VLA-2 (alpha2beta1) integrin promotes rotavirus entry into cells but is not necessary for rotavirus attachment. J Virol 76:1109–1123
24. Ciarlet M, Ludert JE, Iturriza-Gomara M, Liprandi F, Gray JJ, Desselberger U, Estes MK (2002) Initial interaction of rotavirus strains with *N*-acetylneuraminic (sialic) acid residues on the cell surface correlates with VP4 genotype, not species of origin. J Virol 76:4087–4095
25. Clark SM, Roth JR, Clark ML, Barnett BB, Spendlove RS (1981) Trypsin enhancement of rotavirus infectivity: mechanism of enhancement. J Virol 39:816–822
26. Contin R, Arnoldi F, Mano M, Burrone OR (2011) Rotavirus replication requires a functional proteasome for effective assembly of viroplasms. J Virol 85:2781–2792
27. Crawford SE, Labbe M, Cohen J, Burroughs MH, Zhou YJ, Estes MK (1994) Characterization of virus-like particles produced by the expression of rotavirus capsid proteins in insect cells. J Virol 68:5945–5952
28. Cuadras MA, Arias CF, Lopez S (1997) Rotaviruses induce an early membrane permeabilization of MA104 cells and do not require a low intracellular Ca^{2+} concentration to initiate their replication cycle. J Virol 71:9065–9074
29. Cuadras MA, Mendez E, Arias CF, Lopez S (1998) A new cysteine in rotavirus VP4 participates in the formation of an alternate disulfide bond. J Gen Virol 79(pt 11):2673–2677
30. Dector MA, Romero P, Lopez S, Arias CF (2002) Rotavirus gene silencing by small interfering RNAs. EMBO Rep 3:1175–1180
31. Delorme C, Brussow H, Sidoti J, Roche N, Karlsson KA, Neeser JR, Teneberg S (2001) Glycosphingolipid binding specificities of rotavirus: identification of a sialic acid-binding epitope. J Virol 75:2276–2287
32. Diaz-Salinas MA, Romero P, Espinosa R, Hoshino Y, Lopez S, Arias CF (2013) The spike protein VP4 defines the endocytic pathway used by rotavirus to enter MA104 cells. J Virol 87:1658–1663
33. Diaz-Salinas MA, Silva-Ayala D, Lopez S, Arias CF (2014) Rotaviruses reach late endosomes and require the cation-dependent mannose-6-phosphate receptor and the activity of cathepsin proteases to enter the cell. J Virol 88:4389–4402
34. Elbashir SM, Harborth J, Lendeckel W, Yalcin A, Weber K, Tuschl T (2001) Duplexes of 21-nucleotide RNAs mediate RNA interference in cultured mammalian cells. Nature (Lond) 411:494–498
35. Espejo RT, Lopez S, Arias C (1981) Structural polypeptides of simian rotavirus SA11 and the effect of trypsin. J Virol 37:156–160
36. Estes MK, Graham DY, Mason BB (1981) Proteolytic enhancement of rotavirus infectivity: molecular mechanisms. J Virol 39:879–888
37. Estes MK, Greenberg HB (2013) Rotaviruses. In: Knipe DM, Howley PM (eds) Fields virology. Lippincott Williams & Wilkins, Philadelphia, pp 1347–1401
38. Fuentes-Panana EM, Lopez S, Gorziglia M, Arias CF (1995) Mapping the hemagglutination domain of rotaviruses. J Virol 69:2629–2632
39. Glass RI, Parashar U, Patel M, Gentsch J, Jiang B (2014) Rotavirus vaccines: successes and challenges. J Infect 68(suppl 1):S9–18
40. Gonzalez RA, Torres-Vega MA, Lopez S, Arias CF (1998) In vivo interactions among rotavirus nonstructural proteins. Arch Virol 143:981–996
41. Gonzalez RA, Espinosa R, Romero P, Lopez S, Arias CF (2000) Relative localization of viroplasmic and endoplasmic reticulum-resident rotavirus proteins in infected cells. Arch Virol 145:1963–1973
42. Graham KL, Halasz P, Tan Y, Hewish MJ, Takada Y, Mackow ER, Robinson MK, Coulson BS (2003) Integrin-using rotaviruses bind alpha2beta1 integrin alpha2 I domain via VP4 DGE sequence and recognize alphaXbeta2 and alphaVbeta3 by using VP7 during cell entry. J Virol 77:9969–9978
43. Gratia M, Sarot E, Vende P, Charpilienne A, Baron CH, Duarte M, Pyronnet S, Poncet D (2015) Rotavirus NSP3 is a translational surrogate of the poly(A) binding protein-poly(A) complex. J Virol 89:8773–8782

44. Guerrero CA, Zarate S, Corkidi G, Lopez S, Arias CF (2000) Biochemical characterization of rotavirus receptors in MA104 cells. J Virol 74:9362–9371
45. Guerrero CA, Bouyssounade D, Zarate S, Isa P, Lopez T, Espinosa R, Romero P, Mendez E, Lopez S, Arias CF (2002) Heat shock cognate protein 70 is involved in rotavirus cell entry. J Virol 76:4096–4102
46. Gutierrez M, Isa P, Sanchez-San Martin C, Perez-Vargas J, Espinosa R, Arias CF, Lopez S (2010) Different rotavirus strains enter MA104 cells through different endocytic pathways: the role of clathrin-mediated endocytosis. J Virol 84:9161–9169
47. Harb M, Becker MM, Vitour D, Baron CH, Vende P, Brown SC, Bolte S, Arold ST, Poncet D (2008) Nuclear localization of cytoplasmic poly(A)-binding protein upon rotavirus infection involves the interaction of NSP3 with eIF4G and RoXaN. J Virol 82:11283–11293
48. Harding HP, Calfon M, Urano F, Novoa I, Ron D (2002) Transcriptional and translational control in the Mammalian unfolded protein response. Annu Rev Cell Dev Biol 18:575–599
49. Haselhorst T, Blanchard H, Frank M, Kraschnefski MJ, Kiefel MJ, Szyczew AJ, Dyason JC, Fleming F, Holloway G, Coulson BS, von Itzstein M (2007) STD NMR spectroscopy and molecular modeling investigation of the binding of *N*-acetylneuraminic acid derivatives to rhesus rotavirus VP8* core. Glycobiology 17:68–81
50. Hinnebusch AG (2014) The scanning mechanism of eukaryotic translation initiation. Annu Rev Biochem 83:779–812
51. Hu L, Crawford SE, Czako R, Cortes-Penfield NW, Smith DF, Le Pendu J, Estes MK, Prasad BV (2012) Cell attachment protein VP8* of a human rotavirus specifically interacts with A-type histo-blood group antigen. Nature (Lond) 485:256–259
52. Huang P, Xia M, Tan M, Zhong W, Wei C, Wang L, Morrow A, Jiang X (2012) Spike protein VP8* of human rotavirus recognizes histo-blood group antigens in a type-specific manner. J Virol 86:4833–4843
53. Isa P, Lopez S, Segovia L, Arias CF (1997) Functional and structural analysis of the sialic acid-binding domain of rotaviruses. J Virol 71:6749–6756
54. Isa P, Realpe M, Romero P, Lopez S, Arias CF (2004) Rotavirus RRV associates with lipid membrane microdomains during cell entry. Virology 322:370–381
55. Isa P, Arias CF, Lopez S (2006) Role of sialic acids in rotavirus infection. Glycoconj J 23:27–37
56. Jackson RJ, Hellen CU, Pestova TV (2010) The mechanism of eukaryotic translation initiation and principles of its regulation. Nat Rev Mol Cell Biol 11:113–127
57. Kaufman RJ (1999) Stress signaling from the lumen of the endoplasmic reticulum: coordination of gene transcriptional and translational controls. Genes Dev 13:1211–1233
58. Kaufman RJ (2004) Regulation of mRNA translation by protein folding in the endoplasmic reticulum. Trends Biochem Sci 29:152–158
59. Kedersha N, Anderson P (2002) Stress granules: sites of mRNA triage that regulate mRNA stability and translatability. Biochem Soc Trans 30:963–969
60. Kotloff KL, Nataro JP, Blackwelder WC, Nasrin D, Farag TH, Panchalingam S, Wu Y, Sow SO, Sur D, Breiman RF, Faruque AS, Zaidi AK, Saha D, Alonso PL, Tamboura B, Sanogo D, Onwuchekwa U, Manna B, Ramamurthy T, Kanungo S, Ochieng JB, Omore R, Oundo JO, Hossain A, Das SK, Ahmed S, Qureshi S, Quadri F, Adegbola RA, Antonio M, Hossain MJ, Akinsola A, Mandomando I, Nhampossa T, Acacio S, Biswas K, O'Reilly CE, Mintz ED, Berkeley LY, Muhsen K, Sommerfelt H, Robins-Browne RM, Levine MM (2013) Burden and aetiology of diarrhoeal disease in infants and young children in developing countries (the Global Enteric Multicenter Study, GEMS): a prospective, case-control study. Lancet 382:209–222
61. Lappalainen S, Pastor AR, Malm M, Lopez-Guerrero V, Esquivel-Guadarrama F, Palomares LA, Vesikari T, Blazevic V (2015) Protection against live rotavirus challenge in mice induced by parenteral and mucosal delivery of VP6 subunit rotavirus vaccine. Arch Virol 160:2075–2078
62. Li W, Manktelow E, von Kirchbach JC, Gog JR, Desselberger U, Lever AM (2010) Genomic analysis of codon, sequence and structural conservation with selective biochemical-structure mapping reveals highly conserved and dynamic structures in rotavirus RNAs with potential cis-acting functions. Nucleic Acids Res 38:7718–7735

63. Lizano M, Lopez S, Arias CF (1991) The amino-terminal half of rotavirus SA114fM VP4 protein contains a hemagglutination domain and primes for neutralizing antibodies to the virus. J Virol 65:1383–1391
64. Lopez S, Arias CF, Bell JR, Strauss JH, Espejo RT (1985) Primary structure of the cleavage site associated with trypsin enhancement of rotavirus SA11 infectivity. Virology 144:11–19
65. Lopez S, Arias CF, Mendez E, Espejo RT (1986) Conservation in rotaviruses of the protein region containing the two sites associated with trypsin enhancement of infectivity. Virology 154:224–227
66. Lopez S, Arias CF (1987) The nucleotide sequence of the 5′ and 3′ ends of rotavirus SA11 gene 4. Nucleic Acids Res 15:4691
67. Lopez S, Lopez I, Romero P, Mendez E, Soberon X, Arias CF (1991) Rotavirus YM gene 4: analysis of its deduced amino acid sequence and prediction of the secondary structure of the VP4 protein. J Virol 65:3738–3745
68. Lopez S, Arias CF (1993) Sequence analysis of rotavirus YM VP6 and NS28 proteins. J Gen Virol 74(pt 6):1223–1226
69. Lopez S, Arias CF (1993) Protein NS26 is highly conserved among porcine rotavirus strains. Nucleic Acids Res 21:1042
70. Lopez S, Espinosa R, Greenberg HB, Arias CF (1994) Mapping the subgroup epitopes of rotavirus protein VP6. Virology 204:153–162
71. Lopez S, Arias CF (2004) Multistep entry of rotavirus into cells: a Versaillesque dance. Trends Microbiol 12:271–278
72. Lopez S, Arias CF (2006) Early steps in rotavirus cell entry. Curr Top Microbiol Immunol 309:39–66
73. Lopez S, Arias CF (2012) Rotavirus–host cell interactions: an arms race. Curr Opin Virol 2:389–398
74. Lopez S, Sanchez-Tacuba L, Moreno J, Arias CF (2016) Rotavirus strategies against the innate antiviral system. Annu Rev Virol 3(1):591–609
75. Lopez T, Camacho M, Zayas M, Najera R, Sanchez R, Arias CF, Lopez S (2005) Silencing the morphogenesis of rotavirus. J Virol 79:184–192
76. Lopez T, Rojas M, Ayala-Breton C, Lopez S, Arias CF (2005) Reduced expression of the rotavirus NSP5 gene has a pleiotropic effect on virus replication. J Gen Virol 86:1609–1617
77. Lopez T, Silva-Ayala D, Lopez S, Arias CF (2011) Replication of the rotavirus genome requires an active ubiquitin-proteasome system. J Virol 85:11964–11971
78. Lopez T, Silva-Ayala D, Lopez S, Arias CF (2012) Methods suitable for high-throughput screening of siRNAs and other chemical compounds with the potential to inhibit rotavirus replication. J Virol Methods 179:242–249
79. Ludert JE, Feng N, Yu JH, Broome RL, Hoshino Y, Greenberg HB (1996) Genetic mapping indicates that VP4 is the rotavirus cell attachment protein in vitro and in vivo. J Virol 70:487–493
80. Ludert JE, Krishnaney AA, Burns JW, Vo PT, Greenberg HB (1996) Cleavage of rotavirus VP4 in vivo. J Gen Virol 77(pt 3):391–395
81. Martinez MA, Lopez S, Arias CF, Isa P (2013) Gangliosides have a functional role during rotavirus cell entry. J Virol 87:1115–1122
82. Maruri-Avidal L, Lopez S, Arias CF (2008) Endoplasmic reticulum chaperones are involved in the morphogenesis of rotavirus infectious particles. J Virol 82:5368–5380
83. Mendez E, Arias CF, Lopez S (1993) Binding to sialic acids is not an essential step for the entry of animal rotaviruses to epithelial cells in culture. J Virol 67:5253–5259
84. Mendez E, Arias CF, Lopez S (1996) Interactions between the two surface proteins of rotavirus may alter the receptor-binding specificity of the virus. J Virol 70:1218–1222
85. Mendez E, Lopez S, Cuadras MA, Romero P, Arias CF (1999) Entry of rotaviruses is a multistep process. Virology 263:450–459
86. Montero H, Arias CF, Lopez S (2006) Rotavirus nonstructural protein NSP3 is not required for viral protein synthesis. J Virol 80:9031–9038
87. Montero H, Rojas M, Arias CF, Lopez S (2008) Rotavirus infection induces the phosphorylation of eIF2alpha but prevents the formation of stress granules. J Virol 82:1496–1504

88. Mossel EC, Ramig RF (2003) A lymphatic mechanism of rotavirus extraintestinal spread in the neonatal mouse. J Virol 77:12352–12356
89. Nava P, Lopez S, Arias CF, Islas S, Gonzalez-Mariscal L (2004) The rotavirus surface protein VP8 modulates the gate and fence function of tight junctions in epithelial cells. J Cell Sci 117:5509–5519
90. Padilla-Noriega L, Dunn SJ, Lopez S, Greenberg HB, Arias CF (1995) Identification of two independent neutralization domains on the VP4 trypsin cleavage products VP5* and VP8* of human rotavirus ST3. Virology 206:148–154
91. Panas MD, Kedersha N, McInerney GM (2015) Methods for the characterization of stress granules in virus infected cells. Methods 90:57–64
92. Pando V, Isa P, Arias CF, Lopez S (2002) Influence of calcium on the early steps of rotavirus infection. Virology 295:190–200
93. Patton JT, Spencer E (2000) Genome replication and packaging of segmented double-stranded RNA viruses. Virology 277:217–225
94. Patton JT, Vasquez-Del Carpio R, Spencer E (2004) Replication and transcription of the rotavirus genome. Curr Pharm Des 10:3769–3777
95. Patton JT, Vasquez-Del Carpio R, Tortorici MA, Taraporewala ZF (2007) Coupling of rotavirus genome replication and capsid assembly. Adv Virus Res 69:167–201
96. Perez-Vargas J, Romero P, Lopez S, Arias CF (2006) The peptide-binding and ATPase domains of recombinant hsc70 are required to interact with rotavirus and reduce its infectivity. J Virol 80:3322–3331
97. Piron M, Vende P, Cohen J, Poncet D (1998) Rotavirus RNA-binding protein NSP3 interacts with eIF4GI and evicts the poly(A) binding protein from eIF4F. EMBO J 17:5811–5821
98. Poncet D (2003) Translation of rotavirus mRNAs in the infected cell. In: Desselberger U, Gray J (eds) Viral gastroenteritis. Elsevier, Amsterdam, pp 185–205
99. Ramig RF (2004) Pathogenesis of intestinal and systemic rotavirus infection. J Virol 78:10213–10220
100. Realpe M, Espinosa R, Lopez S, Arias CF (2010) Rotaviruses require basolateral molecules for efficient infection of polarized MDCKII cells. Virus Res 147:231–241
101. Rehwinkel J, Reis e Sousa C (2010) RIGorous detection: exposing virus through RNA sensing. Science 327:284–286
102. Rojas M, Arias CF, Lopez S (2010) Protein kinase R is responsible for the phosphorylation of eIF2alpha in rotavirus infection. J Virol 84:10457–10466
103. Rubio RM, Mora SI, Romero P, Arias CF, Lopez S (2013) Rotavirus prevents the expression of host responses by blocking the nucleocytoplasmic transport of polyadenylated mRNAs. J Virol 87:6336–6345
104. Ruiz AM, Lopez IV, Lopez S, Espejo RT, Arias CF (1988) Molecular and antigenic characterization of porcine rotavirus YM, a possible new rotavirus serotype. J Virol 62:4331–4336
105. Ruiz MC, Cohen J, Michelangeli F (2000) Role of Ca^{2+} in the replication and pathogenesis of rotavirus and other viral infections. Cell Calcium 28:137–149
106. Rutkowski DT, Kaufman RJ (2004) A trip to the ER: coping with stress. Trends Cell Biol 14:20–28
107. Sanchez-San Martin C, Lopez T, Arias CF, Lopez S (2004) Characterization of rotavirus cell entry. J Virol 78:2310–2318
108. Sanchez-Tacuba L, Rojas M, Arias CF, Lopez S (2015) Rotavirus controls activation of the 2′-5′-oligoadenylate synthetase/RNase L pathway using at least two distinct mechanisms. J Virol 89:12145–12153
109. Saxena K, Blutt SE, Ettayebi K, Zeng XL, Broughman JR, Crawford SE, Karandikar UC, Sastri NP, Conner ME, Opekun AR, Graham DY, Qureshi W, Sherman V, Foulke-Abel J, In J, Kovbasnjuk O, Zachos NC, Donowitz M, Estes MK (2016) Human intestinal enteroids: a new model to study human rotavirus infection, host restriction, and pathophysiology. J Virol 90:43–56
110. Sen A, Pruijssers AJ, Dermody TS, Garcia-Sastre A, Greenberg HB (2011) The early interferon response to rotavirus is regulated by PKR and depends on MAVS/IPS-1, RIG-I, MDA-5, and IRF3. J Virol 85:3717–3732

111. Silva-Ayala D, Lopez T, Gutierrez M, Perrimon N, Lopez S, Arias CF (2013) Genome-wide RNAi screen reveals a role for the ESCRT complex in rotavirus cell entry. Proc Natl Acad Sci U S A 110:10270–10275
112. Silverman RH (2007) Viral encounters with 2′,5′-oligoadenylate synthetase and RNase L during the interferon antiviral response. J Virol 81:12720–12729
113. Sonenberg N, Hinnebusch AG (2009) Regulation of translation initiation in eukaryotes: mechanisms and biological targets. Cell 136:731–745
114. Svensson L, Desselberger U, Greenberg HB, Estes MK (2016) Introduction. In: Svensson L, Desselberger U, Greenberg HB, Estes MK (eds) Viral gastroenteritis: molecular epidemiology and pathogenesis. Elsevier, London, pp xxi–xxvi
115. Tate JE, Burton AH, Boschi-Pinto C, Parashar UD, World Health Organization-Coordinated Global Rotavirus Surveillance N (2016) Global, regional, and national estimates of rotavirus mortality in children <5 years of age, 2000–2013. Clin Infect Dis 62(suppl 2):S96–S105
116. Torres-Flores JM, Arias CF (2015) Tight junctions go viral! Viruses 7:5145–5154
117. Torres-Flores JM, Silva-Ayala D, Espinoza MA, Lopez S, Arias CF (2015) The tight junction protein JAM-A functions as coreceptor for rotavirus entry into MA104 cells. Virology 475:172–178
118. Torres-Vega MA, Gonzalez RA, Duarte M, Poncet D, Lopez S, Arias CF (2000) The C-terminal domain of rotavirus NSP5 is essential for its multimerization, hyperphosphorylation and interaction with NSP6. J Gen Virol 81:821–830
119. Trujillo-Alonso V, Maruri-Avidal L, Arias CF, Lopez S (2011) Rotavirus infection induces the unfolded protein response of the cell and controls it through the nonstructural protein NSP3. J Virol 85:12594–12604
120. Velazquez FR, Matson DO, Calva JJ, Guerrero L, Morrow AL, Carter-Campbell S, Glass RI, Estes MK, Pickering LK, Ruiz-Palacios GM (1996) Rotavirus infections in infants as protection against subsequent infections. N Engl J Med 335:1022–1028
121. Vesikari T (2016) Rotavirus vaccines and vaccination. In: Svensson L, Desselberger U, Greenberg HB, Estes MK (eds) Viral gastroenteritis: molecular epidemiology and pathogenesis. Elsevier, London, pp 301–328
122. Vijay-Kumar M, Gentsch JR, Kaiser WJ, Borregaard N, Offermann MK, Neish AS, Gewirtz AT (2005) Protein kinase R mediates intestinal epithelial gene remodeling in response to double-stranded RNA and live rotavirus. J Immunol 174:6322–6331
123. Walker CL, Rudan I, Liu L, Nair H, Theodoratou E, Bhutta ZA, O'Brien KL, Campbell H, Black RE (2013) Global burden of childhood pneumonia and diarrhoea. Lancet 381:1405–1416
124. Walsh D, Mohr I (2011) Viral subversion of the host protein synthesis machinery. Nat Rev Microbiol 9:860–875
125. White JP, Lloyd RE (2012) Regulation of stress granules in virus systems. Trends Microbiol 20:175–183
126. Wolf M, Vo PT, Greenberg HB (2011) Rhesus rotavirus entry into a polarized epithelium is endocytosis dependent and involves sequential VP4 conformational changes. J Virol 85:2492–2503
127. Wolf M, Deal EM, Greenberg HB (2012) Rhesus rotavirus trafficking during entry into MA104 cells is restricted to the early endosome compartment. J Virol 86:4009–4013
128. Wollert T, Hurley JH (2010) Molecular mechanism of multivesicular body biogenesis by ESCRT complexes. Nature (Lond) 464:864–869
129. Wu J, Kaufman RJ (2006) From acute ER stress to physiological roles of the unfolded protein response. Cell Death Differ 13:374–384
130. Zambrano JL, Diaz Y, Pena F, Vizzi E, Ruiz MC, Michelangeli F, Liprandi F, Ludert JE (2008) Silencing of rotavirus NSP4 or VP7 expression reduces alterations in Ca^{2+} homeostasis induced by infection of cultured cells. J Virol 82:5815–5824
131. Zambrano JL, Ettayebi K, Maaty WS, Faunce NR, Bothner B, Hardy ME (2011) Rotavirus infection activates the UPR but modulates its activity. Virol J 8:359
132. Zarate S, Espinosa R, Romero P, Guerrero CA, Arias CF, Lopez S (2000) Integrin alpha-2beta1 mediates the cell attachment of the rotavirus neuraminidase-resistant variant nar3. Virology 278:50–54

133. Zarate S, Espinosa R, Romero P, Mendez E, Arias CF, Lopez S (2000) The VP5 domain of VP4 can mediate attachment of rotaviruses to cells. J Virol 74:593–599
134. Zarate S, Cuadras MA, Espinosa R, Romero P, Juarez KO, Camacho-Nuez M, Arias CF, Lopez S (2003) Interaction of rotaviruses with Hsc70 during cell entry is mediated by VP5. J Virol 77:7254–7260
135. Zarate S, Romero P, Espinosa R, Arias CF, Lopez S (2004) VP7 mediates the interaction of rotaviruses with integrin alphavbeta3 through a novel integrin-binding site. J Virol 78:10839–10847

Chapter 3
Calicivirus Biology

Ana Lorena Gutierrez-Escolano

1 Introduction

Caliciviruses are a diverse family constituted of ubiquitous viruses that infect a variety of animals as well as humans. Five genera are classified within the *Caliciviridae* family—*Lagovirus*, *Nebovirus*, *Norovirus*, *Sapovirus*, and *Vesivirus*—with two additional proposed genera, *Recovirus* and *Valovirus* [19, 36]; only noroviruses, sapoviruses, and recovirus [61] have been shown to infect humans [62].

Human noroviruses (HuNoV) can cause acute gastroenteritis in all age groups and represent an important public health problem worldwide; they are considered the leading cause of foodborne gastroenteritis outbreaks [52]. It is estimated that HuNoV are responsible for approximately 23 million total illnesses with a disease burden of 2 billion dollars in the United States alone each year and 200,000 annual deaths of children under the age of 5 years in developing countries [49]. Although the exact burden associated with calicivirus outbreaks in Latin America is hard to estimate, outbreaks in the region associated with health services, closed communities, restaurants, or social events have been well documented. In addition, caliciviruses are a common cause of sporadic acute diarrhea in the region [43]. Finally, following the introduction of rotavirus vaccines, HuNoV have become the predominant gastrointestinal pathogen within pediatric populations in both developed and developing countries including Latin America [50]. In this chapter, research in norovirus and other calicivirus biology and vaccine development carried out in the region is covered. Epidemiological aspects are not covered because two excellent reviews on the subject recently appeared [16, 47].

A.L. Gutierrez-Escolano (✉)
Departamento de Infectómica y Patogénesis Molecular, Centro de Investigación y Estudios Avanzados del Instituto Politécnico Nacional (CINVESTAV-IPN), Mexico City, Mexico
e-mail: alonso@cinvestav.mx

J.E. Ludert et al. (eds.), *Human Virology in Latin America*,
DOI 10.1007/978-3-319-54567-7_3

2 Calicivirus Genomic Organization

Members of the *Caliciviridae* family are non-enveloped icosahedral particles that are 27–40 nm in diameter. The genome consists of a linear, positive-sense, single-stranded RNA approximately 7.4–8.3 kb in length that is covalently linked to a viral-encoded protein (VPg) at the 5′-end [21, 27] and has a polyadenylated tail at the 3′-end. These viruses also produce a subgenomic (sg) RNA during infection that is identical to the 3′-end region of the genome.

The genome contains between two and as many as four open reading frames (ORFs), depending on the genus [6, 13, 23, 63]; the ORF1, located at the 5′-end, encodes for a large polyprotein that is processed by the only viral protease (NS6 or NS6/7) to produce the nonstructural (NS) proteins. Noroviruses and vesiviruses exhibit two additional ORFs, ORF2 and ORF3, that encode for the major (VP1) and minor (VP2) capsid proteins, respectively, although an additional fourth ORF, encoding for the virulence factor 1 (VF1), has been found in MNV [44]. On the other hand, sapoviruses, lagoviruses, and neboviruses contain a genome that is organized in two ORFs: ORF1 encodes for the NS proteins as well as VP1, and ORF2 encodes VP2 [48]. In all cases, ORF1 encodes for a polyprotein, from which the NS proteins are processed, and both structural proteins as well as VF1 are translated from the subgenomic RNA.

Most of these NS proteins have been implicated in the replication complex formation. NS3 is an NTPase; NS5 is the viral protein associated with the genome or VPg, which is involved in the initiation of translation [12, 17, 18, 22] and genome replication [21, 56]; NS6 or NS6/7, which is the only protease described, is responsible for the autocatalytic processing of the ORF1 polyprotein [63]; and N6/7 or NS7 is an RNA-dependent RNA polymerase (RdRP) that synthesizes the plus- and minus-strand viral RNAs [46]. Based on the mature form of the NS7 protein, caliciviruses can be separated into two distinctive groups: (1) the NS6/7 precursor synthesized in cells infected with lagoviruses and noroviruses that suffers further cleavage to produce the fully processed NS6 and NS7 proteins [66] and (2) the NS6/7 precursor synthesized in vesivirus-infected cells, representing a mature and stable bifunctional enzyme that does not undergo additional processing during infection [65].

The major and minor structural proteins that constitute the virus capsid VP1 and VP2 are translated from the subgenomic RNA. The viral capsid is formed by 180 copies of VP1, which spontaneously self-assemble [32]; even though VP2 is dispensable for the self-assembly of empty viral capsids [37, 38], it participates in capsid stability [7, 64].

The additional fourth ORF from the MNV subgenomic RNA, overlapping the VP1 coding region, encodes for the VF1, implicated in the regulation of the immune response and the development of apoptosis [44, 76]. A similar alternative reading frame (ORF3) is found in human sapoviruses; however, the encoded proteins are yet of unknown function [11].

3 Calicivirus Replicative Cycle

The calicivirus replicative cycle initiates with the interaction of the viral particles to the surface of permissive cells. The members of the *Caliciviridae* family recognize cell-surface carbohydrate ligands, glycolipids, and proteins for attachment and entry, including histo-blood group antigens (HBGA) for the HuNoVs [41, 73].

The calicivirus genome translation occurs when the viral RNA has been released into the cytoplasm of the infected cell. In contrast with the majority of positive-sense, single-stranded RNA animal viruses, calicivirus translation depends on the viral protein VPg, which is linked to the 5′-end of the genomic RNA. VPg interacts with some canonical initiation factors (eIFs), such as eIF3, eIF4G, eIF4A, and eIF4E, allowing the recruitment of the pre-initiation complex [12, 21, 22].

The translation of genomic RNA gives rise to a polyprotein that is co- and post-translationally cleaved by the viral protease to produce the NS proteins, which are essential for virus replication. During a calicivirus infection, the NS proteins can be located in the endoplasmic reticulum (ER), as during feline calcivirus (FCV) infection [4], or associated with components of the Golgi apparatus, as seen when HuNoV NS proteins are expressed in cells. During MNV infection, the NS proteins were observed to co-localize with the double-stranded RNA in virus-induced vesicle clusters formed in the cytoplasm of infected cells originating from membranes derived from the secretory pathway [30]. Therefore, NS proteins are important in the induction of intracellular membrane rearrangements that lead to the replication complex formation, where the new genomic and subgenomic RNAs are produced. In these membrane complexes, the subgenomic RNAs, which are also covalently linked to the VPg protein, are also translated to produce the structural proteins VP1 and VP2.

The genomic RNA interacts with the viral replicase NS7 at its 3′-end to generate the negative RNA molecules that form a double-stranded RNA intermediate via a de novo mechanism. The structural proteins NS1/2, VP1, and VP2 can regulate the NS7 replicase activity; particularly, the VP1 produced early during infection promotes the negative-strand RNA synthesis [33, 69]. The negative-stranded RNA also interacts with NS7 at its 3′-end, generating multiple copies of the genomic RNAs via a VPg-primed RNA synthesis at the 3′-end [56], whereas subgenomic RNA is produced by priming at a stem-loop element located near the start site of the subgenomic RNA [81].

Finally, once the subgenomic RNAs are translated to produce VP1 and VP2, 180 copies of VP1 arranged in 90 homodimers form the viral capsid, which is assembled together with the genomic RNA to generate the infectious viral particles by an unknown mechanism. The release of the viral particles from infected cells occurs via a lytic mechanism that involves induction of intrinsic apoptosis, upregulation of caspases [8, 45, 55, 67], and survivin degradation [8, 29]. In this regard, we observed that heat shock treatment of infected cells resulted in delayed cytopathic effect and reduced virus yield, related to the inhibition on caspase 3 and the control of apoptosis [3]. In the presence of caspase inhibitors, MNV infection proceeds via a distinct pathway of rapid cellular necrosis and reduced viral production [20].

4 Virus–Cell Interactions

As in all virus infections, calicivirus protein austerity requires the participation of cellular factors during its replicative cycle. In addition to the eIFs associated with VPg protein, numerous proteinaceous host factors that interact with the genomic RNAs from the HuNoV, FCV, and MNV terminal ends have been identified [25, 26, 78]. The roles of a few of these have been described, such as the polypyrimidine tract-binding (PTB) protein that binds to the 5′-terminal end of the FCV genomic and subgenomic RNAs and inhibits translation as a mechanism to stimulate the synthesis of viral RNA via clearance of ribosomes from viral RNA [34]. On the other hand, we have demonstrated that nucleolin, a phosphoprotein involved in ribosome biogenesis, is a positive regulator of FCV protein synthesis [9, 28]. Efficient viral genome replication of caliciviruses is thought to require the circularization of the RNA genome that occurs via the interaction of complementary sequences present within the 5′- and 3′-ends of NV and MNV genomes that are further stabilized by cellular proteins [40, 57]. For the specific case of MNV, the circularization is promoted by the interaction of heterogeneous nuclear ribonucleoprotein (hnRNP) A1 and the poly(rC)-binding protein 2 (PCBP 2) with both ends of the genome. Another component of the ribonucleoprotein complex formed with the MNV genome is the molecular chaperone Hsp90, which has been described to have a pleiotropic role in the norovirus cell cycle. Both the HuNoV and the MNV capsid proteins require Hsp90 activity for their stability; moreover, targeting Hsp90 in vivo significantly reduces virus replication [79].

The interactions between different viral components and host cell factors are important in most steps of a viral infection. These interactions, which occur throughout the entire viral life cycle, determine the virus–host range, tissue tropism, and viral pathogenesis and drive viral evolution. Therefore, the identification of these host cell factors has provided important information about the viral life cycle and constitutes targets for the development of control strategies [24]. To this regard, it is known that noroviruses bind to histo-blood group antigens (HBGAs; ABH and Lewis) that are complex carbohydrate moieties expressed on red blood cells; on gastrointestinal, genitourinary, and mucosal epithelial cells; and in biological fluids as free oligosaccharides [1, 51]. However, depending on the genotype, they show different binding patterns.

5 Importance of Caliciviruses in Animals

Caliciviruses are also important animal viruses affecting both wild and farm or domestic animals, showing a broad host range spectrum among marine and terrestrial vertebrates, including cetaceans, dogs, cattle, pigs, lions, mink, and monkeys.

Animal caliciviruses can cause a wide spectrum of diseases and lesions such as enteric diseases in calves, pigs, and chicken [80], severe hemorrhagic diseases in rabbits and hares, and upper respiratory and vesicular diseases in felines. In the region nearly 20% of the fecal samples collected in farm pigs have been found positive for either porcine norovirus or sapoviruses, with seroprevalences as high as 80% post weaning [2, 14, 42]. Even though caliciviruses infect a wide range of vertebrates, epidemiological evidence for zoonotic transmission is lacking.

6 Transmission

HuNoV are highly resistant in the environment and can be acquired throughout the fecal–oral route, by consumption of contaminated food or water, and by person-to-person contact. The resistance of the virus in the environment also allows indirect contamination by contact with contaminated surfaces or even via vomit-derived aerosols [43]. Whether transmission via food of animal origin or by contact with diseased animals is possible remains uncertain [5].

7 Clinical and Epidemiological Aspects

Norovirus and sapovirus clinical symptoms are indistinguishable; however, in general, the severity of gastroenteritis is milder in sapovirus infections than that for norovirus and rotavirus infections [48]. Following HuNoV infection, the principal symptoms appear after 1 or 2 days of incubation and include vomiting; watery, non-bloody diarrhea; nausea; abdominal cramps; headache; chills; myalgia; and low-grade fever. HuNoV illness is usually short term and self-limiting; however, for young children, the elderly, and immunocompromised patients, this illness can be serious, leading to severe dehydration, hospitalization, and death. Complications associated with HuNoV infection include infantile convulsion [10] necrotizing enterocolitis (NEC) [68], and, rarely, disseminated disease including multiple organs.

8 Immunity

The relevance of the cellular immunity to protect or resolve HuNoV infections remains controversial; however, antibodies have been shown to be important in protection. Serum HBGA-blocking antibodies can protect norovirus-induced disease [15]; moreover, salivary and circulating memory B cells secreting IgA and

IgG antibodies that primarily recognize VP1 capsid protein correlate with a protective immunity against HuNoV [35, 53]. Duration of HuNoV gastroenteritis immunity has been estimated to be on the scale of months; however, in a recent report based on mathematical studies, duration of immunity was predicted to be from 4.1 to 8.7 years [60]. The seroprevalence for noroviruses is lower in developed than in developing countries, suggesting that the latter populations may be continuously exposed to norovirus from an early age [51]. In this regard, it has been observed that more than 80% of children in Latin America had norovirus antibodies by 2 years of age [31].

9 Treatment, Prevention, and Vaccines

The lack of specific treatment for HuNoV infections limits treatment to supportive care consisting of fluid/electrolyte replenishment for patients with severe dehydration. In patients undergoing immunosuppressive therapy, a temporary hold in immunosuppressive treatment is required to prevent persistent infection [59]. As in most infectious diseases, one of the best prevention strategies for HuNoV infection is hygiene, including frequent hand-washing.

Because HuNoV represent a global health problem with a high estimated morbidity that affects millions of people worldwide, several antivirals and vaccines are under development. However, the lack of a validated cell culture system or small animal model has been a great obstacle for such development [35, 54]. Therefore, efforts on developing candidate vaccines have been based on the use of subviral particles or subunit vaccines, including virus-like particles (VLPs) and P domain complexes that have been shown to be immunogenic [75].

VLPs are advantageous, as they do not contain the viral genome but do contain the viral structural proteins responsible for inducing protective immunity [32, 59]. However, as HuNoV are genetically and antigenically diverse, it is presumed that to induce broad protection against multiple HuNoV strains will be necessary for efficient protection. Moreover, these vaccines may require updating when new pandemic strains emerge [39]. Combination vaccines, including those for other common viral causes of childhood diarrhea, are also under consideration.

Although currently no licensed vaccine for human calicivirus exists, several vaccine candidates are under development, including transgenic plant-based norovirus vaccines [58, 71], norovirus P particles produced in *Escherichia coli* [74], a trivalent norovirus/rotavirus combination vaccine that includes GII.4 and GI.3 VLPs and rotavirus VP6 protein [72], and a bivalent (GI.1 + GII.4) intramuscular VLP vaccine [70, 77]. Another unresolved issue about human calicivirus vaccines is the definition of the vaccine target population, that is, who shall be vaccinated to diminish the risks of severe disease and death caused by HuNoV infections in children, the elderly, and the immunocompromised.

10 Conclusions

Caliciviruses represent a public health problem worldwide. In addition, calicivirus infection affects several animals of veterinary and economic importance. After the introduction of the rotavirus vaccine in the expanded program of immunization in most Latin American countries, caliciviruses, namely, noroviruses, have also emerged as an important etiological agent associated with acute sporadic infantile diarrhea. Therefore, one of the challenges for the region is to understand the importance of caliciviruses as an etiological cause of diarrhea and to expand the local data that will facilitate difficult decisions about possible future vaccine implementation. In addition, basic research to better understand calicivirus biology and to identify host molecules involved in resistance and susceptibility to viral infection is also needed in the quest for strategies of control (Figs. 3.1 and 3.2).

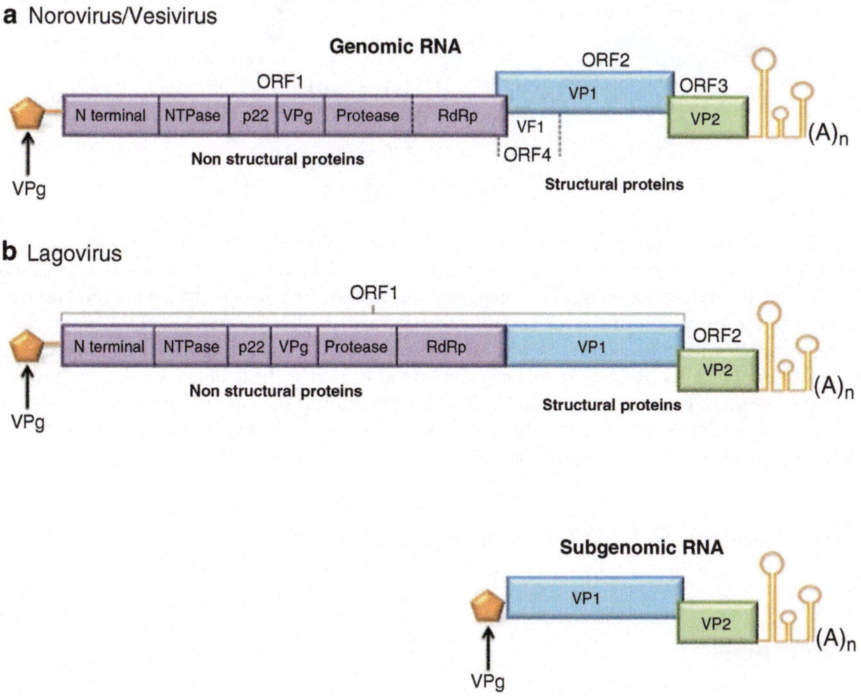

Fig. 3.1 Schematic representation of the genomic organization of three representative genera of the *Caliciviridae* family. The genomic organization of *Norovirus/Vesivirus* (**a**) *and Lagvirus* (**b**) genera contains open reading frame (ORF)1 at the 5′-ends of all the genomes that encode the nonstructural proteins. The structural protein VP1 is either encoded at the 3′-end of ORF1 (*Lagovirus*) or by ORF2 (*Norovirus/Vesivirus),* whereas VP2 is translated from its own ORF. ORF4 is unique to the murine norovirus (MNV). Both VP1 and VP2 structural proteins and VF1 are translated from the subgenomic RNA

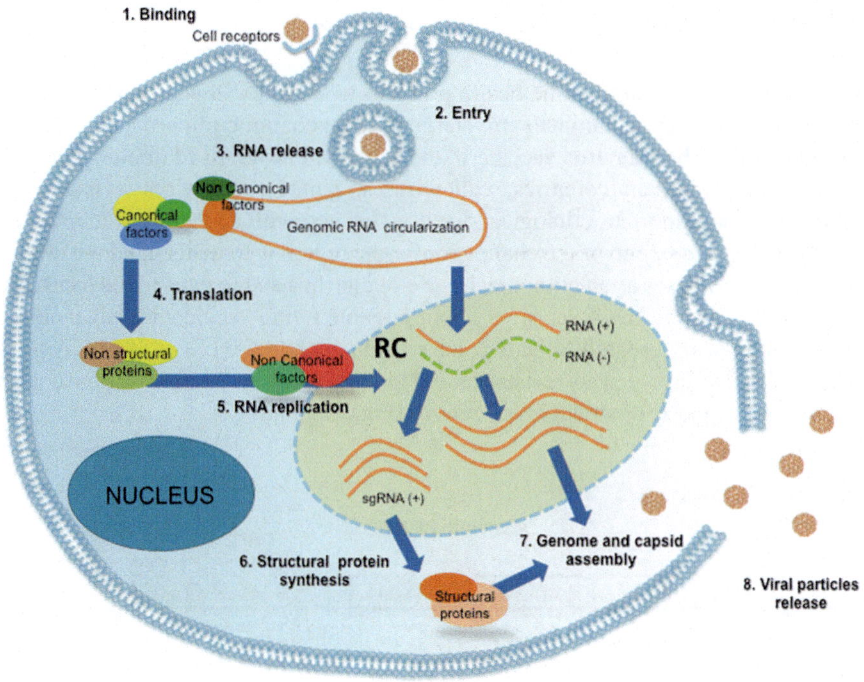

Fig. 3.2 Schematic representation of a typical calicivirus cell cycle. After viral entry into the cell, the RNA is uncoated in the cytoplasm and translated by a VPg-mediated initiation of translation mechanism that requires canonical and noncanonical factors. The nonstructural viral proteins participate in the formation of the replication complex, and the RNA-dependent RNA polymerase (NS7) together with noncanonical factors uses the genomic RNA as a template for the synthesis of negative-strand RNAs [RNA (−)]; this negative strand is used as the template for the synthesis of multiple copies of positive-stranded RNA [RNA (+)] that forms the viral progeny genome and for the subgenomic RNA, which is translated to produce the capsid proteins for RNA encapsidation. After capsid assembly, the viral particles are released from the cell

Acknowledgments CONACYT Proyecto Salud # 261257.

References

1. Ahmed SM, Hall AJ, Robinson AE, Verhoef L, Premkumar P, Parashar UD, Koopmans M, Lopman BA (2014) Global prevalence of norovirus in cases of gastroenteritis: a systematic review and meta-analysis. Lancet Infect Dis 14:725–730
2. Alcala AC, Rodriguez-Diaz J, de Rolo M, Vizzi E, Buesa J, Liprandi F, Ludert JE (2010) Seroepidemiology of porcine enteric sapovirus in pig farms in Venezuela. Vet Immunol Immunopathol 137:269–274
3. Alvarez-Sanchez C, Cancio-Lonches C, Mora-Heredia JE, Santos-Valencia JC, Barrera-Vazquez OS, Yocupicio-Monroy M, Gutierrez-Escolano AL (2015) Negative effect of heat shock on feline calicivirus release from infected cells is associated with the control of apoptosis. Virus Res 198:44–52

4. Bailey D, Karakasiliotis I, Vashist S, Chung LMW, Rees J, McFadden N, Benson A, Yarovinsky F, Simmonds P, Goodfellow I (2010) Functional analysis of RNA structures present at the 3′ extremity of the murine norovirus genome: the variable polypyrimidine tract plays a role in viral virulence. J Virol 84:2870–2859
5. Bank-Wolf BR, Konig M, Thiel HJ (2010) Zoonotic aspects of infections with noroviruses and sapoviruses. Vet Microbiol 140:204–212
6. Rowena B, White PA (2010) Genome oganization and recombination. In: Hansman GS, Jiang XJA, Green KY (eds) Caliciviruses, molecular and cellular virology. Caiser Academic Press, Norfolk, pp 45–63
7. Bertolotti-Ciarlet A, Crawford SE, Hutson AM, Estes MK (2003) The 3′ end of Norwalk virus mRNA contains determinants that regulate the expression and stability of the viral capsid protein VP1: a novel function for the VP2 protein. J Virol 77:11603–11615
8. Bok K, Prikhodko VG, Green KY, Sosnovtsev SV (2009) Apoptosis in murine norovirus-infected RAW264.7 cells is associated with downregulation of survivin. J Virol 83:3647–3656
9. Cancio-Lonches C, Yocupicio-Monroy M, Sandoval-Jaime C, Galvan-Mendoza I, Urena L, Vashist S, Goodfellow I, Salas-Benito J, Gutierrez-Escolano AL (2011) Nucleolin interacts with the feline calicivirus 3′ untranslated region and the protease-polymerase NS6 and NS7 proteins, playing a role in virus replication. J Virol 85:8056–8068
10. Chan CM, Chan C, Ma CK, Chan HB (2011) Norovirus as cause of benign convulsion associated with gastro-enteritis. J Paediatr Child Health 47:373–377
11. Chanit W, Thongprachum A, Khamrin P, Okitsu S, Mizuguchi M, Ushijima H (2009) Intergenogroup recombinant sapovirus in Japan, 2007--2008. Emerg Infect is 15:1084–1087
12. Chaudhry Y, Nayak A, Bordeleau M-E, Tanaka J, Pelletier J, Belsham GJ, Roberts LO, Goodfellow IG (2006) Caliciviruses differ in their functional requirements for eIF4F components. J Biol Chem 281:25315–25325
13. Clarke IN, Lambden PR (1997) The molecular biology of caliciviruses. J Gen Virol 78:291–301
14. Cunha JB, de Mendonca MCL, Miagostovich MP, Leite JPG (2010) Genetic diversity of porcine enteric caliciviruses in pigs raised in Rio de Janeiro State, Brazil. Arch Virol 155:1301–1305
15. Czako R, Atmar RL, Opekun AR, Gilger MA, Graham DY, Estes MK (2012) Serum hemagglutination inhibition activity correlates with protection from gastroenteritis in persons infected with Norwalk virus. Clin Vaccine Immunol 19:284–287
16. da Silva PT, Peiro JR, Mendes LC, Ludwig LF, de Oliveira-Filho EF, Bucardo F, Huynen P, Melin P, Thiry E, Mauroy A (2016) Human norovirus infection in Latin America. J Clin Virol 78:111–119
17. Daughenbaugh KF, Fraser CS, Hershey JWB, Hardy ME (2003) The genome-linked protein VPg of the Norwalk virus binds eIF3, suggesting its role in translation initiation complex recruitment. EMBO J 22:2852–2859
18. Daughenbaugh KF, Wobus CE, Hardy ME (2006) VPg of murine norovirus binds translation initiation factors in infected cells. Virol J 3:33
19. Farkas T, Sestak K, Wei C, Jiang X (2008) Characterization of a rhesus monkey calicivirus representing a new genus of Caliciviridae. J Virol 82:5408–5416
20. Furman LM, Maaty WS, Petersen LK, Ettayebi K, Hardy ME, Bothner B (2009) Cysteine protease activation and apoptosis in murine norovirus infection. Virol J 6:139
21. Goodfellow I (2011) The genome-linked protein VPg of vertebrate viruses: a multifaceted protein. Curr Opin Virol 1:355–362
22. Goodfellow I, Chaudhry Y, Gioldasi I, Gerondopoulos A, Natoni A, Labrie L, Laliberte JF, Roberts L (2005) Calicivirus translation initiation requires an interaction between VPg and eIF4E. EMBO Rep 6:968–972
23. Green KY, Ando T, Balayan MS, Berke T, Clarke IN, Estes MK, Matson DO, Nakata S, Neill JD, Studdert MJ, Thiel HJ (2000) Taxonomy of the caliciviruses. J Infect Dis 181(suppl 2):S322–S330
24. Gutierrez-Escolano AL (2014) Host-cell factors involved in the calicivirus replicative cycle. Future Virol 9:147–160

25. Gutierrez-Escolano AL, Brito ZU, del Angel RM, Jiang X (2000) Interaction of cellular proteins with the 5 ' end of Norwalk virus genomic RNA. J Virol 74:8558–8562
26. Gutierrez-Escolano AL, Vazquez-Ochoa M, Escobar-Herrera J, Hernandez-Acosta J (2003) La, PTB, and PAB proteins bind to the 3′ untranslated region of Norwalk virus genomic RNA. Biochem Biophys Res Commun 311:759–766
27. Herbert TP, Brierley I, Brown TDK (1997) Identification of a protein linked to the genomic and subgenomic mRNAs of feline calicivirus and its role in translation. J Gen Virol 78:1033–1040
28. Hernandez BA, Sandoval-Jaime C, Sosnovtsev SV, Green KY, Gutierrez-Escolano AL (2016) Nucleolin promotes in vitro translation of feline calicivirus genomic RNA. Virology 489:51–62
29. Herod MR, Salim O, Skilton RJ, Prince CA, Ward VK, Lambden PR, Clarke IN (2014) Expression of the murine norovirus (MNV) ORF1 polyprotein is sufficient to induce apoptosis in a virus-free cell model. PLoS One 9:e90679
30. Hyde JL, Sosnovtsev SV, Green KY, Wobus C, Virgin HW, Mackenzie JM (2009) Mouse norovirus replication is associated with virus-induced vesicle clusters originating from membranes derived from the secretory pathway. J Virol 83:9709–9719
31. Jiang X, Matson DO, Velazquez FR, Calva JJ, Zhong WM, Hu J, Ruiz-Palacios GM, Pickering LK (1995) Study of Norwalk-related viruses in Mexican children. J Med Virol 47:309–316
32. Jiang X, Wang M, Graham DY, Estes MK (1992) Expression, self-assembly, and antigenicity of the Norwalk virus capsid protein. J Virol 66:6527–6532
33. Kaiser WJ, Chaudhry Y, Sosnovtsev SV, Goodfellow IG (2006) Analysis of protein–protein interactions in the feline calicivirus replication complex. J Virol 87:363–368
34. Karakasiliotis I, Vashist S, Bailey D, Abente EJ, Green KY, Roberts LO, Sosnovtsev SV, Goodfellow IG (2010) Polypyrimidine tract binding protein functions as a negative regulator of feline calicivirus translation. PLoS One 5:e9562
35. Kocher J, Yuan L (2015) Norovirus vaccines and potential antinorovirus drugs: recent advances and future perspectives. Future Virol 10:899–913
36. L'Homme Y, Sansregret R, Plante-Fortier E, Lamontagne AM, Ouardani M, Lacroix G, Simard C (2009) Genomic characterization of swine caliciviruses representing a new genus of Caliciviridae. Virus Genes 39:66–75
37. Laurent S, Vautherot JF, Madelaine MF, Legall G, Rasschaert D (1994) Recombinant rabbit hemorrhagic-disease virus capsid protein expressed in baculovirus self-assembles into virus-like particles and induces protection. J Virol 68:6794–6798
38. Leite JPG, Ando T, Noel JS, Jiang B, Humphrey CD, Lew JF, Green KY, Glass RI, Monroe SS (1996) Characterization of Toronto virus capsid protein expressed in baculovirus. Arch Virol 141:865–875
39. Lindesmith LC, Ferris MT, Mullan CW, Ferreira J, Debbink K, Swanstrom J, Richardson C, Goodwin RR, Baehner F, Mendelman PM, Bargatze RF, Baric RS (2015) Broad blockade antibody responses in human volunteers after immunization with a multivalent norovirus VLP candidate vaccine: immunological analyses from a phase I clinical trial. PLoS Med 12:e1001807
40. López-Manriquez E, Vashist S, Ueña L, Goodfellow I, Chavez P, Mora-Heredia JE, Cancio-Lonches C, Garrido E, Gutiérrez-Escolano AL (2013) Norovirus genome circularisation and efficient replication is facilitated by binding of PCBPs and hnRNP A1. J Virol 87:11371
41. Marionneau S, Ruvoen N, Le Moullac-Vaidye B, Clement M, Cailleau-Thomas A, Ruiz-Palacois G, Huang P, Jiang X, Le Pendu J (2002) Norwalk virus binds to histo-blood group antigens present on gastroduodenal epithelial cells of secretor individuals. Gastroenterology 122:1967–1977
42. Martinez MA, Alcala AC, Carruyo G, Botero L, Liprandi F, Ludert JE (2006) Molecular detection of porcine enteric caliciviruses in Venezuelan farms. Vet Microbiol 116:77–84
43. Mathijs E, Stals A, Baert L, Botteldoorn N, Denayer S, Mauroy A, Scipioni A, Daube G, Dierick K, Herman L, Van Coillie E, Uyttendaele M, Thiry E (2012) A review of known and hypothetical transmission routes for noroviruses. Food Environ Virol 4:131–152
44. McFadden N, Bailey D, Carrara G, Benson A, Chaudhry Y, Shortland A, Heeney J, Yarovinsky F, Simmonds P, Macdonald A, Goodfellow I (2011) Norovirus regulation of the innate immune

response and apoptosis occurs via the product of the alternative open reading frame 4. PLoS Pathog 7:e1002413
45. Natoni A, Kass GEN, Carter MJ, Roberts LO (2006) The mitochondrial pathway of apoptosis is triggered during feline calicivirus infection. J Gen Virol 87:357–361
46. Neill JD (1990) Nucleotide sequence of a region of the feline calicivirus genome which encodes picornavirus-like RNA-dependent RNA polymerase, cysteine protease and 2C polypeptides. Virus Res 17:145–160
47. O'Ryan M, Riera-Montes M, Lopman B (2016) Norovirus in Latin America: systematic review and meta-analysis. Pediatr Infect Dis J 36(2):127–134
48. Oka T, Wang Q, Katayama K, Saif LJ (2015) Comprehensive review of human sapoviruses. Clin Microbiol Rev 28:32–53
49. Patel MM, Hall AJ, Vinje J, Parashar UD (2009) Noroviruses: a comprehensive review. J Clin Virol 44:1–8
50. Payne DC, Vinje J, Szilagyi PG, Edwards KM, Staat MA, Weinberg GA, Hall CB, Chappell J, Bernstein DI, Curns AT, Wikswo M, Shirley SH, Hall AJ, Lopman B, Parashar UD (2013) Norovirus and medically attended gastroenteritis in U.S. children. N Engl J Med 368:1121–1130
51. Pringle K, Lopman B, Vega E, Vinje J, Parashar UD, Hall AJ (2015) Noroviruses: epidemiology, immunity and prospects for prevention. Future Microbiol 10:53–67
52. Ramani S, Atmar RL, Estes MK (2014) Epidemiology of human noroviruses and updates on vaccine development. Curr Opin Gastroenterol 30:25–33
53. Ramani S, Neill FH, Opekun AR, Gilger MA, Graham DY, Estes MK, Atmar RL (2015) Mucosal and cellular immune responses to Norwalk virus. J Infect Dis 212:397–405
54. Riddle MS, Walker RI (2016) Status of vaccine research and development for norovirus. Vaccine 34:2895–2899
55. Roberts LO, Al-Molawi N, Carter MJ, Kass GEN (2003) Apoptosis in cultured cells infected with feline calicivirus. Apoptosis: from signaling pathways to therapeutic tools. Ann NY Acad Sci 1010:587–590
56. Rohayem J, Robel I, Jager K, Scheffler U, Rudolph W (2006) Protein-primed and de novo initiation of RNA synthesis by norovirus 3Dpol. J Virol 80:7060–7069
57. Sandoval-Jaime C, Gutierrez-Escolano AL (2009) Cellular proteins mediate 5′-3′ end contacts of Norwalk virus genomic RNA. Virology 387:322–330
58. Santi L, Batchelor L, Huang Z, Hjelm B, Kilbourne J, Arntzen CJ, Chen Q, Mason HS (2008) An efficient plant viral expression system generating orally immunogenic Norwalk virus-like particles. Vaccine 26:1846–1854
59. Sarvestani ST, Cotton B, Fritzlar S, O'Donnell TB, Mackenzie JM (2016) Norovirus infection: replication, manipulation of host, and interaction with the host immune response. J Interferon Cytokine Res 36:215–225
60. Simmons K, Gambhir M, Leon J, Lopman B (2013) Duration of immunity to norovirus gastroenteritis. Emerg Infect Dis 19:1260–1267
61. Smits SL, Rahman M, Schapendonk CM, van Leeuwen M, Faruque AS, Haagmans BL, Endtz HP, Osterhaus AD (2012) Calicivirus from novel recovirus genogroup in human diarrhea, Bangladesh. Emerg Infect Dis 18:1192–1195
62. Smits SL, Rahman M, Schapendonk CME, van Leeuwen M, Faruque ASG, Haagmans BL, Endtz HP, Osterhaus ADME (2012) Calicivirus from novel recovirus genogroup in human diarrhea, Bangladesh. Emerg Infect Dis 18:1192–1195
63. Sosnovtsev SV (2010) Proteolytic cleavage and viral proteins. In: Xi JJ, Kim YG, Grant SH (eds) Caliciviruses, molecular and cellular virology, vol 1. Caister Academic, Norfolk
64. Sosnovtsev SV, Belliot G, Chang KO, Onwudiwe O, Green KY (2005) Feline calicivirus VP2 is essential for the production of infectious virions. J Virol 79:4012–4024
65. Sosnovtsev SV, Belliot G, Chang KO, Prikhodko VG, Thackray LB, Wobus CE, Karst SM, Virgin HW, Green KY (2006) Cleavage map and proteolytic processing of the murine norovirus nonstructural polyprotein in infected cells. J Virol 80:7816–7831

66. Sosnovtsev SV, Garfield M, Green KY (2002) Processing map and essential cleavage sites of the nonstructural polyprotein encoded by ORF1 of the feline calicivirus genome. J Virol 76:7060–7072
67. Sosnovtsev SV, Prikhod'ko EA, Belliot G, Cohen JI, Green KY (2003) Feline calicivirus replication induces apoptosis in cultured cells. Virus Res 94:1–10
68. Stuart RL, Tan K, Mahar JE, Kirkwood CD, Andrew Ramsden C, Andrianopoulos N, Jolley D, Bawden K, Doherty R, Kotsanas D, Bradford J, Buttery JP (2010) An outbreak of necrotizing enterocolitis associated with norovirus genotype GII.3. Pediatr Infect Dis J 29:644–647
69. Subba-Reddy CV, Goodfellow I, Kao CC (2011) VPg-primed RNA synthesis of norovirus RNA-dependent RNA polymerases by using a novel cell-based assay. J Virol 85:13027–13037
70. Sundararajan A, Sangster MY, Frey S, Atmar RL, Chen WH, Ferreira J, Bargatze R, Mendelman PM, Treanor JJ, Topham DJ (2015) Robust mucosal-homing antibody-secreting B cell responses induced by intramuscular administration of adjuvanted bivalent human norovirus-like particle vaccine. Vaccine 33:568–576
71. Tacket CO, Mason HS, Losonsky G, Estes MK, Levine MM, Arntzen CJ (2000) Human immune responses to a novel norwalk virus vaccine delivered in transgenic potatoes. J Infect Dis 182:302–305
72. Tamminen K, Lappalainen S, Huhti L, Vesikari T, Blazevic V (2013) Trivalent combination vaccine induces broad heterologous immune responses to norovirus and rotavirus in mice. PLoS One 8:e70409
73. Tan M, Jiang X (2005) Norovirus and its histo-blood group antigen receptors: an answer to a historical puzzle. Trends Microbiol 13:285–293
74. Tan M, Jiang X (2012) Norovirus P particle: a subviral nanoparticle for vaccine development against norovirus, rotavirus and influenza virus. Nanomedicine (Lond) 7:889–897
75. Tan M, Jiang X (2005) The p domain of norovirus capsid protein forms a subviral particle that binds to histo-blood group antigen receptors. J Virol 79:14017–14030
76. Thackray LB, Wobus CE, Chachu KA, Liu B, Alegre ER, Henderson KS, Kelley ST, Virgin HW 4th (2007) Murine noroviruses comprising a single genogroup exhibit biological diversity despite limited sequence divergence. J Virol 81:10460–10473
77. Treanor JJ, Atmar RL, Frey SE, Gormley R, Chen WH, Ferreira J, Goodwin R, Borkowski A, Clemens R, Mendelman PM (2014) A novel intramuscular bivalent norovirus virus-like particle vaccine candidate: reactogenicity, safety, and immunogenicity in a phase 1 trial in healthy adults. J Infect Dis 210:1763–1771
78. Vashist S, Urena L, Chaudhry Y, Goodfellow I (2012) Identification of RNA–protein interaction networks involved in the norovirus life cycle. J Virol 86:11977–11990
79. Vashist S, Urena L, Gonzalez-Hernandez MB, Choi J, de Rougemont A, Rocha-Pereira J, Neyts J, Hwang S, Wobus CE, Goodfellow I (2015) Molecular chaperone Hsp90 is a therapeutic target for noroviruses. J Virol 89:6352–6363
80. Wolf S, Reetz J, Otto P (2011) Genetic characterization of a novel calicivirus from a chicken. Arch Virol 156:1143–1150
81. Yunus MA, Lin X, Bailey D, Karakasiliotis I, Chaudhry Y, Vashist S, Zhang G, Thorne L, Kao CC, Goodfellow I (2015) The murine norovirus core subgenomic RNA promoter consists of a stable stem-loop that can direct accurate initiation of RNA synthesis. J Virol 89:1218–1229

Chapter 4
Astroviruses Biology

Pavel Isa

1 Introduction

Human astroviruses (HAstV) have been found to be among the most important causes of viral diarrhea in children [10, 13, 26]. Epidemiological studies conducted in Argentina, Brazil, Colombia, Chile, Guatemala, Mexico, and Venezuela indicate that in these regions astroviruses are associated with nearly 4% to 7% of severe diarrhea cases in children under 5 years of age, with predominance of the HAstV serotype 1, although co-circulation of other serotypes has been observed [3, 6, 9, 27, 28, 30]. Moreover, the association of turkey astroviruses with the so-called poult enteritis complex has also been reported [4]. Astroviruses are small, non-enveloped viruses with a single-stranded positive-sense RNA genome, grouped into the family *Astroviridae*, which is divided into two genera: *Mamastrovirus*, including viruses infecting mammals, and *Avastrovirus*, including viruses that infect avian species. Human astroviruses (HAstV) are classified into four genotype species, and the classical eight original serotypes are classified now as *Mamastrovirus* genotype 1 [11]. The remaining genotypes are formed by the recently described HAstVs including the MLB, VA, and HMO virus lineages [11], and they are more closely related to animal astroviruses. Recent evidence indicated that these newly described astrovirus serotypes also circulate in the region [16, 32]. Given the importance of astroviruses as a cause of infantile acute diarrhea and the burden caused by diarrheas in Latin America, several research groups in the region have devoted efforts to better understand the astrovirus biology and epidemiology. Herein, the results of some of those efforts are reviewed.

P. Isa (✉)
Instituto de Biotecnología, Universidad Nacional Autónoma de México,
Cuernavaca, Morelos, Mexico
e-mail: pavel@ibt.unam.mx

J.E. Ludert et al. (eds.), *Human Virology in Latin America*,
DOI 10.1007/978-3-319-54567-7_4

2 Astrovirus Replication

As mentioned, the astrovirus genome is composed of a single-stranded positive-sense RNA molecule, which serves as messenger RNA and could be directly translated into the viral polyprotein. The genomic RNA (gRNA) ranges from 6.8 to 7.7 kb in length, depending on the species of isolation [21]. It contains 5′- and 3′-untranslated regions (UTRs) and three open reading frames (ORFs), named ORF1a, 1b, and 2 (Fig. 4.1). In addition, a short ORF X, present in human and some mammalian astrovirus strains and containing between 91 and 122 codons, is predicted [7] (Fig. 4.1). This ORF X is presumably translated in a +1 reading frame from ORF 2; however, there is no experimental evidence of protein synthesis. The astrovirus genome is polyadenylated at the 3′-end and at the 5′-end is linked covalently to VPg, a viral protein that is essential for viral replication [8, 21].

No astrovirus receptors have been identified to date; however, it seems that different serotypes use different cell-surface molecules [2]. The half-time of virus binding to cell surface is about 10 min, and virus decapsidation (analyzed by measurement of release of gRNA) takes place about 130 min later [20]. Astrovirus cell entry is by clathrin-mediated endocytosis and is dependent on the active actin network, endosome acidification, and membrane cholesterol [5, 20]. Given that silencing expression of Rab7, a small GTPase involved in the early-to-late endosome traffic, reduces astrovirus infectivity, it seems that astroviruses need to arrive to late endosomes to enter the cytoplasm. In agreement with this, as mentioned earlier, gRNA is released from particles at about 130 min after infection [20].

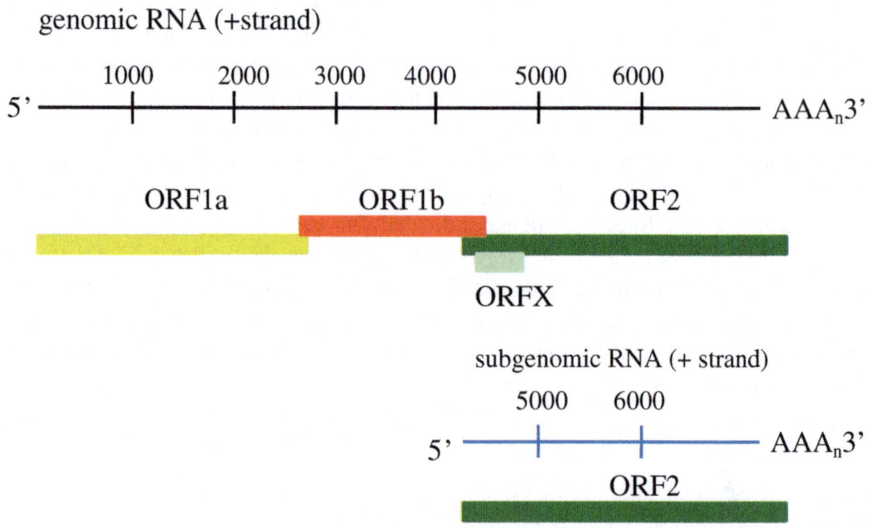

Fig. 4.1 Human astrovirus genome organization. The positive-strand genomic RNA (approximately 6.8 kb) contains three verified open reading frames (ORF): ORF1a, ORF1b, and ORF2. The fourth short ORF (ORFX) has been postulated given its conservation among members of the family. The ORF1b is translated as part of ORF1ab by ribosomal frameshift; ORF2 is translated from subgenomic RNA produced during astrovirus replication cycle

After entering the cells and releasing the gRNA, the protein synthesis starts, with all astrovirus ORFs producing polyproteins, which are proteolytically cleaved to produce the final viral proteins. The nonstructural proteins, nsp1a and nsp1ab, are translated from the 5′-most ORFs (ORF1a and ORF1b) of released gRNA (Fig. 4.2a). These viral proteins use the gRNA as template to synthesize a full-length, negative-sense, antigenomic RNA (agRNA), which in turn is used as template to produce more copies of the full-length gRNA and many copies of a

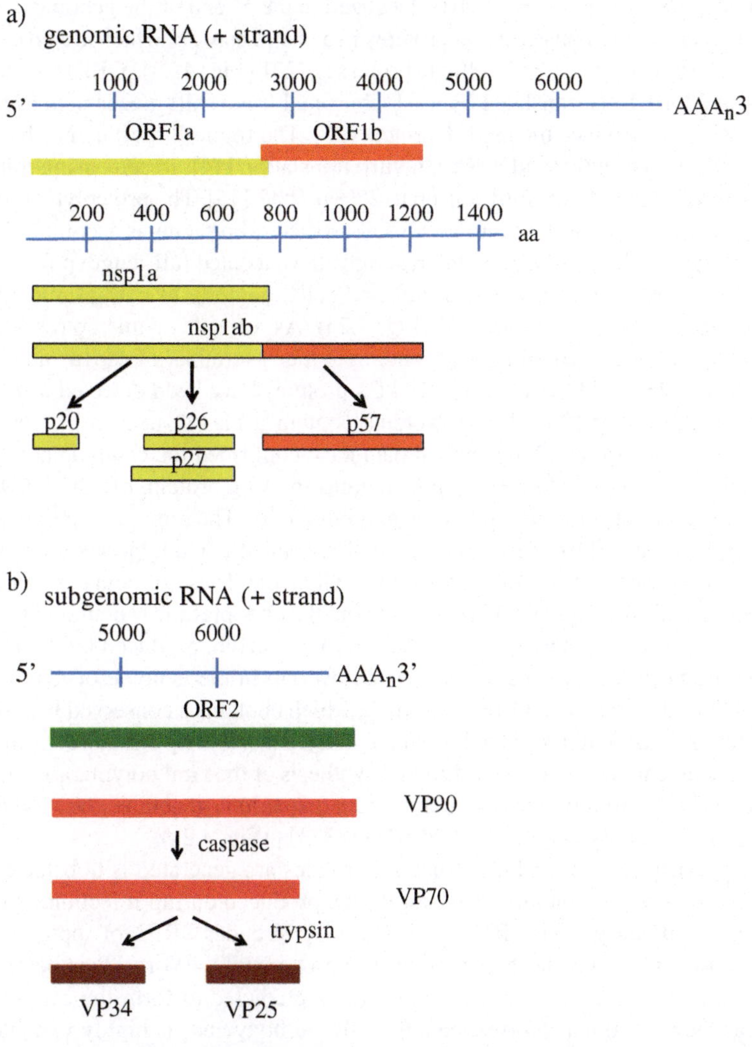

Fig. 4.2 Proteolytic processing of astrovirus proteins. Astrovirus proteins are obtained by proteolytic cleavage of polyprotein precursors. Nonstructural proteins are translated from ORF1a and ORF1ab (**a**); structural proteins are generated from ORF2 (**b**). The polyprotein processing is performed by viral and cellular proteases. In the structural proteins precursor (*VP90*), this polyprotein is cleaved by caspases to generate VP70, which is further processed by trypsin. Characterized final astrovirus proteins are shown at the *bottom*

subgenomic RNA (sgRNA) of about 2.4 kDa [24]. The viral structural proteins (VPs) are synthesized as a polyprotein from ORF2, from sgRNA, which is processed to produce viral structural proteins by cellular and viral proteases [19].

3 Processing of Astrovirus Capsid Polypeptides

As already mentioned, the two ORFs localized in the 5′-end of the genome (ORF1a and 1b) code for the nonstructural proteins nsp1a and nsp1ab, which are proteolytically processed by viral and cellular proteases [23] (Fig. 4.2a). ORF1a codes for nsp1a, and ORF1b is translated by a −1 ribosomal frameshift mechanism as fusion with ORF1a, to produce the nsp1ab protein [23]. The frameshift efficiency has been found to be between 6% and 7% by in vitro translation [14]; in cells using transient expression systems, frameshift was up to 25% to 28% [15]. The proteolytic processing of the initial polyprotein into various nonstructural proteins is accomplished by both viral and cellular proteases. Interestingly, the expected full-length polyproteins nsp1a and nsp1ab are not found in infected cells, pointing to cotranslational processing during their synthesis [23] (Fig. 4.2a). As a result of this cotranslational processing, a 20-kDa protein corresponding to the N-terminal of nsp1a and nsp1ab is produced [23]. Additionally, two 27-kDa proteins have been detected as a result of nsp1a processing [23]. These proteins contain a classic trypsin-like protease motif, and structural and biochemical data have confirmed the catalytic activity of the protein [29]. The following region contains the VPg protein, of 13–15 kDa [8], which is important for viral replication and infectivity. The remaining polypeptides derived from the carboxy-terminal region of nsp1a include a hypervariable region, which is proposed to be relevant for viral replication. These proteins are all phosphorylated and co-localize with the gRNA and the endoplasmic reticulum [12]. The second protein to be translated from the gRNA is nsp1ab, produced by a −1 translational frameshift mechanism. There is an overlap in the genome of mammalian and avian viruses between ORFs 1a and 1b, which contains a conserved heptameric sequence (AAAAAAC) followed by a potential stem-and-loop structure. Both these motifs are essential for the frameshift and synthesis of the viral polymerase encoded in ORF1b [17]. From this second polyprotein (nsp1ab) is generated the viral RNA-dependent RNA polymerase of 57–59 kDa [23, 31] (Fig. 4.2a).

The polypeptide from which structural proteins are generated is translated from a polyadenylated subgenomic RNA (sgRNA), produced during infection, which is 3′ colinear with the genomic RNA [24]. This sgRNA codes ORF2, of approximately 780 amino acids, with the first region of approximately 400 residues being conserved among human astrovirus strains, and is predicted to form the capsid core, whereas the second half (corresponding to the carboxy end) is highly variable and is predicted to form viral spikes. Processing of the ORF2 translation product was studied in detail for human serotype 8 astrovirus YUC 8 [19].

Processing of the ORF2 protein product is schematically shown in Fig. 4.2b. The primary product of ORF2 translation, VP90, is initially cleaved at its carboxy-terminal region to yield the VP70 protein [19, 22]. The intracellular processing of VP90 is mediated by cellular caspases and is associated with the release of human astrovirus [22]. Treatment of purified viral particles with trypsin results in an ordered processing of VP70, to final proteins VP34, VP27, and VP25, and the trypsin processing is associated with an increase in viral infectivity [19]. The trypsin processing responsible for increase of infectivity of YUC8 strain seems to be sequential, with the first cut probably exposing new, trypsin-susceptible sites, suggesting conformational changes associated with trypsin activation.

Astroviruses, as other single-stranded RNA virus families, associate with membranes during their replication. Both structural (VP90) and nonstructural (protease, polymerase) proteins were found to localize with intracellular membranes [18, 25]. The ultrastructural study of membrane rearrangement by electron microscopy showed large groups of viral particles around what appears to be double-membrane O-ring vesicles, inside of which structural proteins and the viral polymerase were detected by immunoelectron microscopy [12, 18].

4 Astrovirus Assembly

Astrovirus assembly has been studied in Caco-2 cells. It was observed that viral particles are formed by VP90, which is then processed by the aforementioned cellular enzymes (caspase 3 and caspase 9) to yield VP70 containing viral particles [22]. Caspase processing of VP90 is gradual, and several intermediate proteins (82, 75, and 72 kDa) are observed. Cleavage of VP90 to VP70 is associated with viral release, through a mechanism so far uncharacterized [1, 22]. Astroviruses need additional proteolytic cleavage to gain infectivity, as viruses grown in the absence of protease contain only VP70 protein, which needs to be processed as described to produce mature viral particles containing the final 25-, 27-, and 34-kDa proteins. This final proteolytic processing is expected to utilize proteases present in the intestinal lumen of the host, although no experimental evidence for this asseveration exists.

Despite progress in characterization of astrovirus biology, many points are still not clear or not well understood. Given the differences observed among serotypes belonging to "classical" *Mamastrovirus* genotype 1, and the recent description of three new astrovirus genotypes, much work still awaits to better understand astrovirus biology.

5 Conclusions

Recent work demonstrates that astroviruses are important pathogens, causing disease in humans and animals. The discovery of new astroviruses from humans that show sequence similarities to strains isolated from other animal species indicates that interspecies transmission could occur. This observation, together with the fact that in some animal species astroviruses cause other than gastrointestinal disease, strengthens the need to further study astrovirus–host cell interaction. It is expected that future characterization of viral particle structure will allow us to understand processes associated with the cell replication cycle, including viral entry, replication, assembly, and egress. It would be important to study and compare these processes among distinct astrovirus groups as there may well be significant differences in replication cycle among them.

References

1. Banos-Lara Mdel R, Mendez E (2010) Role of individual caspases induced by astrovirus on the processing of its structural protein and its release from the cell through a non-lytic mechanism. Virology 401:322–332
2. Brinker JP, Blacklow NR, Herrmann JE (2000) Human astrovirus isolation and propagation in multiple cell lines. Arch Virol 145:1847–1856
3. Cruz JR, Bartlett AV, Herrmann JE, Caceres P, Blacklow NR, Cano F (1992) Astrovirus-associated diarrhea among Guatemalan ambulatory rural children. J Clin Microbiol 30:1140–1144
4. Da Silva SE, Bonetti AM, Petrocelli AT, Ferrari HF, Luvizotto MC, Cardoso TC (2008) Detection of Turkey astrovirus in young poults affected with poult enteritis complex in Brazil. J Vet Med Sci 70:629–631
5. Donelli G, Superti F, Tinari A, Marziano ML (1992) Mechanism of astrovirus entry into Graham 293 cells. J Med Virol 38:271–277
6. Espul C, Martinez N, Noel JS, Cuello H, Abrile C, Grucci S, Glass R, Berke T, Matson DO (2004) Prevalence and characterization of astroviruses in Argentinean children with acute gastroenteritis. J Med Virol 72:75–82
7. Firth AE, Atkins JF (2010) Candidates in Astroviruses, Seadornaviruses, Cytorhabdoviruses and Coronaviruses for +1 frame overlapping genes accessed by leaky scanning. Virol J 7:17
8. Fuentes C, Bosch A, Pinto RM, Guix S (2012) Identification of human astrovirus genome-linked protein (VPg) essential for virus infectivity. J Virol 86:10070–10078
9. Gonzalez GG, Liprandi F, Ludert JE (2011) Molecular epidemiology of enteric viruses in children with sporadic gastroenteritis in Valencia, Venezuela. J Med Virol 83:1972–1982
10. Guerrero ML, Noel JS, Mitchell DK, Calva JJ, Morrow AL, Martinez J, Rosales G, Velazquez FR, Monroe SS, Glass RI, Pickering LK, Ruiz-Palacios GM (1998) A prospective study of astrovirus diarrhea of infancy in Mexico City. Pediatr Infect Dis J 17:723–727
11. Guix S, Bosch A, Pinto RM (2013) Astrovirus taxonomy. In: Schults-Cherry S (ed) Astrovirus research: essential ideas, everyday impacts, future directions. Springer, New York, pp 97–118
12. Guix S, Caballero S, Bosch A, Pinto RM (2004) C-terminal nsP1a protein of human astrovirus colocalizes with the endoplasmic reticulum and viral RNA. J Virol 78:13627–13636
13. Lew JF, Moe CL, Monroe SS, Allen JR, Harrison BM, Forrester BD, Stine SE, Woods PA, Hierholzer JC, Herrmann JE et al (1991) Astrovirus and adenovirus associated with diarrhea in children in day care settings. J Infect Dis 164:673–678
14. Lewis TL, Matsui SM (1995) An astrovirus frameshift signal induces ribosomal frameshifting in vitro. Arch Virol 140:1127–1135
15. Lewis TL, Matsui SM (1996) Astrovirus ribosomal frameshifting in an infection-transfection transient expression system. J Virol 70:2869–2875
16. Lizasoain A, Tort LF, Garcia M, Gomez MM, Leite JP, Miagostovich MP, Cristina J, Colina R, Victoria M (2015) Environmental assessment reveals the presence of MLB-1 human astrovirus in Uruguay. J Appl Microbiol 119:859–867

17. Marczinke B, Bloys AJ, Brown TD, Willcocks MM, Carter MJ, Brierley I (1994) The human astrovirus RNA-dependent RNA polymerase coding region is expressed by ribosomal frameshifting. J Virol 68:5588–5595
18. Mendez E, Aguirre-Crespo G, Zavala G, Arias CF (2007) Association of the astrovirus structural protein VP90 with membranes plays a role in virus morphogenesis. J Virol 81:10649–10658
19. Mendez E, Fernandez-Luna T, Lopez S, Mendez-Toss M, Arias CF (2002) Proteolytic processing of a serotype 8 human astrovirus ORF2 polyprotein. J Virol 76:7996–8002
20. Mendez E, Munoz-Yanez C, Sanchez-San Martin C, Aguirre-Crespo G, Banos-Lara Mdel R, Gutierrez M, Espinosa R, Acevedo Y, Arias CF, Lopez S (2014) Characterization of human astrovirus cell entry. J Virol 88:2452–2460
21. Mendez E, Murillo A, Velazquez-Moctezuma R, Burnham A, Arias C (2013)) Replication cycle of Astroviruses. In: Schults-Cherry S (ed) Astrovirus research: essential ideas, everyday impacts, future directions, vol 1. Springer, New York, pp 19–45
22. Mendez E, Salas-Ocampo MP, Arias C (2004) Caspases mediate processing of the capsid precursor and cell release of human astroviruses. J Virol 78:8601–8608
23. Mendez E, Salas-Ocampo MP, Munguia ME, Arias CF (2003) Protein products of the open reading frames encoding nonstructural proteins of human astrovirus serotype 8. J Virol 77:11378–11384
24. Monroe SS, Jiang B, Stine SE, Koopmans M, Glass RI (1993) Subgenomic RNA sequence of human astrovirus supports classification of *Astroviridae* as a new family of RNA viruses. J Virol 67:3611–3614
25. Murillo A, Vera-Estrella R, Barkla BJ, Mendez E, Arias CF (2015) Identification of host cell factors associated with astrovirus replication in Caco-2 cells. J Virol 89:10359–10370
26. Rodriguez-Baez N, O'Brien R, Qiu SQ, Bass DM (2002) Astrovirus, adenovirus, and rotavirus in hospitalized children: prevalence and association with gastroenteritis. J Pediatr Gastroenterol Nutr 35:64–68
27. Silva PA, Santos RA, Costa PS, Teixeira JM, Giugliano LG, Andreasi MS, Leite JP, Schreier E, Cardoso D (2009) The circulation of human astrovirus genotypes in the Central West Region of Brazil. Mem Inst Oswaldo Cruz 104:655–658
28. Soares CC, Maciel de Albuquerque MC, Maranhao AG, Rocha LN, Ramirez ML, Benati FJ, Timenetsky Mdo C, Santos N (2008) Astrovirus detection in sporadic cases of diarrhea among hospitalized and non-hospitalized children in Rio de Janeiro, Brazil, from 1998 to 2004. J Med Virol 80:113–117
29. Speroni S, Rohayem J, Nenci S, Bonivento D, Robel I, Barthel J, Luzhkov VB, Coutard B, Canard B, Mattevi A (2009) Structural and biochemical analysis of human pathogenic astrovirus serine protease at 2.0 Å resolution. J Mol Biol 387:1137–1152
30. Walter JE, Mitchell DK, Guerrero ML, Berke T, Matson DO, Monroe SS, Pickering LK, Ruiz-Palacios G (2001) Molecular epidemiology of human astrovirus diarrhea among children from a periurban community of Mexico City. J Infect Dis 183:681–686
31. Willcocks MM, Boxall AS, Carter MJ (1999) Processing and intracellular location of human astrovirus non-structural proteins. J Gen Virol 80(pt 10):2607–2611
32. Xavier Mda P, Carvalho Costa FA, Rocha MS, Andrade Jda S, Diniz FK, Andrade TR, Miagostovich MP, Leite JP, Volotao Ede M (2015) Surveillance of human astrovirus infection in Brazil: the first report of MLB1 astrovirus. PLoS One 10:e0135687

Chapter 5
Molecular Epidemiology of Human Rotaviruses in Latin America

Juan I. Degiuseppe, Juan A. Stupka, and Gabriel I. Parra

1 Rotavirus Disease

Rotavirus is the leading cause of acute watery diarrhea in children under 5 years of age worldwide [30]. It causes 125 million cases every year, of which 2 million require hospitalization. It is estimated that 200,000–450,000 deaths annually are associated with rotavirus infections [46, 75, 101].

Rotavirus disease has high economic impact in terms of direct costs to the health system and indirect costs to society. Because the prevalence of rotavirus infection is similar in different countries, this infection is not an accurate indicator of the health status of a population [106]. However, it is associated with higher mortality rates in developing countries: about 90% of rotavirus diarrhea-related deaths occur in low-income and middle- to low-income countries [75].

In Latin America, the latest rotavirus estimates show that it is responsible for 10 million cases of diarrhea, 2 million clinical visits, 75,000 hospitalizations, and between 15,000 and 45,000 deaths each year in children under 5 years [48, 107]. During the past decade most Latin American countries have implemented massive rotavirus vaccination programs, and, consequently, a sustained decline in these numbers has been documented [23, 24, 65, 71, 82, 88], mainly in rotavirus-associated hospitalizations and deaths. In this chapter we briefly describe the current status of the molecular characterization of rotavirus strains circulating in Latin American countries during the past two decades.

J.I. Degiuseppe • J.A. Stupka
Laboratorio Nacional de Referencia para Rotavirus y Norovirus, Instituto Nacional de Enfermedades Infecciosas ANLIS ("Dr. Carlos G. Malbrán"), Buenos Aires, Argentina

G.I. Parra (✉)
Division of Viral Products, Food and Drug Administration, Silver Spring, MD, USA
e-mail: gabriel.parra@fda.hhs.gov

J.E. Ludert et al. (eds.), *Human Virology in Latin America*,
DOI 10.1007/978-3-319-54567-7_5

2 Rotavirus Characteristics

The rotavirus is a non-enveloped virus with a genome composed of 11 segments of double-stranded RNA. These segments encode for six structural proteins (VP1, VP2, VP3, VP4, VP6, VP7) and six nonstructural proteins (NSP1, NSP2, NSP3, NSP4, NSP5, NSP6). During viral replication these 11 segments are packaged within a triple-layered capsid in the cytoplasm of the infected cell. The innermost layer (*core*) of the capsid consists of VP2 protein (Fig. 5.1a, *blue*); the intermediate layer is formed by VP6 protein; and the outermost layer of the capsid is composed of VP4 and VP7 proteins (Fig. 5.1a). Although VP6 constitutes about half the virion structure, VP4 and VP7 are the major targets for specific neutralizing antibodies.

Based on differences in the amino acid sequence of VP6 protein, rotaviruses can be classified into eight groups (A–H); of which only rotavirus group A (RVA), group B (RVB), and group C (RVC) are known to infect humans [30, 63]. RVA are the most predominant (>90% of total isolates worldwide), whereas RVB and RVC have been described to cause outbreaks in adults or are sporadically detected in children with diarrheic symptoms, respectively. In addition to major antigenic differences on VP6, the different rotavirus groups can be identified by the characteristic migration pattern (or electropherotype) of their segmented genome in polyacrylamide gel electrophoresis (Fig. 5.1b). Although minor differences may be detected in the

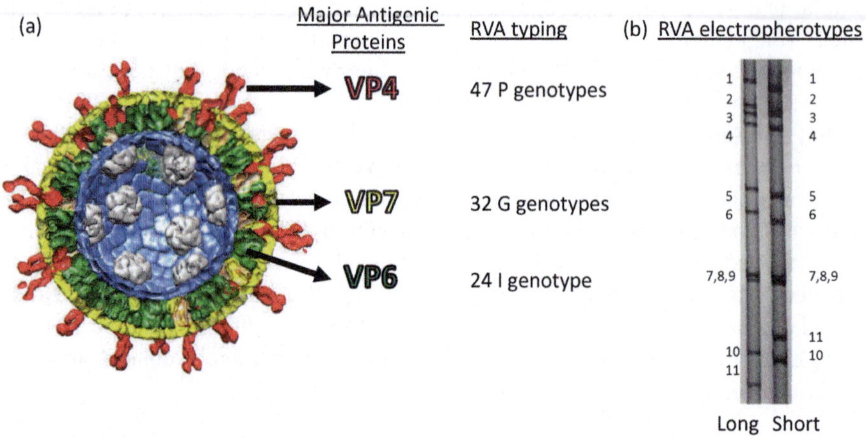

Fig. 5.1 Characterization of rotaviruses. **a** Representation of internal structure of the rotavirus virion. The three major antigenic proteins used for typing are indicated in *red* (*VP4*), *green* (*VP6*), and *yellow* (*VP7*). The current number of genotypes for rotavirus group A (*RVA*) are indicated on the *right side*. The rotavirus virion was modeled using crystal coordinates from VP2, VP6, VP7, and VP4 proteins (Protein Data Bank accession numbers: 3IYU and 3N09) and visualized in Chimera Software. **b** RVA electrophoresis migration patterns (electropherotypes). Strains presenting long and short electropherotypes are shown. In the short electropherotype the gene segment 11 migrates more slowly than gene segment 10

migration patterns of each group, RVA display three major electropherotypes (i.e., long, short, and super-short) depending on the differential migration of the smallest segment. These differences are based on the variable genetic structure of segment 11, where gene rearrangements or AT-rich sequences (of unknown origin) can be detected in the 3′-untranslated region of the segment [30, 34, 60].

Electropherotypes were widely used during the 1980s to distinguish the different rotaviruses in clinical samples; however, this technique was replaced by characterization of the two outermost capsid proteins, VP4 and VP7. These two viral proteins define the G (glycoprotein, VP7) and P (protease sensitive, VP4) serotypes, which serve as the basis of a binary classification system for RVA (Fig. 5.1a) [30]. Because antigenic characterization required specific reagents and a large collection of viruses, not always available in many laboratories, it was replaced by the genetic characterization of the segments encoding the VP7 and VP4 proteins (G and P genotypes, respectively) [21, 33, 39, 43]. Currently, 32 G genotypes and 47 P genotypes of RVA are detected in different species of mammals and birds [85, 103]. Of these, 11 G genotypes and 13 P genotypes have been described as infecting humans. G1, G2, G3, G4, G9, P[4], P[6], and P[8] are the most common genotypes in humans, and although 15 G and P combinations are possible, only 5 of these (G1P[8], G2P[4], G3P[8], G4P[8], and G9P[8]) are predominant in humans [30, 61, 91]. Less common genotypes (G5, G8, and G10) have been reported with high frequencies in certain geographic regions, but the reasons for these different epidemiological patterns are poorly understood [64, 91].

Although it was initially thought that rotaviruses could freely assort their gene segments during coinfection with two strains, the application of genomics and biochemistry to different groups has shown that there are restrictions to the reassortment of genes, which is mostly regulated by interaction between different viral proteins and virus–host interactions [42, 64, 66–69]. Thus, for RVA three characteristic groups of genes (gene constellations or genogroups) have been defined and named according to their reference strain, that is, Wa, DS-1, and AU-1 [74].

Recently, a new classification system including nucleotide sequence analyses of the 11 genomic groups has been implemented for RVA. Hence, in addition to G and P genotypes, genotypes of the other nine genes (Ix-Rx-Cx-Mx-Ax-Nx-Tx-Ex-Hx) are being utilized. In this system, the Wa strain is classified as genotype 1 (G1-P[8]-I1-R1-C1-M1-A1-N1-T1-E1-H1), the DS-1 strain as genotype 2 (G2-P[4]-I2-R2-C2-M2-A2-N2-T2-E2-H2), and the AU-1 strain as genotype 3 (G3-P[9]-I3-R3-C3-M3-A3-N3-T3-E3-H3). The genome constellations of genotypes 1 and 2 are more fit than genotype 3 in the human population and are therefore prevalent. This system provides a universal framework for analyzing inter- and intraspecies evolutionary relationships, genetic reassortment events, and characterization of new strains [61, 64]. In general, there is correlation between the different rotavirus classification systems. For example, RVA displaying a short electropherotype belong to the G2P[4] genotype, and the genotype 2 constellation is observed. However, RVA showing long electropherotypes usually associates with G1P[8], G3P[8], G4P[8], or G9P[8] genotypes, and the genotype 1 genomic constellation [42, 61].

3 Molecular Epidemiology in Latin America

Because of the heavy burden of rotaviruses on the Latin American population, much effort has been dedicated to characterize circulating viruses, both to prepare for introduction of vaccines and for post-vaccination surveillance. Although some countries have better information on the strains circulating in their populations, a search of the literature revealed that 24 Latin American countries have done some type of molecular characterization during the past two decades. Most of this characterization used clinical samples from children (under 5) presenting with gastroenteritis; however, characterization of older children, adults [1, 56, 57], different animal species [5, 49, 69, 70, 79, 84, 96], and environmental samples [31, 73, 102] has also been done to improve understanding of the natural history of this disease. These studies revealed that the most common strains circulating in Latin America are G1P[8], G2P[4], G3P[8], G4P[8], G9P[8], and more recently, G12P[8] [48, 52, 98] (Fig. 5.2). Strains with unusual G and P associations can be detected sporadically in different countries, and some are detected at high levels in the human population, mostly linked to probable

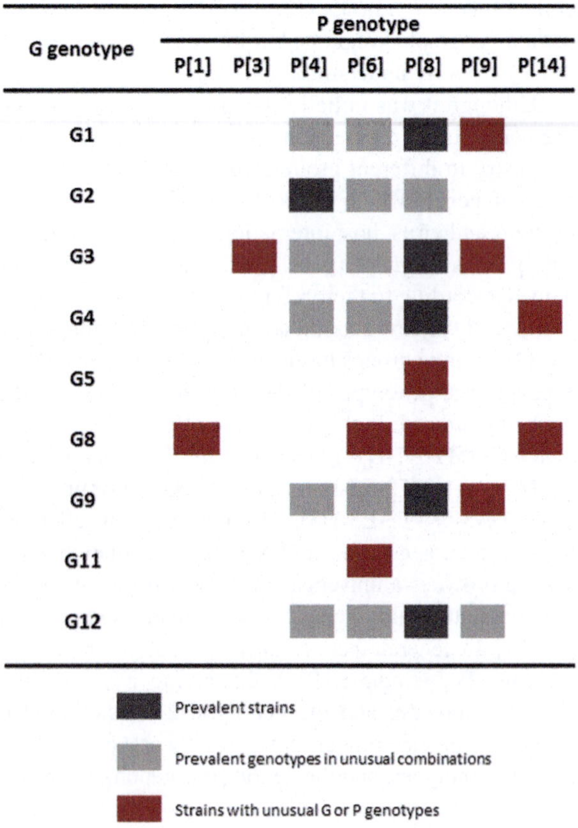

Fig. 5.2 G and P genotype association in rotavirus group A (RVA) detected in Latin America countries

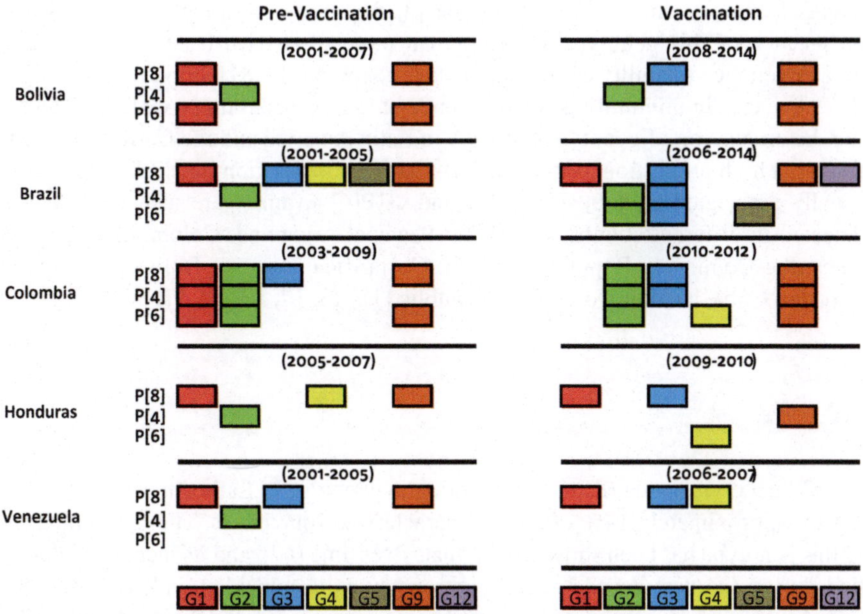

Fig. 5.3 Graphical representation of the different rotavirus group A strains detected in five Latin American countries that have implemented rotavirus vaccination. Periods of surveillance are indicated within *brackets*. Each G genotype is *color coded*, and the corresponding P genotype is indicated on the *left-hand side*

zoonotic events [40, 92]. In Latin America, the "unusual" strains detected in humans have been mostly GxP[6] or GxP[9]; e.g. G1P[9], G3P[9], G4P[6], G9P[6], G12P[6], G12P[9] [11, 19, 51, 56, 58, 90, 93, 97, 104, 105]. In contrast to developed countries, where the G1P[8] strain were the predominant strains each season in the pre-vaccination era, in Latin America the predominant strain changed season to season without any marked epidemiological pattern. Co-circulations and heterogeneous regional distributions were also observed (Fig. 5.3) [26, 76, 78, 91, 98]. For example, different strains predominated in different cities in the same country, thereby complicating our understanding of the epidemiological pattern of rotaviruses.

4 Predominant Genotypes of RVA in Latin America

4.1 G1

G1 is considered the most prevalent circulating RVA genotype worldwide, but its prevalence is significantly higher in developed countries [91]. Phylogenetic analyses based on gene encoding of VP7 by G1 strains have demonstrated that at least six lineages circulate in humans worldwide [3, 81]. The continuous circulation and persistence of this genotype in the human population have been attributed to a greater genetic variability compared to other genotypes [3, 81]; however, the role of these lineages in immunity is not well understood. A recent meta-analysis evaluating the circulation of genotypes of RVA in Latin America and the Caribbean documented a high proportion (~35%) of G1 strains for this region [48]. Although G1 is usually associated with P[8], G1P[4] and G1P[6] strains were also sporadically detected in different countries (G1P[4]: Panama, Argentina, Colombia, Paraguay, Cuba, the Dominican Republic; G1P[6]: Costa Rica, Panama, Bolivia, Colombia, Ecuador, Cuba, and the Dominican Republic [12, 13, 17, 27, 29, 77, 86, 90]).

4.2 G2

The G2 genotype is the second most common genotype in Latin America, accounting for approximately 14% of all the characterized rotaviruses. The predominance of this genotype has been shown to fluctuate over time [47], and its increasing detection during the past decade raised controversy as it coincided with the `massive introduction of the monovalent rotavirus vaccine (G1P[8]) in Brazil [41]. Considering that the circulation of G2P[4] rotaviruses in Paraguay, Argentina, and Colombia increased during the pre-vaccination period, and the high presence of G2P[4] was not sustained in Brazil, this phenomenon seems to be the result of natural fluctuation rather than selective vaccine pressure [56, 76]. Unusual combinations with this G genotype detected in the region were the G2P[6] in Brazil, Colombia, and Argentina, and G2P[8] in Costa Rica, Panama, Argentina, Colombia, Paraguay, Cuba, and Haiti [12, 13, 17, 26, 28, 52, 55, 56, 87, 105].

4.3 G3

The G3 genotype is one of the most frequently detected. This genotype has the broadest host range (humans, pigs, horses, cows, rabbits, cats, dogs, etc.) and consequently the greatest number of combinations with different P-types (P[2], P[3], P[6], P[7], P[8], P[9], P[12], P[14], P[22], P[24]) [30, 91]. Despite its extreme diversity, G3P[8] association is the most prevalent in human infections. G3P[8] was described as circulating at low frequency during the 1990s worldwide, but its reemergence has been observed in the past decade. The increasing frequency of this genotype seems to be associated with a decrease in the prevalence of G1P[8] and G9P[8], all of which present a similar genome constellation (genotype 1). In Latin America, some sporadic detection of G3P[4], G3P[6], G3P[3], and G3P[9] was reported, the last two associated with zoonotic events (Fig. 5.2) [12, 26, 52, 53, 105].

4.4 G4

Although a recent meta-analysis revealed an overall incidence of G4 strains less than 10% in Latin America (the fourth most prevalent genotype after G1, G2, and G9), this genotype has been studied extensively because of its increase in prevalence between the late 1990s and the early 2000s [9, 10, 48, 77, 79]. This increase in incidence, reaching 31% in Argentina, 44% in Paraguay, and 80% in Nicaragua, was attributed to the emergence of strains with two sublineages of VP7 (Ib and Ic). Interestingly, the strains of sublineage Ic presented a sole insertion (Asp76) in the N-terminal portion of the VP7 protein, one of the few limited insertions reported for this protein. After its emergence in Argentina and Paraguay [9, 10, 77] G4 sublineage Ic strains spread to Uruguay, Brazil, Nicaragua, Italy, Ireland, and Japan [8, 14, 97]. G4 rotavirus are common in pigs, usually associated with P[6] and P[9]. Phylogenetic analyses of the VP7 protein showed diversification of G4 into typical "human" and "porcine" strains, the "human" strains associated mostly with P[8] and "porcine" strains with P[6] [45, 79, 97]. G4P[6] strains have been detected in humans with diarrhea in Argentina, Brazil, Colombia, Mexico, Paraguay, Guatemala, Honduras, and Colombia, and showed porcine-like genomes.

4.5 G9

The G9 genotype was first detected in the United States in 1983, and since its detection this genotype has been reported associated with a wide variety of P genotypes [44]. Since the mid-1990s, a significant increase in the frequency of detection has been described globally, initially associated with P[6] and subsequently mostly associated with P[8] [62, 91]. In Latin America, G9P[6] was identified in Argentina and Brazil from 1996 to 1999; G9P[8] was initially identified in Brazil in 1996 [2, 11, 95], and spread throughout the continent. Currently, it is considered the second most prevalent genotype together with G2P[4] rotaviruses. Similarly to G4 rotavirus, phylogenetic analyses of the VP7 encoding gene suggest divergence of G9 viruses into "human-like" and "porcine-like," with the human-like viruses further divided into three lineages (I–III) [44, 94]. Lineage III, comprising more than 90% of all known G9 strains, was able to spread globally in just one decade [62]. Although, the predominant strain is G9P[8], many countries in Latin America have reported rotavirus with G9P[4] (Mexico, Costa Rica, Guatemala, Argentina, Honduras, Colombia, Paraguay, Cuba), G9P[6] (Argentina, Bolivia, Colombia, Ecuador, Peru, Cuba), and G9P[9] (Mexico and Argentina) (Fig. 5.2) [2, 7, 13, 18, 27, 56, 77, 83, 95, 99].

5 Emerging Genotypes

5.1 G5

The G5 genotype is commonly found in pigs, but its circulation has been documented in horses and cattle as well [40]. The first cases of human infection were reported in Brazil, associated with P[8] as a result of zoonotic transmission events [38] that originated the human lineages circulating in Brazil (lineage I), Asia, and Africa (lineage II) [20].

This genotype has been identified in all Brazilian regions and was detected for 14 years (1982–1996). Subsequently, its prevalence decreased sharply, losing its endemic nature, and was detected only sporadically [20, 47]. Although the G5 genotype had a wide dissemination in Brazil during the 1990s, is has not been found to circulate in Latin America since the 2000s [47, 48].

5.2 G8

The G8 genotype is considered a bovine genotype as early reports showed its presence in cattle. Beside cows, this genotype has been found in other species of the Artiodactyla order [50, 80, 91]. The G8 genotype is of great interest for the scientific community as it is highly prevalent in different countries in Africa. In Latin America, it has been detected sporadically as G8P[8] in Central America countries, G8P[6] in Brazil, G8P[14] in Venezuela, and G8P[1] in Paraguay, the latter two associated with interspecies transmission (Fig. 5.2) [47, 59, 72].

5.3 G12

The G12 genotype is considered the latest emerging genotype in humans that has been able to spread worldwide [48, 62]. Its incidence has increased during the past decade, its spread associated with two main lineages of VP7 protein. Strains of lineage II show the P[9] genotype, usually associated with feline strains, whereas lineage III strains show the P[4], P[6], or P[8] genotype (Fig. 5.2). In South America, G12P[9] strains were initially detected in Argentina and Brazil, and later in Paraguay [19, 47, 52, 56, 98]. These strains were found to present a genome very similar to the AU-1 strain (genotype 3). G12P[8] strains were described as circulating in several other regions, but were not detected in South America until 2008, when they were detected with high frequency in Argentina. Complete genome analyses of the Argentine G12P[8] strains demonstrated that they bore genotype 1 genes (i.e., Wa-like). Because the G12P[9] were the first to be introduced into Latin America (more specifically Argentina), it could initially be considered that G12P[8] was a

Table 5.1 Rotavirus full-genome analyses in Latin American countries

Whole Genome in Latinoamerica

Country	Genetic constellation	Aim of the analysis	References
Guatemala	**G8P[14]**-I2-R2-C2-M2-A13-N2-T6-E2-H3	Test zoonotic event	Gautam et al. (2015) Infect Genet Evol [32]
Nicaragua	**G1P[8]**-I1-R1-C1-M1-A1-N2-T1-E1-H1	Vaccination pressure on viral population	Bucardo et al. (2012) Infect Genet Evol [16]
	G12P[8]-I1-R1-C1-M1-A1-N1-T1-E1-H1	Evolutionary origins	Bucardo et al. (2015) Clin Microbiol Infect [15]
Argentina	**G3P[8]**-I1-R1-C1-M1-A1-N1-T1-E1-H1 **G3P[6]**-I2-R2-C2-M2-A2-N2-T2-E2-H2	Evolutionary origins	Degiuseppe et al. (2014) PloS One [26]
	G4P[6]-I1-R1-C1-M1-A8-N1-T7-E1-H1 **G4P[6]**-I1-R1-C1-M1-A8-N1-T1-E1-H1	Test zoonotic event	Degiuseppe et al. (2013) Clin Microbiol Infect [25]
	G12P[8]-I1-R1-C1-M1-A1-N1-T1-E1-H1	Evolutionary origins	Stupka et al. (2012) J Clin Virol [98]
Brazil	**G1P[8]**-I1-R1-C1-M1-A1-N1-T1-E1-H1 **G1P[8]**-I1-R1-C1-M1-A1-N1-T3-E1-H1	Evolutionary origins Vaccination pressure on viral population	Silva et al. (2015) Infect Genet Evol [96]
	G3P[9]-I3-R3-C3-M3-A3-N3-T3-E3-H3 **G9P[9]**-I3-R3-C3-M3-A3-N1-T3-E3-H3 **G9P[9]**-I1-R1-C1-M1-A1-N1-T2-E1-H1 **G1P[9]**-I1-R1-C1-M3-A1-N1-T2-E1-H1 **G1P[9]**-I1-R1-C1-M2-A1-N2-T2-E1-H1 **G3P[9]**-I3-R3-C3-M3-A3-N3-T3-E3-H6 **G12P[9]**-I3-R3-C3-M3-A12-N3-T3-E3-H6	Evolutionary origins Test zoonotic event	Tsugawa et al. (2015) J Gen Virol [104]
	G2P[4]-I2-R2-C2-M2-A2-N2-T2-E2-H2	Evolutionary origins	Gómez et al. (2014) Infect Genet Evol [35]
	G12P[8]-I1-R1-C1-M1-A1-N1-T1-E1-H1 **G12P[9]**-I3-R3-C3-M3-A12-N3-T3-E3-H6	Evolutionary origins	Gómez et al. (2014) Infect Genet Evol [36]
	G1P[8]-I1-R1-C1-M1-A1-N1-T3-E1-H1	Vaccine-related reassortment	Rose et al. (2013) Emerg Infect Dis [89]
	G1P[6]-I1-R1-C1-M1-A1-N1-T1-E1-H1	Evolutionary origins	Gómez et al. (2013) Infect Genet Evol [37]
	G2P[4]-I2-R2-C2-M2-A2-N2-T2-E2-H2 **G1 /G4P[6]**-I1-R1-C1-M1-A1-Nx-T1-E1-H1 **G2P[6]**-Ix-R2-C2-M2-A2-Nx-T2-E1-H2 **G4P[6]**-I1-R1-C1-M1-A8-N1-T7-E1-H1 **G4P[6]**-Ix-R1-C1-M1-A1-N1-T1-E1-H1 **G3P[9]**-Ix-R2-C2-M2-A3-N1-T3-E3-H3	Evolutionary origins Test zoonotic event	Maestri (2012) J Med Virol [54]
Ecuador	**G11P[6]**- I1-R1-C1-M1-A1-N1-T1-E1-H1	Test zoonotic event	Bányai et al. (2009) Arch Virol [6]
Paraguay	**G8P[1]**-I2-R2-C2-M1-Ax-N2-T6-E12-H3	Test zoonotic event	Martinez et al. (2014) Infect Genet Evol [59]
	G4P[6]-I1-R1-C1-M1-A8-N1-T7-E1-H1	Test zoonotic event	Martinez et al. (2014) Infect Genet Evol [58]
Barbados	**G4P[14]**-I1-R1-C1-M1-A8-N1-T1-E1-H1	Test zoonotic event	Tam et al. (2014) Infect Genet Evol [100]

product of reassortment of the G12P[9] VP7 gene into the Wa-like strains circulating at high frequencies (i.e., G3P[8], G9P[8]) (Table 5.1). However, phylogenetic analyses of the VP7 gene sequences revealed that emergence of G12P[8] strains was the result of the introduction of a new strain with a genome constellation well adapted to infect humans [98]. The "versatility" of G12 strains to associate with P[4], P[6], P[8], and P[9] genotypes resembles that of G8 and G9 strains, able to adapt and persist in the human population.

6 Whole-Genome Analyses in Rotaviruses Detected in Latin America

The use of genomic analyses in rotaviruses has become a powerful tool to improve understanding of rotavirus epidemiology and evolution. Although it is still expensive technology for routine use, several laboratories in Latin America have used it to understand the transmission and origin of different rotavirus strains (Table 5.1). Thus, zoonotic events have been documented for feline (G3P[9], G9P[9]), porcine

(G4P[6], G4P[14]), and bovine (G8P[1], G8P[14], G11P[6]) strains [6, 25, 32, 54, 58–59, 100, 104], and understanding has improved regarding the epidemiological origins of G1P[8], G2P[4], G3P[8], G12P[8], and G12P[9] strains [15, 26, 35–7, 54, 96, 98, 104]. In the context of the introduction of massive vaccination, these analyses have also enabled discrimination of wild-type strains from vaccine strains [16, 89]. Several groups have implemented this new methodology in Latin America (Table 5.1).

7 Rotavirus Diversity and Vaccines

To date, two rotavirus vaccines have been licensed and included in national immunization programs in 17 Latin American countries (Fig. 5.4) [22, 23]. Rotarix (GlaxoSmithKline Biologicals, Rixensart, Belgium) is a monovalent vaccine

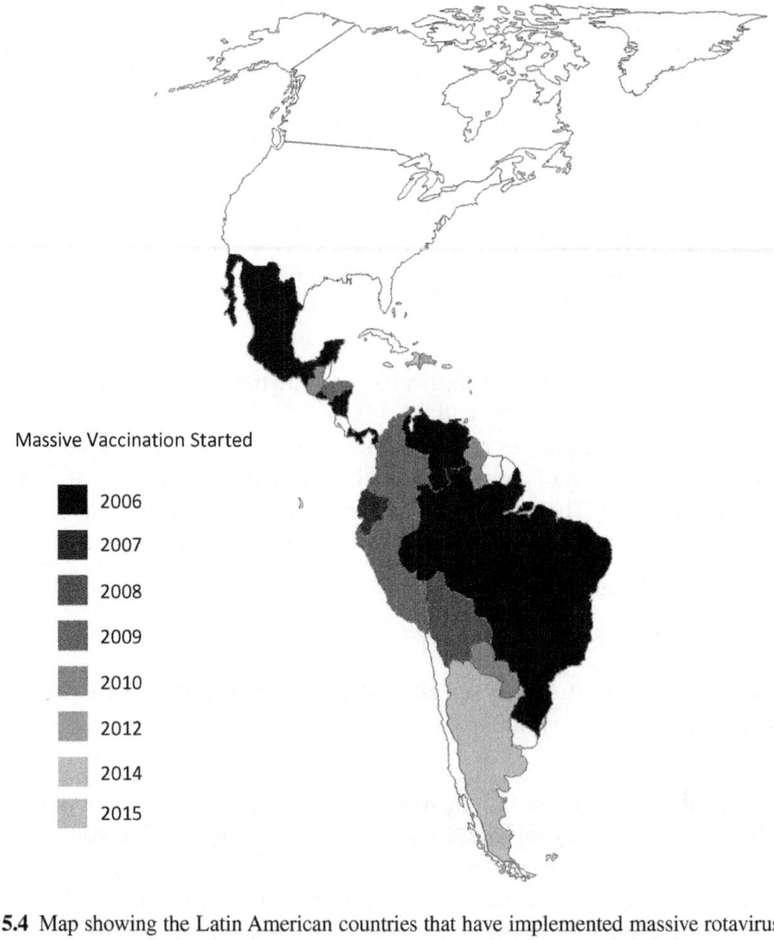

Fig. 5.4 Map showing the Latin American countries that have implemented massive rotavirus vaccination. The years when the vaccination programs started are indicated with different *grey tones*

consisting of an attenuated human G1P[8] rotavirus strain and RotaTeq (Merck and Sanofi Pasteur MSD, NJ, USA) is an oral pentavalent vaccine composed of bovine-human reassortants, with VP7 and VP4 human specificities for genotypes G1–G4 and P[8]. Rotavirus vaccine efficacy varies depending on the geographic region where data are collected. Particularly notorious is the low (<50%) efficacy reported in African and Asian countries [4, 108] where rotavirus diversity is greater compared to developed countries [91]. Despite a large diversity in circulating rotavirus strains, Latin American countries have successfully implemented massive rotavirus vaccination that has reduced the burden of diarrheal disease in the region [23, 24, 65, 71, 82, 88].

Regarding circulating genotypes in countries that have introduced the vaccine, an increase in the proportion of G2P[4] and G3P[8] has been observed in residual rotavirus diarrhea cases. This observation was reported in Bolivia, Brazil, and Colombia (Fig. 5.3). Specifically, an increase in the detection of G2P[4] was reported in countries where the monovalent G1P[8] vaccine was implemented, whereas an increase in G3P[8] was observed in countries where the pentavalent vaccine (G1–G4, P[8]) was implemented. The emergence of unusual genotype combination G9P[4] has been reported in Mexico, Guatemala, and Honduras, which coincided with the decline of G1P[8] and G9P[8] (Fig. 5.3). Although these studies suggested possible genetic drift and selection mechanisms driven by vaccinated populations, studies conducted in the region are insufficient to document pre- and post-vaccination scenarios to determine the effect of the vaccine on viral ecology or whether genetic diversity is influenced by selection mechanisms or natural fluctuation phenomena (Fig. 5.3).

References

1. Amarilla A, Espinola EE, Galeano ME, Farina N, Russomando G, Parra GI (2007) Rotavirus infection in the Paraguayan population from 2004 to 2005: high incidence of rotavirus strains with short electropherotype in children and adults. Med Sci Monit 13:CR333–CR337
2. Araujo IT, Ferreira MS, Fialho AM, Assis RM, Cruz CM, Rocha M, Leite JP (2001) Rotavirus genotypes P[4]G9, P[6]G9, and P[8]G9 in hospitalized children with acute gastroenteritis in Rio de Janeiro, Brazil. J Clin Microbiol 39:1999–2001
3. Arista S, Giammanco GM, De Grazia S, Ramirez S, Lo Biundo C, Colomba C, Cascio A, Martella V (2006) Heterogeneity and temporal dynamics of evolution of G1 human rotaviruses in a settled population. J Virol 80:10724–10733
4. Armah GE, Sow SO, Breiman RF, Dallas MJ, Tapia MD, Feikin DR, Binka FN, Steele AD, Laserson KF, Ansah NA, Levine MM, Lewis K, Coia ML, Attah-Poku M, Ojwando J, Rivers SB, Victor JC, Nyambane G, Hodgson A, Schodel F, Ciarlet M, Neuzil KM (2010) Efficacy of pentavalent rotavirus vaccine against severe rotavirus gastroenteritis in infants in developing countries in sub-Saharan Africa: a randomised, double-blind, placebo-controlled trial. Lancet 376:606–614
5. Badaracco A, Garaicoechea L, Matthijnssens J, Louge Uriarte E, Odeon A, Bilbao G, Fernandez F, Parra GI, Parreno V (2013) Phylogenetic analyses of typical bovine rotavirus genotypes G6, G10, P[5] and P[11] circulating in Argentinean beef and dairy herds. Infect Genet Evol 18:18–30

6. Bányai K, Esona MD, Kerin TK, Hull JJ, Mijatovic S, Vásconez N, Torres C, de Filippis AM, Foytich KR, Gentsch JR (2009) Molecular characterization of a rare, human-porcine reassortant rotavirus strain, G11P[6], from Ecuador. Arch Virol 154(11):1823–1829. Erratum in: Arch Virol. 2010;155(1):147. Foytich, Kimberly R [added]
7. Barril PA, Martinez LC, Giordano MO, Castello AA, Rota RP, Isa MB, Masachessi G, Ferreyra LJ, Glikmann G, Nates SV (2006) Detection of group a human rotavirus G9 genotype circulating in Cordoba, Argentina, as early as 1980. J Med Virol 78:1113–1118
8. Berois M, Libersou S, Russi J, Arbiza J, Cohen J (2003) Genetic variation in the VP7 gene of human rotavirus isolated in Montevideo–Uruguay from 1996–1999. J Med Virol 71:456–462
9. Bok K, Castagnaro N, Borsa A, Nates S, Espul C, Fay O, Fabri A, Grinstein S, Miceli I, Matson DO, Gomez JA (2001) Surveillance for rotavirus in Argentina. J Med Virol 65:190–198
10. Bok K, Matson DO, Gomez JA (2002) Genetic variation of capsid protein VP7 in genotype g4 human rotavirus strains: simultaneous emergence and spread of different lineages in Argentina. J Clin Microbiol 40:2016–2022
11. Bok K, Palacios G, Sijvarger K, Matson D, Gomez J (2001) Emergence of G9 P[6] human rotaviruses in Argentina: phylogenetic relationships among G9 strains. J Clin Microbiol 39:4020–4025
12. Bourdett-Stanziola L, Jimenez C, Ortega-Barria E (2008) Diversity of human rotavirus G and P genotypes in Panama, Costa Rica, and the Dominican Republic. Am J Trop Med Hyg 79:921–924
13. Bourdett-Stanziola L, Ortega-Barria E, Espinoza F, Bucardo F, Jimenez C, Ferrera A (2011) Rotavirus genotypes in Costa Rica, Nicaragua, Honduras and the Dominican Republic. Intervirology 54:49–52
14. Bucardo F, Karlsson B, Nordgren J, Paniagua M, Gonzalez A, Amador JJ, Espinoza F, Svensson L (2007) Mutated G4P[8] rotavirus associated with a nationwide outbreak of gastroenteritis in Nicaragua in 2005. J Clin Microbiol 45:990–997
15. Bucardo F, Mercado J, Reyes Y, González F, Balmaseda A, Nordgren J (2015) Large increase of rotavirus diarrhoea in the hospital setting associated with emergence of G12 genotype in a highly vaccinated population in Nicaragua. Clin Microbiol Infect 21(6):603.e1–603.e7
16. Bucardo F, Rippinger CM, Svensson L, Patton JT (2012) Vaccine-derived NSP2 segment in rotaviruses from vaccinated children with gastroenteritis in Nicaragua. Infect Genet Evol 12(6):1282–1294
17. Caceres DC, Pelaez D, Sierra N, Estrada E, Sanchez L (2006) Burden of rotavirus-related disease among children under five, Colombia, 2004. Rev Panam Salud Publica 20:9–21
18. Castello AA, Arvay ML, Glass RI, Gentsch J (2004) Rotavirus strain surveillance in Latin America: a review of the last nine years. Pediatr Infect Dis J 23:S168–S172
19. Castello AA, Nakagomi T, Nakagomi O, Jiang B, Kang JO, Glass RI, Glikmann G, Gentsch JR (2009) Characterization of genotype P[9]G12 rotavirus strains from Argentina: high similarity with Japanese and Korean G12 strains. J Med Virol 81:371–381
20. da Silva MF, Tort LF, Gomez MM, Assis RM, Volotao Ede M, de Mendonca MC, Bello G, Leite JP (2011) VP7 Gene of human rotavirus A genotype G5: Phylogenetic analysis reveals the existence of three different lineages worldwide. J Med Virol 83:357–366
21. Das BK, Gentsch JR, Cicirello HG, Woods PA, Gupta A, Ramachandran M, Kumar R, Bhan MK, Glass RI (1994) Characterization of rotavirus strains from newborns in New Delhi, India. J Clin Microbiol 32:1820–1822
22. de Oliveira LH, Danovaro-Holliday MC, Matus CR, Andrus JK (2008) Rotavirus vaccine introduction in the Americas: progress and lessons learned. Expert Rev Vaccines 7:345–353
23. de Oliveira LH, Danovaro-Holliday MC, Sanwogou NJ, Ruiz-Matus C, Tambini G, Andrus JK (2011) Progress in the introduction of the rotavirus vaccine in Latin America and the Caribbean: four years of accumulated experience. Pediatr Infect Dis J 30:S61–S66
24. de Palma O, Cruz L, Ramos H, de Baires A, Villatoro N, Pastor D, de Oliveira LH, Kerin T, Bowen M, Gentsch J, Esposito DH, Parashar U, Tate J, Patel M (2010) Effectiveness of

rotavirus vaccination against childhood diarrhoea in El Salvador: case-control study. BMJ 340:c2825

25. Degiuseppe JI, Beltramino JC, Millan A, Stupka JA, Parra GI (2013) Complete genome analyses of G4P[6] rotavirus detected in Argentinean children with diarrhoea provides evidence of interspecies transmission from swine. Clin Microbiol Infect 19:E367–E371
26. Degiuseppe JI, Parra GI, Stupka JA (2014) Genetic diversity of G3 rotavirus strains circulating in Argentina during 1998–2012 assessed by full genome analyses. PLoS One 9:e110341
27. Endara P, Trueba G, Solberg OD, Bates SJ, Ponce K, Cevallos W, Matthijnssens J, Eisenberg JN (2007) Symptomatic and subclinical infection with rotavirus P[8]G9, rural Ecuador. Emerg Infect Dis 13:574–580
28. Esona MD, Buteau J, Lucien MA, Joseph GA, Leshem E, Boncy J, Katz MA, Bowen MD, Balajee SA (2015) Rotavirus group A genotypes detected through diarrheal disease surveillance in Haiti, 2012. Am J Trop Med Hyg 93:54–56
29. Esteban LE, Rota RP, Gentsch JR, Jiang B, Esona M, Glass RI, Glikmann G, Castello AA (2010) Molecular epidemiology of group A rotavirus in Buenos Aires, Argentina 2004-2007: reemergence of G2P[4] and emergence of G9P[8] strains. J Med Virol 82:1083–1093
30. Estes M, Greenberg HB (2013) Rotaviruses. In: Knipe DM, Howley PM, Cohen JI, Griffin DE, Lamb RA, Martin MA, Roizman B, Racaniello VR (eds) Fields virology, vol II, 6th edn. Lippincott/Williams & Wilkins, Philadelphia, pp 1347–1401
31. Fumian TM, Leite JP, Rose TL, Prado T, Miagostovich MP (2011) One year environmental surveillance of rotavirus specie A (RVA) genotypes in circulation after the introduction of the Rotarix(R) vaccine in Rio de Janeiro, Brazil. Water Res 45:5755–5763
32. Gautam R, Mijatovic-Rustempasic S, Roy S, Esona MD, Lopez B, Mencos Y, Rey-Benito G, Bowen MD (2015) Full genomic characterization and phylogenetic analysis of a zoonotic human G8P[14] rotavirus strain detected in a sample from Guatemala. Infect Genet Evol 33:206–211
33. Gentsch JR, Glass RI, Woods P, Gouvea V, Gorziglia M, Flores J, Das BK, Bhan MK (1992) Identification of group A rotavirus gene 4 types by polymerase chain reaction. J Clin Microbiol 30:1365–1373
34. Giambiagi S, Gonzalez Rodriguez I, Gomez J, Burrone O (1994) A rearranged genomic segment 11 is common to different human rotaviruses. Arch Virol 136:415–421
35. Gómez MM, Carvalho-Costa FA, Volotão Ede M, Rose TL, da Silva MF, Fialho AM, Assis RM, de Andrade JS, Sá AC, Zeller M, Heylen E, Matthijnssens J, Leite JP (2014) Prevalence and genomic characterization of G2P[4] group A rotavirus strains during monovalent vaccine introduction in Brazil. Infect Genet Evol 28:486–494
36. Gómez MM, Resque HR, Volotão Ede M, Rose TL, da Silva MF, Heylen E, Zeller M, Matthijnssens J, Leite JP (2014) Distinct evolutionary origins of G12P[8] and G12P[9] group A rotavirus strains circulating in Brazil. Infect Genet Evol 28:385–388
37. Gómez MM, da Silva MF, Zeller M, Heylen E, Matthijnssens J, Ichihara MY, Rose TL, de Mello Volotão E, Leite JP (2013) Phylogenetic analysis of G1P[6] group A rotavirus strains detected in Northeast Brazilian children fully vaccinated with Rotarix™. Infect Genet Evol 19:395–402
38. Gouvea V, de Castro L, Timenetsky MC, Greenberg H, Santos N (1994) Rotavirus serotype G5 associated with diarrhea in Brazilian children. J Clin Microbiol 32:1408–1409
39. Gouvea V, Glass RI, Woods P, Taniguchi K, Clark HF, Forrester B, Fang ZY (1990) Polymerase chain reaction amplification and typing of rotavirus nucleic acid from stool specimens. J Clin Microbiol 28:276–282
40. Gouvea V, Santos N (1999) Rotavirus serotype G5: an emerging cause of epidemic childhood diarrhea. Vaccine 17:1291–1292
41. Gurgel RQ, Cuevas LE, Vieira SC, Barros VC, Fontes PB, Salustino EF, Nakagomi O, Nakagomi T, Dove W, Cunliffe N, Hart CA (2007) Predominance of rotavirus P[4]G2 in a vaccinated population, Brazil. Emerg Infect Dis 13:1571–1573

42. Iturriza-Gomara M, Anderton E, Kang G, Gallimore C, Phillips W, Desselberger U, Gray J (2003) Evidence for genetic linkage between the gene segments encoding NSP4 and VP6 proteins in common and reassortant human rotavirus strains. J Clin Microbiol 41:3566–3573
43. Iturriza-Gomara M, Green J, Brown DW, Desselberger U, Gray JJ (1999) Comparison of specific and random priming in the reverse transcriptase polymerase chain reaction for genotyping group A rotaviruses. J Virol Methods 78:93–103
44. Laird AR, Gentsch JR, Nakagomi T, Nakagomi O, Glass RI (2003) Characterization of serotype G9 rotavirus strains isolated in the United States and India from 1993 to 2001. J Clin Microbiol 41:3100–3111
45. Laird AR, Ibarra V, Ruiz-Palacios G, Guerrero ML, Glass RI, Gentsch JR (2003) Unexpected detection of animal VP7 genes among common rotavirus strains isolated from children in Mexico. J Clin Microbiol 41:4400–4403
46. Lanata CF, Fischer-Walker CL, Olascoaga AC, Torres CX, Aryee MJ, Black RE, Child Health Epidemiology Reference Group of the World Health Organization, UNICEF (2013) Global causes of diarrheal disease mortality in children <5 years of age: a systematic review. PLoS One 8:e72788
47. Leite JP, Carvalho-Costa FA, Linhares AC (2008) Group A rotavirus genotypes and the ongoing Brazilian experience: a review. Mem Inst Oswaldo Cruz 103:745–753
48. Linhares AC, Stupka JA, Ciapponi A, Bardach AE, Glujovsky D, Aruj PK, Mazzoni A, Rodriguez JA, Rearte A, Lanzieri TM, Ortega-Barria E, Colindres R (2011) Burden and typing of rotavirus group A in Latin America and the Caribbean: systematic review and meta-analysis. Rev Med Virol 21:89–109
49. Lorenzetti E, da Silva Medeiros TN, Alfieri AF, Alfieri AA (2011) Genetic heterogeneity of wild-type G4P[6] porcine rotavirus strains detected in a diarrhea outbreak in a regularly vaccinated pig herd. Vet Microbiol 154:191–196
50. Louge Uriarte EL, Badaracco A, Matthijnssens J, Zeller M, Heylen E, Manazza J, Mino S, Van Ranst M, Odeon A, Parreno V (2014) The first caprine rotavirus detected in Argentina displays genomic features resembling virus strains infecting members of the Bovidae and Camelidae. Vet Microbiol 171:189–197
51. Luchs A, Cilli A, Morillo SG, Carmona Rde C, Timenetsky MC (2012) Rare G3P[3] rotavirus strain detected in Brazil: possible human–canine interspecies transmission. J Clin Virol 54:89–92
52. Luchs A, Cilli A, Morillo SG, Gregorio Dde S, de Souza KA, Vieira HR, Fernandes Ade M, Carmona Rde C, Timenetsky MC (2016) Detection of the emerging rotavirus G12P[8] genotype at high frequency in Brazil in 2014: successive replacement of predominant strains after vaccine introduction. Acta Trop 156:87–94
53. Luchs A, do Carmo Sampaio Tavares Timenetsky M (2015) Phylogenetic analysis of human group C rotavirus circulating in Brazil reveals a potential unique NSP4 genetic variant and high similarity with Asian strains. Mol Genet Genomics 290:969–986
54. Maestri RP, Kaiano JH, Neri DL, Soares Lda S, Guerra Sde F, Oliveira Dde S, Farias YN, Gabbay YB, Leite JP, Linhares Ada C, Mascarenhas JD (2012) Phylogenetic analysis of probable non-human genes of group A rotaviruses isolated from children with acute gastroenteritis in Belém, Brazil. J Med Virol 84(12):1993–2002
55. Mandile MG, Esteban LE, Arguelles MH, Mistchenko A, Glikmann G, Castello AA (2014) Surveillance of group A rotavirus in Buenos Aires 2008–2011, long lasting circulation of G2P[4] strains possibly linked to massive monovalent vaccination in the region. J Clin Virol 60:282–289
56. Martinez M, Amarilla AA, Galeano ME, Aquino VH, Farina N, Russomando G, Parra GI (2010) Predominance of rotavirus G2P[4] and emergence of G12P[9] strains in Asuncion, Paraguay, 2006–2007. Arch Virol 155:525–533
57. Martinez M, Farina N, Rodriguez M, Russomando G, Parra GI (2005) Incidence of rotavirus in adults with diarrhea in Asuncion, Paraguay. Rev Argent Microbiol 37:99–101

58. Martinez M, Galeano ME, Akopov A, Palacios R, Russomando G, Kirkness EF, Parra GI (2014) Whole-genome analyses reveals the animal origin of a rotavirus G4P[6] detected in a child with severe diarrhea. Infect Genet Evol 27:156–162
59. Martinez M, Phan TG, Galeano ME, Russomando G, Parreno V, Delwart E, Parra GI (2014) Genomic characterization of a rotavirus G8P[1] detected in a child with diarrhea reveal direct animal-to-human transmission. Infect Genet Evol 27:402–407
60. Matsui SM, Mackow ER, Matsuno S, Paul PS, Greenberg HB (1990) Sequence analysis of gene 11 equivalents from "short" and "super short" strains of rotavirus. J Virol 64:120–124
61. Matthijnssens J, Ciarlet M, McDonald SM, Attoui H, Banyai K, Brister JR, Buesa J, Esona MD, Estes MK, Gentsch JR, Iturriza-Gomara M, Johne R, Kirkwood CD, Martella V, Mertens PP, Nakagomi O, Parreno V, Rahman M, Ruggeri FM, Saif LJ, Santos N, Steyer A, Taniguchi K, Patton JT, Desselberger U, Van Ranst M (2011) Uniformity of rotavirus strain nomenclature proposed by the Rotavirus Classification Working Group (RCWG). Arch Virol 156:1397–1413
62. Matthijnssens J, Heylen E, Zeller M, Rahman M, Lemey P, Van Ranst M (2010) Phylodynamic analyses of rotavirus genotypes G9 and G12 underscore their potential for swift global spread. Mol Biol Evol 27:2431–2436
63. Matthijnssens J, Otto PH, Ciarlet M, Desselberger U, Van Ranst M, Johne R (2012) VP6-sequence-based cutoff values as a criterion for rotavirus species demarcation. Arch Virol 157:1177–1182
64. Matthijnssens J, Van Ranst M (2012) Genotype constellation and evolution of group A rotaviruses infecting humans. Curr Opin Virol 2:426–433
65. McAtee CL, Webman R, Gilman RH, Mejia C, Bern C, Apaza S, Espetia S, Pajuelo M, Saito M, Challappa R, Soria R, Ribera JP, Lozano D, Torrico F (2016) Burden of norovirus and rotavirus in children after rotavirus vaccine introduction, Cochabamba, Bolivia. Am J Trop Med Hyg 94:212–217
66. McDonald SM, Aguayo D, Gonzalez-Nilo FD, Patton JT (2009) Shared and group-specific features of the rotavirus RNA polymerase reveal potential determinants of gene reassortment restriction. J Virol 83:6135–6148
67. McDonald SM, Matthijnssens J, McAllen JK, Hine E, Overton L, Wang S, Lemey P, Zeller M, Van Ranst M, Spiro DJ, Patton JT (2009) Evolutionary dynamics of human rotaviruses: balancing reassortment with preferred genome constellations. PLoS Pathog 5:e1000634
68. McDonald SM, McKell AO, Rippinger CM, McAllen JK, Akopov A, Kirkness EF, Payne DC, Edwards KM, Chappell JD, Patton JT (2012) Diversity and relationships of cocirculating modern human rotaviruses revealed using large-scale comparative genomics. J Virol 86:9148–9162
69. Mino S, Barrandeguy M, Parreno V, Parra GI (2016) Genetic linkage of capsid protein-encoding RNA segments in group A equine rotaviruses. J Gen Virol 97:912–921
70. Molinari BL, Alfieri AF, Alfieri AA (2015) Genetic variability of VP6, VP7, VP4, and NSP4 genes of porcine rotavirus group H detected in Brazil. Virus Res 197:48–53
71. Molto Y, Cortes JE, De Oliveira LH, Mike A, Solis I, Suman O, Coronado L, Patel MM, Parashar UD, Cortese MM (2011) Reduction of diarrhea-associated hospitalizations among children aged <5 years in Panama following the introduction of rotavirus vaccine. Pediatr Infect Dis J 30:S16–S20
72. Montenegro FM, Correia JB, Rodrigues Falbo A, Dove W, Nakagomi T, Nakagomi O, Cuevas LE, Cunliffe NA, Hart CA (2007) Anticipating rotavirus vaccines in Brazil: detection and molecular characterization of emerging rotavirus serotypes G8 and G9 among children with diarrhoea in Recife, Brazil. J Med Virol 79:335–340
73. Moreno S, Alvarado MV, Bermudez A, Gutierrez MF (2009) Phylogenetic analysis indicates human origin of rotavirus and hepatitis A virus strains found in the drinking water of western Colombia. Biomedica 29:209–217
74. Nakagomi O, Nakagomi T, Akatani K, Ikegami N (1989) Identification of rotavirus genogroups by RNA–RNA hybridization. Mol Cell Probes 3:251–261

75. Parashar UD, Burton A, Lanata C, Boschi-Pinto C, Shibuya K, Steele D, Birmingham M, Glass RI (2009) Global mortality associated with rotavirus disease among children in 2004. J Infect Dis 200(suppl 1):S9–S15
76. Parra GI (2009) Seasonal shifts of group A rotavirus strains as a possible mechanism of persistence in the human population. J Med Virol 81:568–571
77. Parra GI, Bok K, Martinez V, Russomando G, Gomez J (2005) Molecular characterization and genetic variation of the VP7 gene of human rotaviruses isolated in Paraguay. J Med Virol 77:579–586
78. Parra GI, Espinola EE, Amarilla AA, Stupka J, Martinez M, Zunini M, Galeano ME, Gomes K, Russomando G, Arbiza J (2007) Diversity of group A rotavirus strains circulating in Paraguay from 2002 to 2005: detection of an atypical G1 in South America. J Clin Virol 40:135–141
79. Parra GI, Vidales G, Gomez JA, Fernandez FM, Parreno V, Bok K (2008) Phylogenetic analysis of porcine rotavirus in Argentina: increasing diversity of G4 strains and evidence of interspecies transmission. Vet Microbiol 126:243–250
80. Parreno V, Bok K, Fernandez F, Gomez J (2004) Molecular characterization of the first isolation of rotavirus in guanacos (*Lama guanicoe*). Arch Virol 149:2465–2471
81. Phan TG, Okitsu S, Maneekarn N, Ushijima H (2007) Evidence of intragenic recombination in G1 rotavirus VP7 genes. J Virol 81:10188–10194
82. Pringle KD, Patzi M, Tate JE, Iniguez Rojas V, Patel M, Inchauste Jordan L, Montesano R, Zarate A, De Oliveira L, Parashar U (2016) Sustained effectiveness of rotavirus vaccine against very severe rotavirus disease through the second year of life, Bolivia 2013–2014. Clin Infect Dis 62(suppl 2):S115–S120
83. Quaye O, McDonald S, Esona MD, Lyde FC, Mijatovic-Rustempasic S, Roy S, Banegas DJ, Quinonez YM, Chinchilla BL, Santiago FG, Lozano HG, Rey-Benito G, de Oliveira LH, Gentsch JR, Bowen MD (2013) Rotavirus G9P[4] in 3 countries in Latin America, 2009–2010. Emerg Infect Dis 19:1332–1333
84. Racz ML, Kroeff SS, Munford V, Caruzo TA, Durigon EL, Hayashi Y, Gouvea V, Palombo EA (2000) Molecular characterization of porcine rotaviruses from the southern region of Brazil: characterization of an atypical genotype G[9] strain. J Clin Microbiol 38:2443–2446
85. https://rega.kuleuven.be/cev/viralmetagenomics/virus-classification/rcwg. 2016. Accessed 17 July 2016
86. Ribas Mde L, Nagashima S, Calzado A, Acosta G, Tejero Y, Cordero Y, Piedra D, Kobayashi N (2011) Emergence of G9 as a predominant genotype of human rotaviruses in Cuba. J Med Virol 83:738–744
87. Ribas Mde L, Tejero Y, Cordero Y, de Los Angeles Leon M, Rodriguez M, Perez-Lastre J, Triana T, Guerra M, Ayllon L, Escalante G, Hadad J (2015) Detection of rotavirus and other enteropathogens in children hospitalized with acute gastroenteritis in Havana, Cuba. Arch Virol 160:1923–1930
88. Richardson V, Parashar U, Patel M (2011) Childhood diarrhea deaths after rotavirus vaccination in Mexico. N Engl J Med 365:772–773
89. Rose TL, Marques da Silva MF, Goméz MM, Resque HR, Ichihara MY, Volotão Ede M, Leite JP (2013) Evidence of vaccine-related reassortment of rotavirus, Brazil, 2008–2010. Emerg Infect Dis 19(11):1843–1846
90. Rivera R, Forney K, Castro MR, Rebolledo PA, Mamani N, Patzi M, Halkyer P, Leon JS, Iniguez V (2013) Rotavirus genotype distribution during the pre-vaccine period in Bolivia: 2007–2008. Int J Infect Dis 17:e762–e767
91. Santos N, Hoshino Y (2005) Global distribution of rotavirus serotypes/genotypes and its implication for the development and implementation of an effective rotavirus vaccine. Rev Med Virol 15:29–56
92. Santos N, Lima RC, Nozawa CM, Linhares RE, Gouvea V (1999) Detection of porcine rotavirus type G9 and of a mixture of types G1 and G5 associated with Wa-like VP4 specificity: evidence for natural human-porcine genetic reassortment. J Clin Microbiol 37:2734–2736

93. Santos N, Soares CC, Volotao EM, Albuquerque MC, Hoshino Y (2003) Surveillance of rotavirus strains in Rio de Janeiro, Brazil, from 1997 to 1999. J Clin Microbiol 41:3399–3402
94. Santos N, Volotao EM, Soares CC, Albuquerque MC, da Silva FM, Chizhikov V, Hoshino Y (2003) VP7 gene polymorphism of serotype G9 rotavirus strains and its impact on G genotype determination by PCR. Virus Res 93:127–138
95. Santos N, Volotao EM, Soares CC, Albuquerque MC, da Silva FM, de Carvalho TR, Pereira CF, Chizhikov V, Hoshino Y (2001) Rotavirus strains bearing genotype G9 or P[9] recovered from Brazilian children with diarrhea from 1997 to 1999. J Clin Microbiol 39:1157–1160
96. Silva FD, Espinoza LR, Tonietti PO, Barbosa BR, Gregori F (2015) Whole-genomic analysis of 12 porcine group A rotaviruses isolated from symptomatic piglets in Brazil during the years of 2012–2013. Infect Genet Evol 32:239–254
97. Stupka JA, Carvalho P, Amarilla AA, Massana M, Parra GI, Argentinean National Surveillance Network for Diarrheas (2009) National Rotavirus Surveillance in Argentina: high incidence of G9P[8] strains and detection of G4P[6] strains with porcine characteristics. Infect Genet Evol 9:1225–1231
98. Stupka JA, Degiuseppe JI, Parra GI, Argentinean National Rotavirus Surveillance Netwwork for Diarrheas (2012) Increased frequency of rotavirus G3P[8] and G12P[8] in Argentina during 2008–2009: whole-genome characterization of emerging G12P[8] strains. J Clin Virol 54:162–167
99. Stupka JA, Parra GI, Gomez J, Arbiza J (2007) Detection of human rotavirus G9P[8] strains circulating in Argentina: phylogenetic analysis of VP7 and NSP4 genes. J Med Virol 79:838–842
100. Tam KI, Roy S, Esona MD, Jones S, Sobers S, Morris-Glasgow V, Rey-Benito G, Gentsch JR, Bowen MD (2014) Full genomic characterization of a novel genotype combination, G4P[14], of a human rotavirus strain from Barbados. Infect Genet Evol 28:524–529
101. Tate JE, Burton AH, Boschi-Pinto C, Steele AD, Duque J, Parashar UD, Network WH-cGRS (2012) 2008 estimate of worldwide rotavirus-associated mortality in children younger than 5 years before the introduction of universal rotavirus vaccination programmes: a systematic review and meta-analysis. Lancet Infect Dis 12:136–141
102. Tort LF, Victoria M, Lizasoain A, Garcia M, Berois M, Cristina J, Leite JP, Gomez MM, Miagostovich MP, Colina R (2015) Detection of common, emerging and uncommon VP4, and VP7 human group A rotavirus genotypes from urban sewage samples in Uruguay. Food Environ Virol 7:342–353
103. Trojnar E, Sachsenroder J, Twardziok S, Reetz J, Otto PH, Johne R (2013) Identification of an avian group A rotavirus containing a novel VP4 gene with a close relationship to those of mammalian rotaviruses. J Gen Virol 94:136–142
104. Tsugawa T, Rainwater-Lovett K, Tsutsumi H (2015) Human G3P[9] rotavirus strains possessing an identical genotype constellation to AU-1 isolated at high prevalence in Brazil, 1997–1999. J Gen Virol 96:590–600
105. Urbina D, Rodriguez JG, Arzuza O, Parra E, Young G, Castro R, del-Portillo P (2004) G and P genotypes of rotavirus circulating among children with diarrhea in the Colombian northern coast. Int Microbiol 7:113–120
106. Velazquez FR, Garcia-Lozano H, Rodriguez E, Cervantes Y, Gomez A, Melo M, Anaya L, Ovalle JC, Torres J, Diaz De Jesus B, Alvarez-Lucas C, Breuer T, Munoz O, Kuri P (2004) Diarrhea morbidity and mortality in Mexican children: impact of rotavirus disease. Pediatr Infect Dis J 23:S149–S155
107. Widdowson MA, Meltzer MI, Zhang X, Bresee JS, Parashar UD, Glass RI (2007) Cost-effectiveness and potential impact of rotavirus vaccination in the United States. Pediatrics 119:684–697
108. Zaman K, Dang DA, Victor JC, Shin S, Yunus M, Dallas MJ, Podder G, Vu DT, Le TP, Luby SP, Le HT, Coia ML, Lewis K, Rivers SB, Sack DA, Schodel F, Steele AD, Neuzil KM, Ciarlet M (2010) Efficacy of pentavalent rotavirus vaccine against severe rotavirus gastroenteritis in infants in developing countries in Asia: a randomised, double-blind, placebo-controlled trial. Lancet 376:615–623

Chapter 6
Environmental Virology

Marize Pereira Miagostovich and Carmen Baur Vieira

1 Introduction

Environmental virology is an extensive area of research; the increasing number of studies in recent years reflects the need to develop more sensitive parameters that can meet the demands of a modern society with growing environmental concerns. Because of its interface with specific public policies such as water resources, sanitation, epidemiological surveillance, and coastal management, this branch of research in virology represents an important tool for improving the quality of life as well as for strengthening environmental monitoring [140]. In 2011, advances were acknowledged and climaxed with the International Society for Food and Environmental Virology foundation [119], whose main goal is to encourage researches associated with viruses transmitted via the environment (water, air, soil, etc.) and food.

Linked to health-environment and health-sanitation concerns, not by chance, progress in the field was achieved during the United Nations International Decade for Action "Water for Life," 2005–2015, when efforts to fulfill international commitments related to the issue of water were stimulated [263]. Despite the increase of knowledge generated worldwide, the overall impact of viral diseases transmitted by environmental routes is still hard to assess. Problems related to subclinical or asymptomatic infections, as well as the difficulties of recovering viruses from different environmental matrices, still influence epidemiological studies, especially when performed in developing countries. Despite all the difficulties and following the global trend, the number of studies in environmental virology in Latin America

M.P. Miagostovich (✉) • C.B. Vieira
Laboratório de Virologia Comparada e Ambiental, Instituto Oswaldo Cruz (IOC), Fundação Oswaldo Cruz (FIOCRUZ), Pavilhão Helio e Peggy Pereira,
Avenida Brasil, 4365, Manguinhos, Rio de Janeiro, RJ 21040-360, Brazil
e-mail: marizepm@ioc.fiocruz.br; c.baur@ioc.fiocruz.br

J.E. Ludert et al. (eds.), *Human Virology in Latin America*,
DOI 10.1007/978-3-319-54567-7_6

increased tenfold compared to the previous decade. In this context, this chapter provides topics in the field focusing on studies of water and foodborne enteric viruses conducted in the region.

1.1 History

Environmental virology as an object of study began in the year 1940 with the recognition of poliovirus (PV) as an enteric infection transmitted through the fecal–oral route and by the attempt to isolate virus by inoculation of sewage in monkeys [153]. In later years, PV was also associated with outbreaks that resulted from consumption of contaminated drinking milk and was considered the main enteric viral agent surveyed in environmental matrices, particularly the result of the development of cell culture medium suitable for viral replication [88].

In the 1950s, a first waterborne outbreak of hepatitis was reported as a result of contamination by wastewater of the water treatment system in New Delhi (India). This outbreak affected about 230,000 people and was caused by the hepatitis E virus (HEV), and its etiological agent was identified years after the episode [24]. Hepatitis A virus (HAV) transmission by consumption of bivalve mollusks in Sweden and later in the United States of America was also described [88].

During the following decade, the improvement of methods for virus concentration from water, oysters, and sediments [92, 118, 155, 156, 222], as well as the association of gastroenteric viruses such as Norwalk (genus *Norovirus*), enteric human adenovirus (HAdV), rotavirus group A (RVA), and human astrovirus (HAstV) with nonbacterial gastroenteritis cases [3, 18, 123, 166, 193], enhanced the studies on environmental virology. However, the greatest impact came from the advent of a highly sensitive molecular tool, notably, the polymerase chain reaction (PCR), that allowed expansion of analysis for those groups of viruses considered fastidious and, especially, for those viruses that did not have cell cultures for viral propagation used routinely in laboratories. Human norovirus (NoV) and HEV still need a three-dimensional (3D) cell culture system for their replication [16, 172].

In Latin America, a study conducted in Brazil was pioneer in demonstrating the contamination of recreational waters by isolating echovirus (E) in beaches located in Guanabara Bay, Rio de Janeiro [112]. Later on, viral contamination in drinking water was first documented in Guadalajara, Mexico, by the detection of RVA and coxsackievirus (CV) in treated drinking water [51]. In the 1990s, the distribution of RVA was demonstrated in raw sewage and creeks in São Paulo, Brazil [150], and in Argentina Caillou et al. [33] evaluated a methodology for enterovirus (EV) recovery from sludge.

1.2 *Global Burden of Diseases and Pathogens*

Waterborne diseases and sanitation-related infections are among the most common causes of death in the world, especially affecting the poorest societies and children less than 5 years old in developing countries. According to the Pan American Health Organization (PAHO), about 4% of the global burden of the disease could be prevented by improving water supply, sanitation, and hygiene. It is estimated that 884 million people in the world—roughly an eighth of the global population—do not have access to safe water. Moreover, 2.6 billion people do not have access to adequate sanitation, and 1.4 million children die every year as a result of waterborne diseases. In Latin American countries, despite improved access to drinking water and sanitation for 93% and 80% of the population, respectively, progress is still slow and is concentrated in urban areas [171].

According to the World Health Organization (WHO), a wide diversity of viruses belonging to different families and genera are well recognized, causing different infectious diseases in humans through exposure to contaminated environments, many of them associated with water- and foodborne outbreaks. HAdV, HAstV, HAV, HEV, RVA, NoV, and EV, including CV and PV, were classified as being of high or moderate importance to health. Viruses are responsible for serious diseases such as hepatitis, encephalitis, meningitis, myocarditis, and cancer, although most of them are associated with cases of acute gastroenteritis [262], the main cause of death in children around the world, second only to pneumonia and neonatal deaths. In 2013, diarrhea was responsible for the death of 578,000 (448,000–750,000) children less than 5 years of age worldwide [137].

Emerging viruses such as aichivirus (AiV), bocavirus (BoV), klassevirus (KV), norovirus genogroup IV (NoV GIV), sapovirus (SAV), and gemycircularvirus (GemyCV) have increased the list of viruses associated with cases of gastroenteritis detected in sewage samples and water contaminated by sewage in Latin America [32, 34, 42, 45, 63, 111, 124, 177, 233, 270, 272]. For foodborne diseases, NoV and HAV are recognized as major etiological agents worldwide, representing a public health problem that is accompanied by considerable economic losses [105].

2 Enteric Viruses

Enteric viruses or viruses of enteric transmission are often the target of investigation in the environment, including all viral groups that are present in the human gastrointestinal tract and which, after fecal–oral transmission, may cause infections in susceptible individuals. These are viruses of icosahedral symmetry, non-enveloped, and highly resistant to unfavorable environmental conditions that share properties, particularly concerning the risk of diseases, as excretion in high concentrations even in asymptomatic cases, lower infectious dose, viral particle stability, and resistance to disinfection procedures [64, 91, 202, 235].

2.1 Source, Route of Exposure, and Transmission

High concentrations of viruses are excreted in the feces and urine of individuals, turning untreated sewage into the largest carrier of those pathogens and the carrier of a wide variety of viruses unknown to us [6, 36].

During the peak of infection, individuals may excrete enteric viruses at high levels (10^{11} virus/g of feces), and according to studies conducted worldwide, they are present in raw and treated sewage samples in concentrations reaching 10^4–10^9 and 10^2–10^7 viruses per liter, respectively [21, 38, 65, 76, 93, 132, 134, 203].

Enteric viruses can be transmitted by water, food, fomites, and human contact. The discharge of viral agents via sewage in natura represents a risk of infection for the population through several routes, including intake of contaminated drinking water and in recreational waters after direct skin contact or by accidental ingestion. They can also be acquired after consumption of shellfish harvested from contaminated waters or vegetables cultivated or irrigated with water contaminated by sewage [24, 202].

Another source of contamination is the transfer of infectious virus particles on animate or inanimate surfaces, such as hands or work surfaces used for food preparation or directly from the hand to the mouth [147, 208]. The surfaces can be contaminated directly by feces, urine, vomit, saliva, blood, and respiratory secretions and by settling of aerosols from infected persons or indirectly by transfer between different types of surfaces [211]. The transmission of NoV by asymptomatic food handlers is well known as an important cause of foodborne outbreaks, especially on cruise ships [15, 165, 249].

2.2 Methods for Viruses Recovered from Environmental Samples

Virus analysis in environmental samples requires several steps, including sampling collection, clarification, concentration, and decontamination/removal of inhibitors, followed by virus detection methods [25]. Natural inhibitors present in environmental samples, such as humic and fulvic acids, heavy metals, and polyphenols, are well recognized by interfering with enzyme amplification of nucleic acids affecting the efficiency of PCR. The use of a process control virus (PCV) before the viral concentration step is an important tool to exclude false-negative reactions [113, 117, 146, 199, 209, 225].

A genetically modified lineage of mengovirus has been proposed in ISO/TS 15216 as PCV, although other viruses, such as feline calicivirus (FCV), murine norovirus 1 (MNV-1), and bacteriophages (MS2, PP7), have also been evaluated in food and water samples [8, 127, 143, 146, 214, 226, 227, 243, 264]. In Latin America, bacteriophage PP7 has been the most used for virus analysis, highlighting its easy propagation for virus stocks [27, 73, 83, 179, 182].

Different methods based on virus properties as adsorption/elution (ionic charge of the viral particle), ultrafiltration (particle size), and ultracentrifugation (density and coefficient of sedimentation) associated with cell culture and molecular assays have been successfully used for recovering virus from environmental matrices [2, 35, 64, 118, 125, 178, 269]. Although viral genome detection does not provide information on infectivity, PCR, for reasons of its high sensitivity, specificity, and speed, has been considered a good technique for environmental monitoring, especially for allowing the cumbersome detection of viruses that do not propagate in cell culture assays [93]. The low stability of RNA free in the environment suggests that molecular methods detect intact viral particles, not free viral genome [37, 94, 269]. However, because infectivity is directly related to human health risk, methodologies able to distinguish infectious and noninfectious particles such as the integrated cell culture–PCR assay (ICC-PCR), viral mRNA detection [41, 130, 194, 195], and enzymatic treatment protocols [133, 169, 237] have been selected for assessing environmental samples. More recently, fluorescent dyes such as ethidium monoazide (EMA) or propidium monoazide (PMA) have also been used to discriminate infectivity in food and water samples [54, 77, 161, 174, 184, 191].

Diverse protocols for detecting viruses in sediments, sewage sludge [13, 92, 100, 101, 130, 136, 209], surfaces [79, 80], and food matrices [7, 26, 29, 70, 135, 141, 160, 207, 226–229] have also been developed. For food matrices, the International Organization for Standardization (ISO) set up methods of elution concentration for recovering and detecting qualitatively and quantitatively NoV and HAV in four classes of food, including drinking water, fruits and vegetables, food surfaces, and bivalve mollusks [120, 121]. More recently, novel methods for molecular detection of viruses using a PCR digital and nanofluidic real-time PCR system have been successfully used to quantify enteric viruses in water and food samples [44, 128, 189].

Studies conducted in Latin America have contributed to the establishment and evaluation of virus recovery methodologies in water matrices [73, 187, 251], sludge [182], and foods such as oysters, cheese, leafy vegetables, and fruits [28, 47, 74, 152, 221] and on different surfaces such as rubber and porous and nonporous formic [83]. They have also demonstrated enteric virus infectivity in freshwater, groundwater, and drinking waters, sediments, swine manure, and biofertilized soil and surfaces [55, 66–69, 81, 82, 159, 250].

2.3 Persistence and Disinfection

Temperature, ultraviolet (UV) light, salts, organic matter, air–water interfaces, and biological factors such as aquatic microflora are important factors for controlling the survival of enteric viruses in water that usually remain stable for months or longer when associated with particulate matter in water and sediments [89].

Processes such as coagulation, flocculation, sedimentation, activated sludge, and filters applied in wastewater treatment plants (WWTPs) are able to remove around 90–99% of the viral load present in wastewater [170, 242]. Further removal of microbiological contaminants is dependent on subsequent disinfection with chlorine tertiary treatment, UV, and ozone [157, 218–220, 234]. Viral load removal efficiency in WWTP depends on the applied and combined processes. Dispersion to aquatic ecosystems in concentrations sufficient to threaten human health is considered an issue related to the difficulty of elimination of viruses [4, 25, 64, 71, 97].

Current studies have demonstrated that viral particles can persist for long periods on surfaces such as medical devices, fomites, and human skin [1, 212, 236], suggesting that they have an important role in the spread of respiratory and enteric viral pathogens in hospitals and children's daycare facilities [39, 78–80, 190, 213]. HAdV in fomites has been described as resistant to hospital disinfection procedures, when recovered from inanimate surfaces from 7 to 90 days, whereas RVA remains viable on inanimate surfaces for periods between 6 and 60 days [131, 206].

In Latin America, a few studies on persistence and/or disinfection have been conducted, mainly to evaluate accumulation and persistence in shellfish, especially in depuration tanks. In Chile, HAV persistence was demonstrated in the mussel *Mytilus chilensis* for about 7 days [56]. Studies on mollusk depuration tanks coupled with UV irradiation revealed the stability of HAdV2, MNV-1, and HAV in seawater samples [48] and demonstrated that 96 h of UV treatment could eliminate HAV and HAdV5 in oysters [43]. Recombinant AdV expressing green fluorescent protein as a model to evaluate disinfection was used to assess the efficiency of free chlorine disinfection in filtered water samples [168] and the effect of UV light in seawater and in shellfish depuration tanks [86].

2.4 Water Quality and Microbial Source Tracking Indicators

Bacterial indicators including *Escherichia coli*, *Enterococcus* sp., and *Clostridium perfringens* have been used for conventional microbiological monitoring of water quality. However, those indicators present some limitations such as sensitivity to inactivation methods when compared to other pathogens as well as the ability to multiply in some environments [23, 104, 122]. Viral water or foodborne diseases have also been reported when those matrices are compliant with the current bacterial standards, generating a global consensus regarding the lack of correlation between bacterial contamination and viral presence [24, 71, 90, 106, 175, 186, 261]. Moreover, those bacteria are not of exclusive fecal origin, becoming unproductive when used to identify sources of contamination and to determine the real risk to public health by contact with these environments [22]. With this background, bacteriophages and human enteric viruses are emerging as alternative indicators for fecal contamination [85, 122, 196, 248], becoming the last one indicated as a promising tool for microbial source tracking (MST) analysis because of its host specificity [22, 64, 116]. Viruses are easily distinguished based on differences in gene sequences common to different species of the same genus. For example, genes of the hexon and fiber of AdV have been used to detect different species [266].

DNA genome viruses such as HAdV and JC polyomavirus (JCPyV) have been described as indicators of fecal contamination of human origin. Factors such as high prevalence, stability in the environment, association to persistent infections, and excretion throughout the year in different geographic regions have contributed to this distinction [21, 22, 91, 109, 149, 178]. For animal species, avian parvovirus, bovine AdV, EV, and PyV, goat and swine AdV, and PyV have been used as a MST tool [40, 114–116, 204, 265, 267, 268].

A few studies conducted in Brazil and Mexico have reported coliphages as good indicators of fecal pollution in seawater and water used for irrigation [12, 31, 57, 95]. The high level of HAdV and JCPyV detection found by environmental surveillance studies conducted in Argentina and Brazil supports the usefulness of viruses to monitor contamination [72, 76, 238, 257] and revealed the human source of contamination in the waters of Negro River in the city of Manaus [205]. In Mexico, *Bacteroidales* 16S rRNA gene sequences were used to indicate fecal contamination in fresh produce [192].

2.5 Data on Environmental Occurrence in Latin America

Distribution and prevalence of viruses of human health concern in water resources is highly variable and dependent on the community's epidemiological profile and geographic conditions, including seasonality, socioeconomic issues, and specific environmental conditions as well as the source of contamination [110].

2.5.1 Sewage, Sediment, and Sludge

Currently, the monitoring of wastewater is considered an important tool for understanding the incidence, distribution, and seasonality of viral agents circulating in certain geographic areas. In some regions, virus identification from raw sewage is a major source of data for epidemiological studies providing phylogenetic analyses [239, 255] and reports of the first description of viruses in the country, as observed for KV in Brazil [34] and for new astrovirus MLB1, canine NoV GVII, and AiV in Uruguay [32, 138, 139]. Table 6.1 summarizes virus data obtained from sewage samples in Latin America.

Environmental surveillance from raw sewage samples was also conducted to monitor the circulation of virus strains prevented by vaccines. For WHO, environmental monitoring of PV is recommended as a complementary surveillance activity of cases of acute flaccid paralysis (AFP), which is currently recognized as one of the main activities included in poliomyelitis eradication efforts. Systematic monitoring of environmental samples aims to evaluate the effects of vaccination programs on the emergence of a vaccine-derived poliovirus (VDPV) as well as to detect the movement of PV through sewage sampling [49, 58, 96, 145, 232, 240, 241].

Table 6.1 Natural occurrence of viruses in raw sewage

Virus	Country	City	Study period	Genotypes	Analyzed samples (% positivity)	Concentration (GC/L)	Reference
Adenovirus	Argentina	Córdoba	2009–2011	NR	30 (86.7)	NR	[62]
	Brazil	Florianópolis	2007–2008	NR	12 (66.6)	1.2E + 04–3.6 + E07	[198]
		Limeira	2004–2005	F	50 (100)	NR	[9]
		Porto Alegre	2009	NR	8 (25)	NR	[245]
		Rio de Janeiro	2005–2008	HAdVC, HAdVD, and HAdVF (40, 41)	12 (58.3)	1.7E + 05–2.3E + 07	[183]
			2009–2010	HAdV41	24 (100)	5.23E + 05–7.04E + 06	[76]
	Venezuela	Caracas	2007–2008	NR	12 (50)	NR	[201]
Astrovirus	Brazil	Rio de Janeiro	2005	NR	24 (16.7)	1.1E + 07[f]	[98]
			2009–2010	HAstV1	24 (29)	6.23E + 02–2.71E + 03	[76]
	Uruguay	Bella Unión, Fray Bentos, Paysandu, and Salto	2011–2012	NR	96 (45)	3.2E + 03–4.3E + 07	[254]
		Bella Unión, Fray Bentos, Paysandu, and Salto	2011–2012	HAstV1 (1a lineage) and HAstV2 (2c and 2d)	96 (44.8)	NR	[138]
		Melo and Treinta y Tres	2012–2013	MLB-1, HAstV1, HAstV2, and HAstV5	20 (20–70)	NR	[139]
	Venezuela	Caracas	2007–2008	HAstV8	12 (66.7)	NR	[201]

Enterovirus	Argentina	Córdoba, Rio Cuarto, Villa Maria, and San Francisco	2005–2006	PV1, PV2, and PV3 Sabin-like	255 (56)	NR	[167]
	Bolivia	Yungas	2012	NR	4 (100)	4.23E + 04–6.2E + 04[a]	[231]
	Brazil	Rio de Janeiro	2011–2012	Echovirus 3, 6, 7, 20, Coxsackievirus B4 and B5, PV 2 and PV3 Sabin-like	31 (87)	NR	[176]
		São Paulo	1999–2001	PV1, PV2, and PV3 Sabin-like	27 (100)	NR	[96]
			2009	NR	24 (100)	0,5E + 00–2.6E + 02[b]	[103]
	Colombia	Cartagena	1991	PV1 Wild-type and vaccine-related PV	42 (21)	NR	[232]
	Cuba	Havana	1997–1998	PV1, PV2, and PV3	56 (50)	NR	[145]
	Mexico	Aguascalientes, Tuxtla Gutiérrez, Benito Juárez, and San Luis Potosí	2010	PV1, PV2, and PV3 Sabin-like and VDPV1 and VDPV2	87 (54)	NR	[58]
		Veracruz	2009	Revertant OPV-1 and nonrevertant OPV-2 and OPV-3	5 (80)	NR	[241]
	Venezuela	Caracas	2007–2008	NR	12 (75)	NR	[201]

(continued)

Table 6.1 (continued)

Virus	Country	City	Study period	Genotypes	Analyzed samples (% positivity)	Concentration (GC/L)	Reference
Hepatitis A virus	Argentina	Córdoba	2009–2010	IA	24 (20.8)	NR	[271]
	Brazil	Florianópolis	2007–2008	NR	12 (8.3)	NR	[198]
		Limeira	2004–2005	NR	50 (48)	NR	[9]
		Rio de Janeiro	2005	IA and IB	25 (32)	1.2E + 05–8.9E + 05	[260]
			2005–2008	IA and IB	12 (33.3)	1.6E + 06–1.2E + 08	[183]
			2009–2010	IA	24 (58)	2.5E + 05–3.7E + 06	[180]
Hepatitis E virus	Argentina	Córdoba	2007, 2009, 2010, and 2011	G3a, G3b, and G3c	48 (6.3)	NR	[144]
	Brazil	Rio de Janeiro	2008	G3b	6 (50)	1.0E + 05	[53][c]
Norovirus GI	Bolivia	Yungas	2012	NR	4 (100)	1E + 03-1E + 06	[231]
	Brazil	Rio de Janeiro	2005	GI.2	24 (4.2)	2.4E + 03	[253]
	Uruguay	Bella Unión, Fray Bentos, Melo, Paysandú, Salto, and Treinta y Tres	2011–2013	GI.1, GI.3, GI.4, GI.5, GI.6, and GI.8	116 (31)	NR	[255]
		Bella Unión, Fray Bentos, Paysandú, and Salto	2011–2012	NR	96 (7.3)	2.9E + 03–3.8E + 07	[254]

Norovirus GII	Argentina	Córdoba	2009	GII.4 2006b and GII.g	7 (42.9)	NR	[19]
	Brazil	Florianópolis	2007–2008	NR	12 (8.3)	NR	[253]
		Rio de Janeiro	2005	GII.3, GII.4, and GGIIb	24 (66.7)	7.29E + 03	[252]
			2005–2008	GII.4	12 (25)	9.7E + 05–1.8E + 06	[183]
			2009–2010	GII.4 and Gll.g	24 (55)	1.5E + 04–2.03E + 05	[76]
	Nicaragua	León	2007–2008	NR	28 (53.6)[d]	1E + 04–1E + 06	[30]
	Uruguay Uruguay	Bella Unión, Fray Bentos, Melo, Paysandú, Treinta y Tres, and Salto	2011–2013	GII.2, GII.3, GII.4, GII.6, GII.13, and GII.17	116 (64.7)	NR	[255]
		Bella Unión, Fray Bentos, Salto, and Paysandú	2011–2012	NR	96 (51)	2.9E + 03–3.8E + 07	[254]
	Venezuela	Caracas	2007–2008	NR	12 (75)	NR	[201]
Norovirus GIV	Brazil	Belém	2008–2010	GIV.1	24 (16.7)	NR	[233]
Polyomavirus	Argentina	Buenos Aires and suburbs	2011 and 2013	BKPyV, JCPyV, MCPyV, and HPyV6[e]	24 (95.8)	NR	[238]
	Brazil	Rio de Janeiro	2005	JCPyV	24 (96)	1.2E + 03–3.2E + 05	[72]
			2009–2010	JCPyV	24 (100)	3.26E + 04–2.11E + 06	[76]

(continued)

Table 6.1 (continued)

Virus	Country	City	Study period	Genotypes	Analyzed samples (% positivity)	Concentration (GC/L)	Reference
Rotavirus	Argentina	Córdoba	2009	G1, G2, G3, G4, G8, and G9	52 (100)	NR	[11]
			2009–2011	G1, G2, G3, G4, G8, G9,P[4], and P[8]	35 (91.4)	2.2E + 02–4.1E + 05	[10]
	Bolivia	Yungas	2012	NR	4 (100)	1E + 05-1E + 06	[231]
	Brazil	Florianópolis	2007–2008	NR	12 (8.3)	NR	[198]
		Porto Alegre	2009	NR	7 (28.6)	NR	[245]
		Rio de Janeiro	2004–2005	G3, G9, P[8], and SGII	24 (45.8)	NR	[61]
			2005–2008	SGI	12 (75)	8.7E + 08–9.6E + 09	[183]
			2009–2010	G2, P[4] and P[6]	24 (100)	2.40E + 05–1.16E + 07	[75]
			NR	G2, P[4] and P[6]	14 (100)	2.5E + 04–1.6E + 07	[73]
		São Paulo	1987–1988	NR	29 (20.6)	NR	[150]
	Nicaragua	León	2007–2008	G2, G4, SGI, and SGII	28 (21.4)	1E + 04-1E + 05	[30]

	Uruguay	Bella Unión, Fray Bentos, Paysandú, and Salto	2011–2012	G2, G3, P[4], P[8], P[10], and P[11]	96 (49)	3.9E + 03–54.3E + 07	[254]
	Uruguay	Bella Unión, Fray Bentos, Melo, Paysandú, Salto, and Treinta y Tres	2011–2013	G1, G2, G3, G12, P[3], P[4], and P[8]	116 (52.6)	NR	[239]
	Venezuela	Caracas	2007–2008	G1, G9, P[4] and P[8]	12 (66.7)	NR	[201]

GC/l (genome copies/l), *NR* not reported
[a]iu/l (infectious units/ml)
[b]PFU/l
[c]Sewage from slaughterhouses
[d]Sewage samples were also positive for NoVGI; however, it is not clear if those samples represent the raw sewage
[e]Also, one sample collected in 2009 resulted positive for HPyV7
[f]G eq/l (genome equivalent/l)

Environmental monitoring for RVA strains was also carried out in countries that include the rotavirus vaccine for immunization schedules. In Brazil, RVA detection from sewage samples obtained in an urban area in Rio de Janeiro demonstrated fluctuation in the RVA genotypes after the introduction of RV1 vaccine in the National Immunization Program, in 2006 [75]. In Nicaragua, the impact of the pentavalent RV5 vaccine was also evaluated by investigating RV dissemination in urban and hospital wastewaters of the city, demonstrating that RVA dissemination decreased in the community 2 years after the introduction of a vaccination program [30]. The importance for RV vaccine strain surveillance in the environment regarding the potential occurrence of unexpected infections and virus genomic reassortments was highlighted by the evidence of its infectivity and high stability in water samples [163].

Studies have also reported enteric viruses from sludge and sediments. HAdV, RVA, NoV, PV, and HAV were detected in sewage sludge samples from WTTPs using conventional activated sludge treatment after anaerobic digestion [181, 182, 215]. HAdV and RVA were detected from sediment samples obtained in the catchment area of Peri Lagoon, Santa Catarina State, Brazil [55].

2.5.2 Surface and Drinking Waters

Viral concentrations reaching levels of 10^8 genome copies/l in fresh and brackish waters indicate how the basic sanitation deficiency in Latin American countries threatens the health of the exposed population, mainly because of the use of these waters for recreational purposes (Table 6.2). For marine recreational waters, studies conducted in Brazil demonstrated RVA, NoV, JCPyV, HAV, and infectious HAdV in the beaches of Florianopolis [164, 198, 253] and Rio de Janeiro [258]. Investigation of enteric viruses during two rainfall events showed a greater permanence of viruses compared with bacterial indicators, highlighting their use as a parameter to determine the microbiological quality of recreational waters. Infectious HAdV, EV, HAV, and NoV were also detected in marine waters in Mexico [59, 210].

HAV outbreaks caused by consumption of drinking water have been described in the region based on epidemiological studies, bacteriological parameters [217, 230], and by direct detection of virus from tap-water samples [259]. Additional studies reported other enteric viruses such as NoV, RVA, EV, HAdV, and torque teno virus (TTV) in tap water [99, 129, 162, 198, 200, 216, 224, 246, 253] and demonstrated the infectivity of HAdV in waters for human consumption using plaque-forming assay and ICC-PCR [67, 69, 87, 159].

2.5.3 Shellfish, Crops, Irrigation Water, and Other Foodstuffs

During the past decades, enteric virus genomes have been detected in shellfish (oysters, clams, and mussels) obtained directly from different aquatic environments such as marine farm crops and mangroves or commercial operations. RVA, HAV,

Table 6.2 Natural occurrence of viruses in surface waters

Virus	Country	City	Water type	Study period	Genotypes	Samples analyzed (% Positivity)	Concentration (GC/L)	Reference
Adenovirus	Brazil	Campo Bom, Esteio, Nova Santa Rita, Parobé, Rolante, Santo Antonio da Patrulha, Taquara, and Três Coroas	River	2011–2013	NR	178 (54)	1.01E + 03–7.98E + 06	[46]
		Campo Bom, Esteio, Parobé, Santo Antônio da Patrulha, Rolante, Taquara, and Três Coroas	Watershed	2011	NR	64 (75)	7.8E + 01–2.1E + 08	[200]
		Caraá, Novo Hamburgo, Taquara, and Sapucaia do Sul	River	NR	NR	4 (25)	NR	[14]
		Concórdia	River	2010–2011	NR	12 (100)	1.3E + 06–3.55E + 06	[87]
		Florianópolis	Creek	2007–2008	NR	12 (75)	2.5E + 04–3.9 + E07	[198]
			Lagoon	2007–2008	NR	12 (75)	4.2E + 04–1.0 + E07	[198]
				2010–2011	HAdV2	12 (75)	1.65E + 06	[67]

(continued)

Table 6.2 (continued)

Virus	Country	City	Water type	Study period	Genotypes	Samples analyzed (% Positivity)	Concentration (GC/L)	Reference
				2010–2011	NR	48 (96)	1.73E + 06–2.41E + 08	[68]
				2011–2012	NR	6 (100)	5.0E + 05–4.1E + 06	[69]
			Lagoon and river	2013	NR	48 (66.7)	6.9E + 05–2.4E + 08	[55]
		Manaus	Stream	2004–2005	HAdV40 and HAdV41	52 (30.8)	NR	[158]
			River	2011–2012	NR	212 (91.9)	2.24E + 02–2.92E + 06	[205, 257]
		Rio de Janeiro	River	2007–2008	NR	12 (16.7)	NR	[258]
				2008–2009	NR	108 (6.5)	NR	[159]
				NR	NR	6 (100)	1.59E + 03–7.11E + 04	[34]
			Lagoon	2007–2008	NR	120 (16.7)	NR	[258]
		Porto Alegre	Lake	2011–2012	NR	18 (77.8)	3.65E + 02–9.04E + 03	[148]
			Stream	2009	NR	14 (21.43)	NR	
	Venezuela	Caracas	River	2007–2008	NR	18 (83)	NR	[201]

Table 6.2 (continued)

Astrovirus	Brazil	Manaus	Stream	2004–2005	HAstV1	52 (15.4)	NR	[158]
	Venezuela	Caracas	River	2007–2008	HAstV8	18 (89)	NR	[201]
Enterovirus	Argentina	Córdoba	River	2010	NR	76 (84.2)	NR	[185]
	Brazil	Caraá, Novo Hamburgo, Taquara, and Sapucaia do Sul	River	NR	NR	4 (25)	NR	[14]
		Porto Alegre	Stream	2009	NR	14 (64.28)	NR	
			Lake	2011–2012	NR	18 (5.6)	1.21E + 04	[148]
	Venezuela	Caracas	River	2007–2008	NR	18 (72)	NR	[201]
Hepatitis A virus	Argentina	Buenos Aires and suburbs and Córdoba	River	2005–2012	IA and IC	264 (16.3)	NR	[20]
		Córdoba	River	2009–2010	IA	31 (16.1)	NR	[271]
	Brazil	Florianópolis	Lagoon	2007–2008	NR	12 (16.6)	NR	[198]
				2010–2011	NR	48 (12)	7.81E + 01–1.33E + 05	[68]
				2011–2012	NR	6 (66.6)	1.2E + 05–2.03E + 05	[69]
			Lagoon and river	2013	NR	48 (29.2)	5.45E + 01–1.11E + 03	[55]

(continued)

Table 6.2 (continued)

Virus	Country	City	Water type	Study period	Genotypes	Samples analyzed (% Positivity)	Concentration (GC/L)	Reference
		Manaus	Stream	2004–2005	IA and IB	52 (92)	6E + 01–5.5E + 03	[50]
		Rio de Janeiro	River	NR	IA	5 (20)	NR	[259]
	Mexico	Huizache Caimanero Lagunary Complex	Estuarine	2006–2007	IB	40 (80)	NR	[108]
	Venezuela	Caracas	Stream	2007–2008	IA	14 (100)	NR	[17]
Hepatitis E virus	Argentina	Córdoba	River	2007, 2009, 2010, and 2011	G3a, G3b, and G3c	31 (3.2)	NR	[144]
Norovirus GI	Brazil	Florianópolis	Creek	2007–2008	GI.2	12 (8.3)	2.6E + 03	[253]
			Lagoon	2007–2008	NR	12 (8.3)	NR	[253]
		Rio de Janeiro	River	2008–2009	NR	108 (2.8)	NR	[159]
	Venezuela	Caracas	River	2007–2008	NR	18 (16.7)	NR	[201]
Norovirus GII	Argentina	Buenos Aires and suburbs, Salta, and Córdoba	River	2005–2011	GII.4, GII.b, GII.2, GII.7, GII.17, GII.e, and GII.g	209 (25.4)	NR	[60]
		Córdoba	River	2009	GII.4 2006b, GII.g	14 (50)	NR	[19]

	Brazil	Florianópolis	Creek	2007–2008	GII.2 and GII.4 2006b	12 (71.7)	1.14E + 02–3.73E + 03	[253]
		Manaus	Stream	2004–2005	NR	52 (5.8)	NR	[158]
			River and stream	2011–2012	NR	212 (7.4)	4.77E + 02–4.77E + 04	[257]
		Rio de Janeiro	Lagoon	2007–2008	NR	120 (17.5)	NR	[258]
			River	NR	NR	6 (33.3)	6.76E + 03–8.57E + 03	[34]
				2007–2008	NR	12 (33.3)	NR	[258]
				2008–2009	NR	108 (9.3)	NR	[159]
	Mexico	Huizache Caimanero Lagunary Complex	Estuarine	2006–2007	GII.4	40 (70)	NR	[108]
	Venezuela	Caracas	River	2007–2008	NR	18 (55.6)	NR	[201]
Norovirus GIV	Brazil	Belém	River	2008–2010	GIV.1	24 (4.2)	NR	[233]
			Bay	2008–2010	GIV.1	24 (4.2)	NR	
			Stream	2008–2010	GIV.1	24 (12.5)	NR	

(continued)

Table 6.2 (continued)

Virus	Country	City	Water type	Study period	Genotypes	Samples analyzed (% Positivity)	Concentration (GC/L)	Reference
Polyomavirus	Argentina	Buenos Aires and suburbs	River	2005, 2006, and 2012	BKPyV, JCPyV, MCPyV, and HPyV6	45 (84)	NR	[238]
	Brazil	Florianópolis	Lagoon	2010–2011	JCPyV	48 (21)	4.1E + 04–6.3E + 06	[68]
		Manaus	River	2011–2012	NR	212 (69.5)	2.32E + 02–7.16E + 05	[205, 257]
		Rio de Janeiro	River	NR	JCPyV	6 (100)	1.58E + 02–2.97E + 04	[34]
Rotavirus	Argentina	Córdoba	River	2010	G1, G2, G3, G4, G8, G9, P[4], P[8], P[10]	76 (48.7)	1.9E + 03–7.9E + 06	[185]
	Brazil	Caraá, Taquara, Novo Hamburgo, and Sapucaia do Sul	River	NR	NR	4 (25)	NR	[14]
		Manaus	River and stream	2011–2012	NR	212 (23.9)	5.66E + 02–1.35E + 05	[257]
			Stream	2004–2005	G1, P[4], P[8], SGII	52 (48)	NR	[61, 158]
		Minas Gerais	Watershed	2011–2012	I1	48 (62.5)	4.74E + 02–6.36E + 04	[5]

		Florianópolis	Creek	2007–2008	NR	12 (58.3)	NR	[198]
			Lagoon	2007–2008	NR	12 (16.6)	NR	[198]
				2010–2011	NR	48 (65)	1.20E + 05–2.50E + 07	[68]
				2011–2012	NR	6 (66.6)	9.5E + 07–3.5E + 08	[69]
			Lagoon and river	2013	NR	48 (33.3)	3.09E + 04–5.4E + 05	[55]
		Rio de Janeiro	Lagoon	2007–2008	NR	120 (25)	NR	[258]
			River	2007–2008	NR	12 (25)	NR	[258]
				NR	NR	6 (66.7)	7.29E + 02–2.7E + 04	[34]
				2008–2009	NR	108 (15.7)	NR	[159]
		São Paulo	Creek	1987–1988	NR	55 (34.5)	NR	[150]
	Venezuela	Caracas	River	2007–2008	G1, G10, P[4], P[8], NSP4A, NSP4B, NSP4C	18 (83)	NR	[201]
Torque teno virus	Brazil	Manaus	River	2004–2005	NR	52 (92)	1.3E + 03–7.46E + 05[a]	[52]
		Porto Alegre	Stream	2009	NR	14 (28.57)	NR	

GC/l (genome copies/l)
NR not reported
[a]G eq/l (genome equivalent/l)

NoV, and HAdV have been detected in *Crassostrea gigas* cultivated in marine farms in Brazil [197, 198, 221, 223] and in samples obtained from the Central Food Supply Station of Mexico City (30%; 10/30) [188]. In Mexico, RVA, HAV, and NoV were also detected from fruits and vegetables in different supermarkets around the country [173, 188]. In Brazilian mangroves, HAdV and RVA genomes were detected by nRT-PCR in 100% of the mussels studied [126].

In Brazil, NoV GI and GII were identified in food samples (herbed butter, cheese, white sauce) served on a cruise ship, indicating these products as sources of the outbreak [165], and in cheeses acquired in markets in Rio de Janeiro [151]. Infectious HAdV was detected in strawberries and in organic products such as lettuce and green onions [142, 152]. HEV RNAs were detected in pork products such as flavored pâté (ham, meat, bacon) [107].

Concerning food safety, irrigation water should also be considered. In the south of Mexico City, RVA and HAstV were detected in approximately 10% of the water samples analyzed that were used for cultivation of flowers and vegetables [57]. In Brazil, HAdV and RVA were detected in groundwater used for irrigation and human consumption [69].

2.5.4 Surfaces

Brazil was the pioneer in Latin America for studying viral contamination on surfaces and fomites. In a hospital intensive care unit in Rio de Janeiro, RVA and HAdV detection and viability were demonstrated from surface samples [81, 82]. In monitoring of surfaces and fomites of two units of a public hospital, including a neonatal intensive care unit and a pediatric ward, the same group showed a significant lower percentage of viruses in the intensive care unit, reflecting concerns for health professionals and families related to hygiene issues [84]. In another study conducted in Sinos Valley region, south of the country, HAdV contamination was observed in 62% (20/32) of samples from surfaces obtained in internship units and in emergency and operating rooms [244].

3 Future Challenges

Prevention of transmission of infectious diseases by exposure to contaminated water remains one of the main tasks for professionals of public healthcare and environmental health and can be accomplished by associating risk analysis studies that estimate the risk and identify vulnerable individuals and its implications in health, places, and routes of exposure to effective management measures [102, 154]. In Latin America, two studies conducted in Argentina and Brazil are pioneers to assess the quantitative risk for RVA infection through contact with surface waters [185, 256]. However, to overcome this challenge, Latin American countries should expand their regulatory actions and networks. The publication of new ordinances that support viral diagnosis from environmental samples when epidemiological data suggest water or foodborne transmission, as well as the establishment of a low-cost method able to be used on a large scale, could accelerate advances in the region. However, it is worth remembering that basic sanitation interventions are the main preventive actions for reducing the diarrheal disease burden [231, 247].

Acknowledgments The authors thank the editors and the institutions that have been supporting projects in environmental virology field at Oswaldo Cruz Institute/Fiocruz, Brazilian National Council for Scientific and Technological Development (CNPq), and Carlos Chagas Filho Foundation for Research Support in the state of Rio de Janeiro (Faperj). M.P.M. (research productivity) and C.B.V. (postdoctoral program) are fellowships of CNPq. This chapter is under the scope of the activities of Fiocruz as a collaborating center of PAHO/WHO of Public and Environmental Health.

References

1. Abad FX, Villena C, Guix S et al (2001) Potential role of fomites in the vehicular transmission of human astroviruses. Appl Environ Microbiol 67:3904–3907. doi:10.1128/AEM.67.9.3904-3907.2001
2. Albinana-Gimenez N, Clemente-Casares P, Bofill-Mas S et al (2006) Distribution of human polyomaviruses, adenoviruses, and hepatitis E virus in the environment and in a drinking-water treatment plant. Environ Sci Technol 40(23):7416–7422
3. Appleton H, Higgins PG (1975) Letter: Viruses and gastroenteritis in infants. Lancet 1(7919):1297
4. Arraj A, Bohatier J, Laveran H et al (2005) Comparison of bacteriophage and enteric virus removal in pilot scale activated sludge plants. J Appl Microbiol 98:516–524. doi:10.1111/j.1365-2672.2004.02485.x
5. Assis AS, Cruz LT, Ferreira AS et al (2015) Relationship between viral detection and turbidity in a watershed contaminated with group A rotavirus. Environ Sci Pollut Res Int 22(9):6886–6897. doi:10.1007/s11356-014-3874-8
6. Aw TG, Howe A, Rose JB (2014) Metagenomic approaches for direct and cell culture evaluation of the virological quality of wastewater. J Virol Methods 210:15–21. doi:10.1016/j.jviromet.2014.09.017
7. Baert L, Uyttendaele M, Debevere J (2008a) Evaluation of viral extraction methods on a broad range of ready-to-eat foods with conventional and real-time RT-PCR for norovirus GII detection. Int J Food Microbiol 123:101–108. doi:10.1016/j.ijfoodmicro.2007.12.020
8. Baert L, Wobus CE, Coillie EV et al (2008b) Detection of murine norovirus 1 by using plaque assay, transfection assay, and real-time reverse transcription-PCR before and after heat exposure. Appl Environ Microbiol 74(2):543–546. doi:10.1128/AEM.01039-07
9. Barrella KM, Garrafa P, Monezi TA et al (2009) Longitudinal study on occurrence of adenoviruses and hepatitis A virus in raw domestic sewage in the city of Limeira, São Paulo. Braz J Microbiol 40(1):102–107. doi:10.1590/S1517-838220090001000017
10. Barril PA, Fumian TM, Prez VE et al (2015) Rotavirus seasonality in urban sewage from Argentina: effect of meteorological variables on the viral load and the genetic diversity. Environ Res 138:409–415. doi:10.1016/j.envres.2015.03.004
11. Barril PA, Giordano MO, Isa MB et al (2010) Correlation between rotavirus A genotypes detected in hospitalized children and sewage samples in 2006, Córdoba, Argentina. J Med Virol 82(7):1277–1281. doi:10.1002/jmv.21800

12. Bartz FE, Hodge DW, Heredia N et al (2016) Somatic coliphage profiles of produce and environmental samples from farms in Northern México. Food Environ Virol 8(3):221–226. doi:10.1007/s12560-016-9240-x
13. Belguith K, Hassen A, Aouni M (2006) Comparative study of four extraction methods for enterovirus recovery from wastewater and sewage sludge. Bioresour Technol 97(3):414–419. doi:10.1016/j.biortech.2005.03.022
14. Bergamaschi B, Rodrigues MT, Silva JV et al (2015) Moving beyond classical markers of water quality: detection of enteric viruses and genotoxicity in water of the Sinos River. Braz J Biol 75(2 suppl):63–67. doi:10.1590/1519-6984.1713
15. Bert F, Scaioli G, Gualano MR et al (2014) Norovirus outbreaks on commercial cruise ships: a systematic review and new targets for the public health agenda. Food Environ Virol 6(2):67–74. doi:10.1007/s12560-014-9145-5
16. Berto A, van der Poel WH, Hakze-van der Honing R et al (2013) Replication of hepatitis E virus in three-dimensional cell culture. J Virol Methods 187(2):327–332. doi:10.1016/j.jviromet.2012.10.017
17. Betancourt WQ, Querales L, Sulbaran YF et al (2010) Molecular characterization of sewage-borne pathogens and detection of sewage markers in an urban stream in Caracas, Venezuela. Appl Environ Microbiol 76(6):2023–2026. doi:10.1128/AEM.02752-09
18. Bishop RF, Davidson GP, Holmes IH et al (1973) Virus particles in epithelial cells of duodenal mucosa from children with acute non-bacterial gastroenteritis. Lancet 2(7841):1281–1283. doi:10.1016/S0140-6736(73)92867-5
19. Blanco Fernández MD, Torres C, Martínez LC et al (2011) Genetic and evolutionary characterization of norovirus from sewage and surface waters in Córdoba City, Argentina. Infect Genet Evol 11(7):1631–1637. doi:10.1016/j.meegid.2011.06.005
20. Blanco Fernández MD, Torres C, Riviello-López G et al (2012) Analysis of the circulation of hepatitis A virus in Argentina since vaccine introduction. Clin Microbiol Infect 18(12):E548–E551. doi:10.1111/1469-0691.12034
21. Bofill-Mas S, Albinana-Gimenez N, Clemente Casares P et al (2006) Quantification and stability of human adenoviruses and polyomavirus JCPyV in wastewater matrices. Appl Environ Microbiol 72(12):7894–7896. doi:10.1128/AEM.00965-06
22. Bofill-Mas S, Rusiñol M, Fernandez-Cassi X et al (2013) Quantification of human and animal viruses to differentiate the origin of the fecal contamination present in environmental samples. Biomed Res Int 2013:192089. doi:10.1155/2013/192089
23. Borchardt MA, Haas NL, Hunt RJ (2004) Vulnerability of drinking-water wells in La Crosse, Wisconsin, to enteric-virus contamination from surface water contributions. Appl Environ Microbiol 70(10):5937–5946. doi:10.1128/AEM.70.10.5937-5946.2004
24. Bosch A (1998) Human enteric viruses in the water environment: a minireview. Int Microbiol 1(3):191–196
25. Bosch A, Guix S, Sano D et al (2008) New tools for the study and direct surveillance of viral pathogens in water. Curr Opin Biotechnol 19(3):295–301. doi:10.1016/j.copbio.2008.04.006
26. Boxman ILA, Tilburg JJHC, Loeke NAJM et al (2007) An efficient and rapid method for recovery of norovirus from food associated with outbreaks of gastroenteritis. J Food Prot 70(2):504–508
27. Brandão ML, Almeida DO, Bispo FC et al (2014a) Assessment of microbiological contamination of fresh, minimally processed, and ready-to-eat lettuces (*Lactuca sativa*), Rio de Janeiro State, Brazil. J Food Sci 79(5):M961–M966. doi:10.1111/1750-3841.12459
28. Brandão MLL, Almeida DO, Marin VA et al (2014b) Recovery of Norovirus from lettuce (*Lactuca sativa*) using an adsorption-elution method with a negatively charged membrane: comparison of two elution buffers. Vigilância Sanitária em Debate: Sociedade, Ciência & Tecnologia 2:58–63
29. Brassard J, Gagné MJ, Généreux M et al (2012) Detection of human food-borne and zoonotic viruses on irrigated, field-grown strawberries. Appl Environ Microbiol 78(10):3763–3766. doi:10.1128/AEM.00251-12

30. Bucardo F, Lindgren PE, Svensson L et al (2011) Low prevalence of rotavirus and high prevalence of norovirus in hospital and community wastewater after introduction of rotavirus vaccine in Nicaragua. PLoS One 6(10):e25962. doi:10.1371/journal.pone.0025962
31. Burbano-Rosero EM, Ueda-Ito M, Kisielius JJ et al (2011) Diversity of somatic coliphages in coastal regions with different levels of anthropogenic activity in São Paulo, Brazil. Appl Environ Microbiol 77(12):4208–4216. doi:10.1128/AEM.02780-10
32. Burutarán L, Lizasoain A, García M et al (2016) Detection and molecular characterization of Aichivirus 1 in wastewater samples from Uruguay. Food Environ Virol 8(1):13–17. doi:10.1007/s12560-015-9217-1
33. Caillou S, Castagnaro N, Naval M (1998) Sewage sludges: application of a technique for recovering enterovirus. Rev Argent Microbiol 30(2):96–99
34. Calgua B, Fumian T, Rusiñol M et al (2013) Detection and quantification of classic and emerging viruses by skimmed-milk flocculation and PCR in river water from two geographical areas. Water Res 47(8):2797–2810. doi:10.1016/j.watres.2013.02.043
35. Calgua B, Mengewein A, Grunert A et al (2008) Development and application of a one-step low cost procedure to concentrate viruses from seawater samples. J Virol Methods 153(2):79–83. doi:10.1016/j.jviromet.2008.08.003
36. Cantalupo PG, Calgua B, Zhao G et al (2011) Raw sewage harbors diverse viral populations. MBio 2(5):e00180–e00111. doi:10.1128/mBio.00180-11
37. Carducci A, Casini B, Bani A et al (2003) Virological control of groundwater quality using biomolecular tests. Water Sci Technol 47(3):261–266
38. Carducci A, Morici P, Pizzi F et al (2008) Study of the viral removal efficiency in a urban wastewater treatment plant. Water Sci Technol 58(4):893–897. doi:10.2166/wst.2008.437
39. Carducci A, Verani M, Lombardi R et al (2011) Environmental survey to assess viral contamination of air and surfaces in hospital settings. J Hosp Infect 77(3):242–247. doi:10.1016/j.jhin.2010.10.010
40. Carratalà A, Rusinol M, Hundesa A et al (2012) A novel tool for specific detection and quantification of chicken/turkey parvoviruses to trace poultry fecal contamination in the environment. Appl Environ Microbiol 78(20):7496–7499
41. Chapron CD, Ballester NA, Fontaine JH et al (2000) Detection of astroviruses, enteroviruses, and adenovirus types 40 and 41 in surface waters collected and evaluated by the information collection rule and an integrated cell culture-nested PCR procedure. Appl Environ Microbiol 66(6):2520–2525. doi:10.1128/AEM.66.6.2520-2525.2000
42. Chiba S, Sakuma Y, Kogasaka R et al (1979) An outbreak of gastroenteritis associated with calicivirus in an infant 307 home. J Med Virol 4(4):249–254
43. Corrêa Ade A, Rigotto C, Moresco V et al. (2012) The depuration dynamics of oysters (*Crassostrea gigas*) artificially contaminated with hepatitis A virus and human adenovirus. Mem Inst Oswaldo Cruz 107(1):11–17. doi:http://dx.doi.org/10.1590/S0074-02762012000100002
44. Coudray-Meunier C, Fraisse A, Martin-Latil S et al (2016) A novel high-throughput method for molecular detection of human pathogenic viruses using a nanofluidic real-time PCR system. PLoS One 11(1):e0147832
45. da Silva Assis MR, Vieira CB, Fioretti JM et al (2016) Detection and molecular characterization of gemycircularvirus from environmental samples in Brazil. Food Environ Virol 8(4):305–309. doi:10.1007/s12560-016-9254-4
46. Dalla Vecchia A, Rigotto C, Staggemeier R et al (2015) Surface water quality in the Sinos River basin, in Southern Brazil: tracking microbiological contamination and correlation with physicochemical parameters. Environ Sci Pollut Res Int 22(13):9899–9911. doi:10.1007/s11356-015-4175-6
47. de Abreu CA, Miagostovich MP (2012) Optimization of an adsorption–elution method with a negatively charged membrane to recover norovirus from lettuce. Food Environ Virol 5: 144–149. doi:10.1007/s12560-013-9113-5
48. de Abreu CA, Souza DSM, Moresco V et al (2012) Stability of human enteric viruses in seawater samples from mollusc depuration tanks coupled with ultraviolet irradiation. J Appl Microbiol 113(6):1554–1563. doi:10.1111/jam.12010

49. de Oliveira Pereira JS, da Silva LR, de Meireles NA et al (2016) Environmental surveillance of polioviruses in Rio de Janeiro, Brazil, in support to the activities of global polio eradication initiative. Food Environ Virol 8(1):27–33. doi:10.1007/s12560-015-9221-5
50. de Paula VS, Diniz-Mendes L, Villar LM et al (2007) Hepatitis A virus in environmental water samples from the Amazon Basin. Water Res 41(6):1169–1176. doi:10.1016/j.watres.2006.11.029
51. Deetz T, Smith E, Goyal M et al (1984) Occurrence of rota- and enteroviruses in drinking and environmental water in a developing nation. Water Res 18(5):567–571. doi:10.1016/0043-1354(84)90205-7
52. Diniz-Mendes L, Paula VS, Luz SL et al (2008) High prevalence of human Torque teno virus in streams crossing the city of Manaus, Brazilian Amazon. J Appl Microbiol 105(1):51–58. doi:10.1111/j.1365-2672.2007.03720.x
53. dos Santos DR, de Paula VS, de Oliveira JM et al (2011) Hepatitis E virus in swine and effluent samples from slaughterhouses in Brazil. Vet Microbiol 149(1-2):236–241. doi:10.1016/j.vetmic.2010.10.024
54. Elizaquível P, Aznar R, Sánchez G (2014) Recent developments in the use of viability dyes and quantitative PCR in the food microbiology field. J Appl Microbiol 116(1):1–13. doi:10.1111/jam.12365
55. Elmahdy EM, Fongaro G, Schissi CD et al (2016) Enteric viruses in surface water and sediment samples from the catchment area of Peri Lagoon, Santa Catarina State, Brazil. J Water Health 14(1):142–154. doi:10.2166/wh.2015.295
56. Enriquez R, Frösner GG, Hochstein-Mintzel V et al (1992) Accumulation and persistence of hepatitis A virus in mussels. J Med Virol 37(3):174–179
57. Espinosa AC, Arias CF, Sánchez-Colón S et al (2009) Comparative study of enteric viruses, coliphages and indicator bacteria for evaluating water quality in a tropical high-altitude system. Environ Health 8:49. doi:10.1186/1476-069X-8-49
58. Esteves-Jaramillo A, Estívariz CF, Peñaranda S et al (2014) Detection of vaccine-derived polioviruses in Mexico using environmental surveillance. J Infect Dis 210(suppl 1):S315–S323. doi:10.1093/infdis/jiu183
59. Félix JL, Fernandez YC, Velarde-Félix JS et al (2010) Detection and phylogenetic analysis of hepatitis A virus and norovirus in marine recreational waters of Mexico. J Water Health 8(2):269–278. doi:10.2166/wh.2009.114
60. Fernández MD, Torres C, Poma HR et al (2012) Environmental surveillance of norovirus in Argentina revealed distinct viral diversity patterns, seasonality and spatio-temporal diffusion processes. Sci Total Environ 437:262–269. doi:10.1016/j.scitotenv.2012.08.033
61. Ferreira FF, Guimarães FR, Fumian TM et al (2009) Environmental dissemination of group A rotavirus: P-type, G-type and subgroup characterization. Water Sci Technol 60(3):633–642. doi:10.2166/wst.2009.413
62. Ferreyra LJ, Giordano MO, Martínez LC et al (2015) Tracking novel adenovirus in environmental and human clinical samples : no evidence of endemic human adenovirus type 58 circulation in Córdoba city, Argentina. Epidemiol Infect 143(7):1427–1431. doi:10.1017/S0950268814002192
63. Fioretti JM, Rocha MS, Fumian TM et al (2016) Occurrence of human sapoviruses in wastewater and stool samples in Rio de Janeiro, Brazil. J Appl Microbiol 121(3):855–862. doi:10.1111/jam.13205
64. Fong TT, Lipp EK (2005) Enteric viruses of humans and animals in aquatic environments: health risks, detection, and potential water quality assessment tools. Microbiol Mol Biol Rev 69(2):357–371. doi:10.1128/MMBR.69.2.357-371.2005
65. Fong TT, Phanikumar MS, Xagorarak I et al (2010) Quantitative detection of human adenoviruses in wastewater and combined sewer overflows influencing a Michigan river. Appl Environ Microbiol 76(3):715–723. doi:10.1128/AEM.01316-09
66. Fongaro G, Hernández M, García-González MC et al (2016) Propidium monoazide coupled with PCR predicts infectivity of enteric viruses in swine manure and biofertilized soil. Food Environ Virol 8(1):79–85. doi:10.1007/s12560-015-9225-1

67. Fongaro G, Nascimento MA, Rigotto C et al (2013) Evaluation and molecular characterization of human adenovirus in drinking water supplies: viral integrity and viability assays. Virol J 10:166. doi:10.1186/1743-422X-10-166
68. Fongaro G, Nascimento MA, Viancelli A et al (2012) Surveillance of human viral contamination and physicochemical profiles in a surface water lagoon. Water Sci Technol 66(12):2682–2687. doi:10.2166/wst.2012.504
69. Fongaro G, Padilha J, Schissi CD et al (2015) Human and animal enteric virus in groundwater from deep wells, and recreational and network water. Environ Sci Pollut Res Int 22(24):20060–20066. doi:10.1007/s11356-015-5196-x
70. Formiga-Cruz M, Hundesa A, Clemente-Casares P et al (2005) Nested multiplex PCR assay for detection of human enteric viruses in shellfish and sewage. J Virol Methods 125(2):111–118. doi:10.1016/j.jviromet.2005.01.009
71. Formiga-Cruz M, Tofiño-Quesada G, Bofill-Mas S et al (2002) Distribution of human virus contamination in shellfish from different growing areas in Greece, Spain, Sweden, and the United Kingdom. Appl Environ Microbiol 68(12):5990–5998. doi:10.1128/AEM.68.12.5990-5998.2002
72. Fumian TM, Guimarães FR, Pereira Vaz BJ et al (2010a) Molecular detection, quantification and characterization of human polyomavirus JC from waste water in Rio De Janeiro, Brazil. J Water Health 8(3):438–445. doi:10.2166/wh.2010.090
73. Fumian TM, Leite JP, Castello AA et al (2010b) Detection of rotavirus A in sewage samples using multiplex qPCR and an evaluation of the ultracentrifugation and adsorption-elution methods for virus concentration. J Virol Methods 170(1-2):42–46. doi:10.1016/j.jviromet.2010.08.017
74. Fumian TM, Leite JP, Marin VA et al (2009) A rapid procedure for detecting noroviruses from cheese and fresh lettuce. J Virol Methods 155(1):39–43. doi:10.1016/j.jviromet.2008.09.026
75. Fumian TM, Leite JP, Rose TL et al (2011) One year environmental surveillance of rotavirus species A (RVA) genotypes in circulation after the introduction of the Rotarix® vaccine in Rio de Janeiro, Brazil. Water Res 45(17):5755–5763. doi:10.1016/j.watres.2011.08.039
76. Fumian TM, Vieira CB, Leite JP et al (2013) Assessment of burden of virus agents in an urban sewage treatment plant in Rio de Janeiro, Brazil. J Water Health 11(1):110–119. doi:10.2166/wh.2012.123
77. Fuster N, Pintó RM, Fuentes C et al (2016) Propidium monoazide RTqPCR assays for the assessment of hepatitis A inactivation and for a better estimation of the health risk of contaminated waters. Water Res 101:226–232. doi:10.1016/j.watres.2016.05.086
78. Gallimore CI, Barreiros MA, Brown DW et al (2004) Noroviruses associated with acute gastroenteritis in a children's day care facility in Rio de Janeiro, Brazil. Braz J Med Biol Res 37(3):321–326. doi:http://dx.doi.org/10.1590/S0100-879X2004000300005
79. Gallimore CI, Taylor C, Gennery AR et al (2006) Environmental monitoring for gastroenteric viruses in a pediatric primary immunodeficiency unit. J Clin Microbiol 44(2):395–399. doi:10.1128/JCM.44.2.395-399.2006
80. Gallimore CI, Taylor C, Gennery AR et al (2005) Use of a heminested reverse transcriptase PCR assay for detection of astrovirus in environmental swabs from an outbreak of gastroenteritis in a pediatric primary immunodeficiency unit. J Clin Microbiol 43(8):3890–3894. doi:10.1128/JCM.43.8.3890-3894.2005
81. Ganime AC, Carvalho-Costa FA, Mendonça MC et al (2012) Group A rotavirus detection on environmental surfaces in a hospital intensive care unit. Am J Infect Control 40(6):544–547. doi:10.1016/j.ajic.2011.07.017
82. Ganime AC, Carvalho-Costa FA, Santos M et al (2014) Viability of human adenovirus from hospital fomites. J Med Virol 86(12):2065–2069. doi:10.1002/jmv.23907
83. Ganime AC, Leite JP, de Abreu CA et al (2015) Evaluation of the swab sampling method to recover viruses from fomites. J Virol Methods 217:24–27. doi:10.1016/j.jviromet.2015.02.009
84. Ganime AC, Leite JP, Figueiredo CE et al (2016) Dissemination of human adenoviruses and rotavirus species A on fomites of hospital pediatric units. Am J Infect Control pii:S0196-6553(16)30226-7. doi:10.1016/j.ajic.2016.04.207

85. Gantzer C, Maul A, Audic JM et al (1998) Detection of infectious enteroviruses, enterovirus genomes, somatic coliphages, and *Bacteroides fragilis* phages in treated wastewater. Appl Environ Microbiol 64(11):4307–4312
86. Garcia LA, Nascimento MA, Barardi CR (2015) Effect of UV light on the inactivation of recombinant human adenovirus and murine norovirus seeded in seawater in shellfish depuration tanks. Food Environ Virol 7(1):67–75. doi:10.1007/s12560-014-9177-x
87. Garcia LA, Viancelli A, Rigotto C et al (2012) Surveillance of human and swine adenovirus, human norovirus and swine circovirus in water samples in Santa Catarina, Brazil. J Water Health 10(3):445–452. doi:10.2166/wh.2012.190
88. Gerba CP (2006) Food virology: past, present, and future. In: Goyal SM (ed) Viruses in food, 1st edn. Springer, New York, pp 1–4
89. Gerba CP (2007) Virus occurrence and survival in the environmental waters. In: Bosch A (ed) Human viruses in water, 1st edn. Elsevier, Amsterdam, pp 91–108
90. Gerba CP, Goyal SM, Labelle RL et al (1979) Failure of indicator bacteria to reflect the occurrence of enteroviruses in marine waters. Am J Public Health 69(11):1116–1119
91. Gerba CP, Gramos DM, Nwachuku N (2002) Comparative inactivation of enteroviruses and adenovirus 2 by UV light. Appl Environ Microbiol 68(10):5167–5169. doi:10.1128/AEM.68.10.5167-5169.2002
92. Gerba CP, Smith EM, Melnick JL (1977) Development of a quantitative method for detecting enteroviruses in estuarine sediments. Appl Environ Microbiol 34(2):158–163
93. Girones R, Ferrús MA, Alonso JL et al (2010) Molecular detection of pathogens in water: the pros and cons of molecular techniques. Water Res 44(15):4325–4339. doi:10.1016/j.watres.2010.06.030
94. Gofti-Laroche L, Gratacap-Cavallier B, Genoulaz O et al (2001) A new analytical tool to assess health associated with the virological quality of drinking water (EMIRA study). Water Sci Technol 43(12):39–48
95. Gortáres-Moroyoqui C-EL, Naranjo JE et al (2011) Microbiological water quality in a large irrigation system: El Vale del Yaqui, Sonora México. J Environ Sci Health A Tox Hazard Subst Environ Eng 46(14):1708–1712. doi:10.1080/10934529.2011.623968
96. Gregio CRV (2006) Caracterização de poliovirus derivado da vacina isolados a partir de amostras ambientais. Dissertation, Oswaldo Cruz Foundation
97. Griffin DW, Donaldson KA, Paul JH et al (2003) Pathogenic human viruses in coastal waters. Clin Microbiol Rev 16:129–143. doi:10.1128/CMR.16.1.129-143.2003
98. Guimarães FR, Ferreira FF, Vieira CB et al (2008) Molecular detection of human astrovirus in an urban sewage treatment plant in Rio de Janeiro, Brazil. Mem Inst Oswaldo Cruz 103(8):819–823. doi:http://dx.doi.org/10.1590/S0074-02762008000800013
99. Gutiérrez MF, Alvarado MV, Martínez E et al (2007) Presence of viral proteins in drinkable water: sufficient condition to consider water a vector of viral transmission? Water Res 41(2):373–378. doi:10.1016/j.watres.2006.09.022
100. Guzmán C, Jofre J, Blanch AR et al (2007a) Development of a feasible method to extract somatic coliphages from sludge, soil, and treated biowaste. J Virol Methods 144(1-2):41–48
101. Guzmán C, Jofre J, Montemayor M et al (2007b) Occurrence and levels of indicators and selected pathogens in different sludges and biosolids. J Appl Microbiol 103(6):2420–2429. doi:10.1111/j.1365-2672.2007.03487.x
102. Haas CN, Rose JB, Gerba CP (1999) Quantitative microbial risk assessment. Wiley, New York
103. Hachich EM, Galvani AT, Padula JA et al (2013) Pathogenic parasites and enteroviruses in wastewater: support for a regulation on water reuse. Water Sci Technol 67(7):1512–1518. doi:10.2166/wst.2013.019
104. Harwood VJ, Levine AD, Scott TM et al (2005) Validity of the indicator organism paradigm for pathogen reduction in reclaimed water and public health protection. Appl Environ Microbiol 71(6):3163–3170. doi:10.1128/AEM.71.6.3163-3170.2005
105. Havelaar AH, Kirk MD, Torgerson PR et al (2015) World Health Organization global estimates and regional comparisons of the burden of foodborne disease in 2010. PLoS Med 12(12):e1001923. doi:10.1371/journal.pmed.1001923

106. He X, Wei Y, Cheng L et al (2012) Molecular detection of three gastroenteritis viruses in urban surface waters in Beijing and correlation with levels of fecal indicator bacteria. Environ Monit Assess 184(9):5563–5570. doi:10.1007/s10661-011-2362-6
107. Heldt FH, Staggmeier R, Gularte JS et al (2016) Hepatitis E virus in surface water, sediments, and pork products marketed in Southern Brazil. Food Environ Virol 8(3):200–205. doi:10.1007/s12560-016-9243-7
108. Hernandez-Morga J, Leon-Felix J, Peraza-Garay F et al (2009) Detection and characterization of hepatitis A virus and Norovirus in estuarine water samples using ultrafiltration: RT-PCR integrated methods. J Appl Microbiol 106(5):1579–1590. doi:10.1111/j.1365-2672.2008.04125.x
109. Hewitt J, Greening GE, Leonard M et al (2013) Evaluation of human adenovirus and human polyomavirus as indicators of human sewage contamination in the aquatic environment. Water Res 47(17):6750–6761. doi:10.1016/j.watres.2013.09.001
110. Hipsey MR, Brookes JD (2013) Pathogen management in surface waters: practical considerations for reducing public health risk. In: Rodriguez-Morales AJ (ed) Current topics in public health. InTech, Rijeka, Croatia, pp 445–475
111. Holtz LR, Finkbeiner SR, Zhao G et al (2009) Klassevirus 1, a previously undescribed member of the family Picornaviridae, is globally widespread. Virol J 6:86. doi:10.1186/1743-422X-6-86
112. Homma A, Schatzmayr HG, Frias AM et al (1975) Viral pollution evaluation of the Guanabara Bay. Rev Inst Med Trop Sao Paulo 17(3):140–145
113. Hoorfar J, Cook N, Malorny B et al (2004) Diagnostic PCR: making internal amplification control mandatory. J Appl Microbiol 96(2):221–222
114. Hundesa A, Bofill-Mas S, Maluquer de Motes C et al (2010) Development of a quantitative PCR assay for the quantitation of bovine polyomavirus as a microbial source-tracking tool. J Virol Methods 163(2):385–389. doi:10.1016/j.jviromet.2009.10.029
115. Hundesa A, Maluquer de Motes C, Albinana-Gimenez N et al (2009) Development of a qPCR assay for the quantification of porcine adenoviruses as an MST tool for swine fecal contamination in the environment. J Virol Methods 158(1-2):130–135. doi:10.1016/j.jviromet.2009.02.006
116. Hundesa A, Maluquer de Motes C, Bofill-Mas S et al (2006) Identification of human and animal adenoviruses and polyomaviruses for determination of sources of fecal contamination in the environment. Appl Environ Microbiol 72(12):7886–7893. doi:10.1128/AEM.01090-06
117. Ijzerman MM, Dahililng DR, Fout GS (1997) A method to remove environmental inhibitors prior to the detection of waterborne enteric viruses by reverse transcription-polymerase chain reaction. J Virol Methods 63(1-2):145–153
118. Ikner LA, Gerba CP, Bright KR (2012) Concentration and recovery of viruses from water: a comprehensive review. Food Environ Virol 4(2):41–67. doi:10.1007/s12560-012-9080-2
119. ISFEV – International Society for Food and Environmental Virology (2016) International Society for Food and Environmental Virology. isfev.org. Accessed 13 June 2016
120. ISO/TS 15216-1:2013 Microbiology of food and animal feed – Horizontal method for determination of hepatitis A virus and norovirus in food using real-time RT-PCR – Part 1: Method for quantification
121. ISO/TS 15216-2:2013(en) Microbiology of food and animal feed — Horizontal method for determination of hepatitis A virus and norovirus in food using real-time RT-PCR – Part 2: Method for qualitative detection
122. Jofre J (2007) Indicators of waterborne enteric viruses. In: Bosch A (ed) Human viruses in water, 1st edn. Elsevier, Amsterdam, pp 227–249
123. Kapikian AZ, Wyatt RG, Dolin R et al (1972) Visualization by immune electron microscopy of a 27-nm particle associated with acute infectious nonbacterial gastroenteritis. J Virol 10:1075–1081
124. Kapoor A, Slikas E, Simmonds P et al (2009) A newly identified bocavirus species in human stool. J Infect Dis 199(2):196–200. doi:10.1086/595831
125. Katayama H, Shimasaki A, Ohgaki S (2002) Development of a virus concentration method and its application to detection of enterovirus and Norwalk virus from coastal seawater. Appl Environ Microbiol 68(3):1033–1039. doi:10.1128/AEM.68.3.1033-1039.2002

126. Keller R, Justino JF, Cassini ST (2013) Assessment of water and seafood microbiology quality in a mangrove region in Vitória, Brazil. J Water Health 11(3):573–580. doi:10.2166/wh.2013.245
127. Kingsley DH, Holliman DR, Calci KR et al (2007) Inactivation of a norovirus by high-pressure processing. Appl Environ Microbiol 73(2):581–585. doi:10.1128/AEM.02117-06
128. Kishida N, Noda N, Haramoto E et al (2014) Quantitative detection of human enteric adenoviruses in river water by microfluidic digital polymerase chain reaction. Water Sci Technol 70(3):555–560. doi:10.2166/wst.2014.262
129. Kluge M, Fleck JD, Soliman MC et al (2014) Human adenovirus (HAdV), human enterovirus (hEV), and genogroup A rotavirus (GARV) in tap water in southern Brazil. J Water Health 12(3):526–532. doi:10.2166/wh.2014.202
130. Ko G, Cromeans TL, Sobsey MD (2003) Detection of infectious adenovirus in cell culture by mRNA reverse transcription-PCR. Appl Environ Microbiol 69(12):7377–7384. doi:10.1128/AEM.69.12.7377-7384.2003
131. Kramer A, Schwebke I, Kampf G (2006) How long do nosocomial pathogens persist on inanimate surfaces? A systematic review. BMC Infect Dis 6:130. doi:10.1186/1471-2334-6-130
132. Kuo HW, Chen LZ, Shih MH (2015) High prevalence of type 41 and high sequence diversity of partial hexon gene of human adenoviruses in municipal raw sewage and activated sludge. J Appl Microbiol 119(4):1181–1195. doi:10.1111/jam.12907
133. Lamhoujeb S, Fliss I, Ngazoa SE et al (2008) Evaluation of the persistence of infectious human noroviruses on food surfaces by using real-time nucleic acid sequence-based amplification. J Appl Environ Microbiol 74(11):3349–3355. doi:10.1128/AEM.02878-07
134. Laverick MA, Wyn-Jones AP, Carter MJ (2004) Quantitative RT-PCR for the enumeration of noroviruses (Norwalk-like viruses) in water and sewage. Lett Appl Microbiol 39(2):127–136. doi:10.1111/j.1472-765X.2004.01534.x
135. Le Guyader FS, Parnaudeau S, Schaeffer J et al (2009) Detection and quantification of noroviruses in shellfish. Appl Environ Microbiol 75(3):618–624. doi:10.1128/AEM.01507-08
136. Lewis GD, Metcalf TG (1988) Polyethylene glycol precipitation for recovery of pathogenic viruses, including hepatitis A virus and human rotavirus, from oyster, water, and sediment samples. Appl Environ Microbiol 54(8):1983–1988
137. Liu L, Oza S, Hogan D et al (2015) Global, regional, and national causes of child mortality in 2000-13, with projections to inform post-2015 priorities: an updated systematic analysis. Lancet 385(9966):430–440. doi:10.1016/S0140-6736(14)61698-6
138. Lizasoain A, Tort LF, García M et al (2015a) Environmental assessment of classical human astrovirus in Uruguay. Food Environ Virol 7:42. doi:10.1007/s12560-015-9186-4
139. Lizasoain A, Tort LF, García M et al (2015b) Environmental assessment reveals the presence of MLB-1human astrovirus in Uruguay. J Appl Microbiol 119(3):859–867. doi:10.1111/jam.12856
140. Machado CJS, Miagostovich MP, Leite JPG et al (2013) Promoção da relação saúde-saneamento-cidade por meio da Virologia Ambiental. Rev Inform Leg 50(199):321–345
141. Mäde D, Trübner K, Neubert E et al (2013) Detection and typing of norovirus from frozen strawberries involved in a large-scale gastroenteritis outbreak in Germany. Food Environ Virol 5:162–168. doi:10.1007/s12560-013-9118-0
142. Marti E, Barardi CR (2016) Detection of human adenoviruses in organic fresh produce using molecular and cell culture-based methods. Int J Food Microbiol 230:40–44. doi:10.1016/j.ijfoodmicro.2016.04.018
143. Martin-Latil S, Hennechart-Collette C, Guillier L et al (2012) Comparison of two extraction methods for the detection of hepatitis A virus in semi-dried tomatoes and murine norovirus as a process control by duplex RT-qPCR. Food Environ Microbiol 31(2):246–253. doi:10.1016/j.fm.2012.03.007
144. Martínez Wassaf MG, Pisano MB, Barril PA et al (2014) First detection of hepatitis E virus in Central Argentina: environmental and serological survey. J Clin Virol 61(3):334–339. doi:10.1016/j.jcv.2014.08.016

145. Más Lago P, Gary HE Jr, Pérez LS et al (2003) Poliovirus detection in wastewater and stools following an immunization campaign in Havana, Cuba. Int J Epidemiol 32(5):772–777. doi:10.1093/ije/dyg185
146. Mattison K, Brassard J, Gagné MJ et al (2009) The feline calicivirus as a sample process control for the detection of food and waterborne RNA viruses. Int J Food Microbiol 132(1):73–77. doi:10.1016/j.ijfoodmicro.2009.04.002
147. Maunula L, Kaupke A, Vasickova P et al (2013) Tracing enteric viruses in the European berry fruit supply chain. Int J Food Microbiol 167:177–185. doi:10.1016/j.ijfoodmicro.2013.09.003
148. Maurer CP, Simonetti AB, Staggemeier R et al (2015) Adenovirus, enterovirus and thermotolerant coliforms in recreational waters from Lake Guaíba beaches, Porto Alegre, Brazil. J Water Health 13(4):1123–1129. doi:10.2166/wh.2015.277
149. McQuaig SM, Scott TM, Lukasik JO et al (2009) Quantification of human polyomaviruses JC virus and BK virus by TaqMan quantitative PCR and comparison to other water quality indicators in water and fecal samples. Appl Environ Microbiol 75(11):3379–3388. doi:10.1128/AEM.02302-08
150. Mehnert DU, Stewien KE (1993) Detection and distribution of rotavirus in raw sewage and creeks in São Paulo, Brazil. Appl Environ Microbiol 59(1):140–143
151. Melgaço FG (2016) Avaliação de metodologia de floculação orgânica para recuperação de vírus entéricos em frutas e queijos. Dissertation, Oswaldo Cruz Foundation
152. Melgaço FG, Victoria M, Corrêa AA et al (2016) Virus recovering from strawberries: evaluation of a skimmed milk organic flocculation method for assessment of microbiological contamination. Int J Food Microbiol 217:14–19. doi:10.1016/j.ijfoodmicro.2015.10.005
153. Melnick JL (1947) Poliomyelitis virus in urban sewage in epidemic and non-epidemic times. Am J Hyg 45(2):240–253
154. Mena KD (2007) Waterborne viruses: assessing the risks. In: Bosch A (ed) Human viruses in water, 1st edn. Elsevier, Amsterdam, pp 163–175
155. Metcalf TG, Melnick JL, Estes MK (1995) Environmental virology: from detection of virus in sewage and water by isolation to identification by molecular biology-a trip of over 50 years. Annu Rev Microbiol 49:461–487. doi:10.1146/annurev.mi.49.100195.002333
156. Metcalf TG, Moulton E, Eckerson D (1980) Improved method and test strategy for recovery of enteric viruses from shellfish. Appl Environ Microbiol 39(1):141–152
157. Mezzanotte V, Antonelli M, Citterio S et al (2007) Wastewater disinfection alternatives: chlorine, ozone, peracetic acid, and UV light. Water Environ Res 79:2373–2379. doi: http://dx.doi.org/10.2175/106143007X183763
158. Miagostovich MP, Ferreira FF, Guimarães FR et al (2008) Molecular detection and characterization of gastroenteritis viruses occurring naturally in the stream waters of Manaus, central Amazonia, Brazil. Appl Environ Microbiol 74(2):375–382. doi:10.1128/AEM.00944-07
159. Miagostovich MP, Guimarães FR, Vieira CB et al (2014) Assessment of water quality in a border region between the Atlantic forest and an urbanised area in Rio de Janeiro, Brazil. Food Environ Virol 6(2):110–115. doi:10.1007/s12560-014-9146-4
160. Morales-Rayas R, Wolffs PFG, Griffiths MW (2009) Anion-exchange filtration and real-time PCR for the detection of a norovirus surrogate in food. J Food Prot 72(10):2178–2183
161. Moreno L, Aznar R, Sánchez G (2015) Application of viability PCR to discriminate the infectivity of hepatitis A virus in food samples. Int J Food Microbiol 201:1–6. doi:10.1016/j.ijfoodmicro.2015.02.012
162. Moreno S, Alvarado MV, Bermúdez A et al (2009) Phylogenetic analysis indicates human origin of rotavirus and hepatitis A virus strains found in the drinking water of western Colombia. Biomedica 29(2):209–217
163. Moresco V, Damazo NA, Barardi CR (2016) Rotavirus vaccine stability in the aquatic environment. J Appl Microbiol 120(2):321–328. doi:10.1111/jam.13021
164. Moresco V, Viancelli A, Nascimento MA et al (2012) Microbiological and physicochemical analysis of the coastal waters of southern Brazil. Mar Pollut Bull 64(1):40–48. doi:10.1016/j.marpolbul.2011.10.026

165. Morillo SG, Luchs A, Cilli A et al (2012) Rapid detection of norovirus in naturally contaminated food: foodborne gastroenteritis outbreak on a cruise ship in Brazil, 2010. Food Environ Virol 4(3):124–129. doi:10.1007/s12560-012-9085-x
166. Morris CA, Flewett TH, Burden AS, Davis H et al (1975) Epidemic viral enteritis in a long-stay children's ward. Lancet 1(7897):4–5
167. Mueller JE, Bessaud M, Huang QS et al (2009) Environmental poliovirus surveillance during oral poliovirus vaccine and inactivated poliovirus vaccine use in Córdoba Province, Argentina. Appl Environ Microbiol 75(5):1395–1401. doi:10.1128/AEM.02201-08
168. Nascimento MA, Magri ME, Schissi CD et al (2015) Recombinant adenovirus as a model to evaluate the efficiency of free chlorine disinfection in filtered water samples. Virol J 12:30. doi:10.1186/s12985-015-0259-7
169. Nuanualsuwan S, Cliver DO (2002) Pretreatment to avoid positive RT-PCR results with inactivated viruses. J Virol Methods 104(2):217–225. doi:10.1016/S0166-0934(02)00089-7
170. Okoh AI, Sibanda T, Gusha SS (2010) Inadequately treated wastewater as a source of human enteric viruses in the environment. Int J Environ Res Public Health 7(6):2620–2637. doi:10.3390/ijerph7062620
171. PAHO – Pan American Health Organization (2016) Pan American Health Organization. http://www.paho.org/hq/index.php?option=com_content&view=article&id=5172&Itemid=40085&lang=es. Accessed 13 June 2016
172. Papafragkou E, Hewitt J, Park GW et al (2013) Challenges of culturing human norovirus in three-dimensional organoid intestinal cell culture models. PLoS One 8(6):e63485. doi:10.1371/journal.pone.0063485
173. Parada-Fabián JC, Juárez-García P, Natividad-Bonifacio I et al (2016) Identification of enteric viruses in foods from Mexico City. Food Environ Virol. doi:10.1007/s12560-016-9244-6
174. Parshionikar S, Laseke I, Fout GS (2010) Use of propidium monoazide in reverse transcriptase PCR to distinguish between infectious and noninfectious enteric viruses in water samples. Appl Environ Microbiol 76:4318–4326. doi:10.1128/AEM.02800-09
175. Payment P, Locas A (2011) Pathogens in water: value and limits of correlation with microbial indicators. Ground Water 49(1):4–11. doi:10.1111/j.1745-6584.2010.00710.x
176. Pereira, JSO (2013) Vigilância ambiental dos poliovírus, no município do Rio de Janeiro, em apoio às atividades de erradicação global da poliomielite. Dissertation, Oswaldo Cruz Foundation
177. Phan TG, Mori D, Deng X et al (2015) Small circular single stranded DNA viral genomes in unexplained cases of human encephalitis, diarrhea, and in untreated sewage. Virology 482:98–104. doi:10.1016/j.virol.2015.03.011
178. Pina S, Puig M, Lucena F et al (1998) Viral pollution in the environment and in shellfish: human adenovirus detection by PCR as an index of human viruses. Appl Environ Microbiol 64(9):3376–3382
179. Poma HR, Rajal VB, Blanco Fernández MD et al (2013) Evaluation of concentration efficiency of the Pseudomonas aeruginosa phage PP7 in various water matrixes by different methods. Environ Monit Assess 185(3):2565–2576. doi:10.1007/s10661-012-2731-9
180. Prado T, Fumian TM, Miagostovich MP et al (2012) Monitoring the hepatitis A virus in urban wastewater from Rio de Janeiro, Brazil. Trans R Soc Trop Med Hyg 106(2):104–109. doi:10.1016/j.trstmh.2011.10.005
181. Prado T, Gaspar AM, Miagostovich MP (2014) Detection of enteric viruses in activated sludge by feasible concentration methods. Braz J Microbiol 45(1):343–349
182. Prado T, Guilayn WCPB, Gaspar AM et al (2013) The efficiency of concentration methods used to detect enteric viruses in anaerobically digested sludge. Mem Inst Oswaldo Cruz 108(1):77–83. doi:10.1590/S0074-02762013000100013
183. Prado T, Silva DM, Guilayn WC et al (2011) Quantification and molecular characterization of enteric viruses detected in effluents from two hospital wastewater treatment plants. Water Res 45(3):1287–1297. doi:10.1016/j.watres.2010.10.012
184. Prevost B, Goulet M, Lucas FS et al (2016) Viral persistence in surface and drinking water: suitability of PCR pre-treatment with intercalating dyes. Water Res 91:68–76. doi:10.1016/j.watres.2015.12.049

185. Prez VE, Gil PI, Temprana CF et al (2015) Quantification of human infection risk caused by rotavirus in surface waters from Córdoba, Argentina. Sci Total Environ 538:220–229. doi:10.1016/j.scitotenv.2015.08.041
186. Pusch D, Oh DY, Wolf S et al (2005) Detection of enteric viruses and bacterial indicators in German environmental waters. Arch Virol 150(5):929–947. doi:10.1007/s00705-004-0467-8
187. Queiroz AP, Santos FM, Sassaroli A et al (2001) Electropositive filter membrane as an alternative for the elimination of PCR inhibitors from sewage and water samples. Appl Environ Microbiol 67(10):4614–4618. doi:10.1128/AEM.67.10.4614-4618.2001
188. Quiroz-Santiago C, Vázquez-Salinas C, Natividad-Bonifacio I et al (2014) Rotavirus G2P[4] detection in fresh vegetables and oysters in Mexico City. J Food Prot 77(11):1953–1959. doi:10.4315/0362-028X.JFP-13-426
189. Rački N, Morisset D, Gutierrez-Aguirre I et al (2014) One-step RT-droplet digital PCR: a breakthrough in the quantification of waterborne RNA viruses. Anal Bioanal Chem 406(3):661–667. doi:10.1007/s00216-013-7476-y
190. Ramani S, Arumugam R, Gopalarathinam N et al (2008) Investigation of the environment and of mothers in transmission of rotavirus infections in the neonatal nursery. J Med Virol 80(6):1099–1105. doi:10.1002/jmv.21177
191. Randazzo W, López-Gálvez F, Allende A et al (2016) Evaluation of viability PCR performance for assessing norovirus infectivity in fresh-cut vegetables and irrigation water. Int J Food Microbiol 229:1–6. doi:10.1016/j.ijfoodmicro.2016.04.010
192. Ravaliya K, Gentry-Shields J, Garcia S et al (2014) Use of Bacteroidales microbial source tracking to monitor fecal contamination in fresh produce production. Appl Environ Microbiol 80(2):612–617. doi:10.1128/AEM.02891-13
193. Retter M, Middleton PJ, Tam JS et al (1979) Enteric adenoviruses: detection, replication, and significance. J Clin Microbiol 10:574–578
194. Reynolds KA, Gerba CP, Abbaszadegan M et al (2001) ICC/PCR detection of enteroviruses and hepatitis A virus in environmental samples. Can J Microbiol 47(2):153–157
195. Reynolds KA, Gerba CP, Pepper IL (1996) Detection of infectious enteroviruses by an integrated cell culture-PCR procedure. Appl Environ Microbiol 62(4):1424–1427
196. Rezaeinejad S, Vergara GG, Woo CH et al (2014) Surveillance of enteric viruses and coliphages in a tropical urban catchment. Water Res 58:122–131. doi:10.1016/j.watres.2014.03.051
197. Rigotto C, Sincero TC, Simões CM et al (2005) Detection of adenoviruses in shellfish by means of conventional-PCR, nested-PCR, and integrated cell culture PCR (ICC/PCR). Water Res 39(2–3):297–304. doi:10.1016/j.watres.2004.10.005
198. Rigotto C, Victoria M, Moresco V et al (2010) Assessment of adenovirus, hepatitis A virus and rotavirus presence in environmental samples in Florianopolis, South Brazil. J Appl Microbiol 109(6):1979–1987. doi:10.1111/j.1365-2672.2010.04827.x
199. Rock C, Alum A, Abbaszadegan M (2010) PCR inhibitor levels in concentrates of biosolid samples predicted by a new method based on excitation-emission matrix spectroscopy. Appl Environ Microbiol 76(24):8102–8109. doi:10.1128/AEM.02339-09
200. Rodrigues MT, Henzel A, Staggemeier R et al (2015) Human adenovirus spread, rainfalls, and the occurrence of gastroenteritis cases in a Brazilian basin. Environ Monit Assess 187(11):720. doi:10.1007/s10661-015-4917-4
201. Rodríguez-Díaz J, Querales L, Caraballo L et al (2009) Detection and characterization of waterborne gastroenteritis viruses in urban sewage and sewage-polluted river waters in Caracas, Venezuela. Appl Environ Microbiol 75(2):387–394. doi:10.1128/AEM.02045-08
202. Rodríguez-Lázaro D, Cook N, Ruggeri FM et al (2012) Virus hazards from food, water and other contaminated environments. FEMS Microbiol Rev 36(4):786–814. doi:10.1111/j.1574-6976.2011.00306.x
203. Rodriguez-Manzano J, Miagostovich M, Hundesa A et al (2010) Analysis of the evolution in the circulation of HAV and HEV in Eastern Spain by testing urban sewage samples. J Water Health 8(2):346e354. doi:10.2166/wh.2009.042
204. Rusiñol M, Carratalà A, Hundesa A et al (2013) Description of a novel viral tool to identify and quantify ovine faecal pollution in the environment. Sci Total Environ 458-460:355–360. doi:10.1016/j.scitotenv.2013.04.028

205. Rusiñol M, Fernandez-Cassi X, Hundesa A et al (2014) Application of human and animal viral microbial source tracking tools in fresh and marine waters from five different geographical areas. Water Res 59:119–129. doi:10.1016/j.watres.2014.04.013
206. Rutala WA, Peacock JE, Gergen MF et al (2006) Efficacy of hospital germicides against adenovirus 8, a common cause of epidemic keratoconjunctivitis in health care facilities. Antimicrob Agents Chemother 50:1419–1424. doi:10.1128/AAC.50.4.1419-1424.2006
207. Rzezutka A, Alotaibi M, D'Agostino M et al (2005) A centrifugation-based method for extraction of norovirus from raspberries. J Food Prot 68(9):1923–1925
208. Rzezutka A, Cook N (2004) Survival of human enteric viruses in the environment and food. FEMS Microbiol Rev 28(4):441–453. doi:http://dx.doi.org/10.1016/j.femsre.2004.02.001
209. Sano D, Fukushi K, Yoshida Y et al (2003) Detection of enteric viruses in municipal sewage sludge by a combination of the enzymatic virus elution method and RT-PCR. Water Res 37:3490–3498. doi:10.1016/S0043-1354(03)00208-2
210. Sassoubre LM, Love DC, Silverman AI et al (2012) Comparison of enterovirus and adenovirus concentration and enumeration methods in seawater from Southern California, USA and Baja Malibu, Mexico. J Water Health 10(3):419–430. doi:10.2166/wh.2012.011
211. Sattar SA (2004) Microbicides and the environmental control of nosocomial viral infections. J Hosp Infect 56(Suppl 2):S64–S69
212. Sattar SA, Lloyd-Evans N, Springthorpe VS et al (1986) Institutional outbreaks of rotavirus diarrhoea: potential role of fomites and environmental surfaces as vehicles for virus transmission. J Hyg (Lond) 96(2):277–289
213. Sattar SA, Springthorpe VS (1996) Transmission of viral infections through animate and inanimate surfaces and infection control through chemical disinfection. In: Hurst CJ (ed) Modeling disease transmission and its prevention by disinfection. Cambridge University Press, Cambridge, pp 224–257
214. Scherer K, Johne R, Schrader C et al (2010) Comparison of two extraction methods for viruses in food and application in a norovirus gastroenteritis outbreak. J Virol Methods 169(1):22–27. doi:10.1016/j.jviromet.2010.06.008
215. Schlindwein AD, Rigotto C, Simões MO et al (2010) Detection of enteric viruses in sewage sludge and treated wastewater effluent. Water Sci Technol 61(2):537–544. doi:10.2166/wst.2010.845
216. Silva HD, Fongaro G, Garcíazapata MT et al (2015) High species C human adenovirus genome copy numbers in the treated water supply of a neotropical area of the central-west region of Brazil. Food Environ Virol 7(3):286–294. doi:10.1007/s12560-015-9192-6
217. Silva LM, Souza EH, Arrebola TM et al (2009) Occurrence of a hepatitis A outbreak in three neighborhoods of the city of Vitória, Espírito Santo State, and its relation with the quality of the water for human consumption. Cien Saude Colet 14(6):2163–2167
218. Simmons FJ, Xagoraraki I (2011) Release of infectious human enteric viruses by full-scale wastewater utilities. Water Res 45:3590–3598. doi:10.1016/j.watres.2011.04.001
219. Simonet J, Gantzer C (2006a) Degradation of the Poliovirus 1 genome by chlorine dioxide. J Appl Microbiol 100(4):862–870. doi:10.1111/j.1365-2672.2005.02850.x
220. Simonet J, Gantzer C (2006b) Inactivation of poliovirus 1 and F-specific RNA phages and degradation of their genomes by UV irradiation at 254 nanometers. Appl Environ Microbiol 72(12):7671–7677. doi:10.1128/AEM.01106-06
221. Sincero TC, Levin DB, Simões CM et al (2006) Detection of hepatitis A virus (HAV) in oysters (*Crassostrea gigas*). Water Res 40(5):895–902. doi:10.1016/j.watres.2005.12.005
222. Sobsey MD, Carrick RJ, Jensen HR (1978) Improved methods for detecting enteric viruses in oysters. Appl Environ Microbiol 36(1):121–128
223. Souza DS, Ramos AP, Nunes FF et al (2012) Evaluation of tropical water sources and mollusks in southern Brazil using microbiological, biochemical, and chemical parameters. Ecotoxicol Environ Saf 76(2):153–161. doi:10.1016/j.ecoenv.2011.09.018
224. Spilki FR, da Luz RB, Fabres RB et al (2014) Detection of human adenovirus, rotavirus and enterovirus in water samples collected on dairy farms from Tenente Portela, Northwest of Rio Grande do Sul, Brazil. Braz J Microbiol 44(3):953–957

225. Stals A, Baert L, Botteldoorn N et al (2009) Multiplex real-time RT-PCR for simultaneous detection of GI/GII noroviruses and murine norovirus 1. J Virol Methods 161(2):247–253. doi:10.1016/j.jviromet.2009.06.019
226. Stals A, Baert L, Keuckelaere AD et al (2011a) Evaluation of a norovirus detection methodology for ready-to-eat foods. Int J Food Microbiol 145(2–3):420–425. doi:10.1016/j.ijfoodmicro.2011.01.013
227. Stals A, Baert L, Van Coillie E et al (2011b) Evaluation of a norovirus detection methodology for soft red fruits. Food Microbiol 28(1):52–58. doi:10.1016/j.fm.2010.08.004
228. Stals A, Baert L, Van Coillie E et al (2012) Extraction of food-borne viruses from food samples: a review. Int J Food Microbiol 153(1–2):1–9. doi:10.1016/j.ijfoodmicro.2011.10.014
229. Summa M, von Bonsdorff CH, Maunula L (2012) Evaluation of four virus recovery methods for detecting noroviruses on fresh lettuce, sliced ham, and frozen raspberries. J Virol Methods 183(2):154–160. doi:10.1016/j.jviromet.2012.04.006
230. Sutmoller F, Gaspar AM, Cynamon SE et al (1982) A water-borne hepatitis A outbreak in Rio de Janeiro. Mem Inst Oswaldo Cruz 77(1):9–17. doi:http://dx.doi.org/10.1590/S0074-02761982000100002
231. Symonds EM, Verbyla ME, Lukasik JO et al (2014) A case study of enteric virus removal and insights into the associated risk of water reuse for two wastewater treatment pond systems in Bolivia. Water Res 65:257–270. doi:10.1016/j.watres.2014.07.032
232. Tambini G, Andrus JK, Marques E et al (1993) Direct detection of wild poliovirus circulation by stool surveys of healthy children and analysis of community wastewater. J Infect Dis 168(6):1510–1514. doi:10.1093/infdis/168.6.1510
233. Teixeira DM, Hernandez JM, Silva LD et al (2016) Occurrence of norovirus GIV in environmental water samples from Belém City, Amazon Region, Brazil. Food Environ Virol 8(1):101–104. doi:10.1007/s12560-015-9220-6
234. Templeton M, Hofmann RA (2004) Ultraviolet disinfection of particle-associated viruses. In: Hahn H, Hoffman E, Odegaard H (eds) Chemical water and wastewater treatment. IWA Publishing, London, pp 109–116
235. Teunis PF, Moe CL, Liu P et al (2008) Norwalk virus: how infectious is it? J Med Virol 80(8):1468–1476. doi:10.1002/jmv.21237
236. Todd EC, Greig JD, Bartleson CA et al (2009) Outbreaks where food workers have been implicated in the spread of foodborne disease. Part 6. Transmission and survival of pathogens in the food processing and preparation environment. J Food Prot 72:202–219
237. Topping JR, Schnerr H, Haines J et al (2009) Temperature inactivation of Feline calicivirus vaccine strain FCV F-9 in comparison with human noroviruses using an RNA exposure assay and reverse transcribed quantitative real-time polymerase chain reaction: a novel method for predicting virus infectivity. Virol Methods 156(1-2):89–95. doi:10.1016/j.jviromet.2008.10.024
238. Torres C, Barrios ME, Cammarata RV et al (2016) High diversity of human polyomaviruses in environmental and clinical samples in Argentina: detection of JC, BK, Merkel-cell, Malawi, and human 6 and 7 polyomaviruses. Sci Total Environ 542(pt A):192–202. doi:10.1016/j.scitotenv.2015.10.047
239. Tort LF, Victoria M, Lizasoain A et al (2015) Detection of common, emerging and uncommon VP4, and VP7 human group A rotavirus genotypes from urban sewage samples in Uruguay. Food Environ Virol 7(4):342–353. doi:10.1007/s12560-015-9213-5
240. Troy SB, Ferreyra-Reyes L, Canizales-Quintero S et al (2012) Real-time polymerase chain reaction analysis of sewage samples to determine oral polio vaccine circulation duration and mutation after Mexican national immunization weeks. J Pediatr Infect Dis Soc 1(3):223–229. doi:10.1093/jpids/pis062
241. Troy SB, Ferreyra-Reyes L, Huang C et al (2011) Use of a novel real-time PCR assay to detect oral polio vaccine shedding and reversion in stool and sewage samples after a Mexican national immunization day. J Clin Microbiol 49(5):1777–1783. doi:10.1128/JCM.02524-10
242. Ueda T, Horan NJ (2000) Fate of indigenous bacteriophage in a membrane bioreactor. Water Res 34(7):2151–2159

243. Uhrbrand K, Myrmel M, Maunula L et al (2010) Evaluation of a rapid method for recovery of norovirus and hepatitis A virus from oysters and blue mussels. J Virol Methods 169(1):70–78. doi:10.1016/j.jviromet.2010.06.019
244. Vanessa dos Santos da Silva J, Henrique de Mello M, Staggemeier R et al (2014) Adenovirus presence in surfaces and equipment from ambulatories, internship units, and operating room in a Brazilian hospital. Am J Infect Control 42(6):693–694. doi:10.1016/j.ajic.2014.02.007
245. Vecchia AD, Fleck JD, Comerlato J et al (2012) Assessment of enteric viruses in a sewage treatment plant located in Porto Alegre, southern Brazil. Braz J Biol 72(4):839–846. doi:http://dx.doi.org/10.1590/S1519-69842012000500009
246. Vecchia AD, Kluge M, dos Santos da Silva JV et al (2013) Presence of Torque teno virus (TTV) in tap water in public schools from Southern Brazil. Food Environ Virol 5(1):41–45. doi:10.1007/s12560-012-9096-7
247. Verbyla ME, Symonds EM, Kafle RC et al (2016) Managing microbial risks from indirect wastewater reuse for irrigation in urbanizing watersheds. Environ Sci Technol 50(13):6803–6813. doi:10.1021/acs.est.5b05398
248. Vergara GG, Goh SG, Rezaeinejad S et al (2015) Evaluation of FRNA coliphages as indicators of human enteric viruses in a tropical urban freshwater catchment. Water Res 79:39–47. doi:10.1016/j.watres.2015.04.022
249. Verhoef L, Depoortere E, Boxman I et al (2008) Emergence of new norovirus variants on spring cruise ships and prediction of winter epidemics. Emerg Infect Dis 14(2):238–243. doi:10.3201/eid1402.061567
250. Viancelli A, Garcia LA, Schiochet M et al (2012) Culturing and molecular methods to assess the infectivity of porcine circovirus from treated effluent of swine manure. Res Vet Sci 93(3):1520–1524. doi:10.1016/j.rvsc.2012.02.005
251. Victoria M, Guimarães F, Fumian T et al (2009) Evaluation of an adsorption-elution method for detection of astrovirus and norovirus in environmental waters. J Virol Methods 156(1–2):73–76. doi:10.1016/j.jviromet.2008.11.003
252. Victoria M, Guimarães FR, Fumian TM et al (2010a) One year monitoring of norovirus in a sewage treatment plant in Rio de Janeiro, Brazil. J Water Health 8(1):158–165. doi:10.2166/wh.2009.012
253. Victoria M, Rigotto C, Moresco V et al (2010b) Assessment of norovirus contamination in environmental samples from Florianópolis City, Southern Brazil. J Appl Microbiol 109(1):231–238. doi:10.1111/j.1365-2672.2009.04646.x
254. Victoria M, Tort LF, García M et al (2014) Assessment of gastroenteric viruses from wastewater directly discharged into Uruguay River, Uruguay. Food Environ Virol 6(2):116–124. doi:10.1007/s12560-014-9143-7
255. Victoria M, Tort LF, Lizasoain A et al (2016) Norovirus molecular detection in Uruguayan sewage samples reveals a high genetic diversity and GII.4 variant replacement along time. J Appl Microbiol 120(5):1427–1435. doi:10.1111/jam.13058
256. Vieira CB (2015) Rastreamento microbiológico de fontes de contaminação humana e animal por marcadores virais e avaliação de risco de infecções por vírus gastroentéricos na bacia do Rio Negro, Manaus, Amazonas. Dissertation, Oswaldo Cruz Foundation
257. Vieira CB, de Abreu CA, de Jesus MS et al (2016) Viruses surveillance under different season scenarios of the Negro River Basin, Amazonia, Brazil. Food Environ Virol 8(1):57–69. doi:10.1007/s12560-016-9226-8
258. Vieira CB, Mendes AC, Guimarães FR et al (2012) Detection of enteric viruses in recreational waters of an urban lagoon in the city of Rio de Janeiro, Brazil. Mem Inst Oswaldo Cruz 107(6):778–784. doi:http://dx.doi.org/10.1590/S0074-02762012000600012
259. Villar LM, de Paula VS, Diniz-Mendes L et al (2006) Evaluation of methods used to concentrate and detect hepatitis A virus in water samples. J Virol Methods 137(2):169–176. doi:10.1016/j.jviromet.2006.06.008
260. Villar LM, de Paula VS, Diniz-Mendes L et al (2007) Molecular detection of hepatitis A virus in a urban sewage in Rio de janeiro, Brazil. Lett Appl Microbiol 45(2):168–173. doi:10.1111/j.1472-765X.2007.02164.x

261. Vivier JC, Ehlers MM, Grabow WO (2004) Detection of enterovirus in treated drinking water. Water Res 38(11):2699–2705
262. WHO – World Health Organization (2011) Guidelines for drinking-water quality. World Health Organization, Geneva
263. WHO – World Health Organization (2016) World Health Organization. http://www.who.int/water_sanitation_health/undwfl/en/. Accessed 13 June 2016
264. Wobus CE, Thackray LB, Virgin HW 4th (2006) Murine Norovirus: a model system to study norovirus biology and pathogenesis. J Virol 80(11):5104–5112. doi:10.1128/JVI.02346-05
265. Wolf S, Hewitt J, Greening GE (2010) Viral multiplex quantitative PCR assays for tracking sources of fecal contamination. Appl Environ Microbiol 76(5):1388–1391. doi:10.1128/AEM.02249-09
266. Wong K, Fong TT, Bibby K et al (2012) Application of enteric viruses for fecal pollution source tracking in environmental waters. Environ Int 45:151–164. doi:10.1016/j.envint.2012.02.009
267. Wong K, Xagoraraki I (2011) Evaluating the prevalence and genetic diversity of adenovirus and polyomavirus in bovine waste for microbial source tracking. Appl Microbiol Biotechnol 90(4):1521–1526. doi:10.1007/s00253-011-3156-z
268. Wong K, Xagoraraki I (2010) Quantitative PCR assays to survey the bovine adenovirus levels in environmental samples. J Appl Microbiol 109(2):605–612. doi:10.1111/j.1365-2672.2010.04684.x
269. Wyn-Jones AP, Sellwood J (2001) Enteric viruses in the aquatic environment. J Appl Microbiol 91(6):945–962. doi:10.1046/j.1365-2672.2001.01470.x
270. Yamashita T, Kobayashi S, Sakae K et al (1991) Isolation of cytopathic small round viruses with BS-C-1 cells from patients with gastroenteritis. J Infect Dis 164(5):954–957. doi:10.1093/infdis/164.5.954
271. Yanez LA, Lucero NS, Barril PA et al (2014) Evidence of hepatitis A virus circulation in central Argentina: seroprevalence and environmental surveillance. J Clin Virol 59(1):38–43. doi:10.1016/j.jcv.2013.11.005
272. Zintz C, Bok K, Parada E et al (2005) Prevalence and genetic characterization of caliciviruses among children hospitalized for acute gastroenteritis in the United States. Genet Evol 5(3):281–290. doi:10.1016/j.meegid.2004.06.010

Chapter 7
Hepatitis A and E in South America: New Challenges Toward Prevention and Control

Marcelo Alves Pinto, Jaqueline Mendes de Oliveira, and Jorge González

1 Introduction

Hepatitis A virus (HAV) and hepatitis E virus (HEV) infections are clinically and epidemiologically similar in many aspects. Both viruses are primarily transmitted by the fecal–oral route and are able to cause large outbreaks via contaminated water and food. HAV and HEV infections often follow a silent clinical course, and symptomatic cases are usually self-limiting. Clinical features range from asymptomatic to acute hepatitis and even to acute liver failure (ALF).

The World Health Organization (WHO) estimates the occurrence of about 1.5 million clinical cases of hepatitis A and 3.4 million symptomatic cases of hepatitis E worldwide every year. HEV genotypes 1 and 2, alone account for approximately 20.1 million HEV infections, 70,000 deaths, and 3,000 stillbirths annually [1, 2].

Concerning HAV infection, improvement in hygiene, sanitation, and socioeconomic conditions, as well as the implementation of universal hepatitis A vaccine programs, are of great impact in the epidemiological patterns of disease, with a shift from high to moderate in several South American countries, including Argentina, Brazil, Chile, and Uruguay. However, the infection rates may vary markedly from country to country or even within the same country [3–7].

HEV infection is now recognized as the major cause of acute viral hepatitis worldwide, mainly in endemic areas of Africa and Asia, where it is associated with large waterborne outbreaks. In the so-called developed world, the burden of

M.A. Pinto • J.M. de Oliveira (✉)
Laboratório de Desenvolvimento Tecnológico em Virologia, Instituto Oswaldo Cruz, Fiocruz, Rio de Janeiro, Brazil
e-mail: jackie@ioc.fiocruz.br; jaque_mendes06@hotmail.com

J. González
Laboratorio Nacional de Referencia INEI – ANLIS Dr Carlos G. Malbran, Instituto de Virologia, Buenos Aires, Argentina

J.E. Ludert et al. (eds.), *Human Virology in Latin America*,
DOI 10.1007/978-3-319-54567-7_7

autochthonous HEV infection has been long underestimated. Nevertheless, it is now clear that the epidemic HEV occurring in endemic regions differs markedly from HEV infection locally acquired in developed countries. The former is caused by genotypes 1 (HEV-1) and 2 (HEV-2), and the latter is caused by genotypes 3 (HEV-3) and 4 (HEV-4). As previously reported, HEV-3 and HEV-4 are of zoonotic nature, and pigs are the major reservoir [8–10].

Finally, it is now accepted that HAV and HEV present considerable differences regarding their natural history and epidemiology. In contrast to HAV, HEV does not cause self-limiting acute hepatitis only. Recent studies have shown that HEV can cause persistent infection that may evolve to chronic hepatitis and fibrosis in immunosuppressed patients, mainly in solid organ transplant (SOT) recipients [11–13]. Still, our understanding of the clinical phenotype of HEV infection in humans is not clearly understood.

2 Hepatitis A Virus

Hepatitis A virus (HAV) is a small non-enveloped virus classified in the genus *Picornavirus* within the family *Picornaviridae*. The viral genome comprises a single-stranded RNA approximately 7.5 kilobases (kb) in length, containing a single open reading frame (ORF) flanked by a short 3′-noncoding region of 40–80 nucleotides followed by a poly-A tail and a typical 5′-noncoding region containing an internal ribosomal entry site (IRES) [14, 15]. The ORF is translated into one single polyprotein of 2225–2227 amino acids [14]. The ORF amino-terminus third (P1) encodes the structural capsid proteins VP2, VP3, and VP1. In contrast to other *Picornaviridae*, the VP4 polypeptide has not yet been demonstrated [16].

The genetic heterogeneity among HAV strains has allowed its classification into genotypes and sub-genotypes by using partial sequences corresponding to different genome regions. In 1992, Robertson and colleagues selected a short segment (168 nucleotides) of the VP1/2A junction region to analyze 152 HAV isolates from the United States of America (U.S.) and Asia [17]. This pioneering study, which did not include samples from other regions of the world such as South America and Africa, resulted in the classification of the virus in seven genotypes (I to VII).

Further, an alternative method using the complete sequence of the VP1 region (900 nucleotides) classified HAV in five genetic groups [18]. This study had not included the JM-55 strain (genotype VI) nor those previously classified as genotype IIIB. Therefore, the compilation of these studies led to the current classification of HAV in six genotypes: three of human origin (I, II, III) and three of simian origin (IV, V, VI), the latter found in some species of Old World monkeys [18, 19].

The geographic distribution of HAV genotypes varies. Genotype I is of global distribution, with a predominance of sub-genotype IA [17, 20]. In North America, as well as in China, Japan, and several countries in Europe, sub-genotypes IA and IB are the most frequent. In South America (except Brazil), sub-genotype IA is of exclusive circulation. Co-circulation of several sub-genotypes has been reported in some regions, such as the simultaneous occurrence of sub-genotypes IA and IB in France and IA and IIIA in India [20]. In Brazil, several studies analyzing sequences from VP1/2A junction revealed co-circulation of sub-genotypes IA and IB in Rio de Janeiro [21–24], Pernambuco [25], Amazonas [26], and Goiás [27].

Only a few complete nucleotide sequences of the HAV genome are currently available. The complete sequence of the HAF-203 strain, isolated in Brazil and adapted in cell culture [28, 29] was classified into the sub-genotype IB [30]. The high homology observed between the HAV strain HAF-203 and HM-175 (99.7%), attenuated HM-175 (99.5%), and MBB (94.8%) is suggestive of a common ancestry [31].

HAV exhibits a considerable diversity of nucleotide sequences among isolates from different regions of the world [17, 18, 32]. However, in relationship to amino acid sequences, a high degree of conservation among samples of human origin is observed, explaining the low antigenic diversity of HAV and therefore the existence of one serotype. Phylogenetic studies using the complete sequence of HAF-203 strain revealed ten nucleotide substitutions associated with amino acid changes, and three resulted in conservative substitutions within genes encoding proteins VP1, 2C, and 3D [30]. It is likely that such mutations have contributed to adaptation of the virus in cell cultures, however, not affecting its phenotype [30], remaining infectious when transmitted experimentally to *Callithrix jacchus* (marmoset) and *Saimiri sciureus* (squirrel monkey) [33, 34].

A high degree of antigenic conservation was also observed among genetically divergent HAV strains, such as the human HM-175 (genotype I) and the simian PA-21 (Panama, genotype III, Panama), recovered from a naturally infected owl monkey (*Aotus trivirgatus*) [35]. Experimental infection in nonhuman primates also showed a high level of cross-protection between those strains [35]. On the other hand, isolates from Old World monkeys, belonging to genotypes IV, V, and VI, showed significant antigenic differences related to amino acid substitutions in regions coding VP3 and VP1 proteins containing the major neutralizing epitopes [36, 37]. In fact, "escape mutants" selected from the HM-175 adapted for cell culture were not neutralized by monoclonal antibodies specific for HAV of human origin [38–40].

3 Replicative Strategies of Hepatitis A Virus

The first step of HAV replication, after uncoating, is the synthesis of its single polyprotein, which is directed by the IRES region, located in the last domain of the 5′-non-coding region. Different from other picornaviruses, HAV displays poor growth in cell culture, which can be explained by its inefficient ribosome entry site (IRES) [41]. Differing from other genera of the *Picornaviridae* family, HAV does not code for the proteases 2A and L, the enzymes responsible for the cleavage of the cellular translation initiation factor eIF4G, thus competing very poorly for the cellular translational machinery [42]. Another difference between HAV and other picornaviruses is its unique codon usage, which has evolved to be complementary to that of human cells and thus results in an increment of rare codons used by HAV [43]. The optimized HAV codon usage contributes to its slow replication and to its low yields, and consequently allows the virus to grow in a quiescent way. Such a unique strategy may also contribute to the modulation of the antiviral cell response [44].

4 Hepatitis E Virus

Hepatitis E virus (HEV) is a small, non-enveloped virus, with a single-stranded positive-sense RNA genome of approximately 7.2 kb organized in three open reading frames. ORF1, at the 5′-end of the genome, encodes nonstructural proteins, such as a methyltransferase (which caps the 5′-end of the genome), a papain-like cysteine protease, a hypervariable region (HVR), and a RNA-dependent RNA polymerase; ORF2, at the 3′-end of the genome, encodes the major viral capsid protein; ORF3, which partially overlaps ORF1 and ORF2, encodes a small regulatory phosphoprotein that is suggested to act as an adapter to link the intracellular transduction pathways [45, 46].

In the early 1980s, HEV was first identified as the etiological agent of enterically transmitted non-A and non-B hepatitis and associated with waterborne epidemics that occurred in New Delhi [47] and in the Kashmir valley [48, 49]. Virus particles were first visualized by immunoelectron microscopy in the feces of an experimentally infected volunteer [50]. The virus was successfully transmitted to cynomolgus monkeys, and its genome was cloned and characterized in 1991 [51, 52]. Subsequent studies confirmed the existence of HEV variants genetically related although epidemiologically distinct. The first HEV strain isolated from an animal host was recovered from a pig in the United States [9]. The perception of pigs as potential reservoirs for HEV opened up a new perspective over its epidemiological pattern, especially in countries that previously reported imported cases only. A number of studies conducted in endemic and nonendemic areas showed the enzootic pattern of HEV transmission among swine and wild animals. Recently, a meta-analysis study assessing HEV prevalence in Europe confirmed the link between exposition to swine and wild animals and higher seroprevalence rates of anti-HEV in comparison with those observed in the general population [10, 53, 54].

HEV was also detected and characterized from other mammals such as wild boar, deer [55–57], cattle [58], camel [59], rabbit [60], rodents [61, 62], ferret [63], mongoose [64], moose [65], minks [66], foxes [67] and bats [68]. Further, the complete genome of the cutthroat trout virus (CTV), isolated in 1988, was related to HEV [69]. Additionally, HEV-related virus has been isolated from chickens and turkeys [70, 71]. A significant prevalence has been observed in chicken flocks in Europe, the United States, and Asia. In chickens, HEV has been associated with big liver and spleen disease (BLS) or hepatitis-splenomegaly (HS) syndrome, ovarian regression, diarrhea, and other pathogenesis [72]. Phylogenetic analysis has shown that avian HEV is genetically related to but distinct from other known HEV strains described. Indeed, avian HEV infection was not successfully reproduced in nonhuman primates [73, 74].

The discovery of HEV-related viruses infecting rabbits [60] and wild boars [75], HEV-like viruses (described in rats, ferrets, and bats), and an even more divergent virus, such as the cutthroat trout virus, propitiated the proposal of a consensus classification. The proposed classification, which established genus and species, respects the different levels of divergence between the cutthroat trout virus and all other herpesviruses. According to the International Committee on Taxonomy of Viruses (ICTV) Hepeviridae Study Group, all HEV are classified in the *Hepeviridae* family with two genera: *Orthohepevirus* with four species (A–D) and *Piscihepevirus* with a single species (A). Strains found in humans, pigs, wild boars, rabbits, cervids, mongooses, and camels are classified into the *Orthohepevirus* A; strains from chickens are classified into the *Orthohepevirus* B; *Orthohepevirus* C encompasses strains found in rats (HEV-C1) and ferrets (HEV-C2); and the bat strain is classified into *Orthohepevirus* D. Also, *Orthohepevirus* A is divided into four genotypes, which are genotypes HEV-1 and HEV-2, including only strains identified in humans, and genotypes HEV-3 and HEV-4, that have been isolated both from humans and from different animal species, and are associated with zoonotic transmission. Genotypes HEV-5 and HEV-6 were found in wild boars in Japan and genotype HEV-7 in dromedary camels in Dubai. The new HEV strains, recently isolated from elks, foxes, and minks, may represent new members of the *Hepeviridae* family [76].

5 Replication of Hepatitis E Virus

Studies on HEV replication strategy are hampered by the limited efficiency of the cell culture models currently available. Besides the successful propagation of HEV in human liver cell lines, HepG2/C3A and PLC/PRF/5 and the human lung carcinoma cell line, A549 [77, 78], knowledge of a replicative cycle is hampered by its inefficient and slow replication. Recently, the 47832c strain, which was isolated from a transplant patient chronically infected with HEV genotype 3, was improved by the generation of the clonal cell line A549/D3 [79]. Improvements in cell line susceptibility have been also described by other authors [80]. Even so, cellular targets to virus entry have already been identified, but a really efficient cell culture system has not yet been reached.

6 Hepatitis A Epidemiology in Argentina

In Argentina, a large nationwide outbreak of hepatitis A occurring in 2003–2004 alerted the health authorities to the need of implementing the national immunization program to include the inactivated hepatitis A vaccine [81]. The outbreak affected mainly children aged 5 to 9 years old living in the northern and western regions of the country. Of note, HAV-associated acute liver failure (ALF) was the major indication for liver transplantation in pediatric patients [82, 83]. Also, previous age-related prevalence studies [84], as well as age-related incidence data reported to the National Surveillance System, supported the implementation of a universal immunization program targeting 12-month-old children.

Routine hepatitis A immunization in Argentina, based on a single-dose regimen of inactivated hepatitis A vaccine, started in June 2005 and reached a national coverage of 95% in 2006. Since then, a very significant decline in the incidence rates has been observed in all age groups as well as in fulminant hepatic failure (FHF) and liver transplantation cases with reduction of 88.1% when compared with the pre-vaccination period [85]. Moreover, 5 years after vaccination, high seroprotection rates (99.7%) observed in children who received the first dose (without booster) confirmed the long-lasting sustained protection after a single-dose regimen [86], thus supporting recommendations in favor of such strategy long used in countries experiencing a shift from high to intermediate levels of HAV infection endemicity. The latest notification data showed a sharp decrease in the incidence of hepatitis A (number of cases per 100,000 inhabitants), ranging from 113.3 in 2004 to 1.4 in 2011 and 0.5 in 2012 [87].

7 Hepatitis A Epidemiology in Brazil

The epidemiology of hepatitis A is changing in Brazil, particularly in urban areas where improved sanitation and living standards contributed to significantly reduce HAV exposition in childhood, with consequent increase of symptomatic disease in adolescents and young adults. In fact, HAV incidence rates have declined in Brazil since the end of the 1990s, according to the National Disease Notification System (SINAN). During the period from 1990 to 2010, nationwide incidence rates dropped from 5.6 to 3.1 cases per 100,000 inhabitants [88].

A Brazilian population-based study conducted in 2005–2009, in a pre-vaccination context, disclosed two epidemiological scenarios, with areas of low endemicity (South and Southeast regions) and intermediate endemicity (North, Northeast, Midwest, and Federal District). Areas of intermediate endemicity showed an anti-HAV prevalence of 68.8%, with higher infection force in subjects 10 to 19 years old, whereas in the South and Southeast regions, seroprevalence was 33.7%, with higher infection force in the 20- to 29-year-old cohort [89].

The national population-based data are consonant with a number of region-wide studies pointing to an age-related increasing exposition to HAV and thereafter indicate that HAV epidemiology in Brazil is shifting from a high to intermediate and low endemicity levels [4, 89–92]. These studies were used as baseline data for implementation of the National Program of Immunizations (PNI) of the Brazilian Ministry of Health with the introduction of the single-dose hepatitis A vaccine for universal immunization of children 12 to 23 months old in July 2014 [93].

8 Hepatitis E Epidemiology in Brazil and in South America

In contrast with hepatitis A, the occurrence of waterborne/foodborne epidemics of HEV infection has not been detected in Brazil or in Argentina [94–96]. By using nucleic acid techniques as a means of identifying sources of virus transmission, only a single locally acquired case of acute hepatitis E in an immunocompetent patient was described in Rio de Janeiro, Brazil [97]. This autochthonous human HEV strain was isolated from a symptomatic patient who seroconverted to anti-HEV IgM and IgG, with HEV RNA detected by PCR, and subsequently classified into the subtype 3b. Phylogenetic and cause–effect analysis linked to zoonotic transmission by ingestion of undercooked pork meat, likely contaminated with HEV-3, which circulates widely among Brazilian pig herds [98–101]. HEV has also been detected in pigs of commercial farms in many provinces in Argentina [96, 102].

Several countries in South America, including Argentina, Brazil, Bolivia, Venezuela, and Uruguay, have reported the detection and characterization of HEV strains. Only three isolates were classified within genotype 1, two from Venezuela and one from Uruguay [103]; the remaining 70 published strains were indigenous isolates belonging to genotype 3 [94, 104–106].

A few sporadic cases of HEV infection have been identified (on the basis of HEV RNA and/or IgM and IgG antibody detection) in Brazilian immunosuppressed patients, such as in parenchymal organ transplant receptors [107–109], and in Argentinean children with acute liver failure [110]. Therefore, acute sporadic cases of HEV infection seem to be infrequent in South America. However, recent studies have reported HEV detection in river water, in raw sewage, and in the effluent of slaughterhouses [96, 106].

9 Pathogenesis of Hepatitis A

Acute HEV infection is clinically indistinguishable from hepatitis A, showing a wide range of clinical manifestations that may vary from asymptomatic or subclinical to a fulminant outcome. In contrast with hepatitis A, some patients become persistently infected and develop chronic hepatitis, progressing to fibrosis and cirrhosis under immunosuppressive conditions [111, 112]. In general, both viruses cause self-limiting hepatitis.

After ingestion, viral particles cross the gut barrier, reaching the liver parenchyma by portal tract replicating in the hepatocytes before being excreted with bile to the intestine. The extrahepatic sites of replication are not well defined; however, HEV replication in the enteric tract and human placenta has been reported for HEV [113–116]. HAV replication in the gastrointestinal tract has yet to be confirmed [117, 118], although it has already demonstrated in the polarized human intestinal epithelial cells [119], kidneys, spleen, and lymph nodes from experimentally infected primates, as well as in the epithelial cells of the intestinal crypts and in cells of the lamina propria in the small intestine of orally inoculated owl monkeys and marmosets [120–123]. In the same study, HAV antigen (HAVAg), besides being detected in the liver, was also detected in the kidneys and spleen but not in pharyngeal tissues of owl monkey [124, 125] that were correlated with the ubiquitous expression of huHAVcr-1-specific receptor in different human cells, working as a functional cellular receptor to HAV [126].

Hepatitis A virus is classically described as a noncytopathic virus in cell line cultures. It has been adapted to diverse primate and nonprimate cell lines [28, 127–130], although HAV does not seem to persist in vivo after an acute HAV infection, as observed with other hepatotropic viruses, such as hepatitis E, B, and C viruses [131]. Recently, HAV cellular receptors, T-cell Ig and mucin domain 1 (TIM-1), were identified, thus clarifying how the HAV genome is released from the viral particle and initiates an infection. In addition, that the virus is released in a non-lytic manner within exosomes was recently reported. This fact has opened up a new paradigm on the transmission of infection mechanisms within infected individuals [132, 133].

The mechanism of HEV replication is not well understood because of the limitations of the cell culture line systems; however, dose-dependent HEV infectivity was confirmed by experimental infection in animal models [134, 135]. The PLC/PRF/5 cell line showed limited permissiveness during long-term culturing for HEV infection [136], and the A549-derived subclonal cell line supports more efficient HEV replication [79]. It was shown that HEV entry depends on endocytic processes associated with dynamin-2, clathrin, membrane cholesterol, actin, and the phosphatidylinositol-3-kinase/Akt pathway in an early post-entry step of viral replication [137]. The HEV-infected A549-derived subclonal cell line allowed understanding of the mechanism by which HEV inhibits interferon (IFN)-α signaling through regulation of STAT1 phosphorylation without showing detectable cytopathic effects [138] and that HEV-1, through ORF3, may transiently activate NF-κB through UPR in early stages and subsequently inhibit TNF-α-induced NF-κB signaling in the late phase so as to create a favorable virus replication environment [139].

10 Pathogenesis of Hepatitis E

Similar to HAV, the mechanism of HEV-induced liver injury is described as immunomediated; however, the severity of HEV infection is dose dependent [135]. In human patients with acute hepatitis E, the characterization of liver-infiltrating lymphocytes pointed predominant CD 8+/CTLs in similarity with other noncytopathic hepatotropic virus (HAV, HBV, HCV). The virus itself is not cytopathic, but liver damage is caused by the host immune reaction [140].

Acute liver failure (ALF) is a clinical syndrome resulting from the massive death of liver cells induced by agents such as viruses, drugs, and autoimmune responses with a high mortality rate. ALF is associated with a critical degree of a liver histological lesion not adequately balanced by a regenerative hepatocellular activity. It is mainly characterized by the abrupt appearance of encephalopathy and coagulation disturbances. Thus, ALF presents extremely high mortality rates, and liver transplantation is the chosen treatment because it is the only promising therapeutic procedure available.

Both HAV and HEV infections may induce ALF. In pregnant women infected with HEV in the first trimester, the elevated mortality rate (30%) was justified by illness progression to liver failure [132]; however, HEV-induced ALF was never confirmed in another country beyond India and Pakistan [133]. We have demonstrated that NKT cells, activated lymphocytes, and an array of cytokines such as interleukin (IL)-10, interferon (IFN)-γ, (IL)-6, and tumor necrosis factor (TNF)-α are released in the liver injury [141]. We also demonstrated that mitochondrial products and ATP work as chemoattractants to granulocytes, contributing to an extensive liver lesion in HAV-induced ALF patients [142].

HEV-induced ALF during pregnancy is associated with high rates of preterm labor and mortality [143]. The precise reason for increased susceptibility to HEV infection during pregnancy is still an enigma [144, 145]. The role of both, mononuclear cells and macrophages, are impaired in pregnant HEV-induced ALF. Reduced toll-like receptor (TLR3 and TLR7) expression and TLR downstream-signaling molecules in pregnant HEV-induced ALF suggest that inadequate triggers of the innate immune responses contribute to the development and severity of the illness [146]. Additionally, HEV ORF3 protein has inhibited the expression of proinflammatory cytokines [TNF-α, interleukin (IL)-1β, IL-6, IL-8, IL-12p40, and IL-18] and chemotactic factors [nitric oxide (NO), interferon-inducible protein-10 (IP-10), macrophage inflammatory protein (MIP)-1α, monocyte chemoattractant protein-1 (MCP-1), granulocyte colony-stimulating factor (G-CSF), and granulocyte macrophage colony-stimulating factor (GM-CSF)] in lipopolysaccharide (LPS)-stimulated human PMA-THP1 cells. Moreover, the inhibition produced corresponding upregulation of IκBα and downregulation of phosphorylated IκB kinase IKKε (p-IKKε) and phosphorylated nuclear factor (NF)-κB (p-NF-κB). These antiinflammatory properties might be of great importance to clarify the role and mechanism of macrophages in chronic HEV infection and cirrhosis [147].

The potential for xenogeneic transmission of HEV from animals to humans via organ, tissue, or cellular transplantation or via ex vivo exposure of humans to biological porcine products for medical purposes should be a public health concern [129]. Furthermore, the use of immunosuppressive drugs may facilitate xenogeneic transmission of swine hepatitis E virus that does not cause detectable clinical symptoms in natural hosts but infects humans and causes hepatitis [130]. Additionally, immunocompromised patients bear the potential risk of developing chronic hepatitis E and short-term cirrhosis [100].

11 Prevention and Control

Socioeconomic development and further improvement in hygiene and sanitation are associated with decreasing infection rates and increasing adult susceptibility levels [4, 148]. The higher correlation between water and infection rate indicates that access to clean water may reduce waterborne outbreaks and direct transmission of

HAV and HEV (genotypes 1 and 2) [148]. Concerning hepatitis E transmission at the population level, prevention is the most effective approach in preventing the disease. Preventive measures to reduce the incidence of the disease include maintaining quality standards for public water supplies and sanitation.

Effective and safe hepatitis A vaccines have been commercially available since 1992 and have been applied either to protect risk groups or to control outbreaks. Groups at increased risk for hepatitis A include men who have sex with men, injection and non-injection drug users, chronic liver disease patients, persons with clotting factor disorders, persons working with nonhuman primates, workers exposed to sewage, and food handlers. Regarding the use of hepatitis A vaccine to control community-wide outbreaks, a single-dose regimen has been successful, mostly in small self-contained communities, where high coverage of multiple age cohorts can be achieved by starting the vaccination early in the course of the outbreak [149].

Currently, WHO recommends that vaccination against HAV should be integrated into the national immunization schedule for children aged 1 year or older if indicated on the basis of incidence of acute hepatitis A, change in endemicity from high to intermediate, and consideration of cost-effectiveness [1]. Ordinarily, long-term protection is achieved after the complete two-dose schedule [149]. However, the inclusion of the single-dose immunization schedule with inactivated hepatitis A vaccine should be considered [1] because long-term immune response has been achieved after the first vaccination (without booster), as observed in Argentina [85, 150], and recommended by the Indian Academy of Pediatrics since 2014 and Israel since 1999 [151, 152].

Moreover, results of a recent study of our research group confirmed that a single dose of inactivated hepatitis A vaccine promotes HAV-specific cellular memory response similar to that induced by a natural infection [153]. Additionally, all available data on monovalent or combined (hepatitis A and B) vaccines suggest that there is no support for a booster when a complete primary vaccination course is offered to immunocompetent subjects.

The role of HEV-related viruses of foodborne transmission by pork and other livestock products, such as game meat, cattle, sheep, and seafood, has also been well established by different reports [66, 154, 155]. Because HEV infection in pigs is subclinical, the diagnosis relies exclusively on laboratory tests, namely, on the detection of viral RNA or specific anti-HEV antibodies [156–158]. Until the present, there is not a well-established diagnostic criterion for HEV detection in livestock.

The hypothesis of massive human exposure to HEV has been raised by the high incidence of hepatitis E virus (HEV) among swine at slaughter age in farms that has been documented in several reports concerning zoonotic HEV genotypes. Thus, it is important to determine measures to reduce HEV spread and infection of slaughter-age pigs within the farms. However, available data about HEV epidemiology on pig farms in South America are poor. Studies about HEV transmission among pigs have suggested that the age in HEV infection is not strictly dependent upon the proportion of piglets with colostrum intake, but is also linked to farm-specific husbandry [159]. In France, the risk of HEV-positive livers was increased by early slaughter, genetic background, lack of hygiene measures and surfaces, and the source of drinking water [160]. Besides hygienic measures, efforts to block HEV spread in the chain production of pork foods and herd vaccination against HEV have been recognized as an efficient method to reduce the proportion of infectious animals at slaughter age, the transmission rate parameter, the susceptibility, the average infectious period, or the combination of these parameters [161].

Comparatively, whether hepatitis E vaccine is long-term protective against heterogeneous HEV infection is still an unsolved issue because of human cases of HEV reinfections in hyperendemic countries [162, 163]. However, a decrease of HEV IgG titers (acquired after natural HEV infection or vaccination) within time was reported in urban and rural children in North India [164]. Efforts to improve the immune response to HEV vaccination have been reported in capsid-designed vaccines. Tsarev and colleagues described a cross-protection between Pakistani and Mexican HEV strains in nonhuman primates vaccinated twice with a 50-μg dose of the recombinant capsid protein (ORF2) [165]. The animals were protected from hepatitis after heterologous genotype challenge with the Mexican strain. In fact, the major anti-HEV antibody response is against conformational epitopes located in a.a. 459–606 of HEV pORF2. All reported neutralization epitopes are present on the dimer domain constructed by this peptide. So far, two recombinant vaccines to HEV are available. The first one, rHEV, is based on a 56-kDa capsid protein and is experimental whereas the other one, HEV 239 designed for humans, is licensed in China since December 2011. HEV 239 vaccine is described as highly immunogenic, but the seroconversion only occurred after the third dose (0, 1, and 6 months). A protective response to HEV infection was evaluated comparing seronegative placebo subjects (evaded naturally acquired immunity) and HEV 239 vaccine-induced immunity and significantly reduced the risk of infection in the vaccinated group [166, 167]. The risk of reinfection observed in the placebo group, which evaded naturally acquired immunity (0.52%; 95% CI, 0.30–0.83), was similar to the risk of breakthrough infection observed in the vaccinated group, which evaded vaccine-induced immunity (0.30%; 95% CI, 0.19–0.44), although GMC among the vaccinated group was substantially higher. Hepatitis E vaccine −239, commercially available in China, has shown high efficacy with sustained protection for more than 4 years [166]. Despite many reports targeting a commercial HEV vaccine applied to humans, there are poor descriptions targeting swine until date. Recently, one study on HEV vaccine evaluation showed rabbits vaccinated with 20 μg of the HEV p179 produced anti-HEV with elevated titers (1:104–1:105), and the animals were completely protected from HEV infection. In this study, when rabbits were vaccinated with 10 μg, anti-HEV was produced with reduced titers (1:103–1:104) and conferred protection against hepatitis E, but two of five rabbits showed fecal virus shedding [167]. Besides the cautions concerning the intake of potentially contaminated meat, antiviral therapy with ribavirin monotherapy or pegylated IFN, associated with a reduction of immunosuppression, has been indicated to reduce viremia in cases of chronic hepatitis, fibrosis, and cirrhosis in immunocompromised patients, and most patients cleared the virus in a few weeks.

In summary, some efforts are key points to reduce the transmission of both HAV and HEV: first, improvement of the sanitary conditions of the human population living in peripheral regions of big cities and slums located in endemic countries; and second, the need for an effective implementation of an HEV vaccination program. Regarding HEV prevention, improvements in sanitary conditions of swine breeding farms and slaughterhouses, including wastewater and sewage treatment, are fundamental to the control of HEV contamination in the chain production of pork foods and HEV investigation in several different swine-derived products to discover the potential source of HEV contamination. In conclusion, the scientific community and public health authorities should consider implementing an effective HEV vaccine for swine.

References

1. WHO (2012) Position paper on hepatitis A vaccines. http://www.who.int/immunization/policy/position_papers/hepatitis_a/en/. Accessed 20 Feb 2017
2. WHO (2015) Hepatitis E vaccine: WHO position paper. Wkly Epidemiol Rec 18:185–200. http://www.who.int/hepatitis/publications/hepE-positionpaper/en/. Accessed 20 Feb 2017
3. Vacchino MN (2008) Incidence of hepatitis A in Argentina after vaccination. J Viral Hepat 15(suppl 2):47–50
4. Vitral CL, Ospina FL, Artimos S, Melgaço JG, Cruz OG, de Paula VS, Luz SB, Freire M, Gaspar LP, Amado LA et al (2012) Declining prevalence of hepatitis A virus antibodies among children from low socioeconomic groups reinforces the need for the implementation of hepatitis A vaccination in Brazil. Mem Inst Oswaldo Cruz 107:652–658
5. Vitral CL, Souto FJ, Gaspar AM (2008) Changing epidemiology of hepatitis A in Brazil: reassessing immunization policy. J Viral Hepat 15(suppl 2):22–25
6. Covarrubías N, Hurtado C, Díaz A, Mezzano G, Brahm J, Venegas M (2015) Hepatitis E virus seroprevalence: a reappraisal. Rev Chil Infectol 32:482–484
7. Quezada A, Baron-Papillon F, Coudeville L, Maggi L (2008) Universal vaccination of children against hepatitis A in Chile: a cost-effectiveness study. Rev Panam Salud Publica 23:303–312
8. Meng XJ, Lindsay DS, Sriranganathan N (2009) Wild boars as sources for infectious diseases in livestock and humans. Philos Trans R Soc Lond Ser B Biol Sci 1530:2697–2707
9. Meng XJ, Purcell RH, Halbur PG, Lehman JR, Webb DM, Tsareva TS, Haynes JS, Thacker BJ, Emerson SU (1997) A novel virus in swine is closely related to the human hepatitis E virus. Proc Natl Acad Sci U S A 94:9860–9865
10. Meng XJ (2010) Hepatitis E virus: animal reservoirs and zoonotic risk. Vet Microbiol 140:256–265
11. Mikolašević I, Sladoje-Martinović B, Orlić L, Milić S, Lukenda V, Župan Z, Štimac D, Rački S (2014) Evaluation of viral hepatitis in solid organ transplantation. Acta Med Croatica 68:151–159
12. Abravanel F, Lhomme S, Rostaing L, Kamar N, Izopet J (2015) Protracted fecal shedding of HEV during ribavirin therapy predicts treatment relapse. Clin Infect Dis 60:96–99
13. Aggarwal R (2011) Clinical presentation of hepatitis E. Virus Res 161:15–22
14. Cohen JI, Ticehurst JR, Purcell RH, Buckler-White A, Baroudy BM (1987) Complete nucleotide sequence of wild-type hepatitis A virus: comparison with different strains of hepatitis A virus and other picornaviruses. J Virol 61:50–59
15. Ticehurst RL, Henry RL (1989) Stage-related behavioural problems in the 1–4 year old child: parental expectations in a child development unit referral group compared with a control group. Aust Paediatr J 25:39–42

16. Martin A, Lemon SM (2006) Hepatitis A virus: from discovery to vaccines. Hepatology 43:S164–S172
17. Robertson BH, Jansen RW, Khanna B, Totsuka A, Nainan OV, Siegl G, Widell A, Margolis HS, Isomura S, Ito K (1992) Genetic relatedness of hepatitis A virus strains recovered from different geographical regions. J Gen Virol 73:1365–1377
18. Costa-Mattioli M, Cristina J, Romero H, Perez-Bercof R, Casane D, Colina R, Garcia L, Vega I, Glikman G, Romanowsky V et al (2002) Molecular evolution of hepatitis A virus: a new classification based on the complete VP1 protein. J Virol 76:9516–9525
19. Costa-Mattioli M, Ferré V, Casane D, Perez-Bercoff R, Coste-Burel M, Imbert-Marcille BM, Andre EC, Bressollette-Bodin C, Billaudel S, Cristina J (2003) Evidence of recombination in natural populations of hepatitis A virus. Virology 311:51–59
20. Cristina J, Costa-Mattioli M (2007) Genetic variability and molecular evolution of hepatitis A virus. Virus Res 127:151–157
21. de Paula VS, Baptista ML, Lampe E, Niel C, Gaspar AM (2002) Characterization of hepatitis A virus isolates from subgenotypes IA and IB in Rio de Janeiro, Brazil. J Med Virol 66:22–27
22. Villar LM, Lampe E, Meyer A, Gaspar AM (2004) Genetic variability of hepatitis A virus isolates in Rio de Janeiro: implications for the vaccination of school children. Braz J Med Biol Res 37:1779–1787
23. Santos DR, Villar LM, Paula VS, Lima GS, Gaspar AM (2008) Hepatitis A virus subgenotypes dissemination during a community outbreak in a surrounding region of Rio de Janeiro. Mem Inst Oswaldo Cruz 103:254–258
24. Amado LA, Villar LM, de Paula VS, Pinto MA, Gaspar AM (2011) Exposure to multiple subgenotypes of hepatitis A virus during an outbreak using matched serum and saliva specimens. J Med Virol 83:768–775
25. Villar LM, Morais LM, Aloise R, Melo MM, Calado IA, Lampe E, Gaspar AM (2006) Co-circulation of genotypes IA and IB of hepatitis A virus in Northeast Brazil. Braz J Med Biol Res 39:873–881
26. De Paula VS, Niel C, Teves SC, Villar LM, Virgolino H, Gaspar AM (2006) Molecular epidemiology of hepatitis A virus in Brazilian Amazon. J Gastroenterol Hepatol 21:1435–1438
27. Fiaccadori FS, Pereira M, Coelho AS, Borges AM, Parente JA, Soares CM, Cardoso D (2008) Molecular characterization of hepatitis A virus isolates from Goiânia, Gioás, Brazil. Mem Inst Oswaldo Cruz 103(8):831–835
28. Gaspar AM, Vitral CL, Yoshida CF, Schatzmayr HG (1992) Primary isolation of a Brazilian strain of hepatitis A virus (HAF-203) and growth in a primate cell line (FRhK-4). Braz J Med Biol Res 25:697–705
29. Gaspar AM, Vitral CL, Yoshida CF, Schatzmayr HG (1993) Fast growth of a Brazilian hepatitis A virus (HAF-203) in a primate cell line. Braz J Med Biol Res 26:203–206
30. Baptista ML, Silva M, Lima MA, Yoshida CF, Gaspar AM, Galler R (2003) Genetic variability of hepatitis A virus strain HAF-203 isolated in Brazil and expression of the VP1 gene in *Escherichia coli*. Mem Inst Oswaldo Cruz 2006(101):759–766
31. Lu L, Ching KZ, de Paula VS, Nakano T, Siegl G, Weitz M, Robertson BH (2004) Characterization of the complete genomic sequence of genotype II hepatitis A virus (CF53/Berne isolate). J Gen Virol 85:2943–2952
32. Costa-Mattioli M, Di Napoli A, Ferré V, Billaudel S, Perez-Bercoff R, Cristina J (2003) Genetic variability of hepatitis A virus. J Gen Virol 84:3191–3201
33. Vitral CL, Marchevsky RS, Yoshida CF, Coelho JM, Gaspar AM, Schatzmayr HG (1995) Intragastric infection induced in marmosets (*Callithrix jacchus*) by a Brazilian hepatitis A virus (HAF-203). Braz J Med Biol Res 28:313–321
34. Pinto MA, Marchevsky RS, Baptista ML, de Lima MA, Pelajo-Machado M, Vitral CL, Kubelka CF, Pissurno JW, Franca MS, Schatzmayr HG et al (2002) Experimental hepatitis A virus (HAV) infection in *Callithrix jacchus*: early detection of HAV antigen and viral fate. Exp Toxicol Pathol 53:413–420
35. Brown EA, Jansen RW, Lemon SM (1989) Characterization of a simian hepatitis A virus (HAV): antigenic and genetic comparison with human HAV. J Virol 63:4932–4937

36. Nainan OV, Margolis HS, Robertson BH, Balayan M, Brinton MA (1991) Sequence analysis of a new hepatitis A virus naturally infecting cynomolgus macaques (*Macaca fascicularis*). J Gen Virol 72:1685–1689
37. Tsarev SA, Emerson SU, Balayan MS, Ticehurst J, Purcell RH (1991) Simian hepatitis A virus (HAV) strain AGM-27: comparison of genome structure and growth in cell culture with other HAV strains. J Gen Virol 72:1677–1683
38. Ping LH, Lemon SM (1992) Antigenic structure of human hepatitis A virus defined by analysis of escape mutants selected against murine monoclonal antibodies. J Virol 66:2208–2216
39. Stapleton JT, Lemon SM (1987) Neutralization escape mutants define a dominant immunogenic neutralization site on hepatitis A virus. J Virol 61:491–498
40. Ping LH, Jansen RW, Stapleton JT, Cohen JI, Lemon SM (1988) Identification of an immunodominant antigenic site involving the capsid protein VP3 of hepatitis A virus. Proc Natl Acad Sci U S A 85:8281–8285
41. Whetter LE, Day SP, Elroy-Stein O, Brown EA, Lemon SM (1994) Low efficiency of the 5′-nontranslated region of hepatitis A virus RNA in directing cap-independent translation in permissive monkey kidney cells. J Virol 68:5253–5263
42. Jünemann C, Song Y, Bassili G, Goergen D, Henke J, Niepmann M (2007) Picornavirus internal ribosome entry site elements can stimulate translation of upstream genes. J Biol Chem 282:132–141
43. Sánchez G, Bosch A, Pintó RM (2003) Genome variability and capsid structural constraints of hepatitis A virus. J Virol 77:452–459
44. Pintó RM, Aragonès L, Costafreda MI, Ribes E, Bosch A (2007) Codon usage and replicative strategies of hepatitis A virus. Virus Res 127:158–163
45. Tam RC, Lau JY, Hong Z (2001) Mechanisms of action of ribavirin in antiviral therapies. Antivir Chem Chemother 12:261–272
46. Chandra V, Taneja S, Kalia M, Jameel S (2008) Molecular biology and pathogenesis of hepatitis E virus. J Biosci 33:451–464
47. Viswanathan R, Sidhu AS (1957) Infectious hepatitis; clinical findings. Indian J Med Res 45:49–58
48. Khuroo MS (1980) Study of an epidemic of non-A, non-B hepatitis. Possibility of another human hepatitis virus distinct from post-transfusion non-A, non-B type. Am J Med 68:818–824
49. Khuroo MS, Duermeyer W, Zargar SA, Ahanger MA, Shah MA (1983) Acute sporadic non-A, non-B hepatitis in India. Am J Epidemiol 118:360–364
50. Balayan MS, Andjaparidze AG, Savinskaya SS, Ketiladze ES, Braginsky DM, Savinov AP, Poleschuk VF (1983) Evidence for a virus in non-A, non-B hepatitis transmitted via the fecal-oral route. Intervirology 20:23–31
51. Emerson SU, Purcell RH (2003) Hepatitis E virus. Rev Med Virol 13:145–154
52. Tam AW, Smith MM, Guerra ME, Huang CC, Bradley DW, Fry KE, Reyes GR (1991) Hepatitis E virus (HEV): molecular cloning and sequencing of the full-length viral genome. Virology 185:120–131
53. Hartl J, Otto B, Madden RG, Webb G, Woolson KL, Kriston L, Vettorazzi E, Lohse AW, Dalton HR, Pischke S (2016) Hepatitis E seroprevalence in Europe: a meta-analysis. Viruses 8. pii: E211
54. Thiry D, Mauroy A, Saegerman C, Thomas I, Wautier M, Miry C, Czaplicki G, Berkvens D, Praet N, van der Poel W et al (2014) Estimation of hepatitis E virus (HEV) pig seroprevalence using ELISA and Western blot and comparison between human and pig HEV sequences in Belgium. Vet Microbiol 172:407–414
55. Takahashi M, Okamoto H (2014) Features of hepatitis E virus infection in humans and animals in Japan. Hepatol Res 44:43–58
56. Rutjes SA, Lodder-Verschoor F, Lodder WJ, van der Giessen J, Reesink H, Bouwknegt M, de Roda Husman AM (2010) Seroprevalence and molecular detection of hepatitis E virus in wild boar and red deer in the Netherlands. J Virol Methods 168:197–206

57. Sonoda H, Abe M, Sugimoto T, Sato Y, Bando M, Fukui E, Mizuo H, Takahashi M, Nishizawa T, Okamoto H (2004) Prevalence of hepatitis E virus (HEV) infection in wild boars and deer and genetic identification of a genotype 3 HEV from a boar in Japan. J Clin Microbiol 42:5371–5374
58. Hu GD, Ma X (2010) Detection and sequences analysis of bovine hepatitis E virus RNA in Xinjiang autonomous region. Bing Du Xue Bao 26:27–32
59. Woo PC, Lau SK, Teng JL, Tsang AK, Joseph M, Wong EY, Tang Y, Sivakumar S, Xie J, Bai R et al (2014) New hepatitis E virus genotype in camels, the Middle East. Emerg Infect Dis 20:1044–1048
60. Zhao C, Ma Z, Harrison TJ, Feng R, Zhang C, Qiao Z, Fan J, Ma H, Li M, Song A et al (2009) A novel genotype of hepatitis E virus prevalent among farmed rabbits in China. J Med Virol 81:1371–1379
61. Li W, Guan D, Su J, Takeda N, Wakita T, Li TC, Ke CW (2013) High prevalence of rat hepatitis E virus in wild rats in China. Vet Microbiol 165:275–280
62. Johne R, Plenge-Bönig A, Hess M, Ulrich RG, Reetz J, Schielke A (2010) Detection of a novel hepatitis E-like virus in faeces of wild rats using a nested broad-spectrum RT-PCR. J Gen Virol 91(pt 3):750–758
63. Raj VS, Smits SL, Pas SD, Provacia LB, Moorman-Roest H, Osterhaus AD, Haagmans BL (2012) Novel hepatitis E virus in ferrets, the Netherlands. Emerg Infect Dis 18:1369–1370
64. Nakamura M, Takahashi K, Taira K, Taira M, Ohno A, Sakugawa H, Arai M, Mishiro S (2006) Hepatitis E virus infection in wild mongooses of Okinawa, Japan: demonstration of anti-HEV antibodies and a full-genome nucleotide sequence. Hepatol Res 34:137–140
65. Lin J, Norder H, Uhlhorn H, Belak S, Widen F (2014) Novel hepatitis E like virus found in Swedish moose. J Gen Virol 95(pt 3):557–570
66. Krog JS, Larsen LE, Schultz AC (2014) Enteric porcine viruses in farmed shellfish in Denmark. Int J Food Microbiol 186:105–109
67. Bodewes R, van der Giessen J, Haagmans BL, Osterhaus AD, Smits SL (2013) Identification of multiple novel viruses, including a parvovirus and a hepevirus, in feces of red foxes. J Virol 87:7758–7764
68. Feng R, Zhao C, Li M, Harrison TJ, Qiao Z, Feng Y, Ma Z, Wang Y (2011) Infection dynamics of hepatitis E virus in naturally infected pigs in a Chinese farrow-to-finish farm. Infect Genet Evol 11:1727–1731
69. Batts W, Yun S, Hedrick R, Winton J (2011) A novel member of the family Hepeviridae from cutthroat trout (*Oncorhynchus clarkii*). Virus Res 158:116–123
70. Hsu IW, Tsai HJ (2014) Avian hepatitis E virus in chickens, Taiwan, 2013. Emerg Infect Dis 20:149–151
71. Peralta B, Biarnés M, Ordóñez G, Porta R, Martín M, Mateu E, Pina S, Meng XJ (2009) Evidence of widespread infection of avian hepatitis E virus (avian HEV) in chickens from Spain. Vet Microbiol 137:31–36
72. Haqshenas G, Shivaprasad HL, Woolcock PR, Read DH, Meng XJ (2001) Genetic identification and characterization of a novel virus related to human hepatitis E virus from chickens with hepatitis-splenomegaly syndrome in the United States. J Gen Virol 82(pt 10):2449–2462
73. Sun ZF, Larsen CT, Huang FF, Billam P, Pierson FW, Toth TE, Meng XJ (2004) Generation and infectivity titration of an infectious stock of avian hepatitis E virus (HEV) in chickens and cross-species infection of turkeys with avian HEV. J Clin Microbiol 42:2658–2662
74. Huang FF, Sun ZF, Emerson SU, Purcell RH, Shivaprasad HL, Pierson FW, Toth TE, Meng XJ (2004) Determination and analysis of the complete genomic sequence of avian hepatitis E virus (avian HEV) and attempts to infect rhesus monkeys with avian HEV. J Gen Virol 85(pt 6):1609–1618
75. Takahashi M, Nishizawa T, Sato H, Sato Y, Jirintai NS, Okamoto H (2011) Analysis of the full-length genome of a hepatitis E virus isolate obtained from a wild boar in Japan that is classifiable into a novel genotype. J Gen Virol 92(pt 4):902–908
76. Smith DB, Simmonds P, Jameel S, Emerson SU, Harrison TJ, Meng XJ, Okamoto H, Van der Poel WH, Purdy MA, Group ICoToVHS (2014) Consensus proposals for classification of the family *Hepeviridae*. J Gen Virol 95(pt 10):2223–2232

77. Okamoto H (2013) Culture systems for hepatitis E virus. J Gastroenterol 48:147–158
78. Johne R, Reetz J, Ulrich RG, Machnowska P, Sachsenröder J, Nickel P, Hofmann J (2014) An ORF1-rearranged hepatitis E virus derived from a chronically infected patient efficiently replicates in cell culture. J Viral Hepatol 21:447–456
79. Schemmerer M, Apelt S, Trojnar E, Ulrich RG, Wenzel JJ, Johne R (2016) Enhanced replication of hepatitis E virus strain 47832c in an A549-derived subclonal cell line. Viruses 8:267
80. Johne R, Trojnar E, Filter M, Hofmann J (2016) Thermal stability of hepatitis E virus as estimated by a cell culture method. Appl Environ Microbiol 82:4225–4231
81. Espul C, Benedetti L, Linares M, Cuello H, Rasuli A (2015) Five-year follow-up of immune response after one or two doses of inactivated hepatitis A vaccine given at 1 year of age in the Mendoza Province of Argentina. J Viral Hepatol 22:453–458
82. Cuarterolo M, Ciocca M, López S, Cervio G, Rojas L, Bianco G, Speranza A, Sasbon J, De Davila MT, Questa H et al (2005) Evolution of children one year post liver transplant. Medicina (B Aires) 65:402–408
83. Munné MS, Vladimirsky S, Otegui L, Brajterman L, Castro R, Soto S, Moreiro R, Ciocca M, Cuarterolo M, Buamscha D, Giannivelli S, Sasbán J, Schlauder G, González JE (2006) Molecular characterization of hepatitis E virus in three acute liver failure cases in children in Argentina. Acta Gastroenterol Latinoam 36:125–130
84. Vizzotti C, González J, Gentile A, Rearte A, Ramonet M, Cañero-Velasco MC, Pérez Carrega ME, Urueña A, Diosque M (2014) Impact of the single-dose immunization strategy against hepatitis A in Argentina. Pediatr Infect Dis J 33:84–88
85. Ott JJ, Wiersma ST (2013) Single-dose administration of inactivated hepatitis A vaccination in the context of hepatitis A vaccine recommendations. Int J Infect Dis 17:e939–e944
86. Gentile A, Ramonet MD, Ciocca M (2013) Hepatitis A immunisation in the Argentinean mandatory schedule. Arch Argent Pediatr 111:155–161
87. BRASIL Boletim Epidemiológico – Hepatites Virais (2012) http://www.riocomsaude.rj.gov.br/Publico/MostrarArquivo.aspx?C=sF6igzRT%2Fp4%3D. Accessed 20 Feb 2017
88. Ximenes RA, Martelli CM, Amaku M, Sartori AM, de Soárez PC, Novacs HM, Pereira LM, Moreira RC, Figueiredo GM, de Azevedo RS et al (2014) Modelling the force of infection for hepatitis A in an urban population-based survey: a comparison of transmission patterns in Brazilian macro-regions. PLoS One 9:e94622
89. Ciaccia MC, Moreira RC, Ferraro AA, Lemos MF, Oba IT, Porta G (2012) Epidemiological and serological aspects of hepatitis A among children and teenagers in the city of Santos: a cross-sectional study. Sao Paulo Med J 130:230–235
90. Markus JR, Cruz CR, Maluf EM, Tahan TT, Hoffmann MM (2011) Seroprevalence of hepatitis A in children and adolescents. J Pediatr 87:419–424
91. Vitral CL, Gaspar AM, Souto FJ (2006) Epidemiological pattern and mortality rates for hepatitis A in Brazil, 1980–2002: a review. Mem Inst Oswaldo Cruz 101:119–127
92. PORTARIA No- 2, DE 18 DE JANEIRO DE (2013) http://www.cgu.gov.br/sobre/legislacao/arquivos/portarias/portaria_cgu_133_2013.pdf. Accessed on 20 Feb 2017
93. Echevarría JM, González JE, Lewis-Ximenez LL, Dos Santos DR, Munné MS, Pinto MA, Pujol FH, Rodríguez-Lay LA (2013) Hepatitis E virus infection in Latin America: a review. J Med Virol 85:1037–1045
94. Vitral CL, da Silva-Nunes M, Pinto MA, de Oliveira JM, Gaspar AM, Pereira RC, Ferreira UM (2014) Hepatitis A and E seroprevalence and associated risk factors: a community-based cross-sectional survey in rural Amazonia. BMC Infect Dis 14:458
95. Martínez Wassaf MG, Pisano MB, Barril PA, Elbarcha OC, Pinto MA, Mendes de Oliveira J, DiGiusto P, Nates SV, Ré VE (2014) First detection of hepatitis E virus in Central Argentina: environmental and serological survey. J Clin Virol 61:334–339
96. Lopes Dos Santos DR, Lewis-Ximenez LL, da Silva MF, de Sousa PS, Gaspar AM, Pinto MA (2010) First report of a human autochthonous hepatitis E virus infection in Brazil. J Clin Virol 47:276–279

97. da Costa Lana MV, Gardinali NR, da Cruz RA, Lopes LL, Silva GS, Caramori Júnior JG, de Oliveira AC, de Almeida SM, Colodel EM, Alfieri AA et al (2014) Evaluation of hepatitis E virus infection between different production systems of pigs in Brazil. Trop Anim Health Prod 46:399–404
98. Gardinali NR, Barry AF, da Silva PF, de Souza C, Alfieri AF, Alfieri AA (2012) Molecular detection and characterization of hepatitis E virus in naturally infected pigs from Brazilian herds. Res Vet Sci 93:1515–1519
99. dos Santos DR, Vitral CL, de Paula VS, Marchevsky RS, Lopes JF, Gaspar AM, Saddi TM, Júnior NC, Guimarães FR, Júnior JG et al (2009) Serological and molecular evidence of hepatitis E virus in swine in Brazil. Vet J 182:474–480
100. de Souza AJ, Gomes-Gouvea MS, Soares Mdo C, Pinho JR, Malheiros AP, Carneiro LA, dos Santos DR, Pereira WL (2012) HEV infection in swine from Eastern Brazilian Amazon: evidence of co-infection by different subtypes. Comp Immunol Microbiol Infect Dis 35:477–485
101. Munné MS, Vladimirsky S, Otegui L, Castro R, Brajterman L, Soto S, Guarnera E, Molina V, Monfellano M, Schlauder GG et al (2006) Identification of the first strain of swine hepatitis E virus in South America and prevalence of anti-HEV antibodies in swine in Argentina. J Med Virol 78:1579–1583
102. Mirazo S, Mainardi V, Ramos N, Gerona S, Rocca A, Arbiza J (2014) Indigenous hepatitis E virus genotype 1 infection, Uruguay. Emerg Infect Dis 20:171–173
103. Dell'Amico MC, Cavallo A, Gonzales JL, Bonelli SI, Valda Y, Pieri A, Segund H, Ibañez R, Mantella A, Bartalesi F et al (2011) Hepatitis E virus genotype 3 in humans and Swine, Bolivia. Emerg Infect Dis 17:1488–1490
104. Gardinali NR, Barry AF, Otonel RA, Alfieri AF, Alfieri AA (2012) Hepatitis E virus in liver and bile samples from slaughtered pigs of Brazil. Mem Inst Oswaldo Cruz 107:935–939
105. dos Santos DR, de Paula VS, de Oliveira JM, Marchevsky RS, Pinto MA (2011) Hepatitis E virus in swine and effluent samples from slaughterhouses in Brazil. Vet Microbiol 149:236–241
106. Passos AM, Heringer TP, Medina-Pestana JO, Ferraz ML, Granato CF (2013) First report and molecular characterization of hepatitis E virus infection in renal transplant recipients in Brazil. J Med Virol 85:615–619
107. Passos-Castilho AM, Porta G, Miura IK, Pugliese RP, Danesi VL, Porta A, Guimarães T, Seda J, Antunes E, Granato CF (2014) Chronic hepatitis E virus infection in a pediatric female liver transplant recipient. J Clin Microbiol 52:4425–4427
108. Hering T, Passos AM, Perez RM, Bilar J, Fragano D, Granato C, Medina-Pestana JO, Ferraz ML (2014) Past and current hepatitis E virus infection in renal transplant patients. J Med Virol 86:948–953
109. Munné MS, Cañero Velasco MC, Moreiro R, Vladimirsky S, Otegui L, Castro R, Brajterman L, Soto S, Mutti J, Nucifora S et al (2006) Duration of viremia and fecal shedding of the virus in hepatitis A infected children. Acta Gastroenterol Latinoam 36:182–189
110. Halac U, Béland K, Lapierre P, Patey N, Ward P, Brassard J, Houde A, Alvarez F (2012) Chronic hepatitis E infection in children with liver transplantation. Gut 61:597–603
111. Kamar N, Abravanel F, Selves J, Garrouste C, Esposito L, Lavayssière L, Cointault O, Ribes D, Cardeau I, Nogier MB et al (2010) Influence of immunosuppressive therapy on the natural history of genotype 3 hepatitis-E virus infection after organ transplantation. Transplantation 89:353–360
112. Cheung MC, Maguire J, Carey I, Wendon J, Agarwal K (2012) Review of the neurological manifestations of hepatitis E infection. Ann Hepatol 11:618–622
113. Williams TP, Kasorndorkbua C, Halbur PG, Haqshenas G, Guenette DK, Toth TE, Meng XJ (2001) Evidence of extrahepatic sites of replication of the hepatitis E virus in a swine model. J Clin Microbiol 39:3040–3046
114. Bose PD, Das BC, Hazam RK, Kumar A, Medhi S, Kar P (2014) Evidence of extrahepatic replication of hepatitis E virus in human placenta. J Gen Virol 95(pt 6):1266–1271
115. Amado LA, Marchevsky RS, de Paula VS, Hooper C, Freire Mda S, Gaspar AM, Pinto MA (2010) Experimental hepatitis A virus (HAV) infection in cynomolgus monkeys (*Macaca fascicularis*): evidence of active extrahepatic site of HAV replication. Int J Exp Pathol 91:87–97

116. Mathiesen LR, Møller AM, Purcell RH, London WT, Feinstone SM (1980) Hepatitis A virus in the liver and intestine of marmosets after oral inoculation. Infect Immun 28:45–48
117. Krawczynski KK, Bradley DW, Murphy BL, Ebert JW, Anderson TE, Doto IL, Nowoslawski A, Duermeyer W, Maynard JE (1981) Pathogenetic aspects of hepatitis A virus infection in enterally inoculated marmosets. Am J Clin Pathol 76:698–706
118. Blank CA, Anderson DA, Beard M, Lemon SM (2000) Infection of polarized cultures of human intestinal epithelial cells with hepatitis A virus: vectorial release of progeny virions through apical cellular membranes. J Virol 74:6476–6484
119. Provost PJ, Villarejos VM, Hilleman MR (1977) Suitability of the rufiventer marmoset as a host animal for human hepatitis A virus. Proc Soc Exp Biol Med 155:283–286
120. Mathiesen LR, Drucker J, Lorenz D, Wagner JA, Gerety RJ, Purcell RH (1978) Localization of hepatitis A antigen in marmoset organs during acute infection with hepatitis A virus. J Infect Dis 138:369–377
121. Karayiannis P, Jowett T, Enticott M, Moore D, Pignatelli M, Brenes F, Scheuer PJ, Thomas HC (1986) Hepatitis A virus replication in tamarins and host immune response in relation to pathogenesis of liver cell damage. J Med Virol 18:261–276
122. Cohen JI, Ticehurst JR, Feinstone SM, Rosenblum B, Purcell RH (1987) Hepatitis A virus cDNA and its RNA transcripts are infectious in cell culture. J Virol 61:3035–3039
123. Provost PJ, Hilleman MR (1979) Propagation of human hepatitis A virus in cell culture in vitro. Proc Soc Exp Biol Med 160:213–221
124. Asher LV, Binn LN, Mensing TL, Marchwicki RH, Vassell RA, Young GD (1995) Pathogenesis of hepatitis A in orally inoculated owl monkeys (*Aotus trivirgatus*). J Med Virol 47:260–268
125. Feigelstock D, Thompson P, Mattoo P, Zhang Y, Kaplan GG (1998) The human homolog of HAVcr-1 codes for a hepatitis A virus cellular receptor. J Virol 72:6621–6628
126. Dotzauer A, Feinstone SM, Kaplan G (1994) Susceptibility of nonprimate cell lines to hepatitis A virus infection. J Virol 68:6064–6068
127. Vallbracht A, Hofmann L, Wurster KG, Flehmig B (1984) Persistent infection of human fibroblasts by hepatitis A virus. J Gen Virol 65:609–615
128. Harmon SA, Summers DF, Ehrenfeld E (1989) Detection of hepatitis A virus RNA and capsid antigen in individual cells. Virus Res 12:361–369
129. Cho MW, Ehrenfeld E (1991) Rapid completion of the replication cycle of hepatitis A virus subsequent to reversal of guanidine inhibition. Virology 180:770–780
130. Rowlands DJ (2015) Human hepatitis A virus is united with a host of relations. Proc Natl Acad Sci U S A 112:15010–15011
131. Wang X, Ren J, Gao Q, Hu Z, Sun Y, Li X, Rowlands DJ, Yin W, Wang J, Stuart DI et al (2015) Hepatitis A virus and the origins of picornaviruses. Nature 517:85–88
132. Seggewiß N, Paulmann D, Dotzauer A (2016) Lysosomes serve as a platform for hepatitis A virus particle maturation and nonlytic release. Arch Virol 161:43–52
133. Tsarev SA, Tsareva TS, Emerson SU, Yarbough PO, Legters LJ, Moskal T, Purcell RH (1994) Infectivity titration of a prototype strain of hepatitis E virus in cynomolgus monkeys. J Med Virol 43:135–142
134. Bouwknegt M, Lodder-Verschoor F, van der Poel WH, Rutjes SA, de Roda Husman AM (2007) Hepatitis E virus RNA in commercial porcine livers in the Netherlands. J Food Prot 70:2889–2895
135. Shiota T, Li TC, Yoshizaki S, Kato T, Wakita T, Ishii K (2015) Establishment of hepatitis E virus infection-permissive and -non-permissive human hepatoma PLC/PRF/5 subclones. Microbiol Immunol 59:89–94
136. Holla P, Ahmad I, Ahmed Z, Jameel S (2015) Hepatitis E virus enters liver cells through a dynamin-2, clathrin and membrane cholesterol-dependent pathway. Traffic 16:398–416
137. Dong C, Zafrullah M, Mixson-Hayden T, Dai X, Liang J, Meng J, Kamili S (2012) Suppression of interferon-α signaling by hepatitis E virus. Hepatology 55:1324–1332
138. Xu J, Wu F, Tian D, Wang J, Zheng Z, Xia N (2014) Open reading frame 3 of genotype 1 hepatitis E virus inhibits nuclear factor-κappa B signaling induced by tumor necrosis factor-α in human A549 lung epithelial cells. PLoS One 9:e100787

139. Drebber U, Odenthal M, Aberle SW, Winkel N, Wedemeyer I, Hemberger J, Holzmann H, Dienes HP (2013) Hepatitis E in liver biopsies from patients with acute hepatitis of clinically unexplained origin. Front Physiol 4:351
140. dos Santos DC, Neves PC, Azeredo EL, Pelajo-Machado M, Martinho JM, Pacheco-Moreira LF, Araújo CC, Cruz OG, de Oliveira JM, Pinto MA (2012) Activated lymphocytes and high liver expression of IFN-γ are associated with fulminant hepatic failure in patients. Liver Int 32:147–157
141. Melgaço JG, Soriani FM, Sucupira PH, Pinheiro LA, Vieira YR, Oliveira JM, Lewis-Ximenez LL, Araújo CCV, Pacheco-Moreira LF, Menezes GB et al (2016) Changes in cellular proliferation and plasma products are associated with liver failure. World J Hepatol 18:1370–1383
142. Kumar A, Beniwal M, Kar P, Sharma JB, Murthy NS (2004) Hepatitis E in pregnancy. Int J Gynaecol Obstet 85:240–244
143. Shalimar ASK (2013) Hepatitis e and acute liver failure in pregnancy. J Clin Exp Hepatol 3:213–224
144. Beniwal M, Kumar A, Kar P, Jilani N, Sharma JB (2003) Prevalence and severity of acute viral hepatitis and fulminant hepatitis during pregnancy: a prospective study from North India. Indian J Med Microbiol 21:184–185
145. Sehgal R, Patra S, David P, Vyas A, Khanam A, Hissar S, Gupta E, Kumar G, Kottilil S, Maiwall R et al (2015) Impaired monocyte-macrophage functions and defective toll-like receptor signaling in hepatitis E virus-infected pregnant women with acute liver failure. Hepatology 62:1683–1696
146. Lei Q, Li L, Cai J, Huang W, Qin B, Zhang S (2016) ORF3 of hepatitis E virus inhibits the expression of proinflammatory cytokines and chemotactic factors in LPS-stimulated human PMA-THP1 cells by inhibiting NF-κB pathway. Viral Immunol 29:105–111
147. Jacobsen KH, Koopman JS (2005) The effects of socioeconomic development on worldwide hepatitis A virus seroprevalence patterns. Int J Epidemiol 34:600–609
148. Fiore AE, Wasley A, Bell BP (2006) Prevention of hepatitis A through active or passive immunization: recommendations of the Advisory Committee on Immunization Practices (ACIP). MMWR Recomm Rep 55((Rr-7)):1–23
149. Vizzotti C, González J, Rearte A, Urueña A, Pérez Carrega M, Calli R, Gentile A, Uboldi A, Ramonet M, Cañero-Velasco M et al (2015) Single-dose universal hepatitis A immunization in Argentina: low viral circulation and high persistence of protective antibodies up to 4 years. J Pediatric Infect Dis Soc 4:e62–e67
150. Rao S, Mao JS, Motlekar S, Fangcheng Z, Kadhe G (2016) A review of immunogenicity and tolerability of live attenuated hepatitis A vaccine in children. Hum Vaccin Immunother 12:3160–3165
151. Barkai G, Belmaker I, Givon-Lavi N, Dagan R (2009) The effect of universal toddlers-only hepatitis A virus vaccination program on seropositivity rate in unvaccinated toddlers: evidence for reduced virus circulation in the community. Pediatr Infect Dis J 28:391–393
152. Melgaco JG, Morgado LN, Santiago MA, Oliveira JM, Lewis-Ximenez LL, Hasselmann B, Cruz OG, Pinto MA, Vitral CL (2015) A single dose of inactivated hepatitis A vaccine promotes HAV-specific memory cellular response similar to that induced by a natural infection. Vaccine 33:3813–3820
153. Shao ZJ, Zheng YJ, Zhang JX, Xiong CL, Lu YH, Ma WT, Xu DZ, Jiang QW (2008) An investigation on hepatitis E virus infection status among livestock in Xi'an area. Zhonghua Liu Xing Bing Xue Za Zhi 29:158–160
154. Colson P, Borentain P, Queyriaux B, Kaba M, Moal V, Gallian P, Heyries L, Raoult D, Gerolami R (2010) Pig liver sausage as a source of hepatitis E virus transmission to humans. J Infect Dis 202:825–834
155. Pezzoni G, Caminiti A, Stercoli L, Grazioli S, Galletti G, Santi A, Tamba M, Brocchi E (2014) Comparison of three in-house ELISAs for the detection of hepatitis E virus infection in pigs under field conditions. J Virol Methods 207:95–103
156. Martin-Latil S, Hennechart-Collette C, Guillier L, Perelle S (2014) Method for HEV detection in raw pig liver products and its implementation for naturally contaminated food. Int J Food Microbiol 176:1–8

157. Van der Poel WH (2014) Food and environmental routes of hepatitis E virus transmission. Curr Opin Virol 4:91–96
158. Andraud M, Casas M, Pavio N, Rose N (2014) Early-life hepatitis E infection in pigs: the importance of maternally-derived antibodies. PLoS One 9:e105527
159. Walachowski S, Dorenlor V, Lefevre J, Lunazzi A, Eono F, Merbah T, Eveno E, Pavio N, Rose N (2014) Risk factors associated with the presence of hepatitis E virus in livers and seroprevalence in slaughter-age pigs: a retrospective study of 90 swine farms in France. Epidemiol Infect 142:1934–1944
160. Backer JA, Berto A, McCreary C, Martelli F, van der Poel WH (2012) Transmission dynamics of hepatitis E virus in pigs: estimation from field data and effect of vaccination. Epidemics 4:86–92
161. Zhang J, Zhang XF, Huang SJ, Wu T, Hu YM, Wang ZZ, Wang H, Jiang HM, Wang YJ, Yan Q et al (2015) Long-term efficacy of a hepatitis E vaccine. N Engl J Med 372:914–922
162. Baylis SA, Crossan C, Corman VM, Blümel J, Scobie L, Dalton HR (2015) Unusual serological response to hepatitis E virus in plasma donors consistent with re-infection. Vox Sang 109:406–409
163. Vivek R, Chandy GM, Brown DW, Kang G (2010) Seroprevalence of IgG antibodies to hepatitis E in urban and rural southern India. Trans R Soc Trop Med Hyg 104:307–308
164. Tsarev SA, Tsareva TS, Emerson SU, Govindarajan S, Shapiro M, Gerin JL, Purcell RH (1997) Recombinant vaccine against hepatitis E: dose response and protection against heterologous challenge. Vaccine 15:1834–1838
165. Zhu FC, Zhang J, Zhang XF, Zhou C, Wang ZZ, Huang SJ, Wang H, Yang CL, Jiang HM, Cai JP et al (2010) Efficacy and safety of a recombinant hepatitis E vaccine in healthy adults: a large-scale, randomised, double-blind placebo-controlled, phase 3 trial. Lancet 376:895–902
166. Li SW, Zhao Q, Wu T, Chen S, Zhang J, Xia NS (2015) The development of a recombinant hepatitis E vaccine HEV 239. Hum Vaccin Immunother 11:908–914
167. Liu P, Du R, Wang L, Han J, Liu L, Zhang Y, Xia J, Lu F, Zhuang H (2014) Management of hepatitis E virus (HEV) zoonotic transmission protection of rabbits against HEV challenge following immunization with HEV 239 vaccine. PLoS One 9:e87600

Part III
Mosquito Borne Viral Diseases

Chapter 8
Dengue Virus and Other *Flaviviruses* (Zika): Biology, Pathogenesis, Epidemiology, and Vaccine Development

Ada M.B. Alves and Rosa M. del Angel

1 Introduction

The *Flaviviridae* family includes a variety of viruses that are distributed worldwide, some of which are associated with high morbidity and mortality. Because there are neither vaccines nor antivirals for most of the *Flavivirus* infections, study of the viral replicative cycle is relevant.

The *Flaviviridae* family comprises three genera: (i) the *Pestivirus*, which infects mammals, including cows and pigs, such as the bovine diarrhea virus 1; (ii) the *Hepacivirus*, which includes only the hepatitis C virus (HCV), an important cause of hepatitis and hepatocellular carcinoma in humans; and (iii) the *Flavivirus*, which contains more than 80 members. A number of *Flaviviruses* are pathogenic to humans and are transmitted via the bite of an arthropod vector (tick or mosquito) to produce an acute cytolytic infection. Examples of flaviviruses affecting humans are yellow fever virus (YFV), dengue virus (DENV), West Nile virus (WNV), Japanese encephalitis virus (JEV), Zika virus, and tick-borne encephalitis virus (TBEV). Most of them cause severe diseases in humans with complex pathologies that on occasions may have fatal results.

The first epidemic of DENV occurred in 1779–1780 in Asia, Africa, and North America. Initially, sporadic outbreaks of the disease were reported, only occurring in its benign form, known as dengue fever. However, after World War II, the infection with DENV spread to different parts of the world, and more than one serotype was

A.M.B. Alves (✉)
Laboratory of Biotechnology and Physiology of Viral Infections, Oswaldo Cruz Institute, Fiocruz, Rio de Janeiro, Brazil
e-mail: ada@ioc.fiocruz.br

R.M. del Angel (✉)
Department of Infectomics and Molecular Pathogenesis, Centro de Investigación y Estudios Avanzados del Instituto Politécnico Nacional (CINVESTAV-IPN), Mexico City, Mexico
e-mail: rmangel@cinvestav.mx

J.E. Ludert et al. (eds.), *Human Virology in Latin America*,
DOI 10.1007/978-3-319-54567-7_8

detected in the same population. This situation increased the number of cases of dengue fever and resulted in the appearance of the most severe form of the infection, known as dengue hemorrhagic fever or severe dengue. Nowadays, one-third of the world's population lives in areas at risk for infection, and more than 100 countries are endemic for dengue, reporting annually more than 400 million cases.

Dengue is endemic in virtually all Latin America (with the exception of Chile and Uruguay). According to the Pan American Health Organization (PAHO) (www.paho.org), in the year 2016, the region reported nearly 650,000 confirmed cases of dengue fever and more than 12,000 cases of severe dengue, with the circulation of all four serotypes. Given the importance of dengue and other *Flavivirus* diseases in the Americas, including the recent emergence of Zika virus, many researchers in the region have devoted their effort to the study and control of this disease. This chapter is aimed to review some of the most relevant findings and contribution made to the biology, epidemiology, and prevention of dengue and also that of Zika disease in the region.

2 Dengue Virus Replicative Cycle

2.1 Early Events

Because DENV is an arbovirus, it can replicate efficiently in mammalian and mosquito hosts. DENV is transmitted to humans by female mosquitoes of the genus *Aedes* during a blood meal. In this process, the virus is inoculated in the human skin where the virus comes in contact with several cell types, including skin-resident dendritic cells and macrophages. These cells are then transported to the regional lymph nodes where the virus infects other macrophages and monocytes, amplifying the infection. Different molecules have been described as DENV receptors in mammalian cells, among them, heparan sulfate [50] proteins associated with CD14 [51]; the lectin DC-SIGN [145, 203]; CLEC5A [49]; the heat shock proteins HSP70, HSP90, and GRP78 [101, 175]; cell receptors such as laminin [213] and mannose receptors; prohibitin [106]; and molecules related to lipid detection such as nLc4, Cer L-3, and TIM and TAM receptors [125]. The variety of cell receptors reported suggest that different receptors are used in different cell types and receptor redundancy use or that DENV uses a receptor complex formed by more than one protein in different steps of viral infection, such as attachment, internalization, and signaling pathway triggering. Later, the virus can be detected in remote lymph nodes to finally induce viremia. Viremia precedes the onset of clinical symptoms, and during this phase, the virus is disseminated to other organs such as the liver, spleen, and kidney.

During the viremia stage, humans can transmit the virus to a healthy mosquito. After the blood meal, DENV can infect epithelial cells of the midgut of the insect. The migration of the virus to the hemocele allows it to reach different organs such as fat bodies, the Malpighian tubules, and finally the salivary glands, where it is ready to be transmitted to humans. Again, glycoproteins of 40 and 45 kDa and polypeptides of 57, 67, and 80 kDa have been identified as DENV putative receptors in mosquito cells, although the final identity of those proteins has been elusive [106, 129, 130, 185, 186, 215, 228].

DENV uses lipid rafts to enter into the host cell, and most of the receptors described for DENV to date are associated with lipid rafts or are recruited to these sections of the membrane at the time of infection [173]. Recent work highlighted the importance of cholesterol and specifically of lipid rafts in the entry and signaling [48, 49] process of DENV in vertebrate cells. The results suggest that the integrity of lipid rafts is required during the infectious process of DENV in monocytes/macrophages in the absence or presence of facilitator antibodies, as well as in mouse neuroblastoma cells N18 [111, 173, 175]. Consistently, a significant reduction in DENV infection was detected in the liver cell line Huh-7 when cells were pretreated with drugs that prevent the formation of these membrane microdomains [199].

Virus attachment is followed by viral entry and decapsidation. Several studies indicate that viral entry occurs by a clathrin-mediated endocytosis [1–3, 138]. The low pH present in late endosomes induces the fusion between viral and endosomal membrane, inducing capsid release into the cytoplasm [33, 134]. Uncoating after viral entry is one of the least studied steps in the *Flavivirus* life cycle, but, recently, it has been described that the capsid is degraded after viral internalization by the host ubiquitin-proteasome system. However, neither the proteasome activity nor capsid degradation is necessary for viral genome release into the cytoplasm, suggesting that DENV capsid degradation is not responsible for genome uncoating. However, DENV genome release requires a non-degradative ubiquitination step [35].

2.2 *Viral Replication*

The dengue virus, as other RNA viruses, modifies the membranes of the endoplasmic reticulum to concentrate all the factors necessary for replication of the viral genome (replicative complex) [140, 166, 202]. The first step in DENV replication is translation of the viral genome, given that the positive polarity of the viral genome allows it to function as mRNA. This RNA contains a cap structure in the 5′-end and lacks a poly A tail in the 3′-end [118]. Translation of the viral genome occurs by a cap-dependent mechanism. However, in conditions where cap-dependent translation is inhibited, the DENV genome can still be translated. Because viral RNA is monocystronic, a polyprotein is synthesized that is cleaved by cellular and viral proteases to generate mature viral proteins. Next, the new synthesized nonstructural proteins initiate viral replication. RNA elements located within the 5′- and 3′-untranslated regions (UTR) regulate translation and genome replication [8, 85]. It has been described that within the 5′-UTR of about 100 nucleotides (nt) there are three regulatory elements: the stem-loop A (SLA), which is the viral polymerase promoter [77, 78, 114]; the stem-loop B (SLB), which contains the 5′ upstream of the AUG region (5′-UAR), complementary to the 3′-UTR (3′-UAR) that mediates viral RNA circularization [11]; and a U-rich spacer which enhances viral replication [114]. The 450-nucleotide-long 3′-UTR contains four domains: domain A1, containing a variable region (VR) [196]; domains A2 and A3, with a dumbbell-like secondary structure functioning as enhancer for viral replication [10, 119]; and domain A4, containing the small hairpin (sHP) and the 3′-stem-loop (3′-SL), all necessary for viral replication [221]. It has been extensively described that long-range RNA–RNA interactions between the 5′-cyclization sequence (5′-CS) with the 3′-CS and 5′-AUR (upstream AUG region) and 3′-UAR are necessary for efficient RNA replication [11]. To initiate viral replication, the viral polymerase NS5 interacts with the 5′-end SLA promoter and then moves to the 3′-end initiation site, located very close because of viral genome cyclization [77]. Recently, specific RNA sequences have been identified in the viral 3′-UTR that are essential for viral replication in mosquito cells but dispensable for replication in mammalian cells [77]. These studies provided direct evidence for host-specific functions of viral RNA elements and raised the question whether viral RNA structures are under specific selective pressures during host adaptation [220].

The size limitations of the viral genome require viruses to depend heavily on host cell factors to complete their replicative cycle successfully. In this regard, a number of viral proteins, including La, PTB, and PABP among others, have been identified to bind specifically to the DENV genome UTRs to effectively modulate the replication process [5, 63, 229, 230].

Other important elements for viral replication are the membranes from the endoplasmic reticulum (ER). During DENV infection, extensive rearrangements of the endoplasmic reticulum occur. Three different substructures have been described: (1) invagination of the ER membrane known as vesicle packets involved in viral replication (VPs), (2) virus "bags" where viral progeny accumulate, and (3) convoluted membranes (CMs) with an unknown function [224]. The nonstructural proteins NS1, NS2A, NS2B, NS3, NS4A, NS4B, and NS5 are located in the invaginations induced in the endoplasmic reticulum.

In mosquito cells, significant differences were observed in the intracellular lipid profile of DENV-infected cells compared to uninfected cells. These new lipids have the ability to alter the structural and functional characteristics of membranes, confirming the idea that infection promotes important rearrangements in membranes through alteration in lipid metabolism [102, 166].

In addition, it has been observed for different *Flavivirus* that if the amount of cholesterol present in these complexes is reduced, the viral genome replication is affected [211]. Furthermore, it has been shown that if the biosynthetic pathway of cholesterol is inhibited (using statins or siRNA against pathway protein cholesterol synthesis), or if capture of cholesterol from the medium (delipidated) is prevented, viral replication in cell lines A549 (lung carcinoma) and K562 (hematopoietic) decreases considerably, showing that cholesterol has a major role in this process [181, 199]. The dependence of DENV replication on cholesterol and lipid metabolism opens possibilities for antiviral treatments. Indeed, compounds that affect lipid metabolism such as nordihydroguaiaretic acid (NDGA) are demonstrated to be an effective inhibitor of DENV replication in vitro [198].

2.3 *Virus Morphogenesis*

The DENV structural proteins C, prM, and E are synthesized in the ER; however, C protein accumulates progressively around cellular organelles named lipid droplets (LDs) during infection [99]. LDs are formed by sphingolipids and cholesterol esters and are located close to the ER. These organelles have been involved in viral assembly. Interestingly, the number of lipid droplets per cell increases after infection, linking lipid droplet metabolism and viral replication. Specific hydrophobic amino acids, located in the center of the capsid protein, have been identified as key elements for lipid droplet association [188]. Surprisingly, the N-terminus of C is necessary for efficient particle formation in mosquito cells, but they are crucial for propagation in human cells, suggesting that this function of C is differentially modulated in different host cells [187].

These findings are consistent with the fact that the pharmacological inhibition of fatty acid synthase (FASN) mediated by C75 promotes a significant inhibition of DENV morphogenesis [170, 188].

Once viral RNA is associated with C, the nascent particle buds into the ER where it acquires the viral membrane containing the C and prM proteins. The immature virions traffic through the trans-Golgi secretory pathway and along this pathway; the prM protein is cleaved by host furin-like proteases to generate mature virions [112, 231]. Along the mature virions, soluble NS1 protein is also secreted to the extracellular milieu. Recently, it was reported that NS1 is secreted also from infected mosquito cells and not only by vertebrate cells, as previously supposed [7].

3 Virus–Host Interactions

As was described earlier, DENV infection can induce a mild disease or can cause a more severe form of the infection called severe dengue. Severe dengue is characterized by the rapid onset of capillary leakage and is accompanied by significant thrombocytopenia and mild-to-moderate liver injury. Hemorrhagic manifestations include bleeding in the skin and gastrointestinal tract. Although the pathogenesis of the severe forms of dengue infection has been broadly studied, the process is not yet fully understood [52, 121, 150]. Secondary infections are recognized as one of the most important risk factors for the development of severe dengue by a complex mechanism known as antibody-dependent enhancement (ADE) of viral infection. It has been postulated that during secondary infection, the antibodies generated during primary infection are able to form virus–antibody complexes that infect Fc-bearing cells such as human monocytic and dendritic cells, through the Fc receptor.

This mechanism is responsible for an increase in the proinflammatory cytokine response, which has the capacity to disturb the apical junction complex in vitro and to cause an increase in vascular permeability in vivo [174]. Indeed, the association between aberrant cytokine levels and dengue severity has long been apparent; past and recent work carried out in Brazil and other parts of the region has contributed greatly to understand the cytokine profiles in sera of patients with dengue and its association with disease severity [22, 76, 91].

Another key element of the antiviral response to DENV is type 1 interferon (IFNα/β). IFN type 1 is secreted very early after DENV infection by mammalian cells, and it has an important function in protection against viral infections [84, 169]. It has been reported that DENV triggers but also counterattacks many of the signaling pathways involved in the induction of a robust IFNα/β response [5, 6, 137, 141, 142, 160, 177, 178]. In concordance with the antiviral activity of IFN type 1, IFN-α levels in patient sera are rapidly modulated after fever onset, and a better clinical condition correlates with higher IFN-α levels, supporting the idea that IFN response has a role in the pathogenesis of DENV [62].

Special mention of the work in dengue conducted in Latin America shall be made to the dengue cohort study being carried out in Nicaragua to study the natural history and transmission of dengue in children. This ambitious study, carried out for more than 10 years, has enrolled thousands of children and has worked in collaboration with public authorities. Among other findings, the Nicaraguan cohort study had provided evidence for the role of neutralizing antibodies in protection against dengue and the role of secondary infections as a risk factor to develop severe dengue [103, 222].

4 Epidemiology of Dengue in Latin America

Dengue reports in the Americas date back to the nineteenth century. In the first moiety of the twentieth century, in Brazil as in other countries of Latin America, the mosquito *Aedes (Ae.) aegypti* was eradicated after a program of the PAHO to control yellow fever, another *Flavivirus* transmitted by the same vector [40]. Unfortunately, this program was discontinued, which led to reinfestation of *Ae. aegypti*, and dengue became one of the most important infectious diseases in Latin America.

As a consequence of this scenario, several efforts have been made to map the epidemiological situation of dengue in the different countries, improve diagnostic tests, identify the circulating serotypes, and better understand the disease with its different forms. The scientific community also focused on studies concerning virus biology, interaction between host and pathogen, and vaccine development. The contribution of Latin America in dengue research was recently analyzed in a bibliometric study, revealing that Brazil was the highest contributor (31.2%), followed by Puerto Rico (12.9%) and Mexico (10.7%) [219].

Epidemiological inquiries have been performed in different regions in Latin America. In 1963 DENV3 was isolated in Jamaica and disseminated to other

Caribbean countries [27]. Later on, the emergence of DENV2 in Cuba in 1981 represented a mark in the epidemiology of dengue in America, with several reported cases of dengue hemorrhagic fever and dengue shock syndrome (DHF/DSS). Investigations revealed that most of the DHF cases presented antibodies against DENV1, which was responsible for the epidemic of 1977 [105]. Also in 1981, DENV4 was introduced in eastern Caribbean islands and spread to the rest of the Caribbean, Mexico, and Central and South America [167]. In 1989 a large dengue epidemic was reported in Venezuela with many cases of DHF [161]. In general, the epidemics of dengue in the Americas occur with recurring peaks of cases at 3- to 5-year intervals. More upsetting is the fact that, over time, peaks become progressively higher, with more cases of reported severe forms of the disease [189]. Nowadays, the four dengue serotypes circulate in several countries, which increases the risk of DHF.

The greatest epidemic in the Americas was in 2013, when almost 2.4 million dengue cases were notified, and approximately 1.6% of them evolved to severe forms of the disease [162]. In that year, Brazil, Mexico, Colombia, and Paraguay had the most important outbreaks, totalizing 83% of the dengue cases in America. Brazil leads the rank of dengue fever cases, with an incidence ranging from 313.8 (in 1998) to 722.4 (in 2013) per 100,000 inhabitants [206].

In Brazil, the first outbreak of dengue with laboratory confirmation was in 1981 in a city in the north of the country, Boa Vista, with simultaneous occurrence of DENV1 and DENV4. This episode was immediately controlled, and the virus did not spread to other regions [156]. Dengue only became a health problem after the epidemic of 1986, when serotype 1 was introduced in the state of Rio de Janeiro and spread to different regions [193]. Since then, dengue has become endemic in Brazil, with explosive epidemics marked by the introduction of DENV2 in 1990, DENV3 in 2000, and DENV4 in 2010 [148, 149, 168, 208]. Epidemiological studies were performed to investigate serotype prevalence and virus distribution, not only by regions but also according to age [29, 30, 95, 104, 204, 207]. These works revealed, for instance, a shift in the age pattern of DHF with more children younger than 15 years being affected [205, 207]. A similar scenario was also observed in Venezuela [210].

More recently, studies have been conducted in many countries in Latin America to map the dengue epidemiological situation in these areas and to access the efficacy of vaccines against it [60]. These studies revealed, for instance, that there is a substantial underreporting of dengue in the epidemiological surveillance systems from Brazil, Colombia, and Mexico [192]. Unapparent infections may also be important in the dengue epidemiology, and scientists have started to investigate this as a possible source for mosquito transmission, which would impact the disease burden [90]. Following the same line, other investigations have been performed to establish the burden of dengue and its economic cost in different countries such as Brazil, Colombia, Nicaragua, Mexico, and even Argentina, where dengue is present but at lower magnitude compared to other regions in America [44, 120, 212, 225, 226].

Other works focused on phylogeny studies, aiming to understand the dynamic of DENV population and the origin of the different virus isolates, by comparing sequences obtained from distinct countries in Latin America and all over the world

[9, 13, 43, 57, 70, 153, 227]. These investigations revealed, for example, that the DENV2 that emerged in Brazil in 1990 continued to circulate until 2003, whereas in 2007 a new DENV2 was isolated that belonged to a different genotype, thus suggesting that this virus did not evolve locally but was rather caused by a new introduction, probably coming from the Caribbean [74]. Similar observation occurred in Peru, also with DENV2 [57].

In parallel with these studies, much effort has been expended to improve the clinical and laboratory dengue diagnostic [4, 16, 34, 56, 59, 116, 157, 171]. Reports have shown, for instance, the incidence of dengue infection by blood transfusion and renal transplant [20, 53]. Recently, it was performed a prospective study with a large cohort of patients from several countries in Latin America and Asia, aiming to differentiate between dengue and other common febrile illness and to identify parameters associated with the progression to severe forms of the disease [100]. These works will certainly help to update guidelines for diagnostics and treatment of dengue all over the world.

Other investigations were performed with postmortem materials of suspected dengue cases to establish/improve diagnostics [19, 46]. Studies with samples from confirmed fatal dengue cases have also been reported [15, 147, 152, 158, 159, 172, 176, 184, 209]. All these works provided important information about the pathogenesis of the disease, especially regarding the severe forms of dengue, and helped to map the target organs/cells for virus replication. It showed, for instance, the commitment of some organs that are not commonly associated with dengue, such as the heart and kidney [159, 172, 209], as well as the reinforcement of the involvement of the central nervous system during the disease [15, 147].

5 Sylvatic Dengue Cycle

DENV exists as sylvatic and urban cycles in Africa and Asia [223]. Investigations to identify a sylvatic cycle for DENV in the Americas or to find evidence of infection in neotropical forest mammals have yielded contradictory results. Molecular and serological evidence for DENV infection in opossums and especially in several species of bats has been reported in studies conducted in French Guiana and Mexico [64, 200]. Yet, other studies conducted with bats also collected in Mexico have failed to find evidence that bats can sustain DENV replication [36, 37].

6 Vaccines

The development of a vaccine against dengue has been pursued since the studies of Sabin in 1945 [183]. One of the major difficulties in this field is to achieve a vaccine against the four dengue serotypes. Studies all over the world showed that infection with one serotype promotes long-term immunity to this serotype, but protection to heterologous infection is only transient. In fact, results revealed an increased risk for severe forms of dengue during a second infection with a heterologous serotype [80]. It seems that the immunity to a specific serotype may induce an uncontrolled immune response during a secondary infection, leading to the DHF/DSS.

One hypothesis to explain the role of immune response in the development of severe forms of the dengue disease is the antibody-dependent enhancement (ADE) of virus replication. In this phenomenon, antibodies generated by the first dengue infection, mainly against the E and prM proteins, bind to the heterologous virus serotype in the second infection but without neutralizing its ability. Instead, the antibody–virus complex can bind to Fc receptors present in monocytes/macrophages and dendritic cells, which are the primary target cells in dengue infection, facilitating, therefore, virus entrance and replication [93]. Another hypothesis, not necessarily exclusionary, is based on the cellular immune response, named original antigenic sin. In this case, cross-reactive memory T cells generated by the first infection are preferentially activated during the second dengue infection. However, these low-affinity T cells are unable to clear infection and can cause an uncontrolled cytokine production (the cytokine storm), which finally results in the plasma leakage that is characteristic of the DHF/DSS [124].

Consequently, it is a consensus that an effective dengue vaccine has to be tetravalent; otherwise, immunization against one serotype could increase the risk for more severe forms of the disease in individuals infected with other serotypes. Different vaccines are now being tested in clinical trials in Latin America. One of these vaccines, produced by Sanofi Pasteur, is based on the recombinant life-attenuated yellow fever (YF)/dengue virus, in which the E and prM genes from YF are substituted by the genes from each DENV serotype. This tetravalent vaccine, CYD, was tested in Asia and Latin America. In Latin America, it was tested in Brazil, Colombia, Peru, Honduras, Mexico, and Puerto Rico [61, 68, 92, 107]. Unfortunately, the efficacy of this vaccine was below expectations, as protection against DENV2 was not achieved in several children. In general, the efficacy of the vaccine was significantly low in children 2 to 5 years of age, especially in individuals with no previous DENV infection. Also, the vaccination regimen involves three doses given with a 6-month interval, which can have logistical difficulties for its administration. This may be a problem in dengue endemic areas because it can leave the population more susceptible to the development of severe disease until the immunological protection is complete. In fact, a statistically significant increase of hospitalization among vaccinated children (2–5 years old) was observed [92]. Because of this, the Sanofi Pasteur vaccine against dengue was recently licensed for commercial use in Brazil and Mexico only in individuals aged from more than 9 years to 45 years, which does not include an important segment of the population at risk for the development of severe dengue [96]. Although results of the clinical trials with this vaccine were quite disappointing, these studies were the most robust performed to date, and they revealed, for instance, that only neutralizing antibodies seem to be not correlated to protection. In fact, vaccinated children presented high levels of neutralizing antibodies to DENV2, but they were not protected against this virus serotype [182].

Another tetravalent vaccine, developed by the National Institute of Health (NIH), is also being tested in Latin America. This vaccine is composed of a mixture of attenuated recombinant virus, obtained by the deletion of 30 nucleotides in the 3′-UTR of DENV1, DENV3, and DENV4 and by the substitution of prM and E genes from DENV4 to those genes from DENV2 [72–74]. This vaccine was tested first in clinical trials in the USA in *Flavivirus*-naïve healthy adults. It was well tolerated, although production of neutralizing antibodies seemed to be associated with the occurrence of rashes. Results suggested that this vaccine could be administered in one single dose, because antibody levels did not increase after the second dose [73]. This vaccine has recently entered in phase 3 clinical tests in some countries in Latin America, especially Brazil in a partnership with Butantan Institute, but results are not yet available.

The Takeda tetravalent dengue vaccine (TDV), in its turn, was constructed by using the life-attenuated DENV2 derived by serial passages of wild-type virus in primary dog kidney cells (PDK). The prM and E genes from this attenuated DENV2 virus, D2 PDK-53, were substituted by those genes from the other virus serotypes [98]. The tetravalent vaccine is composed of the attenuated DENV2 and the three chimeric DENV1, 3, and 4. This vaccine was tested in Puerto Rico and Colombia, and it was well tolerated and immunogenic for all serotypes in volunteers from 1.5 to 45 years of age [197]. In addition, one inactivated vaccine against dengue will soon be tested in Brazil in studies performed by the Institute of Technology in Immunobiologicals (BioManguinhos), from the Oswaldo Cruz Foundation (Fiocruz). This vaccine was developed by GSK, and it will be administered with adjuvants.

Besides the clinical trials of vaccines against dengue, several preclinical investigations have been performed all over the world, including in Latin America. Studies using life-attenuated recombinant virus were also conducted in Brazil, by constructing several chimeric YF/DENV [45, 81, 123]. This strategy was similar to that of the dengue vaccine developed by Sanofi Pasteur, although analyses were only focused on mice and nonhuman primates.

Other strategies used DNA vaccines encoding structural as well as nonstructural dengue proteins. Usually, the vaccines focused on the E protein together with the prM, which works as a chaperonin for the correct folding of the E protein [65, 113, 167]. However, one study from Brazil showed that a DNA vaccine encoding the ectodomain (domains I, II, and III) of the DENV2 E protein, fused to a strong signal peptide, was able to induce high levels of protection in mice challenged with a lethal virus dose [24]. Furthermore, the authors showed that the combination of this DNA vaccine with a chimeric YF/DENV2, constructed in Brazil [45], generated 100% protection in mice with induction of a synergetic neutralizing antibody response [23]. This study also pointed that the cellular immune response elicited by the DNA vaccine was significantly higher when compared to immunization with the chimeric attenuated virus, which may be important for protection and can explain, in part, the low efficiency of the Sanofi Pasteur vaccine. Other investigations focused on DNA vaccines encoding the DENV NS1 or NS3 proteins [54, 55], which also elicited protection. Interestingly, one of these studies showed that the cooperation between CD4+ T cells and antibodies, more than CD8+ T lymphocytes, was crucial for protection induced by a DNA vaccine containing the NS1 gene [89]. Reports of DNA vaccines using the entire E and NS1 gene together or only the domain III of the E protein have also been published [126, 127, 139].

Studies with subunit vaccines have been conducted in different countries in Latin America. Some work has been described in Cuba using the recombinant C protein expressed in *Escherichia coli*, which was tested in mice [86, 109, 115]. The importance of CD4+ and CD8+ T-cell response in the protection elicited by this vaccine was also reported [87]. Other research was based on tetravalent formulations, combining the domain III of the envelope protein to the capsid protein, which similarly induced protection in mice and nonhuman primates [201]. Further, heterologous prime-boost protocols were tested using purified proteins and infective virus [214]. In Mexico, researchers tested the ability of different domains of the envelope protein to induce protection, alone or in combination with the NS1 [83]. Following the same line, studies in Brazil have shown that purified recombinant NS1 protein induced protection in mice in combination with detoxified heat-labile toxin from enterotoxigenic *E. coli* as adjuvant [12]. In Argentina, the E protein was expressed in a plant system to use it as a subunit vaccine or in a diagnostic kit [122]. Another report from Mexico explored the approach of expressing one peptide from the NS3 protein on the surface of *Salmonella* and showed its ability to induce a strong cytotoxic T-cell response [117].

Another difficulty in the development of an effective vaccine against dengue is the lack of an ideal experimental animal model that mimics the disease in all its forms, as we observed in humans. Several studies have been performed with immunodeficient mice, which can develop clinical signs similar to those observed in humans [232, 233]. However, although these animal models are extremely valuable for studying the disease, their use for vaccine tests is controversial because the elicited immune response can differ significantly from that observed in immunocompetent individuals. The most traditional immunocompetent mouse model for testing vaccines against *Flavivirus*, including dengue, is the use of brain-adapted virus inoculated by the intracerebral (i.c.) route. Albeit this is not the natural route of infection in humans; studies with dengue patients have revealed more and more the commitment of the central nervous system during the disease [15, 147]. Also, this model provides a straightforward readout parameter for vaccine testing because virus inoculation is usually lethal. Moreover, recently a study in Brazil was published showing the systemic effect of the virus infection by the i.c. route. Authors detected viremia in these animals, especially in late stages of infection, induction of T-cell responses, and tissue damages in peripheral organs, such as the liver [154]. In another work, researchers investigated the use of immunocompetent mice inoculated with macrophages infected in vitro with a DENV isolate, not laboratory adapted. They observed some aspects of the virus tropism described in humans, with detection of the DENV genome in the same organs [26].

7 Studies with Zika Virus

Apart from the studies with dengue, much research has begun with the Zika virus (ZIKV) in some countries in Latin America, especially in Brazil. The interest toward ZIKV by the scientific community in Latin America is a consequence of the huge health problem we have been experiencing since the beginning of 2015.

This virus also belongs to the *Flaviviridae* family, genus *Flavivirus*. It has a typical *Flavivirus* organization: an enveloped virus with a single-stranded positive-sense RNA genome. The RNA is translated into a polyprotein that is cleaved, generating three structural (C, prM, and E) and seven nonstructural (NS1, NS2A, NS2B, NS3, NS4A, NS4B, and NS5) proteins. The serological diagnosis for ZIKV is not fully conclusive because a significant proportion of tested samples showed cross-reactivity to other viruses, especially DENV. Because of this, laboratory testing generally includes polymerase chain reaction (PCR) assays. Transmission of the ZIKV occurs predominantly via the bites of *Aedes* mosquitoes, mainly *Aedes aegypti*. Moreover, transmission from the mother to the fetus via the placenta was recently established [32, 133]. Sexual transmission has also been reported [69, 97]. More recently, Baca-Carrasco and Velasco Hernández [25] performed a mathematical study to analyze the effects of sexual transmission and migration in the spread of the ZIKV. They concluded that transmission through sexual contact was insufficient to influence the spread of the disease as we observed in Latin America, although it may have affected the magnitude and duration of outbreaks. On the other hand, migration was decisive for the rapid spread of this virus. In addition, the viral RNA and infective particles have been detected in the saliva and urine of ZIKV-infected patients [28, 143] and are now being used as diagnostic tools.

The ZIKV was first isolated in 1947 in a sentinel monkey for monitoring yellow fever in a forest in Uganda named Zika. After the first isolation of the ZIKV, symptomatic cases of infection with this virus were reported in some African and Southeast Asiatic countries. However, little attention was paid to these cases because infection with ZIKV was considered to lead to only mild symptoms. The situation changed after the first recognizable outbreak of ZIKV in Micronesia in 2007, characterized as a DENV-like disease [71]. The second ZIKV outbreak occurred in French Polynesia, affecting approximately 28,000 individuals, in which an increase in the number of cases of Guillain–Barré syndrome (an immunomediated neuropathy that can cause paralysis) was reported [42]. More dramatic was the outbreak of this virus in 2015 in Brazil, after which several cases of microcephaly (occipital frontal circumference below the mean for age and gender, related to serious development problems in the children) was reported in babies whose mothers were infected during pregnancy [94, 132]. In adults, infection usually leads to symptoms as low-grade fever, arthralgia, rash, headache, and myalgia, although most infections are asymptomatic [31, 206].

The autochthonous transmission of ZIKV in Brazil was first reported in the northeast of the country in March 2015, almost simultaneously by two research groups [41, 222]. The first suspicion was that this virus was introduced in Brazil during the World Cup soccer competition in 2014; however, no countries participating in this competition were endemic for ZIKV. Another possibility is that the ZIKV had entered Brazil during the Va'a World Spring Championship canoe in August 2014, when four Pacific countries participated in this event, including French Polynesia [144]. Phylogenetic studies with samples of isolated virus from patients in Brazil also support this hypothesis. In fact, the ZIKV that is circulating in Brazil belongs to the Asian clade and shares 97–100% identity with the virus lineages isolated during the outbreak of 2013 in French Polynesia [38, 88].

After introduction of ZIKV in Brazil, it soon spread throughout Latin America. Colombia was the second country to report circulation of ZIKV in 2015 [21, 39, 146, 191], followed by Mexico, Panama, Haiti, and Puerto Rico [18, 66, 67, 110]. In 2016 many other regions confirmed autochthonous cases of this virus, totalizing 48 countries and territories in the Americas [163].

In Brazil, after the emergence of the ZIKV, it was observed that there was a 20-fold annual increase in the number of microcephaly cases [155, 163, 194]. According to the Brazilian Ministry of Health, between October 2015 and May 2016, a total of 7534 suspected cases of microcephaly and other congenital malformation of the central nervous system (CNS) have been reported [163]. Detection of the virus in pregnant women showing infection symptoms, as well as in amniotic fluid, placenta, and the brains of newborns, intensely reinforced the correlation between infection with ZIKV and malformations of the CNS in newborns, including microcephaly [38, 133, 155]. Mlakar and collaborators [133] published the first data that indicated a strong relationship between ZIKV and microcephaly, describing the case of a Slovenian woman who lived temporarily in the Northeast of Brazil and presented symptoms of the virus infection (febrile illness with rash) at the end of the first trimester of pregnancy. Ultrasonography performed at 29 weeks of gestation revealed microcephaly with calcifications in the fetal brain. The ZIKV RNA was detected in the fetal brain tissue, thus confirming virus transmission from the mother to the fetus. One study with a Brazilian cohort of 88 pregnant women reported that 72 were positive for ZIKV (82%), most of them showing fetal abnormalities [32]. Further, this investigation demonstrated that the fetal abnormalities can happen even when infection with ZIKV occurs after the first trimester of pregnancy. Several other studies also reported the association of ZIKV infection and microcephaly [14, 47, 136, 217]. In one report, the transplacental transmission of ZIKV was evidenced not only by detection of the viral protein and RNA in placental tissues but also by its effects leading to placentitis [151]. The neurotropism of ZIKV was observed by the detection of viral proteins in glial cells and observation of scattered foci of microcalcifications in the fetal brain tissues.

More recently, the term congenital Zika syndrome (CZS) has been preferably used, because it was observed that microcephaly is only one of the clinical signs of this congenital malformation disorder. The clinical features of the CZS are a consequence of direct neurological damages and severe intracranial volume loss. Although the cognitive, sensory, and motor disability components of this syndrome can be shared by other congenital infections, some features seemed to be characteristic: (1) severe microcephaly with partially collapsed skull, (2) thin cerebral cortices with subcortical calcifications, (3) macular scarring and focal pigmentary retinal mottling, (4) congenital contractures, and (5) marked early hypertonia with symptoms of extrapyramidal involvement [135]. Neurological examination of affected infants has shown hypertonia and spasticity, irritability manifested by excessive crying, dysphagia, and, less frequently, hypotonia [194]. A detailed study described the prenatal evolution and perinatal outcomes of 11 neonates who showed developmental abnormalities and neurological damage associated with ZIKV infection in Brazil [128]. The ZIKV was detected in the amniotic fluid, placenta, and cord blood for all patients, as well as from some neonatal tissues collected post mortem. Most of the infants presented with microcephaly, although the authors also found newborns presenting with severe brain lesions with a normal cephalic perimeter. They observed variable injuries as the consequence of brain lesions related with the virus infection, with a common pattern of brain atrophy and changes associated with disturbances in neuronal migration. Some patients showed mild brain atrophy and calcifications, whereas others presented severe malformations, including the absence of the thalamus and lissencephaly [128]. Histopathological and immunohistochemical analysis of tissues from two postmortem babies revealed multiple small foci of calcification and degenerate nerve cells in the brainstem, histiocyte and microglial proliferation, and gliosis, as well as neuronal and axonal degeneration. Ocular abnormalities were also observed, mainly paresis of the oculomotor and abducens muscles with convergent strabismus and loss of photomotor and consensual reflexes. In fact, different studies have been showing that the ZIKA can cause severe injury in the retina [216–218]. The retinal damages include mild to severe macular pigmentary changes and chorioretinal atrophy [216, 217]. Miranda and collaborators [131] described for the first time vascular changes and hemorrhagic retinopathy probably associated with the intrauterine infection with ZIKV. On the other hand, Ventura and collaborators [218] have evaluated the eyes of eight infants whose mothers were infected with ZIKV during pregnancy. Optical coherence tomography technology showed severe involvement of the neurosensory retina, including the internal and external layers and the choroid in most eyes, indicating severe visual impairment in these newborns.

Besides the CZS, infection with the ZIKV is also associated with neurological disorders in adults. Several cases of Guillain–Barré syndrome (GBS) have been reported after infection with ZIKV in Brazil [17, 31, 79, 164, 179]. One study with two cases from Salvador, Bahia, reported the development of ascending paresis after an acute exanthematous illness, evolving later to tetraparesis and cranial nerve palsy, which resolved after intravenous administration of human immunoglobulin [180]. The studies in Brazil supported the association of GBS and ZIKV infection. Furthermore, they served as an alert to other countries in Latin America, where the virus spread recently, of the potential risk not only for CZS in babies but also for neurological commitments in adults and the need for timely detection, diagnosis, and treatment to prevent mortality and long-term sequelae. In fact, from April 2015 to March 2016, a total of 164,237 confirmed or suspected cases of ZIKV disease and 1,474 cases of GBS were reported in Bahia-Brazil, Colombia, the Dominican

Republic, El Salvador, Honduras, Suriname, and Venezuela [190]. Unfortunately, part of these cases progressed to death, as was reported in one study in Colombia, in which 4% of patients with GBS died after respiratory failure and sepsis [165]. Moreover, although epidemiological studies revealed that females had a 75% higher incidence rate of ZIKV disease than males, the greater apparent risk for developing GBS is in males (28% more incidence of GBS among males than among females) [190]. The development of GBS after ZIKV infection probably involves an autoimmune process as described for other viral infections. However, some reports have showed that the development of GBZ after ZIKV infection may follow the pattern of a para-infectious disorder rather than the classic post-infection profile. Actually, a study in Colombia with 66 patients with GBZ revealed that 48% of these individuals had a rapid onset of neurological symptoms without an asymptomatic period after ZIKV infection symptoms (para-infectious) [165]. There are different hypotheses to explain such a scenario. One of them is that the immune molecular mimicry process against the nervous system may initiate before clinical symptoms of the ZIKV infection appear [165]. Additionally, it has been speculated that simultaneous epidemics of DENV and ZIKV may predispose the development of GBS as a result of sequential virus infection and stimulation of the immune system, triggering to an immunopathogenic process [180]. If this is the case, one must pay attention in the developing of vaccines against both *Flavivirus* because exanthematous immunization against one virus may impact in the development of disease caused by another virus.

In an attempt to understand the mechanism by which the ZIKV leads to microcephaly and other malformations in the fetal brain, researchers started to investigate the effect of the virus infection in human neural stem cells, growing as neurospheres and brain organoids [82]. They showed that ZIKV targeted these cells, reducing their viability and growth, which suggests that the virus abrogates the neurogenesis during brain development. However, there are several gaps in this field, and many studies will probably be performed in the future to answer these questions.

In addition, efforts have been made to establish experimental animal models to study the effect of ZIKV infection in several aspects, including intrauterine infections. Most of these studies are based on immunodeficient mice, in particular those lacking type I and II interferon receptors [195]. Similar to the studies with DENV, such a model is useful for investigations concerning the pathogenesis of the Zika, but its use for vaccine tests is controversial. Another investigation was performed with newborn Swiss mice infected with a ZIKV isolated in Brazil [75]. Inoculation of these mice with ZIKV by the intracerebral route led to severe cerebral lesions, with neuronal death, presence of apoptotic bodies, and degeneration of white matter. When the animals were infected by the subcutaneous route, the authors observed moderate cerebral lesions, morphologically similar to those was found in the previous group and additional myelopathy, with architectural loss, marked by neuronal death and apoptotic bodies. In another study, Cugola and collaborators [58] have shown that the association of birth defects, also using a ZIKV isolated in Brazil, depends on the mouse strain. These authors demonstrated that the offspring of immunocompetent pregnant C57BL/6 mice injected intravenously with ZIKV were not infected, indicating that the virus did not cross the placenta barrier. On the other hand, pups from SJL pregnant females infected with ZIKV presented severe intra-uterine growth restriction, resembling that which we observed in humans, including signs of microcephaly.

The establishment of animal models for ZIKV infection has a direct impact in the development of vaccines against this virus. After the dramatic outbreak of such virus in Latin America, much effort has been done toward the development of a vaccine against ZIKV. Several investigations have been conducted by different groups all over the world. One study was performed by Brazilian researchers in collaboration with groups in the U.S., using DNA vaccines or purified inactivated virus [108]. The DNA vaccine based on the full-length prM/E proteins conferred protection against ZIKV in the murine model of SJL mice previously described [58]. However, protection was only measured by absence of viremia and production of antibodies against the E protein in this animal. Other report of DNA vaccines against ZIKV encoding the prM/E proteins was also published [69]. The immunogenicity of such vaccines was tested in mice and nonhuman primates, and protection was evaluated by the lack of viremia in these animals. Clinical trials of some of these DNA vaccines, as well as a ZIKV purified inactivated vaccine, are already ongoing or are about to start in the U.S. [234]. The speed of which these researches have been conducted is a very positive point. However, many studies are still necessary to obtain an efficient and safe vaccine against ZIKV.

8 Conclusions and Future Challenges

Dengue continues to be a major public health problem in the Americas despite efforts and control actions carried out by public health authorities. Research and work carried out in the region have contributed significantly to the understanding of the virus biology, the pathogenesis, the epidemiology, and the diagnosis of this important and burdensome disease. Moreover, researchers in the region have helped in the development of the current licensed dengue vaccine and are also participating in the development of future vaccines and antiviral therapies. Challenges for the future include further understanding of the virus replicative cycle, the pathogenesis of severe dengue, and the immune response to infection to pave the way for effective patient intervention strategies and improved vaccines. Also, a better understanding of the virus–mosquito relationship is needed to implement effective and sustainable mosquito control measurements. Finally, the emergence of Zika virus in the continent, with its severe complications for adults but especially for infected newborns, poses formidable challenges for the region that require urgent attention. In particular, the extensive cross-reactivity between dengue and the Zika virus makes it important to investigate possible effects that vaccination against one *Flavivirus* may have upon the other, in terms of protection and/or pathogenesis, especially in a region where both viruses circulate concomitantly.

References

1. Acosta EG, Castilla V, Damonte EB (2009) Alternative infectious entry pathways for dengue virus serotypes into mammalian cells. Cell Microbiol 11:1533–1549
2. Acosta EG, Castilla V, Damonte EB (2012) Differential requirements in endocytic trafficking for penetration of dengue virus. PLoS One 7:e44835
3. Acosta EG, Castilla V, Damonte EB (2011) Infectious dengue-1 virus entry into mosquito C6/36 cells. Virus Res 160:173–179
4. Acosta PO, Granja F, Meneses CA et al (2014) False-negative dengue cases in Roraima, Brazil: an approach regarding the high number of negative results by NS1 ag kits. Rev Inst Med Trop Sao Paulo 56:447–450
5. Agis-Juarez RA, Galvan I, Medina F et al (2009) Polypyrimidine tract-binding protein is relocated to the cytoplasm and is required during dengue virus infection in Vero cells. J Gen Virol 90:2893–2901
6. Aguirre S, Maestre AM, Pagni S et al (2012) DENV inhibits type I IFN production in infected cells by cleaving human STING. PLoS Pathog 8:e1002934
7. Alcala AC, Medina F, Gonzalez-Robles A et al (2016) The dengue virus non-structural protein 1 (NS1) is secreted efficiently from infected mosquito cells. Virology 488:278–287
8. Alcaraz-Estrada SL, Yocupicio-Monroy M, del Angel RM (2010) Insights into dengue virus genome replication. Future Virol 5:575–592
9. Alfonso HL, Amarilla AA, Goncalves PF et al (2012) Phylogenetic relationship of dengue virus type 3 isolated in Brazil and Paraguay and global evolutionary divergence dynamics. Virol J 9:1–16
10. Alvarez DE, De Lella Ezcurra AL, Fucito S, Gamarnik AV (2005) Role of RNA structures present at the 3'UTR of dengue virus on translation, RNA synthesis, and viral replication. Virology 339:200–212
11. Alvare DE, Lodeiro MF, Luduena SJ, Pietrasanta LI, Gamarnik AV (2005) Long-range RNA-RNA interactions circularize the dengue virus genome. J Virol 79:6631–6643
12. Amorim JH, Diniz MO, Cariri FA et al (2011) Protective immunity to DENV2 after immunization with a recombinant NS1 protein using a genetically detoxified heat-labile toxin as an adjuvant. Vaccine 30:837–845
13. Aquino VH, Amarilla AA, Alfonso HL et al (2009) New genotype of dengue type 3 virus circulating in Brazil and Colombia showed a close relationship to old asian viruses. PLoS One 4:1–8
14. Aragão MFV, van der Linden V, Brainer-Lima AM et al (2016) Clinical features and neuroimaging (CT and MRI) findings in presumed Zika virus related congenital infection and microcephaly: retrospective case series study. BMJ 353:i1901
15. Araújo FM, Araújo MS, Nogueira RM et al (2012) Central nervous system involvement in dengue: a study in fatal cases from a dengue endemic area. Neurology 78:736–472
16. Araújo JM, Gomes GM, Costa Faria NR (2012) Evaluation of a generic RT-nested-PCR for detection of flaviviruses in suspected fatal cases of dengue infection, Rio de Janeiro, Brazil. J Virol Methods 186:167–170
17. Araújo LM, Ferreira ML, Nascimento OJ (2016) Guillain-Barré syndrome associated with the Zika virus outbreak in Brazil. Arq Neuropsiquiatr 74:253–255
18. Araúz D, De Urriola L, Jones J et al (2016) Febrile or exanthematous illness associated with Zika, Dengue, and Chikungunya viruses, Panama. Emerg Infect Dis 22(8):1515–1517
19. Araúz MJ, Ridde V, Hernández LM et al (2015) Developing a social autopsy tool for dengue mortality: a pilot study. PLoS One 10:e0117455
20. Arellanos-Soto D, Cruz V, Mendoza-Tavera N et al (2015) Constant risk of dengue virus infection by blood transfusion in an endemic area in Mexico. Transfus Med 25:122–124
21. Arzuza-Ortega L, Polo A, Pérez-Tatis G et al (2016) Fatal sickle cell disease and Zika virus infection in girl from Colombia. Emerg Infect Dis 22:925–927
22. Azeredo EL, Zagne SM, Santiago MA et al (2001) Characterisation of lymphocyte response and cytokine patterns in patients with dengue fever. Immunobiology 204:494–507

23. Azevedo AS, Gonçalves AJ, Archer M et al (2013) The synergistic effect of combined immunization with a DNA vaccine and chimeric yellow fever/dengue virus leads to strong protection against dengue. PLoS One 8:e58357
24. Azevedo AS, Yamamura AM, Freire MS et al (2011) DNA vaccines against dengue virus type 2 based on truncate envelope protein or its domain III. PLoS One 6:e20528
25. Baca-Carrasco D, Velasco-Hernández JX (2016) Sex, mosquitoes and epidemics: an evaluation of Zika disease dynamics. Bull Math Biol 78:2228–2242
26. Barros VE, dos Santos-Junior NN, Amarilla AA et al (2015) Differential replicative ability of clinical dengue virus isolates in an immunocompetent C57BL/6 mouse model. BMC Microbiol 15:189
27. Belle EA, King SD, Griffiths BB et al (1980) Epidemiological investigation for arboviruses in Jamaica, West Indies. Am J Trop Med Hyg 29:667–675
28. Bingham AM, Cone M, Mock V et al (2016) Comparison of test results for Zika virus RNA in urine, serum, and saliva specimens from persons with travel-associated Zika virus disease – Florida, 2016. MMWR Morb Mortal Wkly Rep 65:475–478
29. Braga C, Albuquerque MF, Cordeiro MT et al (2016) Prospective birth cohort in a hyperendemic dengue area in Northeast Brazil: methods and preliminary results. Cad Saude Publica 32: pii: S0102-311X2016000100601
30. Branco MS, Sousa DM, Monteiro JD et al (2015) Dengue in the State of Rio Grande do Norte, Brazil, 2010-2012. Trop Med Int Health 20:1707–1710
31. Brasil P, Sequeira PC, Freitas AD et al (2016) Guillain-Barré syndrome associated with Zika virus infection. Lancet 387:1482
32. Brasil P, Pereira JP Jr, Raja Gabaglia C et al (2016) Zika virus infection in pregnant women in Rio de Janeiro: preliminary report. N Engl J Med 375:2321–2334
33. Bressanelli S, Stiasny K, Allison SL et al (2004) Structure of a flavivirus envelope glycoprotein in its low-pH-induced membrane fusion conformation. EMBO J 23:728–738
34. Buonora SN, Passos SR, Carmo CN et al (2016) Accuracy of clinical criteria and an immunochromatographic strip test for dengue diagnosis in a DENV 4 epidemic. BMC Infect Dis 16:37
35. Byk LA, Iglesias NG, De Maio FA et al (2016) Dengue virus genome uncoating requires ubiquitination. mBio 7
36. Cabrera-Romo S, Max Ramirez C, Recio-Totoro B et al (2016) No evidence of dengue virus infections in several species of bats captured in central and southern Mexico. Zoonoses Public Health 63:579–583
37. Cabrera-Romo S, Recio-Totoro B, Alcala AC et al (2014) Experimental inoculation of *Artibeus jamaicensis* bats with dengue virus serotypes 1 or 4 showed no evidence of sustained replication. Am J Trop Med Hyg 91:1227–1234
38. Calvet G, Aguiar RS, Melo AS et al. (2016) Detection and sequencing of Zika virus from amniotic fluid of fetuses with microcephaly in Brazil: a case study. Lancet Infect Dis pii: S1473-3099(16)00095-5
39. Camacho E, Paternina-Gomez M, Blanco PJ et al (2016) Detection of autochthonous Zika virus transmission in Sincelejo, Colombia. Emerg Infect Dis 22:927–929
40. Camargo S (1967) History of *Aedes aegypti* eradication in the Americas. Bull World Health Organ 36:602–603
41. Campos GS, Bandeira AC, Sardi SI (2015) Zika virus outbreak, Bahia, Brazil. Emerg Infect Dis 21:1885–1886
42. Cao-Lormeau VM, Roche C, Teissier A et al (2014) Zika virus, French Polynesia, South pacific, 2013. Emerg Infect Dis 20:1085–1086
43. Carneiro AR, Cruz AC, Vallinoto M (2012) Molecular characterisation of dengue virus type 1 reveals lineage replacement during circulation in Brazilian territory. Mem Inst Oswaldo Cruz 107:805–812
44. Castro Rodríguez R, Carrasquilla G, Porras A et al (2016) The burden of dengue and the financial cost to Colombia, 2010-2012. Am J Trop Med Hyg 94:1065–1072
45. Caufour PS, Motta MCA, Yamamura AMY et al (2001) Construction, characterization and immunogenicity of recombinant yellow fever 17D-dengue type 2 viruses. Virus Res 79:1–14

46. Cavalcanti LP, Braga DN, da Silva LM et al (2016) Postmortem diagnosis of dengue as an epidemiological surveillance tool. Am J Trop Med Hyg 94:187–192
47. Cavalheiro S, Lopez A, Serra S et al (2016) Microcephaly and Zika virus: neonatal neuroradiological aspects. Childs Nerv Syst 32:1057–1060
48. Ceballos-Olvera I, Chavez-Salinas S, Medina F et al (2010) JNK phosphorylation, induced during dengue virus infection, is important for viral infection and requires the presence of cholesterol. Virology 396:30–36
49. Chen ST, Lin YL, Huang MT, Wu MF, Cheng SC, Lei HY, Lee CK, Chiou TW, Wong CH, Hsieh SL (2008) CLEC5A is critical for dengue-virus-induced lethal disease. Nature (Lond) 453:672–676
50. Chen Y, Maguire T, Hileman RE et al (1997) Dengue virus infectivity depends on envelope protein binding to target cell heparan sulfate. Nat Med 3:866–871
51. Chen YC, Wang SY, King CC (1999) Bacterial lipopolysaccharide inhibits dengue virus infection of primary human monocytes/macrophages by blockade of virus entry via a CD14-dependent mechanism. J Virol 73:2650–2657
52. Clyde K, Kyle JL, Harris E (2006) Recent advances in deciphering viral and host determinants of dengue virus replication and pathogenesis. J Virol 80:11418–11431
53. Costa SD, da Silva GB Jr, Jacinto CN et al (2015) Dengue fever among renal transplant recipients: a series of 10 cases in a tropical country. Am J Trop Med Hyg 93:394–396
54. Costa SM, Azevedo AS, Paes MV (2007) DNA vaccines against dengue virus based on the ns1 gene: the influence of different signal sequences on the protein expression and its correlation to the immune response elicited in mice. Virology 358:413–423
55. Costa SM, Yorio AP, Gonçalves AJ et al (2011) Induction of a protective response in mice by the dengue virus NS3 protein using DNA vaccines. PLoS One 6:e25685
56. Costa VG, Marques-Silva AC, Moreli ML (2014) A meta-analysis of the diagnostic accuracy of two commercial NS1 antigen ELISA tests for early dengue virus detection. PLoS One 9:e94655
57. Cruz CD, Forshey BM, Juarez DS et al (2013) Molecular epidemiology of American/Asian genotype DENV-2 in Peru. Infect Genet Evol 18:220–228
58. Cugola FR, Fernandes IR, Russo FB et al (2016) The Brazilian Zika virus strain causes birth defects in experimental models. Nature (Lond) 534:267–271
59. Daumas RP, Passos SR, Oliveira RV et al (2013) Clinical and laboratory features that discriminate dengue from other febrile illnesses: a diagnostic accuracy study in Rio de Janeiro, Brazil. BMC Infect Dis 13:77
60. Dayan G, Arredondo JL, Carrasquilla G (2015) Prospective cohort study with active surveillance for fever in four dengue endemic countries in Latin America. Am J Trop Med Hyg 93:18–23
61. Dayan GH, Garbes P, Noriega F et al (2013) Immunogenicity and safety of a recombinant tetravalent dengue vaccine in children and adolescents ages 9–16 years in Brazil. Am J Trop Med Hyg 89:1058–1065
62. De La Cruz Hernandez SI, Puerta-Guardo H, Flores-Aguilar H et al (2014) A strong interferon response correlates with a milder dengue clinical condition. J Clin Virol 60:196–199
63. De Nova-Ocampo M, Villegas-Sepulveda N, del Angel RM (2002) Translation elongation factor-1alpha, La, and PTB interact with the 3' untranslated region of dengue 4 virus RNA. Virology 295:337–347
64. de Thoisy B, Lacoste V, Germain A et al (2009) Dengue infection in neotropical forest mammals. Vector Borne Zoonotic Dis 9:157–170
65. Dhalia R, Maciel M Jr, Cruz FS et al (2009) Membrane and envelope virus proteins co-expressed as lysosome associated membrane protein (LAMP) fused antigens: a potential tool to develop DNA vaccines against flaviviruses. An Acad Bras Cienc 81:663–669
66. Díaz-Quiñonez JA, Escobar-Escamilla N, Wong-Arámbula C et al (2016) Asian genotype Zika virus detected in traveler returning to Mexico from Colombia, October 2015. Emerg Infect Dis 22:937–939

67. Dirlikov E, Ryff KR, Torres-Aponte J et al (2016) Update: Ongoing Zika virus transmission–Puerto Rico, November 1, 2015–April 14. MMWR Morb Mortal Wkly Rep 65:451–455
68. Dorigatti I, Aguas R, Donnelly CA et al (2015) Modelling the immunological response to a tetravalent dengue vaccine from multiple phase-2 trials in Latin America and South East Asia. Vaccine 33:3746–3751
69. Dowd KA, Ko SY, Morabito KM et al (2016) Rapid development of a DNA vaccine for Zika virus. Science 354:237–240
70. Drumond B, Mondini A, Schmidt D et al (2013) Circulation of different lineages of Dengue virus 2, genotype American/Asian in Brazil: dynamics and molecular and phylogenetic characterization. PLoS One 8:e59422
71. Duffy MR, Chen TH, Hancock WT et al (2009) Zika virus outbreak on Yap Island, Federated States of Micronesia. N Engl J Med 360:2536–2543
72. Durbin AP, Kirkpatrick BD, Pierce KK et al (2011) Development and clinical evaluation of multiple investigational monovalent DENV vaccines to identify components for inclusion in a live attenuated tetravalent DENV vaccine. Vaccine 29:7242–7250
73. Durbin AP, Kirkpatrick BD, Pierce KK et al (2016) A 12-month-interval dosing study in adults indicates that a single dose of the National Institutes of Allergy and Infectious Diseases tetravalent dengue vaccine induces a robust neutralizing antibody response. J Infect Dis 214(6):832–835. pii: jiw067
74. Faria NRC, Nogueira RM, de Filippis AMB et al (2013) Twenty years of DENV-2 activity in Brazil: molecular characterization and phylogeny of strains isolated from 1990 to 2010. PLoS Negl Trop Dis 7:e2095
75. Fernandes NC, Nogueira JS, Réssio RA et al. (2016) Experimental Zika virus infection induces spinal cord injury and encephalitis in newborn Swiss mice. Exp Toxicol Pathol pii: S0940-2993(16)30330-X
76. Ferreira RA, de Oliveira SA, Gandini M et al (2015) Circulating cytokines and chemokines associated with plasma leakage and hepatic dysfunction in Brazilian children with dengue fever. Acta Trop 149:138–147
77. Filomatori CV, Iglesias NG, Villordo SM et al (2011) RNA sequences and structures required for the recruitment and activity of the dengue virus polymerase. J Biol Chem 286:6929–6939
78. Filomatori CV, Lodeiro MF, Alvarez DE et al (2006) A 5′ RNA element promotes dengue virus RNA synthesis on a circular genome. Genes Dev 20:2238–2249
79. Fontes CA, Dos Santos AA, Marchiori E (2016) Magnetic resonance imaging findings in Guillain-Barré syndrome caused by Zika virus infection. Neuroradiology 58(8):837–838
80. Fried JR, Gibbons RV, Kalayanarooj S et al (2010) Serotype-specific differences in the risk of dengue hemorrhagic fever: an analysis of data collected in Bangkok, Thailand from 1994 to 2006. PLoS Negl Trop Dis 4:e617
81. Galler R, Marchevsky RS, Caride E et al (2005) Attenuation and immunogenicity of recombinant yellow fever 17D-dengue type 2 virus for rhesus monkeys. Braz J Med Biol Res 38:1835–1846
82. Garcez PP, Loiola EC, Madeiro da Costa R et al (2016) Zika virus impairs growth in human neurospheres and brain organoids. Science 352:816–818
83. García-Machorro J, López-González M, Barrios-Rojas O et al (2013) DENV-2 subunit proteins fused to CR2 receptor-binding domain (P28)-induces specific and neutralizing antibodies to the Dengue virus in mice. Hum Vaccin Immunother 9:2326–2335
84. Garcia-Sastre A, Biron CA (2006) Type 1 interferons and the virus-host relationship: a lesson in detente. Science 312:879–882
85. Gebhard LG, Filomatori CV, Gamarnik AV (2011) Functional RNA elements in the dengue virus genome. Viruses 3:1739–1756
86. Gil L, Izquierdo A, Lazo L et al (2014) Capsid protein: evidences about the partial protective role of neutralizing antibody-independent immunity against dengue in monkeys. Virology 456-457:70–76

87. Gil L, López C, Lazo L et al (2009) Recombinant nucleocapsid-like particles from dengue-2 virus induce protective CD4+ and CD8+ cells against viral encephalitis in mice. Int Immunol 21:1175–1183
88. Giovanetti M, Faria NR, Nunes MR et al (2016) Zika virus complete genome from Salvador, Bahia, Brazil. Infect Genet Evol 41:142–145
89. Gonçalves AJ, Oliveira ER, Costa SM et al (2015) Cooperation between CD4+ T cells and humoral immunity is critical for protection against dengue using aDNA vaccine based on the NS1 antigen. PLoS Negl Trop Dis 9:e0004277
90. Grange L, Simon-Loriere E, Sakuntabhai A et al (2014) Epidemiological risk factors associated with high global frequency of inapparent dengue virus infections. Front Immunol 5:280
91. Guerrero CD, Arrieta AF, Ramirez ND et al (2013) High plasma levels of soluble ST2 but not its ligand IL-33 is associated with severe forms of pediatric dengue. Cytokine 61:766–771
92. Hadinegoro SR, Arredondo-García JL, Capeding MR et al (2015) Efficacy and long-term safety of a dengue vaccine in regions of endemic disease. N Engl J Med 373:1195–1206
93. Halstead SB (2014) Dengue antibody-dependent enhancement: knowns and unknowns. Microbiol Spectr 2
94. Hazin AN, Poretti A, Cruz DD et al (2016) Computed tomographic findings in microcephaly associated with Zika virus. N Engl J Med 374:2193–2195
95. Heringer M, Nogueira RM, de Filippis AM et al (2015) Impact of the emergence and re-emergence of different dengue viruses' serotypes in Rio de Janeiro, Brazil, 2010 to 2012. Trans R Soc Trop Med Hyg 109:268–274
96. Hernández-Ávila M, Lazcano-Ponce E, Hernández-Ávila JE et al (2016) Analysis of the evidence on the efficacy and safety of CYD-TDV dengue vaccine and its potential licensing and implementation through Mexico's Universal Vaccination Program. Salud Publica Mex 58:71–83
97. Hills S, Russell K, Hennessey M et al (2016) Transmission of Zika virus through sexual contact with travelers to areas of ongoing transmission-continental United States, 2016. MMWR Morb Mortal Wkly Rep 65:215–216
98. Huang CY, Butrapet S, Tsuchiya KR et al (2003) Dengue 2 PDK-53 virus as a chimeric carrier for tetravalent dengue vaccine development. J Virol 77:11436–11447
99. Iglesias NG, Mondotte JÁ, Byk LA et al (2015) Dengue virus uses a non-canonical function of the host GBF1-Arf-COPI system for capsid protein accumulation on lipid droplets. Traffic 16:962–977
100. Jaenisch T, Tam DT, Kieu NT et al (2016) Clinical evaluation of dengue and identification of risk factors for severe disease: protocol for a multicentre study in 8 countries. BMC Infect Dis 16(1):20
101. Jindadamrongwech S, Thepparit C, Smith DR (2004) Identification of GRP 78 (BiP) as a liver cell expressed receptor element for dengue virus serotype 2. Arch Virol 149:915–927
102. Junjhon J, Pennington JG, Edwards TJ et al (2014) Ultrastructural characterization and three-dimensional architecture of replication sites in dengue virus-infected mosquito cells. J Virol 88:4687–4697
103. Katzelnick LC, Montoya M, Gresh L et al (2016) Neutralizing antibody titers against dengue virus correlate with protection from symptomatic infection in a longitudinal cohort. Proc Natl Acad Sci U S A 113:728–733
104. Kikuti M, Cunha GM, Paploski IA et al (2015) Spatial distribution of dengue in a Brazilian urban slum setting: role of socioeconomic gradient in disease risk. PLoS Negl Trop Dis 9:e0003937
105. Kourí G, Guzmán MG, Bravo J (1986) Hemorrhagic dengue in Cuba: history of an epidemic. Bull Pan Am Health Organ 20:24–30
106. Kuadkitkan A, Wikan N, Fongsaran C, Smith DR (2010) Identification and characterization of prohibitin as a receptor protein mediating DENV-2 entry into insect cells. Virology 406:149–161

107. Lanata CF, Andrade T, Gil A et al (2012) Immunogenicity and safety of tetrava-lent dengue vaccine in 2–11-year-olds previously vaccinated against yellow fever: randomized, controlled, phase II study in Piura, Peru. Vaccine 30:5935–5941
108. Larocca RA, Abbink P, Peron JP et al (2016) Vaccine protection against Zika virus from Brazil. Nature (Lond) 536:474–478
109. Lazo L, Hermida L, Zulueta A et al (2007) A recombinant capsid protein from Dengue-2 induces protection in mice against homologous virus. Vaccine 25:1064–1070
110. Lednicky J, Beau de Rochars VM, El Badry M et al (2016) Zika virus outbreak in Haiti in 2014: molecular and clinical data. PLoS Negl Trop Dis 10:e0004687
111. Lee CJ, Lin HR, Liao CL, Lin YL (2008) Cholesterol effectively blocks entry of flavivirus. J Virol 82:6470–6480
112. Li Z, Khaliq M, Zhou Z et al (2008) Design, synthesis, and biological evaluation of antiviral agents targeting flavivirus envelope proteins. J Med Chem 51:4660–4671
113. Lima DM, de Paula SO, França RF (2011) A DNA vaccine candidate encoding the structural prM/E proteins elicits a strong immune response and protects mice against dengue-4 virus infection. Vaccine 29:831–838
114. Lodeiro MF, Filomatori CV, Gamarnik AV (2009) Structural and functional studies of the promoter element for dengue virus RNA replication. J Virol 83:993–1008
115. López C, Gil L, Lazo L et al (2009) In vitro assembly of nucleocapsid-like particles from purified recombinant capsid protein of dengue-2 virus. Arch Virol 154:695–698
116. Luna DM, Oliveira MD, Nogueira ML et al (2014) Biosensor based on lectin and lipid membranes for detection of serum glycoproteins in infected patients with dengue. Chem Phys Lipids 180:7–14
117. Luria-Perez R, Cedillo-Barron L, Santos-Argumedo L et al (2007) A fusogenic peptide expressed on the surface of *Salmonella enterica* elicits CTL responses to a dengue virus epitope. Vaccine 25:5071–5085
118. Mackow ER, Makino Y, Zhao BT et al (1987) The nucleotide sequence of dengue type 4 virus: analysis of genes coding for nonstructural proteins. Virology 159:217–228
119. Manzano M, Reichert ED, Polo S et al (2011) Identification of cis-acting elements in the 3′-untranslated region of the dengue virus type 2 RNA that modulate translation and replication. J Biol Chem 286:22521–22534
120. Martelli CMT, Siqueira JB Jr, MPPD P et al (2015) Economic impact of dengue: multicenter study across four Brazilian regions. PLoS Negl Trop Dis 9:e0004042
121. Martina BE, Koraka P, Osterhaus AD (2009) Dengue virus pathogenesis: an integrated view. Clin Microbiol Rev 22:564–581
122. Martínez CA, Topal E, Giulietti AM et al (2010) Exploring different strategies to express Dengue virus envelope protein in a plant system. Biotechnol Lett 32:867–875
123. Mateu GP, Marchevsky RS, Liprandi F et al (2007) Construction and biological properties of yellow fever 17D/dengue type 1 recombinant virus. Trans R Soc Trop Med Hyg 101:289–298
124. Mathew A, Rothman AL (2008) Understanding the contribution of cellular immunity to dengue disease pathogenesis. Immunol Rev 225:300–133
125. Meertens L, Carnec X, Lecoin MP et al (2012) The TIM and TAM families of phosphatidylserine receptors mediate dengue virus entry. Cell Host Microbe 12:544–557
126. Mellado-Sánchez G, García-Cordero J, Luria-Pérez R (2005) DNA priming E and NS1 constructs: homologous proteins boosting immunization strategy to improve immune response against dengue in mice. Viral Immunol 18:709–721
127. Mellado-Sánchez G, García-Machorro J, Sandoval-Montes C et al (2010) A plasmid encoding parts of the dengue virus E and NS1 proteins induces an immune response in a mouse model. Arch Virol 155:847–856
128. Melo AS, Aguiar RS, Amorim MM et al (2016) Congenital Zika virus infection: beyond neonatal microcephaly. JAMA Neurol 73(12):1407–1416. doi:10.1001/jamaneurol.2016.3720
129. Mercado-Curiel RF, Black WC, Munoz ML (2008) A dengue receptor as possible genetic marker of vector competence in *Aedes aegypti*. BMC Microbiol 8:118

130. Mercado-Curiel RF, Esquinca-Aviles HA, Tovar R et al (2006) The four serotypes of dengue recognize the same putative receptors in *Aedes aegypti* midgut and *Ae. albopictus* cells. BMC Microbiol 6:85
131. Miranda HA, Costa MC, Frazão MA et al (2016) Expanded spectrum of congenital ocular findings in microcephaly with presumed Zika infection. Ophthalmology 123:1788–1794
132. Miranda-Filho DB, Martelli CM, Ximenes RA et al (2016) Initial description of the presumed congenital Zika syndrome. Am J Public Health 106:598–600
133. Mlakar J, Korva M, Tul N et al (2016) Zika virus associated with microcephaly. N Engl J Med 374:951–958
134. Modis Y, Ogata S, Clements D, Harrison SC (2004) Structure of the dengue virus envelope protein after membrane fusion. Nature (Lond) 427:313–319
135. Moore C, Staples JE, Dobyns WB et al (2016) Characterizing the pattern of anomalies in congenital Zika syndrome for pediatric clinicians. JAMA Pediatr. doi:10.1001/jamapediatrics.2016.3982
136. Moron AF, Cavalheiro S, Milani H et al (2016) Microcephaly associated with maternal Zika virus infection. Br J Obstet Gynecol 123(8):1265–1269
137. Morrison J, Laurent-Rolle M, Maestre AM et al (2013) Dengue virus co-opts UBR4 to degrade STAT2 and antagonize type I interferon signaling. PLoS Pathog 9:e1003265
138. Mosso C, Galvan-Mendoza IJ, Ludert JE, del Angel RM (2008) Endocytic pathway followed by dengue virus to infect the mosquito cell line C6/36 HT. Virology 378:193–199
139. Mota J, Acosta M, Argotte R et al (2005) Induction of protective antibodies against dengue virus by tetravalent DNA immunization of mice with domain III of the envelope protein. Vaccine 23:3469–3476
140. Mukhopadhyay S, Kuhn RJ, Rossmann MG (2005) A structural perspective of the flavivirus life cycle. Nat Rev Microbiol 3:13–22
141. Munoz-Jordan JL, Fredericksen BL (2010) How flaviviruses activate and suppress the interferon response. Viruses 2:676–691
142. Munoz-Jorda JL, Sanchez-Burgos GG, Laurent-Rolle M, Garcia-Sastre A (2003) Inhibition of interferon signaling by dengue virus. Proc Natl Acad Sci USA 100:14333–14338
143. Musso D, Roche C, Nhan TX et al (2016) Detection of Zika virus in saliva. J Clin Virol 68:53–55
144. Musso D (2015) Zika virus transmission from French Polynesia to Brazil. Emerg Infect Dis 21:1887
145. Navarro-Sanchez E, Altmeyer R, Amara A et al (2003) Dendritic-cell-specific ICAM3-grabbing non-integrin is essential for the productive infection of human dendritic cells by mosquito-cell-derived dengue viruses. EMBO Rep 4:723–728
146. Nishiura H, Mizumoto K, Villamil-Gómez WE et al (2016) Preliminary estimation of the basic reproduction number of Zika virus infection during Colombia epidemic, 2015–2016. Travel Med Infect Dis pii: S1477-8939(16)30008-4
147. Nogueira RM, Filippis AM, Coelho JM et al (2002) Dengue virus infection of the central nervous system (CNS): a case report from Brazil. Southeast Asian J Trop Med Public Health 3:68–71
148. Nogueira RM, Miagostovich MP, de Filippis AM et al (2001) Dengue virus type 3 in Rio de Janeiro, Brazil. Mem Inst Oswaldo Cruz 96:925–926
149. Nogueira RM, Miagostovich MP, Lampe E et al (1990) Isolation of dengue virus type 2 in Rio de Janeiro. Mem Inst Oswaldo Cruz 85:253
150. Noisakran S, Perng GC (2008) Alternate hypothesis on the pathogenesis of dengue hemorrhagic fever (DHF)/dengue shock syndrome (DSS) in dengue virus infection. Exp Biol Med 233:401–408
151. Noronha L, Zanluca C, Azevedo ML et al (2016) Zika virus damages the human placental barrier and presents marked fetal neurotropism. Mem Inst Oswaldo Cruz. pii: S0074–02762016005006103
152. Nunes PC, Paes MV, de Oliveira CA et al (2016) Detection of dengue NS1 and NS3 proteins in placenta and umbilical cord in fetal and maternal death. J Med Virol 88(8):1448–1452

153. Ocazionez-Jiménez RE, Ortiz-Báez AS, Gómez-Rangel SY et al (2013) Dengue virus serotype 1 (DENV-1) from Colombia: its contribution to dengue occurrence in Santander. Biomedica 33(suppl 1):22–30
154. Oliveira ER, Amorim JF, Paes MV et al (2016) Peripheral effects induced in BALB/c mice infected with DENV by the intracerebral route. Virology 489:95–107
155. Oliveira Melo AS, Malinger G, Ximenes R et al (2016) Zika virus intrauterine infection causes fetal brain abnormality and microcephaly: tip of the iceberg? Ultrasound Obstet Gynecol 47:6–7
156. Osanai CH, Travassos da Rosa AP, Tang AT, do Amaral RS, Passos AD et al (1983) Dengue outbreak in Boa Vista, Roraima. Preliminary report. Rev Inst Med Trop Sao Paulo 25:53–54
157. Osorio L, Uribe M, Ardila GI et al (2015) The use of rapid dengue diagnostic tests in a routine clinical setting in a dengue-endemic area of Colombia. Mem Inst Oswaldo Cruz 110:510–516
158. Pagliari C, Quaresma JA, Fernandes ER et al (2014) Immunopathogenesis of dengue hemorrhagic fever: contribution to the study of human liver lesions. J Med Virol 86:1193–1197
159. Pagliari C, Simões Quaresma JA, Kanashiro-Galo L et al (2016) Human kidney damage in fatal dengue hemorrhagic fever results of glomeruli injury mainly induced by IL17. J Clin Virol 75:16–20
160. Pagni S, Fernandez-Sesma A (2012) Evasion of the human innate immune system by dengue virus. Immunol Res 54:152–159
161. PAHO (2013) Number of reported cases of dengue and dengue severe cases in the Americas by country. Available at: http://www.paho.org/hq/index.php?option=com_docman&task=doc_view&gid=24677&Itemid. Accessed 05 Jun 2014
162. PAHO (2016) Regional Zika Epidemiological update (Americas) December 1, 2016. http://www.paho.org/hq/index.php?option=com_content&id=11599&Itemid=41691. Date Accessed 06/12/2016.
163. PAHO (1990) Dengue haemorrhagic fever in Venezuela. Epidemiol Bull 11:7–9
164. Paploski IA, Prates AP, Cardoso CW et al (2016) Time lags between exanthematous illness attributed to Zika virus, Guillain-Barré syndrome, and microcephaly, Salvador, Brazil. Emerg Infect Dis 22:1438–1444
165. Parra B, Lizarazo J, Jiménez-Arango JA et al (2016) Guillain-Barré syndrome associated with Zika virus infection in Colombia. N Engl J Med 375:1513–1523
166. Perera R, Riley C, Isaac G et al (2012) Dengue virus infection perturbs lipid homeostasis in infected mosquito cells. PLoS Pathog 8:e1002584
167. Pérez-Vélez ME, García-Nieves T, Colón-Sánchez C et al (2009) Induction of neutralization antibodies in mice by Dengue-2 envelope DNA vaccines. P R Health Sci J 28:239–250
168. Pinheiro FP, Corber SJ (1997) Global situation of dengue and dengue haemorrhagic fever, and its emergence in the Americas. World Health Stat 50:161–169
169. Pitha PM, Kunzi MS (2007) Type I interferon: the ever unfolding story. Curr Top Microbiol Immunol 316:41–70
170. Poh MK, Shui G, Xie X et al (2012) U18666A, an intra-cellular cholesterol transport inhibitor, inhibits dengue virus entry and replication. Antivir Res 93:191–198
171. Pone SM, Hökerberg YH, de Oliveira R et al (2016) Clinical and laboratory signs associated to severe dengue disease in hospitalized children. J Pediatr (Rio J) 92(5):464–471. 22. pii: S0021-7557(16)30022-5
172. Póvoa TF, Alves AM, Oliveira CA et al (2014) The pathology of severe dengue in multiple organs of human fatal cases: histopathology, ultrastructure and virus replication. PLoS One 9:e83386
173. Puerta-Guardo H, Mosso C, Medina F et al (2010) Antibody-dependent enhancement of dengue virus infection in U937 cells requires cholesterol-rich membrane microdomains. J Gen Virol 91:394–403
174. Puerta-Guardo H, Raya-Sandino A, González-Mariscal L et al (2013) The cytokine response of U937-derived macrophages infected through antibody-dependent enhancement of dengue virus disrupts cell apical-junction complexes and increases vascular permeability. J Virol 87:7486–7501

175. Reyes-Del Valle J, Chavez-Salinas S, Medina F, Del Angel RM (2005) Heat shock protein 90 and heat shock protein 70 are components of dengue virus receptor complex in human cells. J Virol 79:4557–4567
176. Rivera J, Neira M, Parra E et al (2014) Detection of dengue virus antigen in post-mortem tissues. Biomedica 34:514–520
177. Rodriguez-Madoz JR, Belicha-Villanueva A, Bernal-Rubio D et al (2010) Inhibition of the type I interferon response in human dendritic cells by dengue virus infection requires a catalytically active NS2B3 complex. J Virol 84:9760–9774
178. Rodriguez-Madoz JR, Bernal-Rubio D, Kaminski D, Boyd K, Fernandez-Sesma A (2010) Dengue virus inhibits the production of type I interferon in primary human dendritic cells. J Virol 84:4845–4850
179. Rosário MS, Jesus PA, Vasilakis N et al (2016) Guillain-Barré syndrome after Zika virus infection in Brazil. Am J Trop Med Hyg 95:1157–1160
180. Rothwell C, Lebreton A, Young Ng C et al (2009) Cholesterol biosynthesis modulation regulates dengue viral replication. Virology 389:8–19
181. Sabchareon A, Wallace D, Sirivichayakul C et al (2012) Protective efficacy of the recombinant, live-attenuated, CYD tetravalent dengue vaccine in Thai schoolchildren: a randomised, controlled phase 2b trial. Lancet 380:1559–1567
182. Sabin AB, Schlesinger RW (1945) Production of immunity to dengue with virus modified by propagation in mice. Science 101:640–642
183. Sacramento RH, de Melo Braga DN, Sacramento FF et al (2014) Death by dengue fever in a Brazilian child: a case report. BMC Res Notes 7:855
184. Salas-Benito J, Reyes-Del Valle J, Salas-Benito M et al (2007) Evidence that the 45-kD glycoprotein, part of a putative dengue virus receptor complex in the mosquito cell line C6/36, is a heat-shock related protein. Am J Trop Med Hyg 77:283–290
185. Salas-Benito J, del Angel RM (1997) Identification of two surface proteins from C6/36 cells that bind dengue type 4 virus. J Virol 71:7246–7252
186. Sams MM, Mondotte JA, Caramelo JJ, Gamarnik AV (2012) Uncoupling cis-acting RNA elements from coding sequences revealed a requirement of the N-terminal region of dengue virus capsid protein in virus particle formation. J Virol 86:1046–1058
187. Samsa MM, Mondotte JA, Iglesias NG et al (2009) Dengue virus capsid protein usurps lipid droplets for viral particle formation. PLoS Pathog 5:e1000632
188. San Martin JL, Brathwaite O, Zambrano B et al (2010) The epidemiology of dengue in the Americas over the last three decades: a worrisome reality. Am J Trop Med Hyg 82:128–135
189. Santos T, Rodriguez A, Almiron M et al (2016) Zika virus and the Guillain-Barré syndrome: case series from seven countries. N Engl J Med 375:1598–1601
190. Sarmiento-Ospina A, Vásquez-Serna H, Jimenez-Canizales CE (2016) Zika virus associated deaths in Colombia. Lancet Infect Dis. pii: S1473-3099(16)30006-8
191. Sarti E, L'Azou M, Mercado M et al (2016) A comparative study on active and passive epidemiological surveillance for dengue in five countries of Latin America. Int J Infect Dis 44:44–49
192. Schatzmayr HG, Nogueira RMR, Travassos da Rosa APA (1986) An outbreak of dengue virus at Rio de Janeiro. Mem Inst Oswaldo Cruz 81:245–246
193. Schuler-Faccini L, Ribeiro EM, Feitosa IM et al (2016) Possible association between Zika virus infection and microcephaly - Brazil, 2015. MMWR Morb Mortal Wkly Rep 65:59–62
194. Shah A, Kumar A (2016) Zika virus infection and development of a murine model. Neurotox Res 30:131–134
195. Shurtleff AC, Beasley DW, Chen JJ et al (2001) Genetic variation in the 3′ non-coding region of dengue viruses. Virology 281:75–87
196. Sirivichayakul C, Barranco-Santana EA, Esquilin-Rivera I et al (2016) Safety and immunogenicity of a tetravalent dengue vaccine candidate in healthy children and adults in dengue-endemic regions: a randomized, placebo-controlled phase 2 study. J Infect Dis 213:1562–1572
197. Soto-Acosta R, Bautista-Carbajal P, Syed GH et al (2014) Nordihydroguaiaretic acid (NDGA) inhibits replication and viral morphogenesis of dengue virus. Antivir Res 109:132–140

198. Soto-Acosta R, Mosso C, Cervantes-Salazar M et al (2013) The increase in cholesterol levels at early stages after dengue virus infection correlates with an augment in LDL particle uptake and HMG-CoA reductase activity. Virology 442:132–147
199. Sotomayor-Bonilla J, Chaves A, Rico-Chavez O (2014) Dengue virus in bats from southeastern Mexico. Am J Trop Med Hyg 91:129–131
200. Suzarte E, Gil L, Valdés I et al (2015) A novel tetravalent formulation combining the four aggregated domain III-capsid proteins from dengue viruses induces a functional immune response in mice and monkeys. Int Immunol 27:367–379
201. Stiasny K, Heinz FX (2006) Flavivirus membrane fusion. J Gen Virol 87:2755–2766
202. Tassaneetrithep B, Burgess TH, Granelli-Piperno A (2003) DC-SIGN (CD209) mediates dengue virus infection of human dendritic cells. J Exp Med 197:823–829
203. Teixeira MG (2012) Few characteristics of dengue's fever epidemiology in Brazil. Rev Inst Med Trop Sao Paulo 54(suppl 18):S1–S4
204. Teixeira MG, Costa MC, Coelho G, Barreto ML et al (2008) Recent shift in age pattern of dengue hemorrhagic fever, Brazil. Emerg Infect Dis 14:1663
205. Teixeira MG, Costa MCN, Barreto ML et al (2015) Epidemiologia da Dengue. In: Valle D (ed) Dengue: Teorias e práticas, 1st edn. Fiocruz, Rio de Janeiro, pp 293–315
206. Teixeira MG, Siqueira JB, Ferreira GLC et al (2013) Epidemiological trends of dengue disease in Brazil (2000–2010): a systematic literature search and analysis. PLoS Negl Trop Dis 7:e2520
207. Temporão JG, Penna GO, Carmo EH et al (2011) Dengue virus serotype 4, Roraima State, Brazil. Emerg Infect Dis 17:938–940
208. Torres AF, Braga DN, Muniz F et al (2013) Lymphocytic myocarditis at autopsy in patients with dengue fever. Braz J Infect Dis 17:619–621
209. Torres JR, Echezuría L, Fernández M et al (2015) Epidemiology and disease burden of pediatric dengue in Venezuela. J Pediatr Infect Dis Soc 4:288–289
210. Uchil PD, Satchidanandam V (2003) Architecture of the flaviviral replication complex. Protease, nuclease, and detergents reveal encasement within double-layered membrane compartments. J Biol Chem 278:24388–24398
211. Undurraga EA, Betancourt-Cravioto M, Ramos-Castañeda J et al (2015) Economic and disease burden of dengue in Mexico. PLoS Negl Trop Dis 9:e0003547
212. Upanan S, Kuadkitkan A, Smith DR (2008) Identification of dengue virus binding proteins using affinity chromatography. J Virol Methods 151:325–328
213. Valdés I, Hermida L, Gil L, Lazo L et al (2010) Heterologous prime-boost strategy in non-human primates combining the infective dengue virus and a recombinant protein in a formulation suitable for human use. Int J Infect Dis 14:e377–e383
214. Vega-Almeida TO, Salas-Benito M, De Nova-Ocampo MA et al (2013) Surface proteins of C6/36 cells involved in dengue virus 4 binding and entry. Arch Virol 158:1189–1207
215. Ventura CV, Maia M, Bravo-Filho V et al (2016) Zika virus in Brazil and macular atrophy in a child with microcephaly. Lancet 387:228
216. Ventura CV, Maia M, Ventura BV et al (2016) Ophthalmological findings in infants with microcephaly and presumable intra-uterus Zika virus infection. Arq Bras Oftalmol 79:1–3
217. Ventura CV, Ventura LO, Bravo-Filho V et al (2016) Optical coherence tomography of retinal lesions in infants with congenital Zika syndrome. JAMA Ophthalmol 134(12):1420–1427. doi:10.1001/jamaophthalmol.2016.4283
218. Vera-Polania F, Perilla-Gonzalez Y, Martinez-Pulgarin DF et al (2014) Bibliometric assessment of the Latin-American contributions in dengue. Recent Pat Antiinfect Drug Discov 9:195–201
219. Villordo SM, Carballeda JM, Filomatori CV, Gamarnik AV (2016) RNA structure duplications and flavivirus host adaptation. Trends Microbiol 24:270–283
220. Villordo SM, Gamarnik AV (2013) Differential RNA sequence requirement for dengue virus replication in mosquito and mammalian cells. J Virol 87:9365–9372

221. Waggoner JJ, Balmaseda A, Gresh L, Sahoo MK, Montoya M, Wang C, Abeynayake J, Kuan G, Pinsky BA, Harris E (2016) Homotypic dengue virus reinfections in Nicaraguan children. J Infect Dis 214:986–993
222. Weaver SC, Vasilakis N (2009) Molecular evolution of dengue viruses: contributions of phylogenetics to understanding the history and epidemiology of the preeminent arboviral disease. Infect Genet Evol 9:523–540
223. Welsch S, Miller S, Romero-Brey I et al (2009) Composition and three-dimensional architecture of the dengue virus replication and assembly sites. Cell Host Microbe 5:365–375
224. Wenceslao Orellano P, Reynoso JI, Stahl HC et al (2016) Cost-utility analysis of dengue vaccination in a country with heterogeneous risk of dengue transmission. Vaccine 34:616–621
225. Wettstein ZS, Fleming M, Chang AY et al (2012) Total economic cost and burden of dengue in Nicaragua: 1996–2010. Am J Trop Med Hyg 87:616–622
226. Williams M, Mayer SV, Johnson WL et al (2014) Lineage II of Southeast Asian/American DENV-2 is associated with a severe dengue outbreak in the Peruvian Amazon. Am J Trop Med Hyg 91:611–620
227. Yazi Mendoza M, Salas-Bentito JS, Lanz-Mendoza H, Hernandez-Martinez S, del Angel RM (2002) A putative receptor for dengue virus in mosquito tissues: localization of a 45-kDa glycoprotein. Am J Trop Med Hyg 67:76–84
228. Yocupicio-Monroy M, Padmanabhan R, Medina F, del Angel RM (2007) Mosquito La protein binds to the 3′ untranslated region of the positive and negative polarity dengue virus RNAs and relocates to the cytoplasm of infected cells. Virology 357:29–40
229. Yocupicio-Monroy M, Medina F, Reyes-del Valle J, del Angel RM (2003) Cellular proteins from human monocytes bind to dengue 4 virus minus-strand 3′ untranslated region RNA. J Virol 77:3067–3076
230. Yu IM, Zhang W, Holdaway HA et al (2008) Structure of the immature dengue virus at low pH primes proteolytic maturation. Science 319:1834–1837
231. Zanluca C, Melo VC, Mosimann AL et al (2015) First report of autochthonous transmission of Zika virus in Brazil. Mem Inst Oswaldo Cruz 110:569–572
232. Zellweger RM, Shresta S (2014) Mouse models to study dengue virus immunology and pathogenesis. Front Immunol 5:151
233. Zika Virus Net. http://www.zikavirusnet.com/clinical-trials.html. Accessed 07/12/2016

Chapter 9
Alphaviruses in Latin America and the Introduction of Chikungunya Virus

Juan-Carlos Navarro, Jean-Paul Carrera, Jonathan Liria, Albert J. Auguste, and Scott C. Weaver

1 Introduction

The *Togaviridae* is a family of enveloped, single-stranded, plus-strand RNA viruses composed of the genera *Alphavirus* and *Rubivirus*. The rubella virus (the cause of German measles) is the only member of the latter genus [129]. The *Alphaviruses* are arthropod-borne viruses (mainly mosquitoes) with a nearly worldwide geographic distribution, having been reported from all continents except Antarctica and from many islands [16].

The genus *Alphavirus* includes 30 species grouped into 10 complexes based on antigenic and/or genetic similarities [75]. The Barmah Forest, Ndumu, Middelburg, and Semliki Forest complexes occur almost exclusively in the Old World. In the

J.-C. Navarro (✉)
Universidad Internacional SEK, Facultad de Ciencias Naturales y Ambientales, Quito, Ecuador

Universidad Central de Venezuela, Lab. Biología de Vectores-IZET, Caracas, Venezuela
e-mail: juancarlos.navarro@uisek.edu.ec; jcnavac@gmail.com

J.-P. Carrera
Gorgas Memorial Institute of Health Studies, Department of Research in Virology and Biotechnology, Panama City, Panama
e-mail: jpcarrera@gorgas.gob.pa

J. Liria
Universidad Regional Amazónica, IKIAM, Tena, Ecuador
e-mail: jonathan.liria@ikiam.edu.ec

A.J. Auguste • S.C. Weaver
Institute for Human Infections and Immunity and Department of Microbiology and Immunology, University of Texas Medical Branch, Galveston, TX, USA
e-mail: aj1augus@utmb.edu; sweaver@utmb.edu

J.E. Ludert et al. (eds.), *Human Virology in Latin America*,
DOI 10.1007/978-3-319-54567-7_9

New World, the first alphaviruses to be isolated were western equine encephalitis virus (WEEV) in 1930, eastern equine encephalitis virus (EEEV) in 1933, and Venezuelan equine encephalitis virus (VEEV) in 1938 [129]. Other *Alphaviruses* found in Latin America include Mayaro (MAYV), Aura (AURAV), Una (UNAV), Trocara virus (TROV), and the recently introduced chikungunya virus (CHIKV) (Table 9.1). Others found in North America belonging to the WEE complex, Highlands J (HJV), and Fort Morgan viruses (FMV) [129] are recombinants resembling WEEV derived from ancestral EEEV and Sindbis (SINV) from the Old World.

Many of the New World alphaviruses are widely distributed throughout the Americas. WEEV is found from Canada to Argentina, and EEEV and VEEV occur in both North America and South America. Other viruses such as MAYV and AURAV have a more restricted neotropical distribution [106].

Recent phylogenetic analyses depict the evolution of the alphaviruses (Fig. 9.1) based on structural protein genes. Clades or branches are clearly correlated with the antigenic complexes, host/reservoirs, and disease syndromes: the Semliki Forest virus complex is associated with nonhuman primates and human fever/rash/arthralgia (including chikungunya, Mayaro, Ross River, Semliki Forest). The WEEV/EEEV complexes associate with birds, and the VEEV complex is mostly related to rodents as reservoirs, and all three with human and equine encephalitis.

The most recently described alphavirus, Eilat (EILV), a "mosquito-specific" virus from *Anopheles coustani* mosquitoes in Israel [75], is a sister of the WEEV complex (see Fig. 9.1). In contrast to all other mosquito-borne viruses, it is unable to replicate in vertebrate cell lines. EILV has important implications for arbovirus evolution and may help elucidate the viral factors responsible for the virus–cell interactions of pathogenic alphaviruses, facilitate vaccine development, and help develop strategies to control or prevent alphavirus transmission [28, 74, 75, 138].

Figure 9.2 shows a cartoon of alphavirus transmission cycles. All alphaviruses except EILV are zoonotic: mosquitoes (family Culicidae) are the major vector group, especially the genera *Aedes* (chikungunya), *Culex* subgenus *Melanoconion* (VEEV, EEEV, WEEV), and *Haemagogus* and *Sabethes* for Mayaro [26, 72, 84, 112]. In concordance with the phylogenetic tree, enzootic transmission cycles involve nonhuman primates (chikungunya, Mayaro, others that produce arthralgias), birds (EEEV, WEEV), and rodents (VEEV, possibly Madariaga virus). Spillover transmission to humans occurs mainly in rural areas but can become urban for CHIKV and potentially others. Affected animals (including humans) usually generate insufficient viremia to participate in the transmission cycle (i.e., dead-end hosts). VEEV uses equids (horses, donkeys, mules) as amplification hosts (epizootics) to increase spillover to humans (epidemics) [126]. CHIKV is the only alphavirus known to utilize humans as amplification hosts and the urban mosquitoes [*Aedes (Stegomyia) aegypti* or *Aedes (Stegomyia) albopictus*] for transmission, resulting in major epidemics [47, 129]. Mayaro, similar to CHIKV in Africa, uses nonhuman primates as enzootic hosts, but their role is still not well understood [72].

Table 9.1 Alphavirus members from those that occur in the New World, including the new introduction of chikungunya virus

Antigenic complex	Virus	Antigenic subtype	Antigenic variety	Clinical syndrome in humans	Distribution
Eastern equine encephalitis (EEE)	Eastern equine encephalitis virus (EEEV)	I–IV		Febrile illness, encephalitis (none recognized in Latin America)	North, Central, South America
Semliki Forest	Mayaro and Chikungunya virus (CHIKV)			Febrile illness, rash, arthritis	South and Central America
	Una virus (UNAV)			None recognized	South America
Venezuelan equine encephalitis (VEE)	Venezuelan equine encephalitis virus (VEEV)	1	AB	Febrile illness, encephalitis	North, Central, South America
			C	Febrile illness, encephalitis	South America
			D	Febrile illness, encephalitis	South America, Panama
			E	Febrile illness, encephalitis	Central America, Mexico
	Mossos das Pedras virus (MDPV)		F	Febrile illness, encephalitis	Brazil
	Everglades virus (EVEV)	II		Febrile illness, encephalitis	Florida (USA)
	Mucambo virus (MUCV)	III	A	Febrile illness, myalgia	South America, Trinidad
			C	Unknown	Peru
			D	Febrile illness	Peru
	Tonate virus (TONV)	III	B	Febrile illness, encephalitis	Brazil, Colorado (USA)
	Pixuna virus (PIXV)	IV		Febrile illness, myalgia	Brazil
	Cabassou virus (CABV)	V		None recognized	French Guiana
	Rio Negro virus (RNV)	VI		Febrile illness, myalgia	Argentina
Western equine encephalitis	Aura virus (AURAV)			None recognized	South America
	Western equine encephalitis virus (WEEV)	Several		Febrile illness, encephalitis	Western North, South America
	Highlands J virus (HJV)				Eastern North America
	Fort Morgan virus (FMV)	Buggy Creek		None recognized	Western North America
Trocara	Trocara virus (TROV)				South America

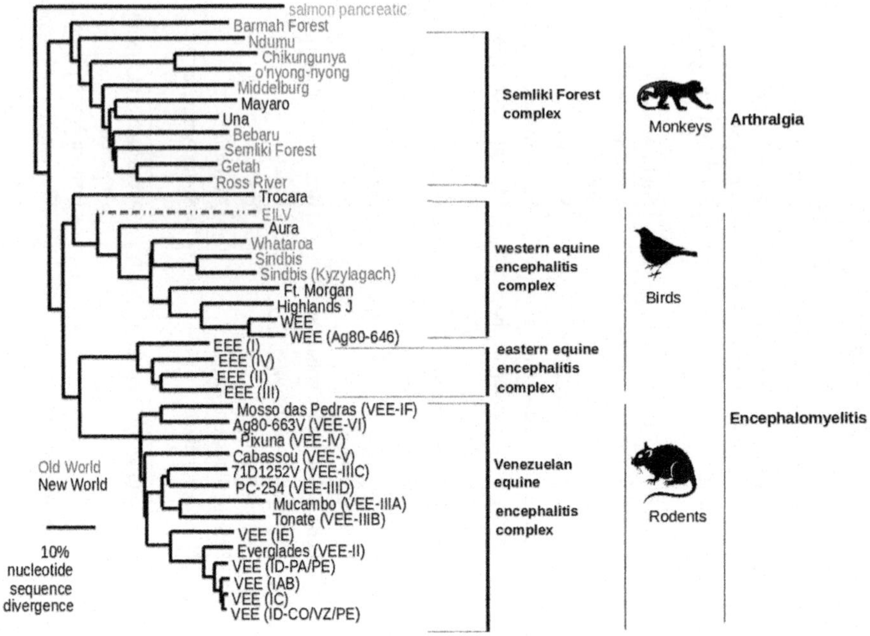

Fig. 9.1 Phylogenetic tree showing the evolutionary hypothesis of *Alphaviruses*. The viruses in *bold* are neotropical and New World viruses; those in *gray* are the Old World viruses (including the mosquito specific virus Eilat). *Lines* on *right* delimit the clades of viruses associated with a serogroup complex, vertebrate hosts, and symptoms (arthralgia and encephalomyelitis). (Modified from Weaver et al. [129])

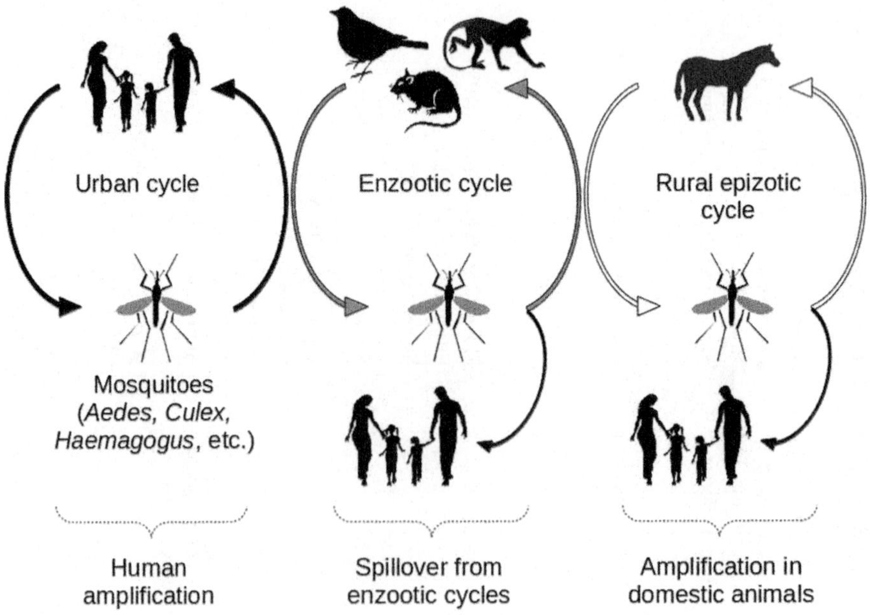

The majority of alphaviruses can cause at least mild febrile disease in humans, with several producing severe, life-threatening diseases; however, many remain poorly studied epidemiologically with unknown public health importance.

Recent studies in several Latin American locations of "dengue-like" illness revealed that alphaviruses such as VEEV, EEEV, MAYV, and CHIKV account for a significant number of cases misdiagnosed clinically as dengue. Moreover, with the recent introduction of the Zika virus, diagnosis based only on signs and symptoms is even more complicated in areas where these viruses are circulating simultaneously.

A better understanding of alphavirus transmission cycles, molecular evolution, vector biology, virus–vector–host coevolution, and rapid diagnostics is needed to prevent alphavirus diseases.

Brief descriptions of the most important alphaviruses in Latin America follow.

2 Alphavirus and Encephalitis

2.1 Venezuelan Equine Encephalitis Virus

The Venezuelan equine encephalitis (VEE) complex includes major human and equine pathogens and consequently is the most thoroughly studied in Latin America [126]. VEEV was recognized first in 1936 and isolated soon thereafter from the brains of fatal equids in Venezuela. This virus is transmitted enzootically between mosquitoes and rodents, and equine-amplified epizootic cycles cause large outbreaks of encephalitis in humans and horses [49, 129]. Outbreaks in Mexico and South America (Colombia, Venezuela, and Peru) demonstrated that VEE is a reemerging disease [44] as a naturally emerging pathogen endemic to South and Central America, Mexico, and Florida [126] circulating among wild rodents and mosquitoes. The introduction of horses, a new and susceptible host, into the Americas during the colonial period triggered outbreaks in these animals and increased the exposure of humans.

Fig. 9.2 Transmission cycles and mechanisms of human infection by alphaviruses. At the *center* is an enzootic cycle, typically involving avian, rodent, or nonhuman primates as amplification or reservoir hosts and mosquito vectors. Humans become infected via direct spillover when they enter enzootic habitats or when amplification results in high levels of circulation. Transmission to humans may involve the enzootic vector or bridge vectors with broader host preferences. *Right panel*: Secondary amplification involving domestic animals can increase circulation around humans, increasing their chance of infection via spillover. In the case of VEEV, mutations that enhance equine viremia are needed for secondary equine amplification. *Left panel*: CHIKV can use humans for amplification, resulting in urban epidemic cycles and massive outbreaks. (Modified from Weaver et al. [129] and Muñoz and Navarro [73])

Systematics and Geographic Distribution

The VEE complex is a sister of the eastern equine encephalitis (EEE) complex [90] and includes six subtypes. Only subtypes IAB and IC are traditionally considered epizootic strains that use equids for amplification via high-titer viremias. Other VEEV subtypes (ID and IE) are considered equine-avirulent, enzootic strains, although IE strains from recent Mexican epizootics appear to be equine-neurovirulent but incapable of generating high-titer equine viremia [35]. The remaining subtypes (II–VI) are also enzootic strains that generally circulate in sylvatic or swamp habitats and are considered incapable of equine amplification. In the United States, these include the Everglades virus in Florida and a variant of Tonate virus, Bijou Bridge, isolated in Colorado from cliff swallow bugs during the 1970s [17].

Geographic phylogenetic correlations of VEEV subtypes ID enzootic lineages or genotypes [1, 13, 93, 97, 124] suggest that geographic barriers explain their current distributions.

The major epidemic/epizootic subtype IAB and IC strains are highly pathogenic for horses, with case-fatality rates of 20% to 80%. The last major VEE outbreak occurred in 1995 in Venezuela and Colombia with 75,000 to 100,000 human cases, more than 300 of them fatal. In 1993, equine disease was associated with VEEV-IE in Mexico, and since 1993, human cases of VEEV ID-associated disease have occurred in Peru [33, 126].

VEEV infection usually causes flu-like symptoms, and encephalitis is rare in adults. Although the case-fatality rate is low (≤1%), neurological disease, including ataxia, disorientation, mental depression, and convulsions, can be detected in up to 14% of infected individuals, mainly children. High seroprevalence has been detected in humans in interepidemic/epizootic periods in Argentina [87, 88]. Neurological sequelae in humans are also common [34, 49, 94].

Evolution of Epizootic Strains from Enzootic Ancestors

Phylogenetic analysis of the VEE complex shows a close evolutionary relationship among IAB, IC, and ID strains and delineates six major lineages of enzootic VEEV, including five ID-like lineages and the subtype IE lineage. All epizootic strains from major outbreaks fall into one of three clades nested within one of these lineages, which is otherwise composed of enzootic ID strains from western Venezuela, Colombia, and northern Peru. These phylogenetic data support the hypothesis that epizootic VEEV strains have arisen on at least four occasions by mutation of enzootic ID strains and changes in host range. In further investigations, the occurrence of two mutations involving charge alterations on the surface of the E2 protein implies alterations in cellular receptor usage that influences pathogenesis as a mechanism of epizootic emergence [38, 91, 124–126].

Epizootic Transmission Cycle

The epizootic transmission cycle of VEEV is fairly well understood [91]. A feature common to all major outbreaks is the role of equids as highly efficient amplification hosts. Although the vertebrate host range of epizootic VEEV strains is wide and includes humans, rodents, bats, dogs, sheep, and some birds, major epidemics in the absence of equine cases have never occurred. Despite the repeated occurrence of epizootics near major cities such as Maracaibo (1995), in western Venezuela, inter-human mosquito-borne transmission has not been detected. However, the potential for urban transmission by a species such as *Aedes aegypti*, which is susceptible to infection after biting humans and exhibits behavioral traits such as multiple host feeding and peri-domesticity that augment its vector competence [43, 54, 80], or the continuous expansion of *Aedes albopictus* should be considered [12, 30, 68, 77], as human populations continue to expand and those of equines decline in rural areas in Latin America [85, 118]. Several small, atypical equine outbreaks were detected in Venezuela in the central llanos (2000–2003) with VEEV sequences showing high genomic stability 10 years after the 1995 outbreak. Cattle sero-surveys indicated the recent circulation of enzootic VEEV strains, and possibly of epizootic strains. Persistence of VEEV subtype IC strains and infection of horses at the end of the rainy season suggested the possibility of an alternative, cryptic transmission cycle involving survival through the dry season of infected vectors or persistently infected vertebrates [76].

Epizootic Vectors

Epizootic strains (subtype IAB and IC) of VEEV are opportunistic in their use of mosquito vectors during outbreaks. Field studies have indicated that more than one principal vector species can be involved in transmission [99, 122, 123, 136]. Although susceptibility to infection is a prerequisite for biological transmission, ecological and behavioral traits can be more important than susceptibility differences in vectorial capacity.

Although several mosquito species have been incriminated as VEEV vectors during epizootics, *Aedes* (*Ochlerotatus*) *taeniorhynchus*, a salt-marsh mosquito, may be the most important epizootic vector. This species is abundant in coastal areas including the Guajira Peninsula (Colombia and Venezuela), where many of the largest outbreaks have occurred, and virus isolations and susceptibility studies have documented its role in transmission [51, 107, 116, 136]. *Culex* (*Deinocerites*) spp. may also be VEEV vectors in coastal areas [37].

Psorophora confinnis and *P. columbiae* were probably important vectors during outbreaks in northern South America and in the 1971 epizootic/epidemic in northern Mexico and Texas [136]. *Aedes* (*Ochlerotatus*) *sollicitans* also exhibited extremely high infection rates in the coastal areas of Mexico and Texas in 1971 [136] and is capable of laboratory transmission following high-titer blood meals [116]. Non-mosquito arthropods (blackflies and ticks) have also been implicated as VEEV vectors but appear to be less important [55–57, 99].

Enzootic Transmission Cycle: Hosts and Vectors

Sylvatic rodents in the genera *Sigmodon*, *Oryzomys*, *Zygodontomys*, *Heteromys*, *Peromyscus*, and *Proechimys* are believed to be the principal reservoir hosts of most enzootic VEE complex viruses because they are frequently infected in nature, have high rates of immunity, and develop moderate- to high-titer viremia [23, 24, 126]. Spiny rats (*Proechimys semispinosus*) and cotton rats (*Sigmodon hispidus*) are the principal reservoir hosts of enzootic subtype ID viruses in Panama and also in Colombia and Venezuela [13, 14, 63, 79, 126]. Comparative studies in Venezuela and Colombia demonstrated a strong correlation between spiny rat (*Proechimys chrysaeolus* in Colombia) populations and levels of VEEV circulation [13]. Other mammals such as opossums (*Didelphis marsupialis*) are also frequently infected, and bats and shorebirds may be involved in the dispersal of enzootic viruses.

The most important enzootic vectors are members of the genus *Culex*, subgenus *Melanoconion*, Spissipes section [9, 25, 31, 71, 78, 101, 123, 126, 134], a diverse and taxonomically difficult group [78, 83, 98, 112]. The Spissipes section [98, 104, 112] includes most vectors of enzootic VEEV and EEEV in Latin America [18, 126], and seven species are proven vectors of VEE complex viruses. Studies of enzootic VEEV ecology have incriminated a single species or multiple species as the principal vector in a given location [9, 13, 31, 69, 123, 134]. A combination of enzootic vectors (transmission within forests) and epizootic vectors (potential exporters of the virus to open agricultural areas) were also studied in enzootic areas of the Catatumbo region in Venezuela [5, 67].

The restriction of most *Melanoconion* arbovirus vectors to the Spissipes section raises the question of what genetic, physiological, or ecological characteristics are shared by the members of this section that predispose them to transmit arboviruses. Recently, the use of ribosomal DNA sequences and phylogenetic methods have revealed evolutionary relationships among the Vomerifer and Pedroi groups of Spissipes [78]. Navarro and Weaver also detected two cryptic potential vector species under *Culex pedroi* supporting the hypothesis of differential vector capacity for VEEV and EEEV [31, 117].

Control and Prevention of VEE Outbreaks

Equine vaccination in enzootic countries, where progenitors of epizootic strains circulate and where recent outbreaks have been documented, can be effective if VEEV circulation is anticipated or recognized quickly during outbreaks. However, governmental responses to epizootics are often slow for reasons of veterinary and public heath surveillance deficiencies. The live-attenuated TC-83 vaccine is the most effective way to prevent and control epizootic VEEV transmission, and it is available throughout most of Latin America. However, some equids in South America are vaccinated with inactivated, multivalent alphavirus vaccines marketed in the United States. Immunity from these vaccines is slower to develop, is less durable, and requires frequent boosters. Therefore, public and veterinary health officials should strongly discourage the use of these inactivated vaccines in regions of Latin America with a history of VEE [126]. The protection of human populations relies principally on personal protection and avoidance of mosquito bites by limiting physical exposure and applying repellants containing the active ingredient diethylmethylbenzamide (DEET). Applying permethrin to clothing to enhance protection of individuals who reside or work near equine herds during epizootics, who contact tropical forest or swamp habitats where enzootic VEEV circulates, or during outbreaks is also effective. The rural–sylvan behavior of enzootic VEEV vectors renders the usual control based on ULV insecticide methods inappropriate and largely ineffective.

2.2 *Eastern and Western Equine Encephalitis Viruses*

2.2.1 Eastern Equine Encephalitis

Eastern equine encephalitis virus (EEEV) (*Alphavirus*, *Togaviridae*) is an single-stranded RNA mosquito-borne zoonotic pathogen transmitted throughout the Americas [105, 135]. In North America, EEEV circulates principally along the east cost of United States and Canada, and related strains have been also detected in northern Mexico and the Caribbean region. In North America, sporadic human cases averaging approximately five or six per year occur in swamp habitats where the enzootic cycle is involved mainly among birds of the Passeriformes order and the ornithophilic mosquito *Culiseta melanura* and *Culex* subgenus *Melanoconion* species in the southeast [105, 135]. EEEV infections in domestic animals are common; the case-fatality rates in both human and equine cases average about 50% to 70% or more.

South American eastern equine encephalitis virus (EEEV/SA) was first characterized in Argentina in 1933 [70]. New evidence shows that North American EEEV (EEEV/NA) and EEEV/SA variants have developed differences in the ecological, epidemiological, pathogenic, antigenic, and genetic profiles that allowed the classification of EEE/SA into a new species called *Madariaga virus* (MADV), named after the place of its first collection in Argentina. Consequently, MADV is composed of three distinct genetic lineages: one that circulates in Guatemala, Brazil, and Peru; a second lineage in Argentina, Brazil, Colombia, Ecuador, Guyana, Panama, Peru, Venezuela, and Trinidad; and a third represented by a single location isolated in Brazil [8].

Early reports suggest that MADV virus was human avirulent and causes equine epizootics with high mortality averaging approximately 70%; despite human exposure during epizootics, cases have never been detected in Panama and Argentina despite active surveillance [27, 70, 96]. Studies in Peru revealed that although the isolation of MADV in mosquitoes known to feed on humans was not uncommon, no MADV was isolated in acutely febrile patients [3]. Only three human cases of MADV infection had been recognized in Brazil (one) and in Trinidad (two) [21, 32] before the first documented epidemic was detected in Panama during 2010 [18].

Vector and Host

Culex (Melanoconion) taeniopus and *Cx. (Mel.) pedroi* are recognized as the main enzootic vectors of MADV in Central and South America, respectively [108, 117]. Forest-dwelling rodents and marsupials have been implicated as possible hosts based on serosurveys, although birds may also serve as hosts in the Amazon and southern regions of South America. Lizards have also been suggested as MADV hosts in Panama [22]. Recent serological studies in Panama suggest that the short-tailed cane mouse (*Zygodontomys brevicauda*) is a host for MADV, and humans active in pastures and farms where this rodent is abundant are at increased risk of infection [121].

Epidemiology
Human MADV infections in Panama are recognized after equine cases have been detected early in the rainy season (May–June). The distribution of cases is typically clustered in the Province of Darien close to the Colombian border. Severe cases are observed principally in children with a median age of 5.1 years, with a case-fatality rate around 10% [18].

With the exception of Panama, there is no report of human disease outbreaks attributed to MADV, and the lack of human cases in the rest of Latin America may result from (a) cross-protective immunity by heterologous alphavirus antibodies such as VEEV or (b) intrinsic characteristics of MADV strains, such as the inability to evade the interferon response [2, 3]. In addition to previous studies, in Panama recent evidence supports the effect of cross-protective immunity, as areas of high VEEV transmission appear to have reduced MADV transmission [121].

Clinical Characteristics
Symptomatic MADV cases present a prodromal phase, principally with fever and headache; vomiting and diarrhea occur less frequently. A neurological stage follows accompanied by disorientation, somnolence, seizures, and coma. Patients typically show elevated white cell counts and protein elevation in the cerebrospinal fluid (CSF). Severe cases can develop long-term neurological sequelae including seizures, hemiparesis, psychomotor retardation, and coma [18, 61].

Laboratory Diagnosis
Laboratory diagnosis is a challenge in endemic countries, where a high level of training to perform the viral isolation and serological assays is required. Co-circulation of multiple alphaviruses and antibody cross-reactions are common in Latin America. Furthermore, in endemic regions where multiple alphaviruses such as VEEV and EEEV (MADV) that cause similar clinical presentations are circulating, the interpretation of laboratory results is complex. The alphavirus IgM antibody response lasts around 2 to 3 months, and multiple diagnostic tools such as viral isolation, antibody tests, and viral RNA detection should be implemented [18, 62]. Although incidental laboratory infections with MADV have not been reported, diagnostic confirmation in endemic regions may require the use of live VEEV and EEEV in biosafety level 3 containment. In this case, neutralization tests for MADV and VEEV can be performed with EEEV chimeric viruses and TC-83 live-attenuated vaccine VEEV strain in biosafety level 2 facilities with results similar to those obtained with the wild-type strains [48, 121].

2.2.2 Western Equine Encephalitis

Western equine encephalitis virus (WEEV) is a recombinant alphavirus descended from Sindbis- and EEEV-like ancestors [40]. WEEV causes sporadic epizootics in the western United States and Canada, associated with increased rainfall in early spring followed by warmer-than-normal temperatures. *Culex tarsalis* is recognized as the principal vector in North America, and *Aedes albifasciatus* has been implicated in South America; passerine birds are the main enzootic hosts.

Circulation of WEEV in Latin America has been recognized during epizootics in Argentina, Uruguay, and Cuba [128]. A human fatal case was reported in Uruguay during 2009. However, the available evidence suggests that WEEV circulation is declining, with the last human case in North America reported during 1994, and the last detection in mosquito pools in 2008 [15].

3 Alphavirus and Arthralgias

3.1 Mayaro Virus

Mayaro virus (MAYV) is a unique, exceptional New World alphavirus, and, before the introduction of chikungunya virus (CHIKV) in 2013, represented the sole arthralgic alphavirus endemic to the Western Hemisphere. MAYV was first detected in forest workers in the county of Mayaro, Trinidad, in 1954 [6]. Since then, there is evidence of MAYV infection, either by serological detection or virus isolation, in several regions of Latin America, including Brazil, Colombia, Ecuador, Peru, Surinam, Bolivia, French Guiana, Trinidad, and Venezuela [4, 10, 11, 33, 39, 46, 50, 59, 66, 72, 81, 109–111, 113, 120, 137]. Although MAYV has not recently emerged sufficiently to result in major outbreaks, several acute undifferentiated febrile illness surveillance studies have clearly shown that the virus commonly infects humans, and that those infected most are infected as a result of occupational exposure [33, 120].

MAYV causes sporadic outbreaks, which have been localized primarily to regions of Brazil [11, 53, 64, 82, 120, 137], but has also been detected in Bolivia and Peru [33]. A pediatric infection was recently detected in Haiti, suggesting local enzootic (although no wild monkeys are present) or endemic circulation [139]. Typically, infection with MAYV is not fatal but cases usually present with fever, headache, retro-orbital pain, myalgia, vomiting, diarrhea, rash, and often persistent (<1 year) severe arthralgia [41, 100]. In this regard, MAYV may be more incapacitating than other common arboviruses such as dengue virus (DENV). Given the low economic impact of MAYV, there is very little vaccine development effort, the exception being an IRES-based attenuated live-attenuated vaccine [133]. The largest documented MAYV outbreak occurred in Belterra, Brazil, in 1978, with approximately 790 persons possibly affected and 55 confirmed cases: the virus was isolated from 43 cases [53, 86]. Since then, outbreaks have been very limited, until recently in Venezuela in 2010, where there was an outbreak of 77 suspected cases, of which 6 were detected by virus isolation from acute-phase sera [10].

Given the extent of the most recent MAYV outbreak in Venezuela, it is important to consider the possibility that *Aedes (Steg) aegypti* was involved in virus transmission, in addition to its enzootic vectors. *Ae. aegypti*-vectored arboviruses are among the most important arboviral pathogens, and previous studies suggest this is a moderately competent MAYV vector [60]. *Ae. albopictus* is a competent vector for several arboviruses [131], and in contrast to *Ae. aegypti*, it is found in periurban areas and more temperate regions. *Ae. albopictus* competency for MAYV has been examined experimentally, and results suggest it is not a competent vector unless bloodmeals are taken from highly viremic mice (>7 logs). If MAYV were to adapt for more efficient *Ae. aegypti* or *Ae. albopictus* transmission, it can present a significant global threat [140].

Phylogenetic studies of MAYV sequences show that the viruses can be further delineated into three genotypes, designated genotype D, L, and N [10, 89]. Genotype D includes isolates from Trinidad, Brazil, French Guiana, Surinam, Peru, and Bolivia; genotype L contains isolates from Brazil, but it is unclear if this genotype is still in circulation because it has not been detected since 1991; and genotype N consists of a single strain isolated from Peru in 2010 [10]. The single genotype N sequence is intermediary in phylogeny between genotypes D and L. It would be interesting to determine the prevalence of this strain in Peru and fully characterize its pathogenicity in mice relative to the other genotypes.

There is no evidence that the MAYV phylogeny is temporally structured, but it appears to be influenced to some extent by geography. Genotype D strains can be further delineated into smaller clades based on the geographic region of collection [10, 89]. Given this geographically structured phylogeny, we cannot exclude the possibility that there may be potential restrictions associated with vector competence, vector distributions, or alternative vertebrate amplification hosts that might affect this apparent population subdivision. However, sampling bias and the difficulty associated with isolating MAYV should be considered when proposing these conclusions. Additionally, recent studies provide evidence that MAYV strains concurrently circulating within Venezuela are undergoing regionally independent evolution, suggesting the absence of a single panmictic viral population, at least in Venezuela [10].

The MAYV ecological niche model based on localities of virus isolations [10, 89], vector distributions, and 19 climatic variables [58] predicts the suitable MAYV habitats in the following order of importance: eco-regions, followed by *Haemagogus (Hg.) leucocelaenus*, *Hg. celeste*, and *Hg. clarki* distributions. Figure 9.3 presents a

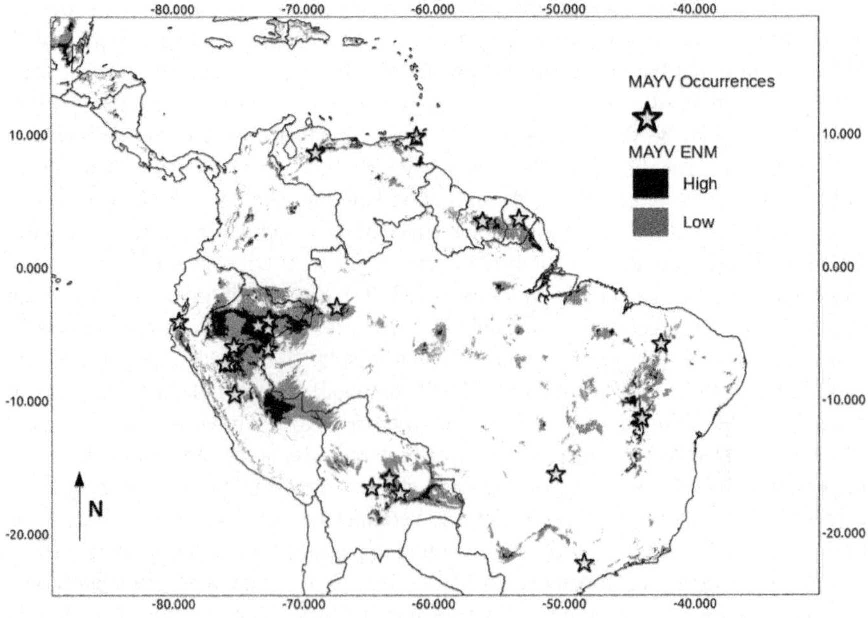

Fig. 9.3 Ecological niche model for Mayaro viruses (MAYV) based on virus isolation localities, seropositivity records, primary vector distribution (*Haemagogus* spp.), and 19 climatic variables

map of the predicted geographic distribution of the virus. MAYV presents six distributional patterns across South America related to terrestrial eco-regions showing the high and low probabilities of predicted areas.

Previous findings suggest that MAYV circulates between canopy-dwelling *Haemagogus* mosquitoes and nonhuman primates [45, 72, 89], but very little is known about the enzootic transmission cycles that perpetuate MAYV.

The MAYV genome is highly conserved (i.e., 96.4%–100% nucleotide sequence identities and 97.7%–100% amino acid sequence identities), across the complete genome, among genotype D strains [10]. Given this level of sequence conservation, it is unlikely that there may be significant phenotypic variation among strains, but recent work has shown the presence of five positively selected sites across the genome and nonsynonymous mutations that delineate various genotypes as well as the Venezuelan 2010 outbreak strains [10]. Reverse genetic studies are necessary to determine if any of these substitutions can cause phenotypic alterations. These studies can also be used to compare virulence among historical and contemporaneous isolates.

3.2 Chikungunya Virus

Chikungunya virus (CHIKV) is an alphavirus in the family *Togaviridae* and a relative of other neotropical viruses such as Venezuelan, western, and eastern equine encephalitis (although most strains in Latin America are part of the species *Madariaga* virus) viruses, as well as Mayaro (MAYV) and Una viruses, its closest relatives in the New World [132].

As does MAYV, CHIKV causes an acute febrile disease typically accompanied by severe arthralgia that can persist for years [130]. However, differing from MAYV wherein human infections are thought to result mainly from direct spillover of enzootic strains, CHIKV causes disease via direct spillover as well as by entering a human–mosquito–human cycle in urban areas, typically involving transmission by the anthropophilic mosquito *Aedes ae.* and recently also by *Ae. albopictus*; this leads to major epidemics involving millions of persons with efficient spread via infected air travelers during recent outbreaks. Although CHIKV is rarely fatal, newborns infected during birth, as well as the elderly, especially those with complicating, underlying medical conditions, can have severe neurological disease,. However, in addition to CHIKV being a direct cause of extensive morbidity in all age groups because of its typically high attack rates, the debilitating and often chronic arthralgia results in extensive economic impacts when infected persons cannot work or care for their families [102].

Chikungunya virus is believed to have originated in sub-Saharan Africa in enzootic cycles involving nonhuman primates and sylvatic *Aedes* spp. mosquito vectors, and these cycles continue in many regions of that continent. The history of CHIKV in Latin America probably began centuries ago when sailing ships carried it from Africa to port cities around the world, including the Caribbean and Latin America [127]. In fact, the term "dengue" may have originally described CHIKV infections, with the terminology becoming confused over the centuries with what is now known as dengue fever [42]. During modern scientific history since CHIKV was first isolated and associated with febrile illness in 1952 [65, 95], CHIKV is believed to have emerged from the enzootic African cycle to initiate urban transmission on several occasions, beginning about a century ago when the Asian lineage was introduced into South and Southeast Asia and caused outbreaks first recognized in 1958 (Fig. 9.4) [127]. This Asian lineage, transmitted primarily by *Ae. aegypti*, has continued to cause sporadic outbreaks in Asia and Oceania ever since. The next major emergence of a strain into a stable urban cycle began in 2004 when an outbreak began in coastal Kenya [20] and spread into the Indian Ocean basin as well as into Asia to infect millions of persons. Following the importation of this Indian Ocean lineage (IOL) strain by tens of thousands of infected travelers, outbreaks were also detected in Italy [92] and France [36]. However, despite importations into permissive (naïve human populations and abundant *Ae. aegypti*) dengue-endemic regions of the Americas, no local transmission was detected in the Western Hemisphere until late 2013, when an Asian lineage strain was implicated in human infections on the island of St. Martin in the Caribbean. Subsequently, this strain spread to nearly all Caribbean islands and throughout tropical and subtropical regions of Latin America during 2014, with continued circulation in many regions as of 2016. Then, in 2014, another CHIKV strain was introduced into northeastern Brazil directly from Africa (a member of the East/Central/South African, or ECSA, lineage). The distributions of the two CHIKV strains (Asian and ECSA lineages) are not completely known, but the Asian strain has been detected by sequencing in the Caribbean, Central America, Mexico, Florida in the United States (briefly following introductions in 2014), and northern South America, whereas the ECSA strain has not been reported outside Brazil. Determination of the geographic ranges of the two strains could be important because many ECSA strains have the ability to adapt for more efficient transmission by *Ae. albopictus* via mutations in the E1 and E2 envelope glycoprotein genes [114], although Asian lineages are epistatically constrained from such adaptation [115]. However, the ECSA strain circulating in Brazil and possibly beyond may have a different epistatic constraint based on a different E2 residue (position 211) [115]. Reverse genetic studies are needed to more definitively assess this adaptive potential of the Brazilian ECSA strain because the ability to use *Ae. albopictus* as an efficient vector could allow CHIKV to extend its geographic range into rural and temperate regions of Latin America.

Control of CHIKV in Latin America represents the same challenges imposed by dengue and now Zika viruses. Until a vaccine can be licensed (and several promising candidates are in late preclinical or early clinical stages of testing), vector control represents the only means of preventing infection and limiting spread. Although CHIKV has already spread extensively to many regions of Latin America and the Caribbean, and high seroprevalence [52, 103] as well as a drop in reported cases since 2014 (PAHO data) suggest that the epidemic has peaked in many regions, CHIKV infections continue to occur and outbreaks have not been reported in some areas with a history of dengue, suggesting continued spread. Unfortunately, past failures with the control of *Ae. aegypti* because of the wide range of challenges posed by this species do not bode well for this approach to CHIKV control [29]. The presence of MAYV in many parts of South America could also influence further

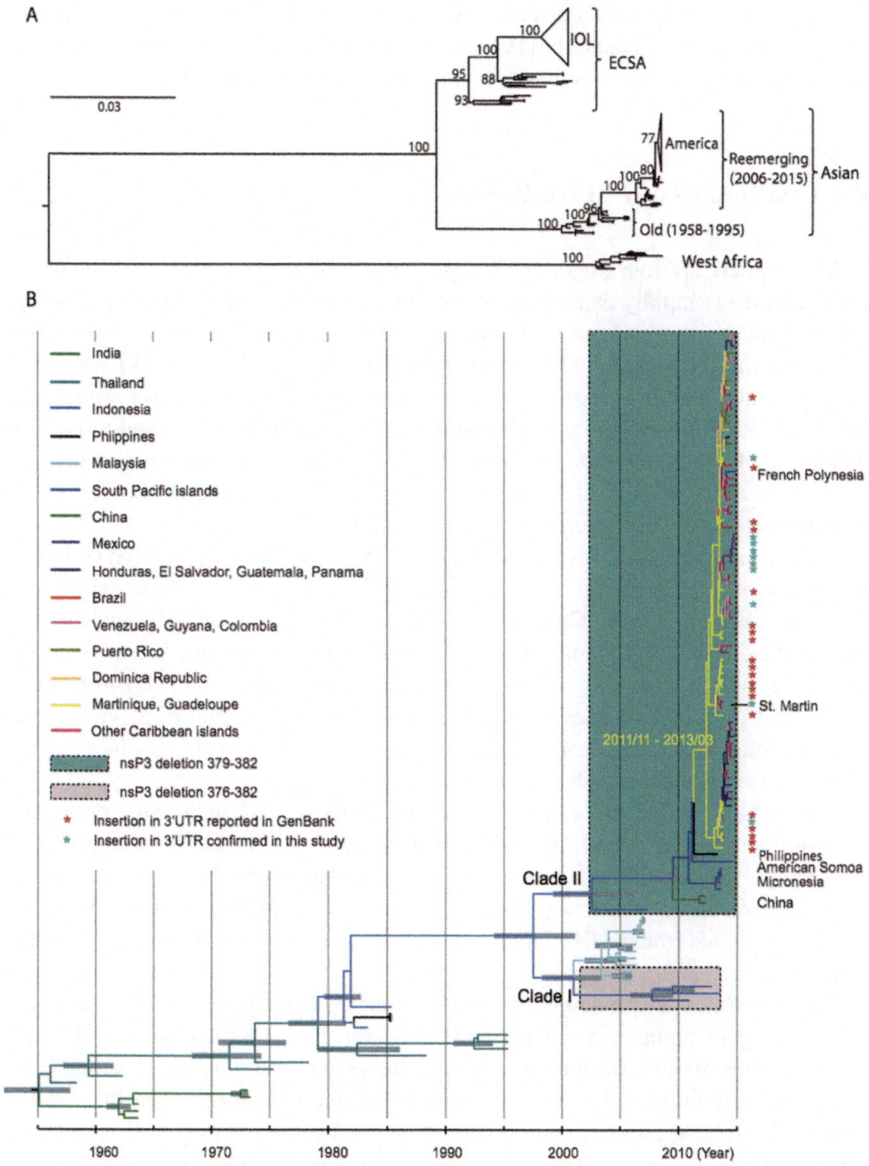

Fig. 9.4 Evolution of CHIKV. **a** Maximum-likelihood tree of major CHIKV lineages based on concatenated open reading frames (ORFs). Branch lengths reflect genetic distance. Bootstrap values are labeled for major lineages and clades. **b** Maximum clade credibility (MCC) tree of the Asian CHIKV lineage based on the Skyride population model using all Asian lineage strains. Branch length is scaled to the sampling and divergence time, and the branches are color coded for sample location. *Node bars* representing 95% highest probability density (HPD) value of the node height are shown only for those with a posterior probability of 90 or higher. (Modified from Chen et al. [19])

CHIKV spread and possibly disease manifestations. These two alphaviruses exhibit some antigenic cross-reactivity [16] that could reduce viremia to affect pathogenesis and possibly vector transmission, which should be evaluated in the near future.

4 Challenges of Future Research

In VEEV, there are four major challenges that we believe can be solved using new approaches: (1) rapidly estimating the origin of a newly discovered VEEV strain; (2) estimating its equine and/or human amplification and thus epidemic potential; (3) predicting the human virulence phenotype of a newly discovered VEEV strain. Phylogenetic relationships of a diverse collection of VEEV strains have proved useful for identification of the genetic features leading to epidemic spread to humans and livestock of this zoonotic pathogen. Also, (4) search for synapomorphic genetic, physiological, or ecological factors shared by Spissipes mosquitoes could explain their important role in transmitting arboviruses [34, 126].

Meanwhile, in EEEV several advances in the understanding of MADV pathogenesis have been achieved in the recent year. However, the available evidence is limited to in vitro studies. Genetic determinants of virulence are still unclear: animal models have failed to reproduce the natural history of disease, although cotton rats seem to be a promising model for evaluation of this question [7]. Basic epidemiological investigations are needed to understand the potential of MADV emergence in other Latin American countries; evaluation of cross-protective immunity is also important for vaccine design.

For MAYV and CHIKV, the degree and longevity of such cross-protection between both viruses should be further assessed not only to assist with predicting interactions that might limit circulation and spread but also to determine if an effective CHIKV vaccine could also limit disease caused by MAYV, which is probably indistinguishable from CHIKV infection and appears to be grossly underreported in Latin America [119].

Further work is warranted to truly understand the ecology of MAYV as a potentially emergent alphavirus. Of particular interest are (i) which vectors maintain enzootic transmission, (ii) what species are competent bridge vectors that facilitate transmission to humans, (iii) which nonhuman primate species serve as the primary amplification host, and last (iv) do other canopy-dwelling vertebrates, rodents, or birds have a role in transmission, or act as dead-end hosts only. These questions can be addressed through experimental infections in the laboratory or via field studies aimed at virus isolation and serological detection of MAYV among vectors and vertebrates in known endemic areas.

Acknowledgments We thank the Editors for the invitation to contribute this chapter. Projects Fonacit MC-2008000911-4 (Venez), P011617-UISEK (ECU).

Competing Interest The authors declare no conflict of interest.

References

1. Aguilar PV, Adams AP, Suárez V, Beingolea L, Vargas J, Manock S, Freire J, Espinoza WR, Felices V, Diaz A, Liang X, Roca Y, Weaver SC, Kochel TJ (2009) Genetic characterization of Venezuelan equine encephalitis virus from Bolivia, Ecuador and Peru: identification of a new subtype ID lineage. PLoS Negl Trop Dis 3:e514. doi:10.1371/journal.pntd.0000514
2. Aguilar PV, Paessler S, Carrara A-S, Baron S, Poast J, Wang E, Moncayo AC, Anishchenko M, Watts D, Tesh RB, Weaver SC (2005) Variation in interferon sensitivity and induction among strains of eastern equine encephalitis virus. J Virol 79:11300–11310
3. Aguilar PV, Robich RM, Turell MJ, O'Guinn ML, Klein TA, Huaman A, Guevara C, Rios Z, Tesh RB, Watts DM, Olson J, Weaver SC (2007) Endemic eastern equine encephalitis in the Amazon region of Peru. Am J Trop Med Hyg 76:293–298
4. Aitken TH, Downs WG, Anderson CR, Sence L, Casals J (1960) Mayaro virus isolated from a Trinidadian mosquito, Mansonia Venezuelensis. Science 131:986
5. Alfonzo D, Grillet ME, Liria J, Navarro J-C, Weaver SC, Barrera R (2005) Ecological characterization of the aquatic habitats of mosquitoes (Diptera: Culicidae) in enzootic foci of Venezuelan equine encephalitis virus in western Venezuela. J Med Entomol 42:278–284. doi:10.1603/0022-2585(2005)042[0278:ECOTAH]2.0.CO;2
6. Anderson CR, Downs WG, Wattley GH, Ahin NW, Reese AA (1957) Mayaro virus: a new human disease agent. II. Isolation from blood of patients in Trinidad, B.W.I. Am J Trop Med Hyg 6:1012–1016
7. Arrigo NC, Adams AP, Watts DM, Newman PC, Weaver SC, Paige Adams A, Watts DM, Newman PC, Weaver SC (2010) Cotton rats and house sparrows as hosts for north and south american strains of eastern equine encephalitis virus. Emerg Infect Dis 16:1373–1380. doi:10.3201/eid1609.100459
8. Arrigo NC, Adams AP, Weaver SC (2010) Evolutionary patterns of eastern equine encephalitis virus in north versus South America suggest ecological differences and taxonomic revision. J Virol 84:1014–1025. doi:10.1128/JVI.01586-09
9. Auguste AJ, Adams AP, Arrigo NC, Martinez R, Travassos da Rosa APA, Adesiyun AA, Chadee DD, Tesh RB, Carrington CVF, Weaver SC (2010) Isolation and characterization of sylvatic mosquito-borne viruses in Trinidad: enzootic transmission and a new potential vector of Mucambo virus. Am J Trop Med Hyg 83:1262–1265. doi:10.4269/ajtmh.2010.10-0280
10. Auguste AJ, Liria J, Forrester NL, Giambalvo D, Moncada M, Long KC, Morón D, de Manzione N, Tesh RB, Halsey ES, Kochel TJ, Hernandez R, Navarro J, Weaver SC, de Manzione N, Tesh RB, Halsey ES, Kochel TJ, Hernandez R, Navarro J, Weaver SC (2015) Evolutionary and ecological characterization of Mayaro virus strains isolated during an outbreak, Venezuela, 2010. Emerg Infect Dis J 21:1742–1750. doi:10.3201/eid2110.141660
11. Azevedo RSS, Silva EVP, Carvalho VL, Rodrigues SG, Nunes Neto JP, Monteiro HAO, Peixoto VS, Chiang JO, Nunes MRT, Vasconcelos PFC, Nunes-Neto JP, Monteiro HAO, Peixoto VS, Chiang JO, Nunes MRT, Vasconcelos PFC (2009) Mayaro fever virus, Brazilian amazon. Emerg Infect Dis 15:1830–1832. doi:10.3201/eid1511.090461
12. Bargielowski IE, Lounibos LP, Shin D, Smartt CT, Carrasquilla MC, Henry A, Navarro JC, Paupy C, Dennett JA (2015) Widespread evidence for interspecific mating between *Aedes aegypti* and *Aedes albopictus* (Diptera: Culicidae) in nature. Infect Genet Evol 36:6–11. doi:10.1016/j.meegid.2015.08.016
13. Barrera R, Ferro C, Navarro J-C, Freier J, Liria J, Salas R, Ahumada M, Vasquez C, Gonzalez M, Kang W, Boshell J, Weaver SC (2002) Contrasting sylvatic foci of Venezuelan equine encephalitis virus in northern South America. Am J Trop Med Hyg 67:324–334
14. Barrera R, Torres N, Freier JE, Navarro JC, García CZ, Salas R, Vasquez C, Weaver SC (2001) Characterization of enzootic foci of Venezuelan equine encephalitis virus in western Venezuela. Vector Borne Zoonotic Dis 1:219–230. doi:10.1089/153036601753552585
15. Bergren NA, Auguste AJ, Forrester NL, Negi SS, Braun WA, Weaver SC (2014) Western equine encephalitis virus: evolutionary analysis of a declining alphavirus based on complete genome sequences. J Virol 88:9260–9267. doi:10.1128/JVI.01463-14

16. Calisher C, Karabatsos N (1988) Arbovirus serogroups: definition and geographic distribution. In: Montar T (ed) The arboviruses: epidemiology and ecology, vol I. CRC Press, Boca Raton, pp 19–57
17. Calisher CH, Lazuick JS, Justines G, Francy DB, Monath TP, Gutierrez E, Haven N (1981) Viruses isolated from Aedomyia squamipennis collected in Panama, Ecuador and Argentina: establishment of the Gamboa Serogroup. Am J Trop Med Hyg 30:219–223
18. Carrera J-P, Forrester N, Wang E, Vittor AY, Haddow AD, López-Vergès S, Abadía I, Castaño E, Sosa N, Báez C, Estripeaut D, Díaz Y, Beltrán D, Cisneros J, Cedeño HG, Travassos da Rosa AP, Hernandez H, Martínez-Torres AO, Tesh RB, Weaver SC (2013) Eastern equine encephalitis in Latin America. N Engl J Med 369:732–744. doi:10.1056/NEJMoa1212628
19. Chen R, Puri V, Fedorova N, Lin D, Hari KL, Jain R, Rodas JD, Das SR, Shabman RS, Weaver SC (2016) Comprehensive genome scale phylogenetic study provides new insights on the global expansion of chikungunya virus. J Virol 90:10600–10611. doi:10.1128/JVI.01166-16
20. Chretien J-P, Anyamba A, Bedno SA, Breiman RF, Sang R, Sergon K, Powers AM, Onyango CO, Small J, Tucker CJ, Linthicum KJ (2007) Drought-associated chikungunya emergence along coastal East Africa. Am J Trop Med Hyg 76:405–407
21. Corniou B, Ardoin P, Bartholomew C, Ince W, Massiah V (1972) First isolation of a South American strain of eastern equine virus from a case of encephalitis in Trinidad. Trop Geogr Med 24:162–167
22. Craighead JE, Shelokov A, Peralta PH (1962) The lizard: a possible host for eastern equine encephalitis virus in Panama. Am J Hyg 76:82–87
23. Deardorff ER, Estrada-Franco JG, Freier JE, Navarro-Lopez R, Travassos Da Rosa A, Tesh RB, Weaver SC, Da Rosa AT, Tesh RB, Weaver SC (2011) Candidate vectors and rodent hosts of Venezuelan equine encephalitis virus, Chiapas, 2006–2007. Am J Trop Med Hyg 85:1146–1153. doi:10.4269/ajtmh.2011.11-0094
24. Deardorff ER, Forrester NL, Travassos-da-Rosa AP, Estrada-Franco JG, Navarro-Lopez R, Tesh RB, Weaver SC (2009) Experimental infection of potential reservoir hosts with Venezuelan equine encephalitis virus, Mexico. Emerg Infect Dis 15:519–525. doi:10.3201/eid1504.081008
25. Deardorff ER, Weaver SC (2010) Vector competence of *Culex* (*Melanoconion*) *taeniopus* for equine-virulent subtype IE strains of Venezuelan equine encephalitis virus. Am J Trop Med Hyg 82:1047–1052. doi:10.4269/ajtmh.2010.09-0556
26. Díaz-González EE, Kautz TF, Dorantes-Delgado A, Malo-García IR, Laguna-Aguilar M, Langsjoen RM, Chen R, Auguste DI, Sánchez-Casas RM, Danis-Lozano R, Weaver SC, Fernández-Salas I (2015) First report of *Aedes aegypti* transmission of chikungunya virus in the Americas. Am J Trop Med Hyg 93:1325–1329. doi:10.4269/ajtmh.15-0450
27. Dietz WH, Galindo P, Johnson KM (1980) Eastern equine encephalomyelitis in Panama: the epidemiology of the 1973 epizootic. Am J Trop Med Hyg 29:133–140
28. Erasmus JH, Needham J, Raychaudhuri S, Diamond MS, Beasley DWC, Morkowski S, Salje H, Fernandez Salas I, Kim DY, Frolov I, Nasar F, Weaver SC (2015) Utilization of an Eilat virus-based chimera for serological detection of chikungunya infection. PLoS Negl Trop Dis 9:e0004119. doi:10.1371/journal.pntd.0004119
29. Fernández-Salas I, Danis-Lozano R, Casas-Martínez M, Ulloa A, Bond JG, Marina CF, Lopez-Ordóñez T, Elizondo-Quiroga A, Torres-Monzón JA, Díaz-González EE (2015) Historical inability to control *Aedes aegypti* as a main contributor of fast dispersal of chikungunya outbreaks in Latin America. Antivir Res 124:30–42. doi:10.1016/j.antiviral.2015.10.015
30. Fernandez Z, Moncayo AC, Carrara AS, Forattini OP, Weaver ASC (2003) Vector competence of rural and urban strains of *Aedes* (*Stegomyia*) *albopictus* (Diptera: Culicidae) from São Paulo state, Brazil for IC, ID, and IF subtypes of Venezuelan equine encephalitis virus. J Med Entomol 40:522–527
31. Ferro C, Boshell J, Moncayo AC, Gonzalez M, Ahumada ML, Kang W, Weaver SC (2003) Natural enzootic vectors of Venezuelan equine encephalitis virus, Magdalena Vally, Colombia. Emerg Infect Dis 9:49–54. doi:10.3201/eid0901.020136

32. Alice FJ (1956) Infeccao humana pelo virus "leste" da encefalite equina. Bol Inst Biol da Bahia 3:3–9
33. Forshey BM, Guevara C, Laguna-Torres VA, Cespedes M, Vargas J, Gianella A, Vallejo E, Madrid C, Aguayo N, Gotuzzo E, Suarez V, Morales AM, Beingolea L, Reyes N, Perez J, Negrete M, Rocha C, Morrison AC, Russell KL, Blair PJ, Olson JG, Kochel TJ, Blair PJ, Olson JG, Kochel TJ (2010) Arboviral etiologies of acute febrile illnesses in western South America, 2000–2007. PLoS Negl Trop Dis 4:2000–2007. doi:10.1371/journal.pntd.0000787
34. Gardner SN, McLoughlin K, Be NA, Allen J, Weaver SC, Forrester N, Guerbois M, Jaing C (2016) Characterization of genetic variability of Venezuelan equine encephalitis viruses. PLoS One 11:e0152604. doi:10.1371/journal.pone.0152604
35. Gonzalez-Salazar D, Estrada-Franco JG, Carrara AS, Aronson JF, Weaver SC (2003) Equine amplification and virulence of subtype IE Venezuelan equine encephalitis viruses isolated during the 1993 and 1996 Mexican epizootics. Emerg Infect Dis 9:161–168. doi:10.3201/eid0902.020124
36. Grandadam M, Caro V, Plumet S, Thiberge JM, Souarès Y, Failloux AB, Tolou HJ, Budelot M, Cosserat D, Leparc-Goffart I, Desprès P (2011) Chikungunya virus, southeastern France. Emerg Infect Dis 17:910–913. doi:10.3201/eid1705.101873
37. Grayson MA, Galindo P (1972) Experimental transmission of Venezuelan equine encephalitis virus by *Deinocerites pseudes* Dyar and Knab, 1909. J Med Entomol 9:196–200
38. Greene IP, Paessler S, Austgen L, Anishchenko M, Brault AC, Bowen RA, Weaver SC (2005) Envelope glycoprotein mutations mediate equine amplification and virulence of epizootic venezuelan equine encephalitis virus. J Virol 79:9128–9133. doi:10.1128/JVI.79.14.9128-9133.2005
39. Groot H, Morales A, Vidales H (1961) Virus isolations from forest mosquitoes in San Vicente de Chucuri, Colombia. Am J Trop Med Hyg 10:397–402
40. Hahn CS, Lustig S, Strauss EG, Strauss JH (1988) Western equine encephalitis virus is a recombinant virus. Proc Natl Acad Sci U S A 85:5997–6001
41. Halsey ES, Siles C, Guevara C, Vilcarromero S, Jhonston EJ, Ramal C, Aguilar PV, Ampuero JS (2013) Mayaro virus infection, Amazon Basin region, Peru, 2010-2013. Emerg Infect Dis 19:1839–1842. doi:10.3201/eid1911.130777
42. Halstead SB (2015) Reappearance of chikungunya, formerly called dengue, in the Americas. Emerg Infect Dis 21:557–561. doi:10.3201/eid2104.141723
43. Harrington LC, Edman JD, Scott TW (2001) Why do female *Aedes aegypti* (Diptera: Culicidae) feed preferentially and frequently on human blood? J Med Entomol 38:411–422. doi:10.1603/0022-2585-38.3.411
44. Hawley RJ, Eitzen EM (2001) Biological weapons—a primer for microbiologists. Annu Rev Microbiol 55:235–253. doi:10.1146/annurev.micro.55.1.235
45. Hoch AL, Peterson NE, LeDuc JW, Pinheiro FP (1981) An outbreak of Mayaro virus disease in Belterra, Brazil. III. Entomological and ecological studies. Am J Trop Med Hyg 30:689–698
46. Izurieta RO, Macaluso M, Watts DM, Tesh RB, Guerra B, Cruz LM, Galwankar S, Vermund SH (2011) Hunting in the rainforest and Mayaro virus infection: an emerging alphavirus in Ecuador. J Glob Infect Dis 3:317–323. doi:10.4103/0974-777X.91049
47. Johansson MA, Powers AM, Pesik N, Cohen NJ, Erin Staples J (2014) Nowcasting the spread of chikungunya virus in the Americas. PLoS One 9:e104915. doi:10.1371/journal.pone.0104915
48. Johnson BW, Kosoy O, Wang E, Delorey M, Russell B, Bowen RA, Weaver SC (2011) Use of sindbis/eastern equine encephalitis chimeric viruses in plaque reduction neutralization tests for arboviral disease diagnostics. Clin Vaccine Immunol 18:1486–1491. doi:10.1128/CVI.05129-11
49. Johnson KM, Martin DH (1974) Venezuelan equine encephalitis. Adv Vet Sci Comp Med 18:79–116
50. Karbaat J (1965) Arthropod-borne virus infections in Dutch soldiers stationed in Surinam. Trop Geogr Med 17:187–189

51. Kramer LD, Scherer WF (1976) Vector competence of mosquitoes as a marker to distinguish central American and Mexican epizootic from enzootic strains of Venezuelan encephalitis virus. Am J Trop Med Hyg 25:336–346
52. Kuan G, Ramirez S, Gresh L, Ojeda S, Melendez M, Sanchez N, Collado D, Garcia N, Mercado JC, Gordon A, Balmaseda A, Harris E (2016) Seroprevalence of anti-chikungunya virus antibodies in children and adults in Managua, Nicaragua, after the first chikungunya epidemic, 2014-2015. PLoS Negl Trop Dis 10:e0004773. doi:10.1371/journal.pntd.0004773
53. LeDuc JW, Pinheiro FP, Travassos da Rosa AP (1981) An outbreak of Mayaro virus disease in Belterra, Brazil. II. Epidemiology. Am J Trop Med Hyg 30:682–688
54. Liebman KA, Stoddard ST, Reiner RC, Perkins TA, Astete H, Sihuincha M, Halsey ES, Kochel TJ, Morrison AC, Scott TW (2014) Determinants of heterogeneous blood feeding patterns by *Aedes aegypti* in Iquitos, Peru. PLoS Negl Trop Dis 8:e2702. doi:10.1371/journal.pntd.0002702
55. Linthicum KJ, Gordon SW, Monath TP (1992) Comparative infections of epizootic and enzootic strains of Venezuelan equine encephalomyelitis virus in *Amblyomma cajennense* (Acari: Ixodidae). J Med Entomol 29:827–831
56. Linthicum KJ, Logan TM (1994) Laboratory transmission of Venezuelan equine encephalomyelitis virus by the tick *Hyalomma truncatum*. Trans R Soc Trop Med Hyg 88:126. doi:10.1016/0035-9203(94)90536-3
57. Linthicum KJ, Logan TM, Bailey CL, Gordon SW, Peters CJ, Monath TP, Osorio J, Francy DB, McLean RG, Leduc JW et al (1991) Venezuelan equine encephalomyelitis virus infection in and transmission by the tick *Amblyomma cajennense* (Arachnida: Ixodidae). J Med Entomol 28:405–409
58. Liria J, Navarro JC (2010) Modelo de nicho ecológico en *Haemagogus* Williston (Diptera: Culicidae), vectores del virus de la fiebre amarilla. Rev Biomédica 21:149–161
59. Llagonne-Barets M, Icard V, Leparc-Goffart I, Prat C, Perpoint T, André P, Ramière C (2016) A case of Mayaro virus infection imported from French Guiana. J Clin Virol 77:66–68. doi:10.1016/j.jcv.2016.02.013
60. Long KC, Ziegler SA, Thangamani S, Hausser NL, Kochel TJ, Higgs S, Tesh RB (2011) Experimental transmission of Mayaro virus by *Aedes aegypti*. Am J Trop Med Hyg 85:750–757. doi:10.4269/ajtmh.2011.11-0359
61. Luciany K, Abadía I, Cisneros J, Guerra I, Garcıa M, Martínez-Torres AO, Estripeaut D, Carrera J-P (2015) Case report : madariaga virus infection associated with a case of acute disseminated encephalomyelitis. Am J Trop Med Hyg 92:1130–1132. doi:10.4269/ajtmh.14-0845
62. Luciany K, Abadía II, Cisneros J, Guerra I, Garcıa M, Martínez-Torres AO, Estripeaut D, Carrera J-P, Garcı M, Luciany K, Abadía II, Cisneros J, Guerra I, Garcıa M, Martínez-Torres AO, Estripeaut D, Carrera J-P, Luciani K, Abadía II, Martínez-Torres AO, Cisneros J, Guerra I, García M, Estripeaut D, Carrera J-P (2015) Case report : madariaga virus infection associated with a case of acute disseminated encephalomyelitis. Am J Trop Med Hyg 92:1130–1132. doi:10.4269/ajtmh.14-0845
63. Grayson MA, Galindo P, Grayson MA, Galindo P (1969) Ecology of Venezuelan equine encephalitis virus in Panama. Am Vet Med Assoc 155:2141–2145
64. do Carmo Alves Martins V, de Souza Bastos M, Ramasawmy R, de Figueiredo RP, JBL G, WSM B, Nogueira ML, Nozawa S, Naveca FG, LTM F, MPG M (2014) Clinical and virological descriptive study in the 2011 outbreak of dengue in the Amazonas, Brazil. PLoS One 9:e100535. doi:10.1371/journal.pone.0100535
65. Mason PJ, Haddow AJ (1957) An epidemic of virus disease in Southern Province, Tanganyika territory, in 1952–1953. Trans R Soc Trop Med Hyg 51:238–240. doi:10.1016/0035-9203(57)90022-6
66. Medina G, Garzaro DJ, Barrios M, Auguste AJ, Weaver SC, Pujol FH (2015) Genetic diversity of Venezuelan alphaviruses and circulation of a Venezuelan equine encephalitis virus subtype IAB strain during an interepizootic period. Am J Trop Med Hyg 93:7–10. doi:10.4269/ajtmh.14-0543

67. Mendez W, Liria J, Navarro JC, Garcia CZ, Freier JE, Salas R, Weaver SC, Barrera R (2001) Spatial dispersion of adult mosquitoes (Diptera: Culicidae) in a sylvatic focus of Venezuelan equine encephalitis virus. J Med Entomol 38:813–821
68. Miller MJ, Loaiza JR (2015) Geographic expansion of the invasive mosquito *Aedes albopictus* across Panama—implications for control of dengue and chikungunya viruses. PLoS Negl Trop Dis 9:e0003383. doi:10.1371/journal.pntd.0003383
69. Mitchell CJ, Monath TP, Sabattini MS, Cropp CB, Daffner JF, Calisher CH, Jakob WL, Christensen HA (1985) Arbovirus investigations in Argentina, 1977–1980. II. Arthropod collections and virus isolations from argentine mosquitoes. Am J Trop Med Hyg 34:945–955
70. Monath TP, Sabattini MS, Mitchell CJ, Bowen GS (1985) Arbovirus investigations in Argentina, 1977–1980. Surveys 34:966–975
71. Monath TP, Sabattini MS, Pauli R, Daffner JF, Mitchell CJ, Bowen GS, Cropp CB (1985) Arbovirus investigations in Argentina, 1977-1980. IV. Serologic surveys and sentinel equine program. Am J Trop Med Hyg 34:966–975
72. Muñoz M, Navarro JC (2012) Mayaro: a re-emerging arbovirus in Venezuela and Latin America. Biomedica (Bogotá) 32:286–302. doi:10.1590/S0120-41572012000300017
73. Muñoz M, Navarro JC (2012) Virus Mayaro : un arbovirus reemergente en Venezuela y Latinoamérica. Biomédica (Bogotá) 32:286–302
74. Nasar F, Gorchakov RV, Tesh RB, Weaver SC (2015) Eilat virus host range restriction is present at multiple levels of the virus life cycle. J Virol 89:1404–1418. doi:10.1128/JVI.01856-14
75. Nasar F, Palacios G, Gorchakov RV, Guzman H, Da Rosa APT, Savji N, Popov VL, Sherman MB, Lipkin WI, Tesh RB, Weaver SC (2012) Eilat virus, a unique alphavirus with host range restricted to insects by RNA replication. Proc Natl Acad Sci U S A 109:14622–14627. doi:10.1073/pnas.1204787109
76. Navarro J-C, Medina G, Vasquez C, Coffey LL, Wang E, Suárez A, Biord H, Salas M, Weaver SC (2005) Postepizootic persistence of Venezuelan equine encephalitis virus, Venezuela. Emerg Infect Dis 11:1907–1915. doi:10.3201/eid1112.050533
77. Navarro J-C, Quintero L, Zorrilla A, González R (2013) Molecular tracing with mitochondrial ND5 of the invasive mosquito *Aedes* (*Stegomyia*) *albopictus* (Skuse) in Northern South America. J Entomol Zool Stud 1(4):32–39
78. Navarro J-C, Weaver SC (2004) Molecular phylogeny of the Vomerifer and pedroi groups in the Spissipes section of the subgenus *Culex* (*Melanoconion*). J Med Entomol 41:575–581
79. Navarro J, Medina G, Vasquez C, Coffey LL, Wang E, Suárez A, Biord H, Salas M, Weaver SC (2005) Postepizootic persistence of Venezuelan equine encephalitis. Emerg Infect Dis 11:1907–1915
80. Ortiz DI, Kang W, Weaver SC (2008) Susceptibility of *Ae. aegypti* (Diptera: Culicidae) to infection with epidemic (subtype IC) and enzootic (subtypes ID, IIIC, IIID) Venezuelan equine encephalitis complex alphaviruses. J Med Entomol 45:1117–1125. doi:10.1603/0022-2585(2008)45[1117:SOAADC]2.0.CO;2
81. Pauvolid-Correa A, Tavares FN, Alencar J, Silva Jdos S, Murta M, Serra-Freire NM, Pellegrin AO, Gil-Santana H, Guimaraes AE, Silva E (2010) Preliminary investigation of Culicidae species in south Pantanal, Brazil and their potential importance in arbovirus transmission. Rev Inst Med Trop Sao Paulo 52:17–24
82. Pauvolid-Corrêa A, Juliano RS, Campos Z, Velez J, Nogueira RMR, Komar N (2015) Neutralising antibodies for Mayaro virus in Pantanal, Brazil. Mem Inst Oswaldo Cruz 110:125–133. doi:10.1590/0074-02760140383
83. Pecor JE, Jones J, Turell MJ, Fernandez R, Carbajal F, O'Guinn M, Sardalis M, Watts D, Zyzak M, Calampa C, Klein TA (2000) Annotated checklist of the mosquito species encountered during arboviral studies in Iquitos, Peru (Diptera: Culicidae). J Am Mosq Control Assoc 16:210–218
84. Petersen LR, Powers AM (2016) Chikungunya: epidemiology. F1000Research 5:1–8. doi:10.12688/f1000research.7171.1
85. Pfeffer M, Dobler G (2010) Emergence of zoonotic arboviruses by animal trade and migration. Parasit Vectors 3:35. doi:10.1186/1756-3305-3-35

86. Pinheiro FP, Freitas RB, Travassos da Rosa JF, Gabbay YB, Mello WA, LeDuc JW (1981) An outbreak of Mayaro virus disease in Belterra, Brazil. I. Clinical and virological findings. Am J Trop Med Hyg 30:674–681
87. Pisano MB, Oria G, Beskow G, Aguilar J, Konigheim B, Cacace ML, Aguirre L, Stein M, Contigiani MS (2013) Venezuelan equine encephalitis viruses (VEEV) in Argentina: serological evidence of human infection. PLoS Negl Trop Dis 7:e2551. doi:10.1371/journal.pntd.0002551
88. Pisano MB, Ré VE, Díaz LA, Farías A, Stein M, Sanchez-Seco MP, Tenorio A, Almirón WR, Contigiani MS (2010) Enzootic activity of pixuna and Rio Negro viruses (Venezuelan equine encephalitis complex) in a neotropical region of Argentina. Vector Borne Zoonotic Dis 10:199–201. doi:10.1089/vbz.2008.0156
89. Powers AM, Aguilar PV, Chandler LJ, Brault AC, Meakins TA, Watts D, Russell KL, Olson J, Vasconcelos PFC, Da Rosa AT, Weaver SC, Tesh RB (2006) Genetic relationships among Mayaro and Una viruses suggest distinct patterns of transmission. Am J Trop Med Hyg 75:461–469. 75/3/461 [pii]
90. Powers AM, Brault AC, Shirako Y, Strauss EG, Kang W, Strauss JH, Weaver SC (2001) Evolutionary relationships and systematics of the alphaviruses. J Virol 75:10118–10131. doi:10.1128/JVI.75.21.10118-10131.2001
91. Powers AM, Oberste MS, Brault AC, Rico-Hesse R, Schmura SM, Smith JF, Kang W, Sweeney WP, Weaver SC (1997) Repeated emergence of epidemic/epizootic Venezuelan equine encephalitis from a single genotype of enzootic subtype ID virus. J Virol 71:6697–6705
92. Rezza G, Nicoletti L, Angelini R, Romi R, Finarelli A, Panning M, Cordioli P, Fortuna C, Boros S, Magurano F, Silvi G, Angelini P, Dottori M, Ciufolini M, Majori G, Cassone A (2007) Infection with chikungunya virus in Italy: an outbreak in a temperate region. Lancet 370:1840–1846. doi:10.1016/S0140-6736(07)61779-6
93. Rico-Hesse R, Weaver SC, de Siger J, Medina G, Salas RA (1995) Emergence of a new epidemic/epizootic Venezuelan equine encephalitis virus in South America. Proc Natl Acad Sci U S A 92:5278–5281. doi:10.1073/pnas.92.12.5278
94. Ronca SE, Dineley KT, Paessler S (2016) Neurological sequelae resulting from encephalitic alphavirus infection. Front Microbiol 7:959. doi:10.3389/fmicb.2016.00959
95. Ross RW (1956) The Newala Epidemic. III. The virus: isolation, pathogenic properties and relationship to the epidemic. J Hyg (Lond) 54:177–191. doi:10.1017/S0022172400044442
96. Sabattini MS, Daffner JFMT, Al E (1991) Localized eastern equine encephalitis in Santiago del Estero Province, Argentina, without human infection. Med Buenos Aires 51:3–8
97. Salas RA, Garcia CZ, Liria J, Barrera R, Navarro JC, Medina G, Vasquez C, Fernandez Z, Weaver SC (2001) Ecological studies of enzootic Venezuelan equine encephalitis in north-central Venezuela, 1997-1998. Am J Trop Med Hyg 64:84–92
98. Sallum MA, Forattini OP (1996) Revision of the Spissipes section of Culex (Melanoconion) (Diptera:Culicidae). J Am Mosq Control Assoc 12:517–600
99. Sanmartin C, Mackenzie RB, Trapido H, Barreto P, Mullenax CH, Gutierrez E, Lesmes C (1973) Venezuelan equine encephalitis in Colombia, 1967. Bol Oficina Sanit Panam 74:108–137
100. Santiago FW, Halsey ES, Siles C, Vilcarromero S, Guevara C, Silvas JA, Ramal C, Ampuero JS, Aguilar PV (2015) Long-term arthralgia after Mayaro virus infection correlates with sustained pro-inflammatory cytokine response. PLoS Negl Trop Dis 9:e0004104. doi:10.1371/journal.pntd.0004104
101. Scherer WF, Weaver SC, Taylor CA, Cupp EW, Dickerman RW, Rubino HH (1987) Vector competence of *Culex* (*Melanoconion*) *taeniopus* for allopatric and epizootic Venezuelan equine encephalomyelitis viruses. Am J Trop Med Hyg 36:194–197
102. Schilte C, Staikovsky F, Couderc T, Madec Y, Carpentier F, Kassab S, Albert ML, Lecuit M, Michault A (2013) Chikungunya virus-associated long-term arthralgia: a 36-month prospective longitudinal study. PLoS Negl Trop Dis. doi:10.1371/journal.pntd.0002137
103. Simmons G, Brès V, Lu K, Liss NM, Brambilla DJ, Ryff KR, Bruhn R, Velez E, Ocampo D, Linnen J, Latoni G, Petersen LR, Williamson PC, Busch MP (2016) High incidence of chikungunya virus and frequency of viremic blood donations during epidemic, Puerto Rico, USA, 2014. Emerg Infect Dis 22:1221–1228. doi:10.3201/eid2207.160116

104. Sirivanakarn S (1982) A review of the systematics and a proposed scheme of internal classification of the new world subgenus *Melanoconion* of *Culex* (Diptera, Culicidae). Mosq Syst 14:265–333
105. Smith DW, Mackenzie JS, Weaver SC (2009) Alphaviruses. In: Richman DD, Whitley RJ, Hayden FG (eds) Clinical virology. ASM Press, Washington, DC
106. Strauss JH, Strauss EG (1994) The alphaviruses: gene expression, replication, and evolution. Microbiol Rev 58:491–562
107. Suarez O, Bergold G (1968) Investigations of an outbreak of Venezuelan equine encephalitis in towns of eastern Venezuela. Am J Trop Med Hyg 17:875–880
108. Sunthorn S, Galindo P (1967) The isolation of eastern equine encephalitis from *Culex (Melanoconion) taeniopus* Dyar and Knab in Panama. Mosq News 27:74–76
109. Talarmin A, Chandler LJ, Kazanji M, de Thoisy B, Debon P, Lelarge J, Labeau B, Bourreau E, Vié JC, Shope RE, Sarthou JL (1998) Mayaro virus fever in French Guiana: isolation, identification, and seroprevalence. Am J Trop Med Hyg 59:452–456
110. Terzian ACB, Auguste AJ, Vedovello D, Ferreira MU, da Silva-Nunes M, Speranca MA, Suzuki RB, Juncansen C, Araujo JP, Weaver SC, Nogueira ML (2015) Isolation and characterization of Mayaro virus from a human in Acre, Brazil. Am J Trop Med Hyg 92:401–404. doi:10.4269/ajtmh.14-0417
111. de Thoisy B, Gardon J, Salas RA, Morvan J, Kazanji M (2003) Mayaro virus in wild mammals, French Guiana. Emerg Infect Dis 9:1326–1329. doi:10.3201/eid0910.030161
112. Torres-Gutierrez C, Sallum MAM (2015) Catalog of the subgenus *Melanoconion* of *Culex* (Diptera: Culicidae) for South America. Zootaxa 4028:1–50
113. Torres JR, Russell KL, Vasquez C, Barrera R, Tesh RB, Salas R, Watts DM (2004) Family cluster of Mayaro fever, Venezuela. Emerg Infect Dis 10:1304–1306. doi:10.3201/eid1007.030860
114. Tsetsarkin KA, Chen R, Yun R, Rossi SL, Plante KS, Guerbois M, Forrester N, Perng GC, Sreekumar E, Leal G, Huang J, Mukhopadhyay S, Weaver SC (2014) Multi-peaked adaptive landscape for chikungunya virus evolution predicts continued fitness optimization in *Aedes albopictus* mosquitoes. Nat Commun 5:4084. doi:10.1038/ncomms5084
115. Tsetsarkin KA, Chen R, Leal G, Forrester N, Higgs S, Huang J, Weaver SC (2011) Chikungunya virus emergence is constrained in Asia by lineage-specific adaptive landscapes. Proc Natl Acad Sci U S A 108:7872–7877. doi:10.1073/pnas.1018344108
116. Turell MJ, Ludwig GV, Beaman JR (1992) Transmission of Venezuelan equine encephalomyelitis virus by *Aedes sollicitans* and *Aedes taeniorhynchus* (Diptera: Culicidae). J Med Entomol 29:62–65
117. Turell MJ, O'Guinn ML, Jones JW, Sardelis MR, Dohm DJ, Watts DM, Fernandez R, Travassos da Rosa A, Guzman H, Tesh R, Rossi CA, Ludwig V, Mangiafico JA, Kondig J, Wasieloski LP, Pecor J, Zyzak M, Schoeler G, Mores CN, Calampa C, Lee JS, Klein TA (2005) Isolation of viruses from mosquitoes (Diptera: Culicidae) collected in the Amazon Basin region of Peru. J Med Entomol 42:891–898. doi:10.1603/0022-2585(2005)042[0891:IOVFMD]2.0.CO;2
118. Vasconcelos PF, Travassos da Rosa AP, Rodrigues SG, Travassos da Rosa ES, Dégallier N, Travassos da Rosa JF (2001) Inadequate management of natural ecosystem in the Brazilian Amazon region results in the emergence and reemergence of arboviruses. Cad saude publica / Minist da Saude, Fund Oswaldo Cruz, Esc Nac Saude Publica 17(suppl):155–164. doi:10.1590/S0102-311X2001000700025
119. Vasconcelos PFC, Calisher CH (2016) Emergence of human arboviral diseases in the Americas, 2000-2016. Vector Borne Zoonotic Dis 16:1–7. doi:10.1089/vbz.2016.1952
120. Vieira C, Silva D, Barreto E, Siqueira C, Colombo T, Ozanic K, Schmidt D, Drumond B, Mondini A, Nogueira ML, Bronzoni R (2015) Detection of Mayaro virus infections during a dengue outbreak in Mato Grosso, Brazil. Acta Trop 147:12–16. doi:10.1016/j.actatropica.2015.03.020
121. Vittor AY, Armien B, Gonzalez P, Carrera J-P, Dominguez C, Valderrama A, Glass GE, Beltran D, Cisneros J, Wang E, Castillo A, Moreno B, Weaver SC (2016) Epidemiology of emergent Madariaga encephalitis in a region with endemic Venezuelan equine encephalitis:

initial host studies and human cross-sectional study in Darien, Panama. PLoS Negl Trop Dis 10:e0004554. doi:10.1371/journal.pntd.0004554
122. Walder R, Suarez OM, Calisher CH (1984) Arbovirus studies in the Guajira region of Venezuela: activities of eastern equine encephalitis and Venezuelan equine encephalitis viruses during an interepizootic period. Am J Trop Med Hyg 33:699–707
123. Walder R, Suarez OM, Calisher CH (1984) Arbovirus studies in southwestern Venezuela during 1973-1981. II. Isolations and further studies of Venezuelan and eastern equine encephalitis, Una, Itaqui, and Moju viruses. Am J Trop Med Hyg 33:483–491
124. Wang E, Barrera R, Boshell J, Ferro C, Freier JE, Navarro JC, Salas R, Vasquez C, Weaver SC (1999) Genetic and phenotypic changes accompanying the emergence of epizootic subtype IC Venezuelan equine encephalitis viruses from an enzootic subtype ID progenitor. J Virol 73:4266–4271
125. Wang E, Bowen RA, Medina G, Powers AM, Kang W, Chandler LM, Shope RE, Weaver SC (2001) Virulence and viremia characteristics of 1992 epizootic subtype IC Venezuelan equine encephalitis viruses and closely related enzootic subtype ID strains. Am J Trop Med Hyg 65:64–69
126. Weaver SC, Ferro C, Barrera R, Boshell J, Navarro JC (2004) Venezuelan equine encephalitis. Annu Rev Entomol 49:141–174. doi:10.1146/annurev.ento.49.061802.123422
127. Weaver SC, Forrester NL (2015) Chikungunya: evolutionary history and recent epidemic spread. Antivir Res 120:32–39. doi:10.1016/j.antiviral.2015.04.016
128. Weaver SC, Kang W, Shirako Y, Rumenapf T, Strauss EG, Strauss JH (1997) Recombinational history and molecular evolution of western equine encephalomyelitis complex alphaviruses. J Virol 71:613–623
129. Weaver SC, Klimstra WB, Ryman KD(2009) Togaviruses: general features. In: Mahy BWJ, van Regenmortel MHV (eds) Desk encyclopedia of plant and fungal virology. Elsevier/ Academic Press, New York. doi: 10.1016/B978-012374410-4.00627-0
130. Weaver SC, Lecuit M (2015) Chikungunya virus infections. N Engl J Med 373:93–95. doi:10.1056/NEJMc1505501
131. Weaver SC, Reisen WK (2010) Present and future arboviral threats. Antivir Res 85:328–345. doi:10.1016/j.antiviral.2009.10.008
132. Weaver SC, Winegar R, Manger ID, Forrester NL (2012) Alphaviruses: population genetics and determinants of emergence. Antivir Res 94:242–257. doi:10.1016/j.antiviral.2012.04.002
133. Weise WJ, Hermance ME, Forrester N, Adams AP, Langsjoen R, Gorchakov R, Wang E, Alcorn MDH, Tsetsarkin K, Weaver SC (2014) A novel live-attenuated vaccine candidate for Mayaro fever. PLoS Negl Trop Dis 8:e2969. doi:10.1371/journal.pntd.0002969
134. Yanoviak SP, Aguilar PV, Lounibos LP, Weaver SC (2005) Transmission of a Venezuelan equine encephalitis complex alphavirus by *Culex (Melanoconion) gnomatos* (Diptera: Culicidae) in northeastern Peru. J Med Entomol 42:404–408
135. Zacks MA, Paessler S (2010) Encephalitic alphaviruses. Vet Microbiol 140:281–286. doi:10.1016/j.vetmic.2009.08.023
136. Zehmer RB, Dean PB, Sudia WD, Calisher CH, Sather GE, Parker RL (1974) Venezuelan equine encephalitis epidemic in Texas, 1971. Health Serv Rep 89:278–282
137. Zuchi N, da Silva Heinen LB, dos MAM S, Pereira FC, Slhessarenko RD (2014) Molecular detection of Mayaro virus during a dengue outbreak in the state of Mato Grosso, central-West Brazil. Mem Inst Oswaldo Cruz 109:820–823
138. Erasmus JH, Auguste AJ, Kaelber JT, Luo H, Rossi SL, Fenton K, Leal G, Kim DY, Chiu W, Wang T, Frolov I, Nasar F, Weaver SC (2017) A chikungunya fever vaccine utilizing an insect-specific virus platform. Nat Med 23:192–199. PMCID: PMCPMC5296253
139. Lednicky J, De Rochars VM, Elbadry M, Loeb J, Telisma T, Chavannes S, Anilis G, Cella E, Ciccozzi M, Okech B, Salemi M, Morris JG Jr (2016) Mayaro virus in child with acute febrile illness, haiti, 2015. Emerg Infect Dis 22:2000–2002
140. Laboratory studies of a Brazilian strain of Aedes albopictus as a potential vector of Mayaro and Oropouche viruses. J Am Mosq Control Assoc (1991) 7(1):89–93

Part IV
Hemorragic, Skin and Respiratory Viral Diseases

Chapter 10
Arenaviruses and Hemorrhagic Fevers: From Virus Discovery to Molecular Biology, Therapeutics, and Prevention in Latin America

Víctor Romanowski, Matías L. Pidre, Mario E. Lozano, and Sandra E. Goñi

1 Introduction and New Arenavirus Taxonomy

This chapter focuses on arenavirus studies carried out in past decades in Latin America. Some information comes also from reports of international collaborative research and study groups in which scientists from this region have participated.

The *Arenaviridae* family presently includes more than 30 viral species. The number of new arenaviruses isolated and characterized in the past few years has grown dramatically and led to the establishment of two genera: *Mammarenavirus* (known to infect mammals) and *Reptarenavirus* (identified in snakes) [71] (Table 10.1).

The mammarenaviruses (referred to as "arenaviruses" in the literature before 2016) are generally associated with infection in rodents and are divided into two major groups on the basis of serological cross-reactivity, phylogeny, and geographic site of isolation: the Old World (OW) complex (Africa, Europe, and Asia) and the larger New World (NW) complex (Americas), subdivided into clades. *Lymphocytic choriomeningitis virus* (LCMV), the type species, infects the common house mouse *(Mus musculus)*, a fact that explains its global distribution. In contrast, all other

V. Romanowski (✉) • M.L. Pidre
Instituto de Biotecnología y Biología Molecular (IBBM, UNLP-CONICET), Facultad de Ciencias Exactas, Universidad Nacional de la Plata, Buenos Aires, Argentina
e-mail: victor@biol.unlp.edu.ar

M.E. Lozano • S.E. Goñi
Instituto de Microbiología Básica y Aplicada (IMBA), Laboratorio de Ingeniería Genética y Biología Celular y Molecular, Área de Virosis Emergentes y Zoonóticas (IMBA, LIGBCM-AVEZ), Departamento de Ciencia y Tecnología (DCyT), Universidad Nacional de Quilmes, Buenos Aires, Argentina

J.E. Ludert et al. (eds.), *Human Virology in Latin America*,
DOI 10.1007/978-3-319-54567-7_10

Table 10.1 *Arenaviridae* taxonomy

Genus	Group[a]	Species	Virus [abbreviation]	Clade[b]
Mammarenavirus	Old World (OW) Arenaviruses	*Ippy mammarenavirus*	Ippy [IPPYV]	
		Lassa mammarenavirus	Lassa [LASV]	
		Lujo mammarenavirus	Lujo [LUJV]	
		Luna mammarenavirus	Luna [LUAV]	
		Lunk mammarenavirus	Lunk [LNKV]	
		Lymphocytic choriomeningitis mammarenavirus	Lymphocytic choriomeningitis [LCMV]	
		Merino Walk mammarenavirus	Merino Walk [MRWV]	
		Mobala mammarenavirus	Mobala [MOBV]	
		Mopeia mammarenavirus	Mopeia [MPOV]; Morogoro [MORV]	
		Gairo mammarenavirus	Gairo [GAIV]	
		Mariental mammarenavirus	Mariental [MRLV]	
		Okahandja mammarenavirus	Okahandja [OKAV]	
		Wēnzhōu mammarenavirus	Wēnzhōu [WENV]	
	New World (NW) Arenaviruses[a]	*Pichindé mammarenavirus*	Pichindé [PICHV]	A
		Pirital mammarenavirus	Pirital [PIRV]	A
		Paraná mammarenavirus	Paraná [PRAV]	A
		Flexal mammarenavirus	Flexal [FLEV]	A
		Allpahuayo mammarenavirus	Allpahuayo [ALLV]	A
		Junín mammarenavirus	Junín [JUNV]	B
		Amaparí mammarenavirus	Amaparí [AMAV]	B
		Cupixi mammarenavirus	Cupixi [CUPXV]	B
		Chaparé mammarenavirus	Chaparé [CHPV]	B
		Guanarito mammarenavirus	Guanarito [GTOV]	B

		Machupo mammarenavirus	Machupo [MACV]	B
		Sabiá mammarenavirus	Sabiá [SBAV]	B
		Tacaribe mammarenavirus	Tacaribe [TCRV]	B
		Latino mammarenavirus	Latino [LATV]	C
		Oliveros mammarenavirus	Oliveros [OLVV]	C
		Whitewater Arroyo mammarenavirus	Big Brushy Tank [BBRTV], Catarina [CTNV], Skinner Tank [SKTV], Tonto Creek [TTKV], Whitewater Arroyo [WWAV]	D (rec)
		Tamiami mammarenavirus	Tamiami [TMMV]	D (rec)
		Bear Canyon mammarenavirus	Bear Canyon [BCNV]	D (rec)
Reptarenavirus		*Alethinophid 1 reptarenavirus*	Golden Gate [GOGV]	
		Alethinophid 2 reptarenavirus	ROUT [ROUTV]; University of Helsinki [UHV]	
		Alethinophid 3 reptarenavirus	CAS [CASV]	

The table has been adapted from http://www.ictvonline.org; by Radoshitzky et al. [71]

[a]Old World (OW) and New World (NW) arenavirus groups are based on serological cross-reactivity and geographic site of virus isolation

[b]NW arenavirus, also known as Tacaribe complex, is clustered in four clades: A, B, C, and D (rec). The S RNA segment of clade D viruses contains a clade A-like N gene and a clade B-like GPC gene (S RNA recombination); the L RNA belongs to clade A

The species were defined using pairwise sequence comparison (PASC), a web tool for analysis of pairwise identity distribution within viral families [10], available at http://www.ncbi.nlm.nih.gov/sutils/pasc/

Black Mesa virus, Collierville virus, Gbagroube virus, Jirandogo virus, Kodoko virus, Ocozocoautla de Espinosa virus, Orogrande virus, Pinhal virus, Real de Catorce virus, and the unnamed North American arenaviruses are considered tentative members of the family until more data become available Bao et al. [10]

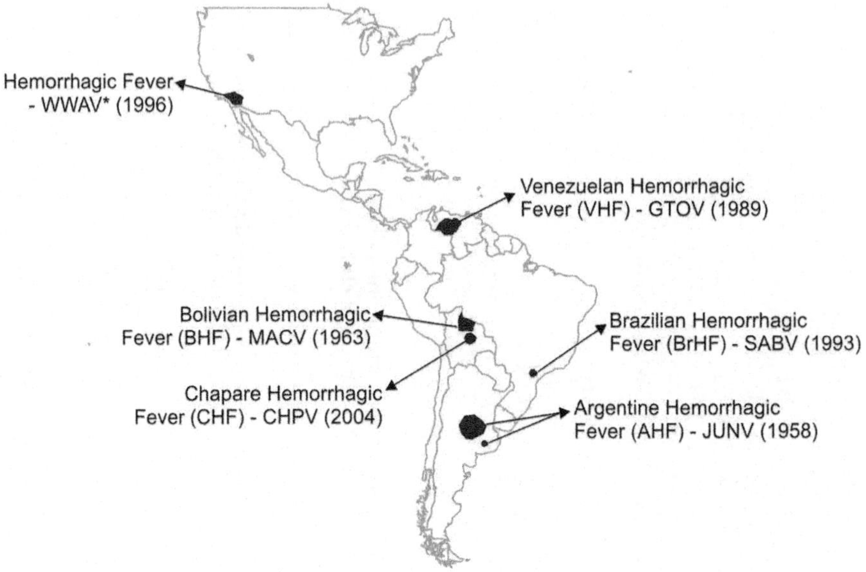

Fig. 10.1 New World arenaviruses that cause hemorrhagic fevers. Endemic areas for American human pathogenic arenaviruses. The year of isolation/description of these viruses is indicated in parentheses. WWAV* includes several different North American arenaviruses, some of which are associated with cases of hemorrhagic fever with liver failure and belong to the species *Whitewater Arroyo mammarenavirus*

arenaviruses show restricted geographic distribution coincident with the location of habitats of natural reservoir hosts. In rodents, arenaviruses usually establish an asymptomatic chronic infection. This fact has a correlation in vitro, where arenaviruses can produce persistent infections without associated cytopathic effects. However, occasionally some arenaviruses may be transmitted to humans through contact with urine or blood-contaminated materials and produce severe hemorrhagic fever. Included are Lassa (LASV) and Lujo (LUJV) viruses, in West Africa, and Junín (JUNV), Machupo (MACV), Guanarito (GTOV), Sabiá (SABV), Chaparé (CHPV), and Whitewater Arroyo (WWAV) viruses, in the Americas (Fig. 10.1).

At present, arenaviruses cause as many as 500,000 zoonotic infections per year in endemic areas of Africa and South America that can lead to severe and lethal hemorrhagic fever symptoms. Human pathogenic arenaviruses are considered potential biological weapons.

Because of the number of cases of human disease in Latin America and the availability of locally and internationally published reports, this chapter discusses JUNV in more detail and refers to other arenaviruses when appropriate. By no means should this chapter be considered a thorough compilation of the research on arenaviruses conducted in Latin America; it is rather an overview of the diverse published studies in this field and a short perspective on future developments.

2 Virion Structure and Genome Organization

Members of the *Mammarenavirus* genus share the following features: virions are pleomorphic (mostly spherical), 50–300 nm in diameter (mean, ~120 nm); they contain several copies of circular nucleocapsids and include a variable number of ribosomes [74, 77]. They acquire a lipid envelope with club-shaped projections, 8–10 nm in length (spikes), during the budding process from the host cell membrane at the end of the infectious cycle.

Their genome consists of two single-stranded RNA segments: small (S) and large (L), about 3.5 kb and 7.5 kb in length, respectively. Each segment has two nonoverlapping open reading frames (ORF) of opposite sense, which was the origin of the term *ambisense* to describe this type of coding strategy [9]. The sizes of the gene products indicated in the following text are those of JUNV.

The ORFs of opposite polarity are separated in both RNAs by a noncoding intergenic region predicted to fold into a stable secondary structure [31]. The L segment codes for both the 94 amino acid (aa) zinc-binding Z-matrix protein that drives virus budding (~11 kDa), as well as for the RNA-dependent RNA polymerase L (2210 aa; ~250 kDa). The S RNA codes for both a nucleocapsid protein N (564 aa) as well as the glycoprotein precursor GPC (485 aa) [31, 77]. GPC is synthesized as a single polypeptide chain and is post-translationally cleaved to yield mature virion glycoproteins G1 (192 aa) and G2 (235 aa) and a stable signal peptide SSP (58 aa) [87]. G1/G2/SSP trimers form the spikes decorating the virus surface.

G1 is located at the top of the spike and mediates virus interaction with host cell-surface receptors, and G2 is similar to others class I viral fusion proteins [70]. SSP is generated by signal peptidase cleavage but, in contrast to conventional signal peptides, is stable, unusually long (58 aa vs. the usual 15–25 aa), and myristoylated; it contains two hydrophobic segments that span the lipid bilayer with both N- and C-termini residing in the cytosol; and contributes to G2 fusion activity through its C-terminal region [86, 87].

The nucleocapsid protein N is the most abundant virion protein, followed by G2, G1, Z, and L (~1500, 650, 650, 450, and 30 molecules per virion, respectively, as calculated per "old style" methods) [84]. The RNA–N interactions and zinc-binding capacity have been identified in computational and experimental studies [65, 80, 81].

3 Virus Entry and Cell Tropism

Specific virus interaction with receptor molecules on the cell membrane is a crucial step in the infectious process: it drives subsequent entry into the host cell, making cell receptors the major determinants of viral cell tropism, host range, and pathogenesis. Until 2005, little was known about the mechanism by which JUNV entered host cells. By using pseudo-typed retroviruses, several laboratories confirmed that JUNV, as well as other clade B NW arenaviruses, did not interact with α-dystroglycan, the known receptor for Old World (OW) arenaviruses, to enter the cells [72]. The next big breakthrough came 1 year later when using a proteomic pull-down approach, applying a recombinant receptor-binding G1 moiety of Machupo virus (MACV) as bait, transferrin receptor 1 (TfR1) was identified as the first known JUNV, MACV, GTOV, and SABV cell receptor [70]. A recent study has shown that a neutralizing monoclonal antibody directed toward G1 maps to the same site that makes contact with hTRf1, suggesting that this is the basis for immune therapy success [50].

Although the hTfR1 is definitely a major receptor that allows JUNV infection, there is information on additional or alternative cell-surface molecules that seem to promote virus entry. Dendritic cell-specific intercellular adhesion molecule-3-grabbing nonintegrin (DC-SIGN) is a type II transmembrane lectin receptor. DC-SIGN is abundantly expressed on immature dendritic cells (iDCs), one of the principal targets of JUNV. Previous reports showed that nonpermissive cells lacking TfR1 became significantly more susceptible for JUNV infection when transfected with a plasmid DNA construct expressing DC-SIGN receptor. In addition, pretreatment of these genetically modified cells with anti-DC-SIGN or mannan reduces the infection with JUNV. This work validates a direct cell-to-cell transmission of JUNV and supports the role of type C lectins in viral adsorption, internalization, and intracellular transport [25, 38].

After binding to its receptor, a virus can resort to different internalization mechanisms. Briefly, there are three general mechanisms for viral internalization: clathrin-mediated endocytosis, caveolar/raft pathway, and cholesterol-dependent endocytosis. It has been demonstrated in Vero cells that clathrin-mediated endocytosis is the main route used by JUNV and involves the cytoskeleton and a host of cellular proteins [55–57]. Finally, direct evidence of JUNV cell entry was obtained using transmission electron microscopy [55]. Pichinde virus (PICV), another NW arenavirus, has also been shown to enter cells through a clathrin-dependent endocytic pathway, trafficked through the dynamin 2 endocytic pathway in which the virus travels through Rab5-mediated early endosomes and Rab7-mediated late endosomes [83]. Similar results have been obtained for JUNV [57]. JUNV internalization leads to PI3K/Akt signaling pathway activation [42], which requires both intact actin and a dynamic microtubule network [56]. An alternative virus internalization pathway has been recently described for TCRV [73].

Later, the fusion of the viral envelope with the endosomal membrane is essential for the progression of the infection cycle. Although G1 interacts with the TfR1, G2 is responsible for the fusion process. These studies revealed a crucial role of the SSP in pH-dependent membrane fusion [87]. It has been shown that sera from AHF patients inhibit fusion activity in an in vitro system [16].

4 RNA Transcription and Replication

The 3′- and 5′-terminal sequences of 19-nucleotide RNA segments are complementary and very well conserved in all arenaviruses. The base complementarity of these termini is probably the molecular basis for the circular conformation of the nucleocapsids that has been observed [29]. These termini are essential for replication and transcription and are believed to function as a binding site for viral polymerases (reviewed in [2]).

Genome replication and transcription take place in the cytoplasm of infected cells and require that viral proteins combine with viral RNA to form ribonucleoprotein (RNP) complexes. The L protein mediates viral transcription and replication using RNPs as templates. In TCRV, N and L proteins together with virus RNA are the minimal components of RNP complexes and are sufficient for genome replication and transcription [44]. Both N and L proteins are necessary and sufficient for these early steps in vivo in reconstructed JUNV transcription-replication [2]. During genome replication, full-length copies of genomic S and L RNAs are synthesized, generating the corresponding antigenomic S and L RNAs. In response to the *ambisense* coding strategy, both genomic and antigenomic RNAs serve as template for viral mRNA transcription (Fig. 10.2). Transcripts contain a cap but are not polyadenylated. The 3′-end sequences of the subgenomic mRNAs fall within the intergenic, suggesting that the stem-loop structure is involved in transcription termination regulated by interaction with N [36, 80]. However, elements regulating termination have not yet been well defined.

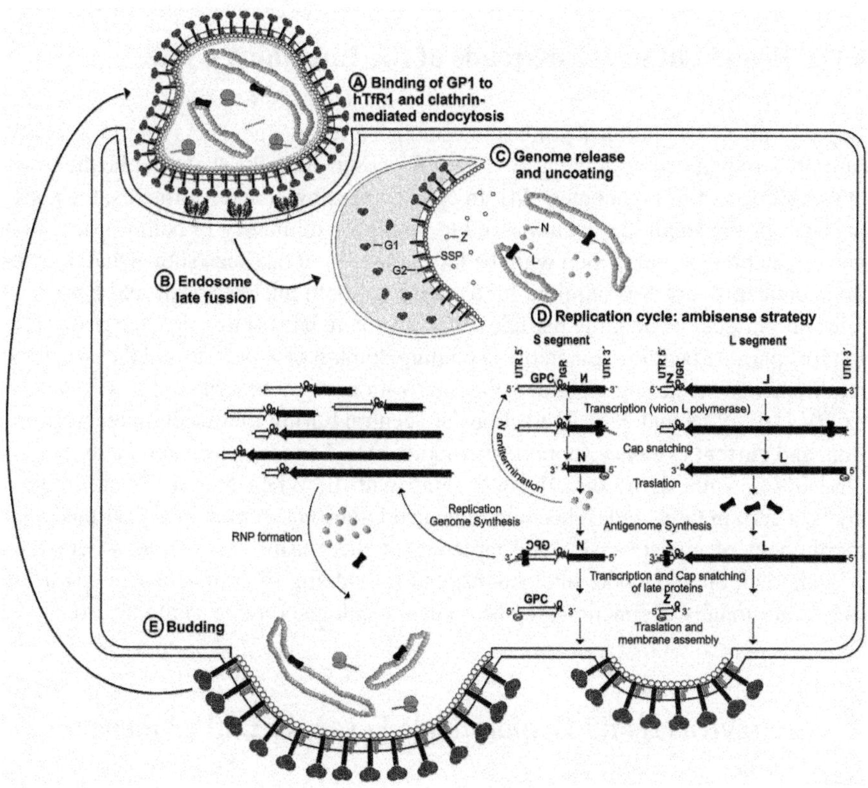

Fig. 10.2 Schematic of the JUNV infection cycle. Virion adsorption to the cell surface is mediated by G1 (head of trimeric GP complex) interaction with the hTfR1 cell receptor. After receptor-mediated endocytosis, the drop in pH triggers conformational changes in G2 that drive virion and endosome membrane fusion. Uncoating releases the viral nucleoprotein, which serves as template for transcription and replication of the ambisense genomic RNAs. Translation of GPC mRNA and processing of the precursor via the secretory pathway yields the GP complexes (trimers of G1 + G2 + SSP) inserted in the cell membrane. The Z protein drives the budding process by curving the GP-containing membrane GP patches and interacting with the newly formed nucleocapsids. No nuclear phase is required for arenavirus replication

5 Glycoprotein Processing and Envelope Assembly

GPC is processed to yield SSP + G1 + G2 to form trimeric spikes protruding on the virus surface. SSP is required for transport of the G1–G2 precursor protein GPC from the endoplasmic reticulum (ER). After cleavage, JUNV SSP is retained and positioned in the GP complex through interaction with a zinc-binding domain in the cytoplasmic tail of G2. The G1–G2 precursor is cleaved by the cellular SKI-1/S1P protease in the Golgi compartment to form the mature G1 and G2 subunits. As with other class I viral fusion proteins, proteolytic cleavage of the GPC precursor is required to render the GP complex competent for membrane fusion [2, 87]. Other protein–protein interactions are necessary to package the genome and induce budding to generate virions. In addition, it has been shown that GP complexes become localized to cholesterol-rich lipid microdomains [18].

6 Z Protein at the Crossroads of the Infectious Cycle

López et al. (2001) initiated studies to establish a reverse genetic system for TCRV (a close nonpathogenic relative of JUNV) focused initially at the replication and encapsidation of minigenomes [44]. In cells expressing N and L proteins, the coexpression of the small Z protein proved to be highly inhibitory to both transcription and replication via interaction with the L protein [39]. It has been shown that interaction between Z and N is required for assembly of both nucleocapsids and glycoproteins into infectious budding particles [15]. Z protein has been assigned a major role in virus particle budding. Later, the L-binding domain of Z protein and the structural requirements mediating Z homo-oligomerization were described for JUNV and TCRV [46]. N–N and N–L interactions are central during transcription and replication, and current evidence supports the notion that Z operates as a key modulator of viral RNA synthesis by directly interacting with L. When N and GP accumulate above a certain threshold, Z becomes engaged in virion assembly via Z–N-mediated recruitment of nucleocapsids and targeting of the plasma membrane, where Z–Z oligomerization and Z–G2 interactions lead to budding of complete virus particles [45]. Additionally, Z function may be related to cell response to viral infection.

7 Arenaviruses and Hemorrhagic Fevers in Latin America

Five NW arenaviruses are known to naturally cause severe febrile disease in humans in Latin America: the Guanarito (GTOV), Junín (JUNV), Machupo (MACV), Sabiá (SABV), and Chaparé (CHPV) viruses [17, 20, 66, 79]. The diseases range from sporadic cases to small outbreaks to hyperendemic episodes. Humans usually become infected with arenaviruses by inhalation of virus in aerosolized droplets of rodent excreta.

In contrast, other arenaviruses are not associated with human disease, such as Mopeia virus (MOPV) in West Africa or Tacaribe (TCRV), Pichinde (PICV), and Oliveros (OLVV), in the Americas.

Argentine hemorrhagic fever (AHF) is the best studied South American hemorrhagic fever (HF) and is similar to others in clinical presentation. It is a severe viral hemorrhagic syndrome endemic to the agricultural plains of central Argentina. Its incidence is mainly seasonal [24, 75].

The clinical symptoms of AHF include hematological, neurological, cardiovascular, renal, and immunological alterations. This emerging disease was first recognized in 1955, and its etiological agent was characterized and designated Junín virus (JUNV) for the geographic site where it was first isolated [66, 67]. JUNV is a rodent-borne virus and belongs to the clade B New World (NW) arenavirus within the *Arenaviridae* family [77].

The population of humans at risk is composed mainly of agricultural workers who become infected by inhaling aerosols of rodent excreta, although viral entry may occur by other routes, such as the conjunctival membranes, other mucous membranes, ingestion, and direct contact with damaged skin [60]. Transmission between humans has been reported even though AHF is usually not contagious from human to human. In patients with AHF, the viremia is present during the entire acute febrile period. Moreover, the virus was occasionally isolated from oral swabs, urine, and breast milk from infected subjects. Sexual transmission of JUNV was reported from convalescent men to women [13].

Since its emergence in the 1950s, annual epidemics of the disease have been recorded. The initially high case fatality rate of the disease was markedly reduced, first with adequate supportive measures and, more significantly, with the use of immune plasma [24].

Former endemic hotspots are currently cooling off; however, there is a steady and progressive geographic expansion of the endemic region into north-central Argentina, and currently almost 5 million people are considered to be at risk of contracting AHF [24].

A collaborative effort conducted by the U.S. and Argentine governments in the 1980s led to the production of a live attenuated Junín virus vaccine [51]. The availability of the live attenuated vaccine has contributed to a substantial reduction in the number of AHF cases in recent years [24].

JUNV may enter the body through the skin, respiratory tract, or gastrointestinal mucosa. After replication, generalized dissemination occurs, but gross pathology changes are nonspecific [34]. Capillary dilatation ensues with perivascular erythrocyte diapedesis and bleeding; minor edema of the vascular wall has also been observed. Erythroblastopenia with morphologically abnormal erythroid and leukopoietic cell lines and normal megakaryocytes has been described in bone marrow, as well as severe meningeal edema and hemorrhages in Virchow–Robin spaces in the central nervous system (CNS) (reviewed in [54]).

Decreased T- and B-lymphocyte counts and a diminished response to mitogens are expressions of immunosuppression during the acute phase of the disease. Low numbers of null, B, and T cells, as well as a lower T4/T8 ratio, have been observed during the acute phase of AHF. Null and T8 cell numbers improve after immune plasma infusion, and all cell subsets return to normal in early convalescence. It has been proposed that circulating monocytes (macrophages) are targets for JUNV replication, contributing to viral spread during the acute AHF [54]. However, at this stage as well as in early convalescence, patient peripheral-blood mononuclear cells may exert antibody-dependent cell cytotoxicity, suggesting JUNV replication in macrophages does not affect their killing capacity [54].

During the first week after symptom onset, AHF patients show very high serum interferon-alpha (IFN-α) titers. Even though these values slowly normalize during the second week of illness in survivors, they remain elevated in severe cases. Interferon levels at admission correlate with outcome and are significantly lower in patients who survive [41]. Other cytokines described as significantly elevated in the serum of acute AHF patients include tumor necrosis factor (TNF)-α, interleukin (IL)-6, IL-8, and IL-10 [53], although their individual function in disease pathogenesis has not been studied. For more details on human disease findings, readers should consult the review by Marta et al. [54].

Pathological lesions in fatal AHF include generalized vasocongestion with multiple hemorrhages in the gastrointestinal mucosa and different organs, such as the liver, kidney, and lungs, as well as in subcutaneous tissue. The highest virus titers are found in the spleen, lymph nodes, and lungs, and high levels of viral antigen are found in cells of the monocyte/macrophage lineage in peripheral blood, lymphatic tissue, lung, and liver [54].

8 Expansion of Agriculture and Emergence of AHF

This disease is endemic, with annual outbreaks from the end of summer until midwinter, coincident with the harvest of maize and with the increase in the population of the wild rodents *Calomys musculinus, Calomys laucha*, *Akodon azarae*, and *Oryzomys flavescens* [60]. It was assumed that appearance of AHF disease in the mid-1950s was caused by human changes made in natural habitats in relationship to agricultural practices. Those environmental modifications are thought to have favored the growth of the *C. musculinus* population and facilitated its contact with humans. The epidemiological features of AHF are determined by the natural cycle of JUNV and by the behavior of the rodent reservoirs [60]. AHF mainly affects rural workers from the agricultural region known as the humid pampa, in central-east Argentina [24]. There are, however, urban cases in which the origin of infection is not easy to establish.

9 Rodent Reservoirs

All arenaviruses pathogenic for humans are rodent viruses. Although each arenavirus can infect many species of rodents, in every geographic site there is one species that is the principal reservoir because of higher population density and the prevalence and characteristics of infection. *Calomys musculinus* (family Muridae, subfamily Sigmodontinae) has been identified as the principal reservoir of JUNV, although virus has also been isolated from the organs and body fluids of other rodents captured in the endemic area, including *Calomys laucha* and *Akodon azarae*, and occasionally from *Mus musculus*, *Necromys benefactus*, and *Oligoryzomys flavescens* [76].

Some of these animals develop an acute disease with antibody response and clearance of the virus, whereas others develop a persistent infection, with low titers or absence of antibodies, chronic viremia, and shedding of virus in urine, feces, and saliva [85]. The chronically infected rodents are usually asymptomatic and exhibit normal behavior. Field studies of natural populations demonstrated that infection with JUNV among *C. musculinus* was more frequent among males than females and was positively correlated with age and the presence of wounds and scars [60]. JUNV among rodents may be transmitted via aerosols and bites, as well as sexually.

10 Screening and Discovery of Arenaviruses in the Americas

Serological screening for arenavirus infection in wild rodents and patients, combined with reverse transcriptase (RT)-polymerase chain reaction (PCR), sequencing, and virus isolation, has been the basis for discovery and definition of new species or virus variants [33, 49, 59].

New monoclonal antibodies (mAbs) against JUNV N were obtained by Nakauchi et al. [63]. Three epitopes comprising residues 12–17 (WTQSLR), 72–79 (KEVDRLMS), and 551–558 (PPSLLFLP) are recognized by different mAbs with different degrees of specificity, that is, ranging from broadly reactive with South American arenaviruses to JUNV specific [63]. RT-PCR-based methods have been described using both *Arenaviridae* family-specific and species-specific methods that can be applied to detect arenaviruses in rodents captured in the field [37, 48].

Comparison of endpoint antibody titers to WWAV and AMAV in individual blood samples from nearly 5000 rodents indicated that the Tacaribe complex viruses that are enzootic in New Mexico, Texas, and Mexico are antigenically diverse [58]. In particular, the samples from Chiapas (Mexico) showed a strong reaction to AMAV antigen. Analyses of nucleotide and amino acid sequence data indicated that the deer mice were infected with a novel Tacaribe serocomplex virus (proposed name: Ocozocoautla de Espinosa virus, OCEV), which is phylogenetically closely related to Tacaribe serocomplex viruses that cause hemorrhagic fever in humans in South America (clade B) [14]. Hypothetically, OCEV or an arenavirus phylogenetically closely related to OCEV was the etiological agent in the hemorrhagic fever epidemic in Chiapas in 1967 and presently is the cause of a human disease that is clinically indistinct from dengue hemorrhagic fever and other severe febrile illnesses endemic to Chiapas. Moreover, it has been speculated that these findings support the notion that epidemics of highly lethal hemorrhagic fever(s) in the highlands of Mexico in the sixteenth century were caused by arenavirus(es) native to Mesoamerica [1, 52].

More recently, aiming at the identification of the natural rodent reservoir for SABV, a broadly cross-reactive enzyme-linked immunosorbent assay (ELISA) was used to screen for antibody-positive animals. RT-PCR amplification provided evidence of a new arenavirus (proposed name: Pinhal virus) of the lineage C and no evidence of involvement in human disease [11].

11 Clinical Presentation

The AHF incubation period ranges from 6 to 12 days, ending with the onset of fever, usually associated with a flu-like syndrome that may include myalgia, arthralgia, headache, relative bradycardia, conjunctivitis, nausea, vomiting, and diarrhea, with little central nervous system (CNS) or hematological involvement during the first week. The early symptoms of AHF differ from those of acute respiratory infections by an almost constant absence of sore throat, cough, or nasal congestion. At the end of the first week of evolution, oliguria and different degrees of dehydration are present; neurological symptoms are common, and, in female patients, mild to moderate metrorrhagia is always present, being in some cases the first symptom of this disease [34].

In the second week of the disease, about 75% of infected individuals begin to improve, whereas the remaining 25% manifest neurological disorders or severe bleeding. Overlapping shock and bacterial infections appear 6 to 12 days after the onset of symptoms. Fever persists, and petechiae in the oral mucosa and the axillary region as well as gingival bleeding can be observed. Less common and more severe hemorrhagic signs may be present including hematemesis, melena, hemoptysis, epistaxis, hematomas, metrorrhagia, and hematuria. CNS involvement can also be present during the second week in the form of hyporeflexia and mental confusion. When severe, this phase can progress to include areflexia, muscular hypotonia, ataxia, increased irritability, and tremors, followed by delirium, generalized seizures, and coma [34, 61].

Clinically apparent disease occurs in almost two thirds of infected subjects. The fatality rate is as high as 30% among untreated patients. Immune plasma therapy reduces mortality to less than 1%, although this specific therapy is effective only when started during the first week of illness.

In early convalescence, 10% of cases treated with immune plasma from convalescent patients develop a late neurological syndrome (LNS). The LNS occurs after a period free of manifestations, differs from the neurological symptoms of the acute period of AHF, and is characterized by fever syndrome and manifestations from the cerebellar trunk [21]. Patients have a prolonged convalescence. Temporary loss of hair is common; many patients experience fatigue, irritability, and memory changes, but these symptoms are temporary and disappear gradually.

During the second week of illness, patients who are improving start to produce antibodies against JUNV as well as cellular immune response to clear up the virus. Moreover, robust titers of neutralizing antiviral antibodies can be detected in immune plasma from convalescent patients [reviewed in [24]].

12 Diagnosis

Reporting of AHF disease is mandatory in Argentina. At present, the AHF diagnosis to establish specific therapy is based on clinical and laboratory data. During the early phase of the illness, the clinical manifestations of AHF are nonspecific and can be confused with several acute febrile conditions. Therefore, if platelet counts less than 100,000/mm^3 in combination with white blood cell counts less than 2,500/mm^3 are detected, when screening patients in endemic areas these criteria can be considered potentially useful to identify individuals at risk [reviewed in [24]].

Seroconversion occurs only late in the course of infection; serological tests are not useful markers in the early stages of the disease. Neutralizing anti-JUNV antibodies (Abs) consisting mainly of the IgG1 subtype are usually present from day 12 on (reviewed in [24]). Serological diagnosis can be done by complement fixation, indirect immunofluorescent antibody assays, neutralization tests, and ELISA. The sensitivity and specificity of ELISA make it the routine method of choice for the etiological diagnosis of reported cases retrospectively and for the surveillance of the zoonosis [62]. More recently, a more accurate ELISA was developed employing recombinant JUNV N protein [82]. Immunohistochemistry is used to examine organ specimens from autopsy and confirm etiology.

During the acute phase of infection, virus titers in blood are low. Therefore, JUNV antigen detection is not a method of choice for early diagnosis until more sensitive techniques become available. During this phase, virus isolation can be performed from whole blood or peripheral blood mononuclear cells (PBMCs), a useful (reliable, but lengthy and cumbersome) tool to retrospectively confirm the clinical diagnosis in addition to serological tests [7].

Because immune plasma therapy is able to reduce mortality when introduced during the first 8 days of infection, the availability of rapid and early diagnostic tests is fundamental. In this context, a RT-PCR-based assay has been established for rapid diagnosis [47] and has also been successfully applied to establish an etiological diagnostic in subjects who died before the appearance of the specific antibodies. At present, the RT-PCR analysis to detect JUNV genome seems to be the most sensitive, rapid, and early test for the specific diagnosis of the infection [40].

13 AHF Vaccine

A scientific collaboration between the U.S. and Argentine governments allowed the development of a live attenuated Junín virus vaccine, Candid#1 [51].

Nucleotide and amino acid sequence alignments, performed on Candid#1 and XJ ancestor strains, revealed several nucleotide substitutions throughout the GPC and L genes [5, 31, 32]. One of the changes proposed to affect infectivity was found in the G2 protein and later confirmed to be responsible for the attenuated phenotype by reverse genetics [3]. Candid#1 turned out to be safe, immunogenic, and effective in preventing AHF in preclinical studies in mice, guinea pigs, and rhesus monkeys. Guinea pigs and rhesus monkeys inoculated with increasing doses of Candid#1 developed neutralizing antibodies and became JUNV resistant if inoculated with highly virulent strains. Candid#1 also protected these animals against MACV, the etiological agent of Bolivian hemorrhagic fever [23]. These studies also showed the absence of neurovirulence, neurotropism, or hemorrhagic manifestations and the stability of the attenuated strain.

In phase III clinical trials conducted in the period 1988–1990, Candid#1 showed a protective efficacy ≥84% and no serious adverse effects. As expected, immune response to Candid#1 boosts preexisting immunity to JUNV but is not changed by previous exposure to *Lymphocytic choriomeningitis virus* (LCMV) [24].

The live attenuated JUNV vaccine, Candid#1, has proven effective during the past two decades in more than 100,000 persons [24]. The vaccine has been recently produced in Argentina and tested in a compatible clinical study with 946 healthy volunteers who participated to support the comparability of Candid#1 vaccine manufacturing in the U.S. and Argentina [6]. Results presented by Enria et al. (2010) showed that the vaccine produced in Argentina is equivalent to that manufactured in the U.S., both in ability to immunize against JUNV (immunogenicity ≥95.5%) and in the lack of promoting any serious adverse effects [22]. Candid#1 is the first effective vaccine against arenaviruses; it is effective to protect against AHF, promoting humoral- and cell-mediated responses, and, since January 2007, in Argentina, is part of the National Immunization Program in the AHF risk area. The availability of such live attenuated vaccine has led to a substantial reduction in the incidence of AHF disease [22, 23].

14 Prognosis and Treatment

Without treatment, more than 80% of patients improve after the second week, although bacterial infection is a frequent complication.

Significant improvement in clinical AHF management has been achieved using immune plasma from convalescents, with mortality rates dropping from almost 30% to less than 1%. Additional administration of ribavirin may enhance these results even further (reviewed in [24]). Approximately 10% of cases treated with immune plasma develop late neurological syndrome (LNS). After a symptom-free period, LNS onset is characterized by fever, cerebellar signs, and cranial nerve palsies. LNS has never been registered among AHF patients recovering without specific treatment.

Current anti-arenaviral therapy is limited to an off-label use of ribavirin (1-β-D-ribofuranosyl-1,2,4-triazole-3-carboxamide), which has had only mixed success in the treatment of severe infections and is associated with significant toxicity in humans [24].

15 Conclusions and Future Challenges

During the past years, impressive progress has been made in developing rapid diagnostic tools and preventive and therapeutic approaches supported by our increasing understanding of the basic molecular and cellular biology of JUNV.

Rapid and specific diagnostic tools have been developed that can be applied both for early detection of JUNV in AHF suspected patients and in epidemiological surveillance studies, including samples from field rodents.

A robust reverse genetic system for JUNV combined with new detailed knowledge on virus–host interactions has the potential to be utilized in a rational design of novel live attenuated virus vaccines with precisely engineered disruptions of pathogenic properties. The genomic sequence of Candid #1 can be used as starting information to precisely engineer a prototypic live attenuated vaccine incorporating these and other attenuating features, thereby improving the vaccine identity, efficacy, and safety [3]. Other approaches to safe and effective vaccines are being explored, including recombinant vectors expressing selected arenaviral proteins to generate immunogens based on live recombinant viruses, subunit vaccines, or virus-like particles (VLPs) [40].

The VLPs can be generated from cells transfected with a JUNV Z expression vector in the absence of any other viral protein. In view of the capacity of Z protein to support fusions at the C-terminus without compromising its membrane budding properties, Borio et al. speculated on the possibility of using it as a vehicle of specific antigens to be included in eVLPs [12].

In particular, the development of a reverse genetic system for JUNV represented an important breakthrough and provided a powerful tool to precisely address questions regarding the biology and pathogenicity of JUNV (and other NW arenaviruses) [3, 4, 45].

Current studies are focused on the ability of arenaviruses to subvert the host cell innate antiviral defenses, the impact of arenavirus infection on the differentiation and function of cells targeted by hemorrhagic arenaviruses in vivo, including APCs such as macrophages and DCs, endothelial cells, and megakaryocytes involved in platelet formation [30, 64, 68, 69]. At the same time, novel animal models will provide important new information about the interaction of hemorrhagic arenaviruses with the host adaptive immune system, in particular, virus-induced immunosuppression, and understanding of the terminal hemorrhagic shock syndrome.

To study the direct effects of virus replication and gene expression that may be responsible for the perturbation of endothelial cell function, future research involving cell culture models for human endothelium that allow detailed analysis of virus-induced cell biological and biochemical alterations will have great importance.

Research on early molecular events of JUNV infection involving viral glycoprotein spikes and cell-surface receptors as well as virion and cell membrane fusion provided the basis for the development of novel therapeutic strategies [19, 35]. Other potential therapeutic targets being explored are specific steps for virus entry, processing, and replication [27, 78]. In particular, the multifunctional Z protein has been explored as target for antiviral compounds including siRNA [8, 26, 28]. Host cell factors have been also proposed as antiviral targets in arenavirus infection [43].

In addition to small antiviral molecules, immune therapy can be regarded as an alternative for AHF patients. Based on the success of immune therapy in controlling AHF mortality, it is possible to design a strategy replacing convalescent immune plasma, which is in short supply, with controlled humanized neutralizing monoclonal antibodies appropriately tested for efficacy [50, 88].

Further studies on new potential treatments are needed to block viral replication without causing toxicity and to prevent the increased vascular permeability that is responsible for hypotension and shock.

References

1. Acuna-Soto R, Romero LC et al (2000) Large epidemics of hemorrhagic fevers in Mexico 1545–1815. Am J Trop Med Hyg 62(6):733–739
2. Albarino CG, Bergeron E et al (2009) Efficient reverse genetics generation of infectious Junin viruses differing in glycoprotein processing. J Virol 83(11):5606–5614
3. Albarino CG, Bird BH et al (2011) The major determinant of attenuation in mice of the Candid1 vaccine for argentine hemorrhagic fever is located in the G2 glycoprotein transmembrane domain. J Virol 85(19):10404–10408
4. Albarino CG, Bird BH et al (2011) Reverse genetics generation of chimeric infectious Junin/Lassa virus is dependent on interaction of homologous glycoprotein stable signal peptide and G2 cytoplasmic domains. J Virol 85(1):112–122
5. Albarino CG, Ghiringhelli PD et al (1997) Molecular characterization of attenuated Junin virus strains. J Gen Virol 78(Pt 7):1605–1610
6. Ambrosio A, Saavedra M et al (2011) Argentine hemorrhagic fever vaccines. Hum Vaccin 7(6):694–700
7. Ambrosio AM, Enria DA et al (1986) Junin virus isolation from lympho-mononuclear cells of patients with Argentine hemorrhagic fever. Intervirology 25(2):97–102
8. Artuso MC, Ellenberg PC et al (2009) Inhibition of Junin virus replication by small interfering RNAs. Antivir Res 84(1):31–37
9. Auperin DD, Romanowski V et al (1984) Sequencing studies of pichinde arenavirus S RNA indicate a novel coding strategy, an ambisense viral S RNA. J Virol 52(3):897–904
10. Bao Y, Chetvernin V et al (2014) Improvements to pairwise sequence comparison (PASC): a genome-based web tool for virus classification. Arch Virol 159(12):3293–3304
11. Bisordi I, Levis S et al (2015) Pinhal virus, a new arenavirus isolated from *Calomys tener* in Brazil. Vector Borne Zoonotic Dis 15(11):694–700
12. Borio CS, Bilen MF et al (2012) Antigen vehiculization particles based on the Z protein of Junin virus. BMC Biotechnol 12(1):1–9
13. Briggiler AM, Enria DA et al (1987) Contagio interhumano e infección clfnica con virus Junin (V J) en matrimonios residentes en el area endémica de Fiebre Hemorráigica Argentina (FHA). Medicina (B Aires) 47:565
14. Cajimat MN, Milazzo ML et al (2012) Ocozocoautla de espinosa virus and hemorrhagic fever, Mexico. Emerg Infect Dis 18(3):401–405
15. Casabona JC, Levingston Macleod JM et al (2009) The RING domain and the L79 residue of Z protein are involved in both the rescue of nucleocapsids and the incorporation of glycoproteins into infectious chimeric arenavirus-like particles. J Virol 83(14):7029–7039
16. Castilla V, Contigiani M et al (2005) Inhibition of cell fusion in Junin virus-infected cells by sera from argentine hemorrhagic fever patients. J Clin Virol 32(4):286–288
17. Coimbra TLM, Nassar ES et al (1994) New arenavirus isolated in Brazil. Lancet 343(8894):391–392
18. Cordo SM, Valko A et al (2013) Membrane localization of Junin virus glycoproteins requires cholesterol and cholesterol rich membranes. Biochem Biophys Res Commun 430(3):912–917
19. Crispin M, Zeltina A et al (2016) Native functionality and therapeutic targeting of arenaviral glycoproteins. Curr Opin Virol 18:70–75
20. Delgado S, Erickson BR et al (2008) Chapare virus, a newly discovered arenavirus isolated from a fatal hemorrhagic fever case in Bolivia. PLoS Pathog 4(4):e1000047

21. Enria D, Briggiler AM et al (1998) An overview of the epidemiological, ecological and preventive hallmarks of Argentine hemorrhagic fever (Junin virus). Bull Inst Pasteur 96:12
22. Enria DA, Ambrosio AM et al (2010) Candid#1 vaccine against Argentine hemorrhagic fever produced in Argentina. Immunogenicity and safety. Medicina 70(3):215–222
23. Enria DA, Barrera Oro JG (2002) Junin virus vaccines. Curr Top Microbiol Immunol 263:239–261
24. Enria DA, Briggiler AM et al (2008) Treatment of argentine hemorrhagic fever. Antivir Res 78(1):132–139
25. Forlenza MB, Roldán JS et al (2011) Interacción del virus Junín con lectinas de tipo C y mecanismos endocíticos involucrados. In: X Congreso Argentino de Virología, pp 162–163
26. Garcia CC, Djavani M et al (2006) Arenavirus Z protein as an antiviral target: virus inactivation and protein oligomerization by zinc finger-reactive compounds. J Gen Virol 87(pt 5):1217–1228
27. García CC, Sepúlveda CS et al (2011) Novel therapeutic targets for arenavirus hemorrhagic fevers. Future Virol 6(1):27–44
28. Garcia CC, Topisirovic I et al (2010) An antiviral disulfide compound blocks interaction between arenavirus Z protein and cellular promyelocytic leukemia protein. Biochem Biophys Res Commun 393(4):625–630
29. Ghiringhelli PD, Rivera-Pomar RV et al (1991) Molecular organization of Junin virus S RNA: complete nucleotide sequence, relationship with other members of the Arenaviridae and unusual secondary structures. J Gen Virol 72(9):2129–2141
30. Gomez RM, Pozner RG et al (2003) Endothelial cell function alteration after Junin virus infection. Thromb Haemost 90(2):326–333
31. Goni SE, Iserte JA et al (2006) Genomic features of attenuated Junin virus vaccine strain candidate. Virus Genes 32(1):37–41
32. Goni SE, Iserte JA et al (2010) Molecular analysis of the virulence attenuation process in Junin virus vaccine genealogy. Virus Genes 40(3):320–328
33. Goni SE, Stephan BI et al (2011) Viral diversity of Junin virus field strains. Virus Res 160(1–2):150–158
34. Harrison LH, Halsey NA et al (1999) Clinical case definitions for argentine hemorrhagic fever. Clin Infect Dis 28(5):1091–1094
35. Helguera G, Jemielity S et al (2012) An antibody recognizing the apical domain of human transferrin receptor 1 efficiently inhibits the entry of all new world hemorrhagic fever arenaviruses. J Virol 86(7):4024–4028
36. Iapalucci S, Lopez N et al (1991) The 3′ end termini of the Tacaribe arenavirus subgenomic RNAs. Virology 182(1):269–278
37. Iserte JA, Stephan BI et al (2013) Family-specific degenerate primer design: a tool to design consensus degenerated oligonucleotides. Biotechnol Res Int 2013:383646
38. Iula LJ, Martínez MG et al (2011) La expresión del receptor celular DC-SIGN aumenta la capacidad infectiva del virus Junín. Interacción directa del receptor con la glicoproteína viral. In: X Congreso Argentino de Virología, p 42
39. Jacamo R, Lopez N et al (2003) Tacaribe virus Z protein interacts with the L polymerase protein to inhibit viral RNA synthesis. J Virol 77(19):10383–10393
40. Kerber R, Reindl S et al (2015) Research efforts to control highly pathogenic arenaviruses: a summary of the progress and gaps. J Clin Virol 64:120–127
41. Levis SC, Saavedra MC et al (1985) Correlation between endogenous interferon and the clinical evolution of patients with argentine hemorrhagic fever. J Interferon Res 5(3):383–389
42. Linero FN, Scolaro LA (2009) Participation of the phosphatidylinositol 3-kinase/Akt pathway in Junin virus replication in vitro. Virus Res 145(1):166–170
43. Linero FN, Sepulveda CS et al (2012) Host cell factors as antiviral targets in arenavirus infection. Viruses 4(9):1569–1591
44. Lopez N, Jacamo R et al (2001) Transcription and RNA replication of tacaribe virus genome and antigenome analogs require N and L proteins: Z protein is an inhibitor of these processes. J Virol 75(24):12241–12251

45. Loureiro ME, D'Antuono A et al (2012) Uncovering viral protein-protein interactions and their role in arenavirus life cycle. Viruses 4(9):1651–1667
46. Loureiro ME, Wilda M et al (2011) Molecular determinants of arenavirus Z protein homo-oligomerization and L polymerase binding. J Virol 85(23):12304–12314
47. Lozano ME, Enria D et al (1995) Rapid diagnosis of argentine hemorrhagic fever by reverse transcriptase PCR-based assay. J Clin Microbiol 33(5):1327–1332
48. Lozano ME, Posik DM et al (1997) Characterization of arenaviruses using a family-specific primer set for RT-PCR amplification and RFLP analysis. Its potential use for detection of uncharacterized arenaviruses. Virus Res 49(1):79–89
49. Machado AM, Figueiredo GG et al (2010) Standardization of an ELISA test using a recombinant nucleoprotein from the Junin virus as the antigen and serological screening for arenavirus among the population of Nova Xavantina, state of Mato Grosso. Rev Soc Bras Med Trop 43(3):229–233
50. Mahmutovic S, Clark L et al (2015) Molecular basis for antibody-mediated neutralization of New World hemorrhagic fever mammarenaviruses. Cell Host Microbe 18(6):705–713
51. Maiztegui JI, McKee KT Jr et al (1998) Protective efficacy of a live attenuated vaccine against argentine hemorrhagic fever. AHF study group. J Infect Dis 177(2):277–283
52. Marr JS, Kiracofe JB (2000) Was the huey cocoliztli a haemorrhagic fever? Med Hist 44(3):341–362
53. Marta RF, Montero VS et al (1999) Proinflammatory cytokines and elastase-alpha-1-antitrypsin in Argentine hemorrhagic fever. Am J Trop Med Hyg 60(1):85–89
54. Marta RF, Montero VS et al (1998) Systemic disorders in argentine haemorrhagic fever. Bull Inst Pasteur 96:115–124
55. Martinez MG, Cordo SM et al (2007) Characterization of Junín arenavirus cell entry. J Gen Virol 88(6):1776–1784
56. Martinez MG, Cordo SM et al (2008) Involvement of cytoskeleton in Junin virus entry. Virus Res 138(1–2):17–25
57. Martinez MG, Forlenza MB et al (2009) Involvement of cellular proteins in Junin arenavirus entry. Biotechnol J 4(6):866–870
58. Milazzo ML, Barragan-Gomez A et al (2010) Antibodies to Tacaribe serocomplex viruses (family *Arenaviridae*, genus *Arenavirus*) in cricetid rodents from New Mexico, Texas, and Mexico. Vector Borne Zoonotic Dis 10(6):629–637
59. Milazzo ML, Cajimat MN et al (2015) Epizootiology of Tacaribe serocomplex viruses (*Arenaviridae*) associated with neotomine rodents (Cricetidae, Neotominae) in southern California. Vector Borne Zoonotic Dis 15(2):156–166
60. Mills JN, Ellis BA et al (1994) Prevalence of infection with Junin virus in rodent populations in the epidemic area of Argentine hemorrhagic fever. Am J Trop Med Hyg 51(5):554–562
61. Molinas FC, de Bracco MM et al (1981) Coagulation studies in Argentine hemorrhagic fever. J Infect Dis 143(1):1–6
62. Morales MA, Calderon GE et al (2002) Evaluation of an enzyme-linked immunosorbent assay for detection of antibodies to Junin virus in rodents. J Virol Methods 103(1):57–66
63. Nakauchi M, Fukushi S et al (2009) Characterization of monoclonal antibodies to Junin virus nucleocapsid protein and application to the diagnosis of hemorrhagic fever caused by South American arenaviruses. Clin Vaccine Immunol CVI 16(8):1132–1138
64. Negrotto S, Mena HA et al (2015) Human plasmacytoid dendritic cells elicited different responses after infection with pathogenic and nonpathogenic Junin virus strains. J Virol 89(14):7409–7413
65. Parisi G, Echave J et al (1996) Computational characterisation of potential RNA-binding sites in arenavirus nucleocapsid proteins. Virus Genes 13(3):247–254
66. Parodi AS, Greenway DJ et al (1958) Sobre la etiología del brote epidémico de Junín. Dia Med 30:2300–2301
67. Parodi AS, Rugiero HR et al (1959) Isolation of the Junin virus (epidemic hemorrhagic fever) from the mites of the epidemic area (*Echinolaelaps echidninus* Barlese). Prensa Med Argent 46:2242–2244

68. Peña Cárcamo JR (2016) Viperina, un factor de restricción en la infección con el arenavirus Junín. PhD thesis, Universidad de Buenos Aires, Argentina
69. Pozner RG, Ure AE et al (2010) Junin virus infection of human hematopoietic progenitors impairs in vitro proplatelet formation and platelet release via a bystander effect involving type I IFN signaling. PLoS Pathog 6(4):e1000847
70. Radoshitzky SR, Abraham J et al (2007) Transferrin receptor 1 is a cellular receptor for New World haemorrhagic fever arenaviruses. Nature (Lond) 446(7131):92–96
71. Radoshitzky SR, Bao Y et al (2015) Past, present, and future of arenavirus taxonomy. Arch Virol 160(7):1851–1874
72. Reignier T, Oldenburg J et al (2006) Receptor use by pathogenic arenaviruses. Virology 353(1):111–120
73. Roldan JS, Martinez MG et al (2016) Human transferrin receptor triggers an alternative Tacaribe virus internalization pathway. Arch Virol 161(2):353–363
74. Romanowski V, Bishop DH (1983) The formation of arenaviruses that are genetically diploid. Virology 126(1):87–95
75. Romanowski V, Pidre ML et al (2013) Argentine hemorrhagic fever. In: Singh SK (ed) Viral hemorrhagic fevers. CRC Press, Boca Raton [http://www.crcpress.com/]/Taylor & Francis Group, pp 317–337
76. Sabattini MS, Gonzalez de Rios LE et al (1977) Infección natural y experimental de roedores con virus Junin. Medicina (B Aires) 37:149–159
77. Salvato MS, Clegg JCS et al (2012) Family *Arenaviridae*. In: Virus taxonomy. Elsevier, San Diego, pp 715–723
78. Sepulveda CS, Garcia CC et al (2012) Inhibition of Junin virus RNA synthesis by an antiviral acridone derivative. Antiviral Res 93:16–22
79. Tesh RB, Jahrling PB et al (1994) Description of Guanarito virus (*Arenaviridae*: arenavirus), the etiologic agent of Venezuelan hemorrhagic fever. Am J Trop Med Hyg 50(4):452–459
80. Tortorici MA, Albarino CG et al (2001) Arenavirus nucleocapsid protein displays a transcriptional antitermination activity in vivo. Virus Res 73(1):41–55
81. Tortorici MA, Ghiringhelli PD et al (2001) Zinc-binding properties of Junin virus nucleocapsid protein. J Gen Virol 82(pt 1):121–128
82. Ure AE, Ghiringhelli PD et al (2008) Argentine hemorrhagic fever diagnostic test based on recombinant Junin virus N protein. J Med Virol 80(12):2127–2133
83. Vela EM, Colpitts TM et al (2008) Pichinde virus is trafficked through a dynamin 2 endocytic pathway that is dependent on cellular Rab5- and Rab7-mediated endosomes. Arch Virol 153(7):1391–1396
84. Vezza AC, Gard GP et al (1977) Structural components of the arenavirus Pichinde. J Virol 23(3):776–786
85. Vitullo AD, Hodara VL et al (1987) Effect of persistent infection with Junin virus on growth and reproduction of its natural reservoir, *Calomys musculinus*. Am J Trop Med Hyg 37(3):663–669
86. York J, Agnihothram SS et al (2005) Genetic analysis of heptad-repeat regions in the G2 fusion subunit of the Junin arenavirus envelope glycoprotein. Virology 343(2):267–274
87. York J, Romanowski V et al (2004) The signal peptide of the Junin arenavirus envelope glycoprotein is myristoylated and forms an essential subunit of the mature G1–G2 complex. J Virol 78(19):10783–10792
88. Zeitlin L, Geisbert JB et al (2016) Monoclonal antibody therapy for Junin virus infection. Proc Natl Acad Sci U S A 113(16):4458–4463

Chapter 11
Hantavirus: General Features and Present Situation in Latin America

Adriana Delfraro, Sonia M. Raboni, and Claudia Nunes Duarte dos Santos

1 Introduction

Several factors have contributed to the emergence and reemergence of viruses of medical importance: the increase in displacement of individuals in all areas of the globe, demographic increase and geographic expansion, destruction of biodiversity, and global warming are some examples. Additionally, the genome of most of the roboviruses (rodent-borne viruses) consists of RNA molecule(s), which present high mutation and recombination rates. There is also the possibility of viral genomic segments reassortment between different viruses (in the case of segmented genomes) that are pivotal events for viral evolution but can also increase the risk of the emergence of more adapted and virulent strains.

Hantaviruses are members of the genus *Hantavirus*, family *Bunyaviridae* [7], which contains more than 350 members and represents a major class of zoonotic pathogens that cause two severe diseases in humans: hemorrhagic fever with renal syndrome (HFRS) in Eurasia and hantavirus cardiopulmonary syndrome (HCPS) in the Americas. The name *Hantavirus* is after a river's name in Korea where the *Hantaan virus* (HTNV) was first identified in a rodent by Ho-Wang Lee and colleagues in the 1970s [48].

A. Delfraro
Sección Virología, Facultad de Ciencias, Universidad de la República, Montevideo, Uruguay

S.M. Raboni
Laboratório de Virologia, Universidade Federal de Paraná, Curitiba, PR, Brazil

C.N.D. dos Santos (✉)
Laboratório de Virologia Molecular, Instituto Carlos Chagas, Fiocruz PR, Curitiba, PR, Brazil
e-mail: clsantos@fiocruz.br

J.E. Ludert et al. (eds.), *Human Virology in Latin America*,
DOI 10.1007/978-3-319-54567-7_11

The transmission of hantavirus to humans occurs through inhalation of aerosols generated from feces, urine, and saliva of infected rodents, but other small mammals such as shrews, moles, and bats can also harbor the virus [30, 36, 98]. The rodent-borne hantaviruses persistently infect rodents from the family *Muridae* and subfamilies *Arvicolinae* (Europe), *Murinae* (Europe and Asia), and *Sigmodontinae/Neotominae* (America) [71]. In general, each hantavirus is predominantly associated with a rodent-host species in a given geographic region (reviewed in Plyusnin 2002), although few cases of interhuman transmission have been described [64, 78, 80, 81].

HCPS is an immunopathology of rapid progression that begins with a mild fever and can progress to noncardiogenic pulmonary edema and shock. HCPS was initially described in 1993 in a cluster of patients presenting an acute respiratory distress disorder in the southwestern region of the United States [76]. The hantavirus implicated in this outbreak was named Sin Nombre (SNV), harbored by the wild rodent reservoir *Peromyscus maniculatus*. In the Americas, hantaviruses are emergent, and more than 40 genotypes have been described, indicating that these agents are widely dispersed, and nearly half of them are pathogenic to humans [31].

Mortality rates related to hantavirus infection vary from 0.1% to 40%, depending on the specific virus involved. The emergence of hantavirus in human populations is correlated with rodent population density, which depends on several environmental factors, such as precipitation, temperature, habitat quality, and food availability, or, alternatively, anthropogenic behavior in the rodent–host environment [93]. Although hantavirus disease has been recognized for more than four decades, there is still no specific therapy available, and medical treatment is mainly palliative. The Syrian golden hamster recapitulates the human clinical picture and is a valuable model to study vital pathogenesis, viral evolution, and antiviral strategies for prevention and prophylactics of HCPS [37].

2 Virion Structure and Replicative Cycle

The viral particles are spherical with a diameter of 80–120 nm, with an envelope (derived from Golgi membranes of the host cell) and containing equimolar amounts of three negative single-stranded RNA genome segments designated large (L), medium (M), and small (S) [74]. The three RNA segments are coated with the nucleoprotein (N), forming ribonucleoproteins (RNPs) [20, 45].

The L segment of approximately 6.6 kb encodes a 250-kDa RNA-dependent RNA polymerase (viral replicase) that is associated to each viral segment in the virions and is required to initiate viral replication in the host cell cytosol. Comparative analyses of nucleotide and amino acid sequences of the L segment of different hantaviruses showed a conserved primary structure, despite the marked variability of its nucleotide sequence.

The M segment, about 3.7 kb, has a single open reading frame and encodes a precursor glycoprotein (GPC) that is processed by a cellular protease into two envelope glycoproteins, Gn and Gc (formerly named G1 and G2), 70 kDa and ~50 kDa, respectively. Gn and Gc glycosylation takes place in the Golgi complex during the maturation process [21]. Structural studies have demonstrated that Gn and Gc proteins form spikes protruding from the viral membrane and that these Gn/Gc glycoprotein heterodimers may interact with cellular receptors. Gn and Gc proteins present large globular and hydrophobic transmembrane domains and a C-terminal cytoplasmic tail. Because bunyaviruses lack a classical matrix protein, it is suggested that the Gn/Gc tail would interact with the N protein [22, 24]. The hantavirus glycoproteins are involved in the modulation of host innate immune response and virulence [59, 66].

The S segment, approximately 1.09 kb, encodes the nucleoprotein (N) with a molecular mass of approximately 50 kDa, which associates with the viral RNA segments to form helicoidal filamentous structures called ribonucleocapsids (RNPs) [20, 98, 108]. Among the structural proteins, the nucleocapsid is the most abundant and antigenic protein in the virus particle and accumulates in the cytoplasm of infected cells early during infection. Immunodominant and cross-reactive epitopes at the N-terminal domain of the N protein make it suitable to be used as a recombinant antigen for diagnosis purposes. Hantavirus-infected patients present a very short-term viremia and have detectable IgM and IgG antibodies against the nucleocapsid antigen at the onset of clinical symptoms [11, 79]; thus, serological tests are often used for the detection of these antibody classes [109].

N is a multifunctional protein involved in the initiation of transcription and translation of the viral genome, binding selectively viral RNA (vRNA) and viral positive sense antigenomic RNA (cRNA), leading to the encapsidation of newly synthesized vRNA and direct virus assembly, and interacts with the MxA proteins interfering with the host type I interferon response [74].

The 5′- and 3′-noncoding regions (NCRs) of the viral genomic segments present complementary sequences at their ends that are paired to form "panhandle"-like structures, which presumably account for the circular form of the RNAs observed by electron microscopy. These complementary regions are likely to have an important role in replication, providing signals for recognition by RdRp synthesis or for the packaging of the viral genome [20] (Fig. 11.1).

The virus enters the host cell through interaction of viral glycoproteins with specific cell receptors and subsequent endocytosis. Gavrilovskaya et al. [28] described that the use of vitronectin, a protein that binds to β3 integrin, inhibits the entry of the *Sin Nombre virus* (SNV) and *New York virus* (NYV) in Vero E6 cells, indicating that this would be a receptor on the host cell involved in viral penetration [28]. Potential receptors for Old World hantaviruses also include β1, β2, and β3 integrins (reviewed in Albornoz et al. [2]). In the acidic compartment of the late endosomes, RNPs are released in the cytoplasm after viral glycoprotein and endosome membrane fusion. Subsequently, the virion supplied RNA-dependent RNA polymerase-mediated primary transcription of negative-strand RNA in the cytoplasm, following viral mRNA translation; transcription shifts from mRNA to positive-strand complementary RNA (cRNA) and de novo negative-strand viral RNA synthesis with the concomitant formation of ribonucleoprotein structures. Transcription and translation can also take place in the endoplasmic reticulum–Golgi compartment. The RdRp possesses transcriptase, polymerase, and endonuclease activity [23]. Its endonuclease activity is involved in cap-snatching cytoplasmic cellular mRNAs to prime viral mRNA

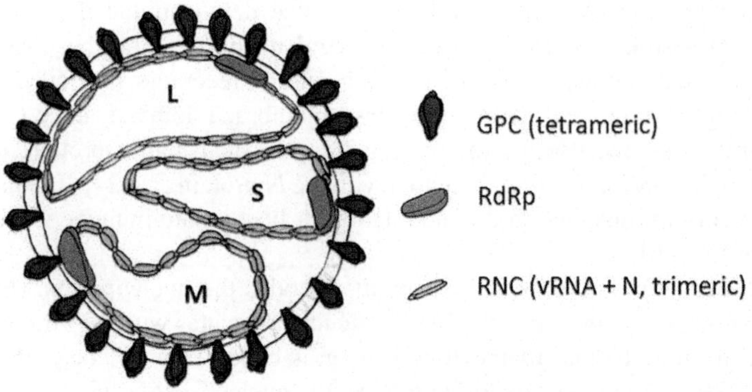

Fig. 11.1 Schematic representation of the hantavirus virion. The viral particles are spherical with a diameter of 80–120 nm, with an envelope derived from Golgi membranes of the host cell. The GPC glycoprotein forms spikes protruding from the envelope. The genome is composed of three segments of negative single-stranded RNA designated large (L), medium (M), and small (S). Each RNA segment is coated with the nucleoprotein (N) and is associated with RNA-dependent RNA polymerase (viral replicase)

synthesis. The GPC synthesis (derived from M segment mRNA) occurs on endoplasmic reticulum (ER) membrane-bound ribosomes, with the simultaneous primary glycosylation of envelope proteins. Gn and Gc glycoproteins are translocated to the Golgi complex, and new virus particles are formed by a budding process at smooth-surface vesicles in the Golgi and are liberated from the cell by exocytosis [98].

3 Origin, Evolution, and Classification of Hantaviruses

For several years, hantaviruses were only associated with rodents (*Murinae* and *Cricetidae*), with the exception only of *Thottapalayam virus*, which was isolated in 1964 from the Asian house shrew *Suncus murinus* [13]. From 2007 to present, more than 20 new hantaviruses were discovered in association with insectivore mammals such as shrews and moles (order Eulipotyphla; families Soricidae and Talpidae) [5, 29, 30, 43, 52, 99]. To add more complexity to the picture, recent findings report new hantaviruses associated with bats (order Chiroptera; families Vespertilionidae, Rhinolophidae, Nycteridae, and Hipposideridae) in Asia and Africa [4, 102, 113, 114]. Up to now, no human disease has been associated with these newfound hantaviruses.

Evolutive studies on the origin of placental mammals propose that the superorder Laurasiatheria (which include bats, shrews, and moles) diverged from Euarchontoglires (where order Rodentia is placed) at 100 million years ago [8, 68]. The discovery of new hantaviruses in such divergent reservoir hosts leads to revising the origin of the *Hantavirus* genus and the main forces driving its evolution.

Early studies on hantaviruses and their rodent hosts showed a high degree of congruence in their respective phylogenies, reinforcing the idea that coevolution between hosts and viruses was the main force influencing hantavirus evolution.

Basically, each hantavirus was carried by a unique rodent species, and in turn, genetically similar viruses were hosted by closely related rodents. Incongruent topologies in phylogenies were explained by the occurrence of sporadic host switching [38, 82]. From 2007 to present, the finding of a growing number of new hantaviruses harbored by insectivores questioned this paradigm. Co-phylogenetic reconciliation analyses and estimations on the evolutionary rates showed that divergence times between hantavirus lineages from rodents, insectivores, and bats were far more recent than the divergence times between their respective hosts. Further, host switching events appeared more frequently than expected under a co-evolutionary theory. So, as a result of these approaches, the similarities between hantaviruses and their mammalian hosts phylogenies may be the result of preferential host switching followed by local adaptation instead of co-evolution [89, 90].

Nowadays the debate remains open. New phylogenetic approaches and more comprehensive analyses lead to reevaluating the role of co-divergence in hantavirus evolution. Sequence analyses of complete and partial L genes from all known rodent, insectivore, and bat hantaviruses together with phylogenetic fossil host hypothesis testing showed that mammals in the superorder Laurasiatheria could have been the potential hosts of ancestral hantaviruses at most basal tree nodes. According to these inferences, hantaviruses from Muridae and Cricetidae rodents appear as paraphyletic groups, originated by two independent host switches from hantaviruses carried by laurasiatherian mammals. Overall, the main hantavirus groups show typical systematics of co-speciation, where virus phylogeny primarily resembles the phylogeny of host mammals and only secondarily their geographic dispersion [83, 114, 116].

Another point of controversy is the criterion for species delimitation in hantaviruses. According to the International Committee on Taxonomy of Viruses (ICTV), a hantavirus species should fulfill the following criteria: a hantavirus species should (i) occupy a unique ecological niche (i.e., a clear association of a new hantavirus with a different primary rodent reservoir species or subspecies), (ii) have at least a 7% difference in the amino acid sequences of the complete nucleocapsid (N) and glycoprotein precursor (GPC) proteins to all known species, (iii) show an at least fourfold difference in a two-way cross-neutralization test, and (iv) show the absence of genetic reassortment with other species in nature [81, 82].

For the majority of known hantaviruses it has been difficult to fulfill the four criteria. Genetic reassortment of closely related hantaviruses has been reported in vitro, but also in nature [34, 46, 92, 117]. Additionally, several South American hantaviruses have been detected in more than one rodent species, and the species involved may vary according to the geographic areas studied. As an example, the genetically related Juquitiba, Araucaria, and Itapúa hantaviruses were found in five different species (*Oligoryzomys nigripes*, *Oxymycterus judex*, *Oxymycterus nasutus*, *Akodon montensis*, *and Akodon paranaensis*) in South Brazil, Paraguay, and Uruguay [15, 19, 86]. This finding also raises difficulties in determining the primary reservoir host for a given hantavirus and the role of other sympatric rodents in maintaining the virus in the environment.

Another difficulty is the requirement to perform cross-neutralization test assays to define species, given that hantaviruses have proven to be very hard to isolate in tissue culture. The majority of the rodent- and insectivore-borne and all the bat-borne hantaviruses have not been isolated. In fact, most of them were identified through phylogenetic analyses on (mostly) partial sequences of the N, GPC, or L protein genes.

The last ICTV report for the *Hantavirus* genus recognizes only 24 species, 23 of which are rodent borne and 1 shrew borne, so the majority of the hantaviruses remain taxonomically unclassified [1]. Recently, another scheme for the demarcation of hantavirus species has been suggested. Maes et al. propose an amino acidic sequence distance >10% for N protein or >12% for GPC to limit species and an amino acidic sequence distance >24% for N protein or >32% for GPC to delimitate hantavirus groups [62].

4 Hantavirus Studies in Latin America

HCPS is a serious health problem in Latin America. Growing urbanization, together with the expansion of agriculture and cattle-breeding areas into natural ecosystems, has increased the chances of close contact between infected rodents and humans. Several studies have pointed out that habitat fragmentation caused by human activities tends to reduce rodent diversity, which in turn results in an increase of hantavirus prevalence in endemic areas [47, 58, 103, 104].

The rodent family Cricetidae is the second largest mammalian family (more than 500 species), including all the New World mice, and it is divided into three subfamilies: Sigmodontinae (predominantly South American), Neotominae (almost exclusively North American), and the Arvicolinae. All these groups are morphologically and ecologically diverse. Sigmodontinae rodents of South America present high diversity because of recent invasion of the continent followed by a rapid adaptive radiation [101]. In accordance with this diversity, many different hantavirus lineages were characterized in Latin America (Fig. 11.2); most of them are associated with Sigmodontinae rodents and its three main tribes: Akodontini, Oryzomyini, and Phyllotini.

The first HCPS outbreak reported in Latin America occurred in 1993. The three cases were diagnosed in the rural locality of Juquitiba, São Paulo State, Brazil. Further molecular studies identified the virus responsible for the outbreak; this new hantavirus was named Juquitiba [41, 110].

In 1994–1995, several outbreaks of HCPS were reported in Central and Southern Argentina. Genome amplification, sequencing, and phylogenetic analyses allowed the identification of two novel hantaviruses: Andes and Lechiguanas. Soon after that, rodent trapping in the likely places of exposure for human cases led to the identification of the long-tailed rice rat and the yellow pigmy rice rat (*Oligoryzomys longicaudatus* and *O. flavescens*, respectively) as the primary reservoir hosts for Andes and Lechiguanas hantavirus [50, 51, 55]. Isolation of the aforementioned hantavirus from rodent tissues and the molecular cloning and protein expression of Araucaria and Araraquara N protein were of capital importance to locally develop diagnostic tools. These techniques allowed detecting antibodies to autochthonous hantavirus in human and rodent samples with better sensitivity and specificity [50,

Fig. 11.2 Hantavirus lineages characterized in Latin America. Hantaviruses are transmitted directly to humans by small mammals, which are their natural reservoir. In general, each hantavirus has a unique rodent host. The majority of the Latin American hantaviruses are associated with rodents of the subfamily Sigmodontinae and its three main tribes: Akodontini, Oryzomyini, and Phyllotini

57, 77, 87]. From 1996 and on, outbreaks were reported in Argentina, Chile, Brazil, Uruguay, and Paraguay, and numerous field studies were carried out to determine the reservoir hosts in each geographic area [25, 40, 86, 88, 106].

Currently, almost all Latin American countries have reported HCPS cases or outbreaks. Countries with no reports on HCPS cases or rodent survey are Ecuador, Guyana, Surinam, Cuba, Nicaragua, El Salvador, Guatemala, and Belize. Since 1993, about 4000 accumulated cases have been reported in the Americas, and almost half of them occurred in Brazil [26].

Up to today, 28 hantavirus lineages are present in Latin America (Table 11.1). Most of them have been identified through genome amplification plus sequencing and phylogeny, based on viral S and/or M segments [10, 18, 27, 49, 69, 70, 85, 94, 96]. In turn, field studies allowed the identification of 14 rodent-only viral lineages that still have not been associated with human disease [16, 44, 54, 65, 72, 75, 95, 107, 112].

Except for Andes, Lechiguanas, and Maciel hantavirus, viral isolation was not achieved for these hantaviruses, hampering the determination of their taxonomic status according to the ICTV criteria. Despite this, availability of new sequence information on Latin American hantavirus lineages together with accurate rodent

Table 11.1 Hantavirus lineages present in Latin America

Virus lineage	Reservoir host	Geographic distribution	Human disease	References
Andes	*Oligoryzomys longicaudatus, Abrothrix longipilis*	Argentina, Chile	HCPS	[50, 55, 106]
Anajatuba	*Oligoryzomys fornesi*	Northern Brazil	HCPS	[95, 107]
Araraquara-Paranoa	*Bolomys lasiurus, Calomys tener*	Brazil	HCPS	[25, 26]
Bermejo-Ñeembucu	*Oligoryzomys flavescens*	Argentina, Paraguay	HCPS	[15, 50, 51]
Calabazo	*Zygodontomys brevicauda*	*Panamá*	Unknown	[112]
Caño Delgadito	*Sigmodon alstoni*	Venezuela	Unknown	[27]
Carrizal	*Reithrodontomys sumichrasti*	Mexico	Unknown	[44]
Castelo dos Sonhos	*Oligoryzomys utiaritensis, Oligoryzomys moojeni*	Brazil	HCPS	[96]
Catacamas	*Oryzomys couesi*	Honduras	Unknown	[69]
Choclo	*Oligoryzomys fulvescens*	Panamá	HCPS	[112]
El Moro Canyon	*Reithrodontomys megalotis*	Mexico	Unknown	[44]
Huitzilac	*Reithrodontomys megalotis*	Mexico	Unknown	[44]
Jabora	*Akodon montensis, Akodon serrensis*	Southern Brazil, Paraguay	Unknown	[15, 86]
Juquitiba-Araucaria	*Oligoryzomys nigripes, Oxymycterus judex, Oxymycterus nasutus, Akodon montensis, Akodon paranaensis*	Brazil, Uruguay, Paraguay	HCPS	[15, 19, 86, 88]
Laguna Negra	*Calomys laucha, Calomys callosus*	Argentina, Bolivia, Brazil, Paraguay	HCPS	[40, 49, 85]
Lechiguanas	*Oligoryzomys flavescens*	Argentina, Uruguay	HCPS	[18, 50, 51]
Maciel	*Necromys obscurus*	Argentina	Unknown	[50]
Maripa	*Oligoryzomys fulvescens, Zygodontomys brevicauda*	French Guiana	HCPS	[65]
Maporal	*Oligoryzomys fulvescens, Oligoryzomys delicatus*	Venezuela	HCPS?[a]	[70]
Montano	*Peromyscus beatae*	Mexico	Unknown	[44]

Necocli	*Zygodontomys cherriei*	Colombia	Unknown	[72]
Oran	*Oligoryzomys chacoensis*	Northern Argentina	HCPS	[50]
Pergamino	*Akodon azarae*	Argentina	Unknown	[50]
Playa de Oro	*Oryzomys couesi, Sigmodon mascotensis*	Mexico	Unknown	[16]
Rio Mamore	*Oligoryzomys microtis*	Peru, Bolivia	HCPS	[94]
Rio Mearim	*Holochilus sciureus*	Northern Brazil	Unknown	[95]
Rio Segundo	*Reithrodontomys mexicanus*	Costa Rica	Unknown	[44]
Tunari	Undetermined	Bolivia	HCPS	[26]

[a]No cases of HCPS caused by Maporal hantavirus were reported in the literature; however, experimental infection in the Syrian hamster showed this virus is capable of reproducing HCPS

Source: Milazzo et al. (2002) *J Infect Dis* 186(10):1390–1395)

identification will contribute to clarifying the virus reservoir–host relationships and the epidemiology of this relevant zoonosis.

5 Pathogenesis and Implication on Treatment

Hantavirus infection induces serious microvascular leakage, and clinically, patients present with hemorrhages, hemoconcentration, and hypotension [35]. Hantavirus infections are associated with two diseases: hemorrhagic fever with renal syndrome (HFRS) and hantavirus cardiopulmonary syndrome (HCPS), which present similar pathological findings, being the consequence of both innate and adaptive and humoral and cellular immune mechanisms, but the contribution of these factors to disease development remains indeterminate [60].

Previous studies showed that CD4+ and CD8+ T-cell responses to hantavirus infection in humans are associated with immunoprotection, including the magnitude of the cellular immune response and the frequency of the CD8+ T-cell response, which were much higher in patients with mild/moderate HFRS than in those with severe/critical disease at the acute stage of the disease [53, 105]. Moreover, *Hantavirus* glycoprotein produces a strong CD4+ T-cell response, which elicits greater defense against the infection and is inversely correlated with plasma viral load and disease outcome [56]. Conversely, a report has shown the importance of monocyte activation to hantavirus infection. Increased monocyte counts have been positively correlated with elevated plasma-soluble CD14 levels, as well as increased tumor necrosis factor (TNF)-α and soluble CD163 levels, which are associated with severe acute kidney injury in HFRS patients [105].

Primarily, it was supposed that the endothelial cell (EC) dysfunction demonstrated in HFRS and HCPS patients was related to a strong cellular immune response, elicited by cytotoxic CD8+ T and NK cells, although no clear endothelial cell damage has been clearly observed. Recent studies in ANDV-infected Syrian hamsters showed that depletion of T cells did not impact disease onset or outcome of HCPS [33, 84]. The infected EC is probably protected from cytotoxic lymphocyte-mediated killing through an inhibition of apoptosis induction in infected cells in combination with an increased expression of HLA class [12]. Contrasting with previous findings, in a recently reported macaque model for Sin Nombre hantavirus disease, an association between the expansion of T-lymphocyte-activated and disease severity was observed [91]. The importance and function of T cells activated in human hantavirus infections are still poorly understood, and further studies are needed.

The increased vascular permeability occurs without any cytopathic evidence, suggesting that the pathogenesis is associated with immunopathological mechanisms. It results from the interaction between EC surface-receptor and cytoplasmic signaling responses such as pro-inflammatory cytokine: interleukin (IL)-6, IL-8, IL-33, interferon (IFN)-γ, and tumor necrosis factor-α (TNF-α) secreted by activated innate immune cells, which are upregulated in hantavirus infections [9, 105, 115], as well as EC interactions with immune cells [17].

In hantavirus acute infection, a significant upregulation of cytokines has been shown, some associated with leukocyte migration and repair of lung tissue, and other factors linked to increasing the endothelial monolayer permeability that facilitates leukocyte transendothelial migration. In addition, a downregulation of cytokines also can be observed and has been associated with platelet numbers, reduction, and dysfunction [73].

Endothelial cells (ECs) are fundamental to preserve the fluid barrier and have a vital role in maintaining the balance between tissues and vascular compartments, through a complex system of surface–receptor interactions. Microvascular and lymphatic EC surface receptors and the endothelial glycocalyx are keys to fluid management and vascular homeostasis [6]. Furthermore, lymphatic tissues and lymphatic endothelial cells (LECs) are exclusively regulated by cell-surface receptors and are responsible for edema regulation by clearing fluid from tissues, and LECs are sentinel antigen-presenting cells that determine tolerance and viral clearance [17]. Studies showed that hantavirus infection is associated with LEC dysfunction and, consequently, alteration of liquid pulmonary clearance in patients with HCPS [61].

Nonpathogenic hantaviruses use $\alpha v\beta 1$ integrin receptors, whereas pathogenic hantavirus binds and inactivates $\alpha v\beta 3$ integrin conformers, which usually form complexes with VEGF receptors, leading to dysregulation of VEGF-induced permeability; this may lead to impairment of vascular endothelial cadherin expression and subsequent loss of endothelial barrier function [17, 63].

Currently, there are no antiviral drugs or immunotherapeutic or effective post-exposure prophylactics available for hantavirus infection [3, 32], and only supportive interventions, such as early diagnosis and aggressive support in a cardiac and pulmonary intensive care unit, are recommended [42]. In vitro studies have demonstrated ribavirin as a drug capable of preventing hantavirus infection, including in vivo studies carried out in golden Syrian hamsters which showed that drug administration before or up to 3 days after infection was able to prevent infection in all animals [97]. However, despite randomized studies conducted in HRFS individuals who showed response to ribavirin when it was administrated up to 7 days of infection, in HCPS patients no benefit of treatment could be observed, probably in consequence of the long incubation period of the disease, which can last up to 5 weeks [63].

The use of molecules to block receptor- and pathway-specific regulation of VEGFR2 or downstream signaling pathway responses that control EC barrier functions has also been evaluated as a therapeutic intervention. However, the effectiveness of this approach in patients who are already symptomatic remains to be determined [63].

Based on the immune response exacerbation observed during the course of hantavirus infection, studies using corticosteroids to prevent severe HCPS in acute disease were performed, but no benefits were demonstrated and it has not been recommended [111]. An alternative approach to HCPS treatment was the passive administration of neutralizing antibodies by serum transfusion from convalescent patients. Preliminary results carried out in Chile have indicated a reduced mortality in the treated patients [111], although the final evaluation of this study is not yet available.

In conclusion, during hantavirus infection, several factors contribute simultaneously, culminating with vascular leakage, and the underlying mechanisms associated with disease pathogenesis are not completely understood. Prospective studies with animal models are crucial to evaluate individual immune cell types involved in the disease process. Knowing their different mechanisms of action is fundamental to the development of drugs that will act on specific targets, seeking to reduce disease severity and, consequently, to reduce mortality rates.

6 Prevention

Small interfering RNA (siRNA) against *Andes virus* (ANDV) genes has been tested as a potential antiviral strategy [14]. Although with promising results, such therapy would take years to be available for human use.

A phase III inactivated viral vaccine against the Old World hantavirus Hantaan has been tested in healthy adults, but the results of the plaque reduction neutralization test (PRNT) after a three-dose vaccination are modest [100]. Some DNA vaccines were developed and tested in animal models, but the effects are still pending [39, 67].

Despite the long time since the association of hantavirus infection and HFRS and HCPS in humans has been known, no specific treatment and prophylaxis have been obtained, although efforts have been committed. Several aspects of hantavirus biology and pathogenesis remain unclear. As an example, it has been assumed that the primary target cells during HCPS infection were human alveolar macrophages. Recently, Hammerbeck and colleagues have demonstrated that depletion of alveolar macrophages in golden Syrian hamsters does not prevent hantavirus disease [32]. These findings open new perspectives on hantavirus infection strategies on human beings.

References

1. Adams MJ, Lefkowitz EJ, King AMQ, Bamford DH, Breitbart M, Davison AJ, Ghabrial SA, Gorbalenya AE, Knowles NJ, Krell P, Lavigne R, Prangishvili D, Sanfaçon H, Siddell SG, Simmonds P, Carstens EB (2015) Ratification vote on taxonomic proposals to the International Committee on Taxonomy of Viruses (2015). Arch Virol 160:1837–1850. doi:10.1007/s00705-015-2425-z
2. Albornoz A, Hoffmann A, Lozach P-Y, Tischler N (2016) Early bunyavirus-host cell interactions. Viruses 8:143. doi:10.3390/v8050143
3. Angulo J, Pino K, Echeverría-chagas N, Marco C, Martínez-valdebenito C, Galeno H, Villagra E, Vera L, Lagos N, Becerra N, Mora J, Bermúdez A, Cárcamo M, Díaz J, Miquel JF, Ferrés M, López-lastra M (2015) Association of single-nucleotide polymorphisms in IL28B, but not TNF-α, with severity of disease caused by Andes virus. Clin Infect Dis 61:62–69. doi:10.1093/cid/civ830
4. Arai S, Nguyen ST, Boldgiv B, Fukui D, Araki K, Dang CN, Ohdachi SD, Nguyen NX, Pham TD, Boldbaatar B, Satoh H, Yoshikawa Y, Morikawa S, Tanaka-Taya K, Yanagihara

R, Oishi K (2013) Novel bat-borne hantavirus, Vietnam. Emerg Infect Dis 19:1159–1161. doi:10.3201/eid1907.121549
5. Arai S, Ohdachi SD, Asakawa M, Kang HJ, Mocz G, Arikawa J, Okabe N, Yanagihara R (2008) Molecular phylogeny of a newfound hantavirus in the Japanese shrew mole (*Urotrichus talpoides*). Proc Natl Acad Sci U S A 105:16296–16301. doi:10.1073/pnas.0808942105
6. Baumgartner-Parzer SM, Waldhäusl WK (2001) The endothelium as a metabolic and endocrine organ: its relation with insulin resistance. Exp Clin Endocrinol Diabetes 109:S166–S179. doi:10.1055/s-2001-18579
7. Beaty B, Calisher C (1991) Bunyaviridae—natural history. Curr Top Microbiol Immunol 169:27–78
8. Bininda-emonds ORP, Cardillo M, Jones KE, Macphee RDE, Beck RMD, Grenyer R, Price SA, Vos RA, Gittleman JL, Purvis A (2007) The delayed rise of present-day mammals. Nature (Lond) 446:507–512. doi:10.1038/nature05634
9. Bondu V, Schrader R, Gawinowicz M, McGuire P, Lawrence D, Hjelle B, Buranda T (2015) Elevated cytokines, thrombin and PAI-1 in severe HCPS patients due to Sin Nombre virus. Viruses 7:559–589. doi:10.3390/v7020559
10. de Borba L, Delfraro A, Raboni SM, dos Santos CND (2013) First evidence of asymptomatic infection related to the araucaria (Juquitiba-like) hantavirus. BMJ Case Rep 2013:2–4
11. Borges AA, Campos GM, Moreli ML, Souza RLM, Aquino VH, Saggioro FP, Figueiredo LTM (2006) Hantavirus cardiopulmonary syndrome: immune response and pathogenesis. Microbes Infect 8:2324–2330. doi:10.1016/j.micinf.2006.04.019
12. Braun M, Bjorkstrom NK, Gupta S, Sundstrom K, Ahlm C, Klingstrom J, Ljunggren HG (2014) NK cell activation in human hantavirus infection explained by virus-induced IL-15/IL15Ralpha expression. PLoS Pathog 10:e1004521. doi:10.1371/journal.ppat.1004521
13. Carey DE, Reuben R, Panicker KN, Shope RE, Myers RM (1971) Thottapalayam virus: a presumptive arbovirus isolated from a shrew in India. Indian J Med Res 59:1758–1760
14. Chiang CF, Albariño CG, Lo MK, Spiropoulou CF (2014) Small interfering RNA inhibition of Andes virus replication. PLoS One. doi:10.1371/journal.pone.0099764
15. Chu YK, Goodin D, Owen RD, Koch D, Jonsson CB (2009) Sympatry of 2 hantavirus strains, Paraguay, 2003–2007. Emerg Infect Dis 15:1977–1980. doi:10.3201/eid1512.090338
16. Chu YK, Owen RD, Sanchez-Hernandez C, Romero-Almaraz Mde L, Jonsson CB (2008) Genetic characterization and phylogeny of a hantavirus from western Mexico. Virus Res 131:180–188
17. Dalrymple NA, Mackow ER (2014) Virus interactions with endothelial cell receptors: implications for viral pathogenesis. Curr Opin Virol 7:134–140. doi:10.1016/j.coviro.2014.06.006
18. Delfraro A, Clara M, Tome L, Achaval F, Levis S, Calderon G, Enria D, Lozano M, Russi J, Arbiza J (2003) Yellow pigmy rice rat (*Oligoryzomys flavescens*) and hantavirus pulmonary syndrome in Uruguay. Emerg Infect Dis 9:846–852
19. Delfraro A, Tomé L, D'Elía G, Clara M, Achával F, Russi JC, Arbiza Rodonz JR (2008) Juquitiba-like hantavirus from 2 nonrelated rodent species, Uruguay. Emerg Infect Dis 14:1447–1451. doi:10.3201/eid1409.080455
20. Elliott RM (1990) Molecular biology of the Bunyaviridae. J Gen Virol 71:501–522. doi:10.1099/0022-1317-71-3-501
21. Elliott RM, Brennan B (2014) Emerging phleboviruses. Curr Opin Virol 5:50–57. doi:10.1016/j.coviro.2014.01.011
22. Estrada DF, Boudreaux DM, Zhong D, St Jeor SC, De Guzman RN (2009) The hantavirus glycoprotein G1 tail contains dual CCHC-type classical zinc fingers. J Biol Chem 284:8654–8660. doi:10.1074/jbc.M808081200
23. Fernández-García Y, Reguera J, Busch C, Witte G, Sánchez-Ramos O, Betzel C, Cusack S, Günther S, Reindl S (2016) Atomic structure and biochemical characterization of an RNA endonuclease in the N terminus of Andes virus L protein. PLoS Pathog 12:e1005635. doi:10.1371/journal.ppat.1005635

24. Fernando Estrada D, Conner M, St. Jeor SC, De Guzman RN (2011) The structure of the hantavirus zinc finger domain is conserved and represents the only natively folded region of the Gn cytoplasmic tail. Front Microbiol. doi:10.3389/fmicb.2011.00251
25. Figueiredo GG, Borges AA, Campos GM, Machado AM, Saggioro FP, Sabino Junior Gdos S, Badra SJ, Ortiz AA, Figueiredo LT (2010) Diagnosis of hantavirus infection in humans and rodents in Ribeirao Preto, State of Sao Paulo, Brazil. Rev Soc Bras Med Trop 43:348–354
26. Figueiredo LTM, de Souza WM, Ferres M, Enria DA (2014) Hantaviruses and cardiopulmonary syndrome in South America. Virus Res 187:43–54. doi:10.1016/j.virusres.2014.01.015
27. Fulhorst CF, Monroe MC, Salas RA, Duno G, Utrera A, Ksiazek TG, Nichol ST, de Manzione NM, Tovar D, Tesh RB (1997) Isolation, characterization and geographic distribution of Cano Delgadito virus, a newly discovered south American hantavirus (family Bunyaviridae). Virus Res 51:159–171
28. Gavrilovskaya IN, Shepley M, Shaw R, Ginsberg MH, Mackow ER (1998) Beta3 integrins mediate the cellular entry of hantaviruses that cause respiratory failure. Proc Natl Acad Sci U S A 95:7074–7079. doi:10.1073/pnas.95.12.7074
29. Gu SH, Dormion J, Hugot J-P, Yanagihara R (2014) High prevalence of Nova hantavirus infection in the European mole (*Talpa europaea*) in France. Epidemiol Infect 142:1167–1171. doi:10.1017/S0950268813002197
30. Guo WP, Lin XD, Wang W, Tian JH, Cong ML, Zhang HL, Wang MR, Zhou RH, Wang JB, Li MH, Xu J, Holmes EC, Zhang YZ (2013) Phylogeny and origins of hantaviruses harbored by bats, insectivores, and rodents. PLoS Pathog 9(2):e1003159. doi:10.1371/journal.ppat.1003159
31. Guterres A, de Oliveira RC, Fernandes J, Schrago CG, de Lemos ERS (2015) Detection of different south American hantaviruses. Virus Res 210:106–113. doi:10.1016/j.virusres.2015.07.022
32. Hammerbeck CD, Brocato RL, Bell TM, Schellhase CW, Mraz SR, Queen LA, Hooper JW (2016) Depletion of alveolar macrophages does not prevent hantavirus disease pathogenesis in golden Syrian hamsters. J Virol 90:6200–6215. doi:10.1128/JVI.00304-16
33. Hammerbeck CD, Hooper JW (2011) T cells are not required for pathogenesis in the Syrian hamster model of hantavirus pulmonary syndrome. J Virol 85:9929–9944. doi:10.1128/JVI.05356-11
34. Handke W, Oelschlegel R, Franke R, Wiedemann L, Krüger DH, Rang A (2010) Generation and characterization of genetic reassortants between Puumala and Prospect Hill hantavirus in vitro. J Gen Virol 91:2351–2359. doi:10.1099/vir.0.021139-0
35. Hepojoki J, Vaheri A, Strandin T (2014) The fundamental role of endothelial cells in hantavirus pathogenesis. Front Microbiol 5:1–7. doi:10.3389/fmicb.2014.00727
36. Holmes EC, Zhang Y-Z (2015) The evolution and emergence of hantaviruses. Curr Opin Virol 10:27–33. doi:10.1016/j.coviro.2014.12.007
37. Hooper JW, Larsen T, Custer DM, Schmaljohn CS (2001) A lethal disease model for hantavirus pulmonary syndrome. Virology 289:6–14. doi:10.1006/viro.2001.1133
38. Jackson AP, Charleston MA (2004) A cophylogenetic perspective of RNA-virus evolution. Mol Biol Evol 21:45–57. doi:10.1093/molbev/msg232
39. Jiang DB, Sun YJ, Cheng LF, Zhang GF, Dong C, Jin BQ, Song CJ, Ma Y, Zhang FL, Yang K (2015) Construction and evaluation of DNA vaccine encoding hantavirus glycoprotein N-terminal fused with lysosome-associated membrane protein. Vaccine 33:3367–3376. doi:10.1016/j.vaccine.2015.05.007
40. Johnson AM, Bowen MD, Ksiazek TG, Williams RJ, Bryan RT, Mills JN, Peters CJ, Nichol ST (1997) Laguna Negra virus associated with HPS in western Paraguay and Bolivia. Virology 238:115–127
41. Johnson AM, de Souza LT, Ferreira IB, Pereira LE, Ksiazek TG, Rollin PE, Peters CJ, Nichol ST (1999) Genetic investigation of novel hantaviruses causing fatal HPS in Brazil. J Med Virol 59:527–535
42. Jonsson CB, Hooper J, Mertz G (2008) Treatment of hantavirus pulmonary syndrome. Antivir Res 78:162–169. doi:10.1016/j.antiviral.2007.10.012

43. Kang HJ, Arai S, Hope AG, Cook JA, Yanagihara R (2010) Novel Hantavirus in the flat-skulled shrew (*Sorex roboratus*). Vector Borne Zoonotic Dis 10:593–597. doi:10.1089/vbz.2009.0159
44. Kariwa H, Yoshida H, Sánchez-Hernández C, Romero-Almaraz Mde L, Almazán-Catalán JA, Ramos C, Miyashita D, Seto T, Takano A, Totani M, Murata R, Saasa N, Ishizuka M, Sanada T, Yoshii K, Yoshimatsu K, Arikawa J, Takashima I (2012) Genetic diversity of hantaviruses in Mexico: identification of three novel hantaviruses from Neotominae rodents. Virus Res 163:486–494. doi:10.1016/j.virusres.2011.11.013
45. Khaiboullina SF, Morzunov SP, St Jeor SC (2005) Hantaviruses: molecular biology, evolution and pathogenesis. Curr Mol Med 5:773–790. doi:10.2174/156652405774962317
46. Klempa B, Schmidt HA, Ulrich R, Kaluz S, Labuda M, Meisel H, Hjelle B, Krüger DH (2003) Genetic interaction between distinct Dobrava hantavirus subtypes in *Apodemus agrarius* and *A. flavicollis* in nature. J Virol 77:804–809. doi:10.1128/JVI.77.1.804
47. Langlois JP, Fahrig L, Merriam G, Artsob H (2001) Landscape structure influences continental distribution of hantavirus in deer mice. Landsc Ecol 16:255–266. doi:10.1023/A:1011148316537
48. Lee HW, Lee PW, Johnson KM (1978) Isolation of the etiologic agent of Korean hemorrhagic fever. J Infect Dis 137:298–308. doi:10.1016/j.jviromet.2009.04.006
49. Levis S, Garcia J, Pini N, Calderon G, Ramirez J, Bravo D, St Jeor S, Ripoll C, Bego M, Lozano E, Barquez R, Ksiazek TG, Enria D (2004) Hantavirus pulmonary syndrome in northwestern Argentina: circulation of Laguna Negra virus associated with *Calomys callosus*. Am J Trop Med Hyg 71:658–663
50. Levis S, Morzunov SP, Rowe JE, Enria D, Pini N, Calderon G, Sabattini M, St Jeor SC (1998) Genetic diversity and epidemiology of hantaviruses in Argentina. J Infect Dis 177:529–538
51. Levis S, Rowe JE, Morzunov S, Enria DA, St Jeor S (1997) New hantaviruses causing hantavirus pulmonary syndrome in Central Argentina. Lancet 349:998–999. doi:10.1016/S0140-6736(05)62895-4
52. Ling J, Sironen T, Voutilainen L, Hepojoki S, Niemimaa J, Isoviita VM, Vaheri A, Henttonen H, Vapalahti O (2014) Hantaviruses in Finnish soricomorphs: evidence for two distinct hantaviruses carried by *Sorex araneus* suggesting ancient host-switch. Infect Genet Evol 27:51–61. doi:10.1016/j.meegid.2014.06.023
53. Liu B, Ma Y, Zhang Y, Zhang C, Yi J, Zhuang R, Yu H, Yang A, Zhang Y (2015) CD8 low CD100− T cells identify a novel CD8 T cell subset associated with viral control during human Hantaan virus infection. J Virol 89:11834–11844. doi:10.1128/JVI.01610-15.Editor
54. Londoño AF, Díaz FJ, Agudelo-Flórez P, Levis S, Rodas JD (2011) Genetic evidence of hantavirus infections in wild rodents from northwestern Colombia. Vector Borne Zoonotic Dis 11:701–708. doi:10.1089/vbz.2010.0129
55. López N, Padula P, Rossi C, Lázaro ME, Franze-Fernández MT (1996) Genetic identification of a new hantavirus causing severe pulmonary syndrome in Argentina. Virology 220:223–226. doi:10.1006/viro.1996.0305
56. Ma Y, Yuan B, Zhuang R, Zhang Y, Liu B, Zhang C (2015) Hantaan virus infection induces both Th1 and ThGranzyme B+ cell immune responses that associated with viral control and clinical outcome in humans. PLoS Pathogen 11(4):e1004788. doi:10.1371/journal.ppat.1004788
57. Machado AM, Machado ARSR, Moreli ML, Ribeiro BM, Figueiredo LTM, Wolff JLC (2011) Expression of recombinant Araraquara hantavirus nucleoprotein in insect cells and its use as an antigen for immunodetection compared to the same antigen expressed in *Escherichia coli*. Virol J 8:218. doi:10.1186/1743-422X-8-218
58. Mackelprang R, Dearing MD, St Jeor S (2001) High prevalence of sin Nombre virus in rodent populations, Central Utah: a consequence of human disturbance? Emerg Infect Dis 7:480–482. doi:10.3201/eid0703.017328
59. Mackow ER, Dalrymple NA, Cimica V, Matthys V, Gorbunova E, Gavrilovskaya I (2014) Hantavirus interferon regulation and virulence determinants. Virus Res 187:65–71. doi:10.1016/j.virusres.2013.12.041

60. Mackow ER, Gavrilovskaya IN (2009) Hantavirus regulation of endothelial cell functions. Thromb Haemost 102(6):1030–1041. doi:10.1160/TH09-09-0640
61. Mackow ER, Gorbunova EE, Dalrymple NA, Gavrilovskaya IN (2013) Role of vascular and lymphatic endothelial cells in hantavirus pulmonary syndrome suggests targeted therapeutic approaches. Lymphat Res Biol 11:128–135. doi:10.1089/lrb.2013.0006
62. Maes P, Klempa B, Clement J, Matthijnssens J, Gajdusek DC, Krüger DH, Van Ranst M (2009) A proposal for new criteria for the classification of hantaviruses, based on S and M segment protein sequences. Infect Genet Evol 9:813–820. doi:10.1016/j.meegid.2009.04.012
63. Manigold T, Vial P (2014) Human hantavirus infections: epidemiology, clinical features, pathogenesis and immunology. Swiss Med Wkly 44:w13937. doi:10.4414/smw.2014.13937
64. Martinez VP, Bellomo C, San Juan J, Pinna D, Forlenza R, Elder M, Padula PJ (2005) Person-to-person transmission of Andes virus. Emerg Infect Dis 11:1848–1853. doi:10.3201/eid1112.050501
65. Matheus S, Lavergne A, de Thoisy B, Dussart P, Lacoste V (2012) Complete genome sequence of a novel hantavirus variant of Rio Mamore virus, Maripa virus, from French Guiana. J Virol 86:5399–5399. doi:10.1128/JVI.00337-12
66. Matthys VS, Cimica V, Dalrymple NA, Glennon NB, Bianco C, Mackow ER (2014) Hantavirus GnT elements mediate TRAF3 binding and inhibit RIG-I/TBK1-directed beta interferon transcription by blocking IRF3 phosphorylation. J Virol 88:2246–2259. doi:10.1128/JVI.02647-13
67. Mccoy JR, Mendoza JM, Spik KW, Badger C, Gomez AF, Schmaljohn CS, Sardesai NY, Broderick KE (2015) A multi-head intradermal electroporation device allows for tailored and increased dose DNA vaccine delivery to the skin. Hum Vaccines Immunother 11:746–754. doi:10.4161/21645515.2014.978223
68. Meredith RW, Janečka JE, Gatesy J, Ryder OA, Fisher CA, Teeling EC, Goodbla A, Eizirik E, Simão TLL, Stadler T, Rabosky DL, Honeycutt RL, Flynn JJ, Ingram CM, Steiner C, Williams TL, Robinson TJ, Burk-Herrick A, Westerman M, Ayoub NA, Springer MS, Murphy WJ (2011) Impacts of the cretaceous terrestrial revolution and KPg extinction on mammal diversification. Science 334:521–524. doi:10.1126/science.1211028
69. Milazzo ML, Cajimat MNB, Hanson JD, Bradley RD, Quintana M, Sherman C, Velásquez RT, Fulhorst CF (2006) Catacamas virus, a hantaviral species naturally associated with *Oryzomys couesi* (Coues' oryzomys) in Honduras. Am J Trop Med Hyg 75:1003–1010
70. Milazzo ML, Duno G, Utrera A, Richter MH, Duno F, de Manzione N, Fulhorst CF (2010) Natural host relationships of hantaviruses native to western Venezuela. Vector Borne Zoonotic Dis 10:605–611. doi:10.1089/vbz.2009.0118
71. Mills J, Amman B, Glass G (2010) Ecology of hantaviruses and their hosts in North America. Vector Borne Zoonotic Dis 10:563–574
72. Montoya-Ruiz C, Cajimat MNB, Milazzo ML, Diaz FJ, Rodas JD, Valbuena G, Fulhorst CF (2015) Phylogenetic relationship of Necoclí virus to other south American hantaviruses (Bunyaviridae: hantavirus). Vector Borne Zoonotic Dis 15:438–445. doi:10.1089/vbz.2014.1739
73. Morzunov SP, Khaiboullina SF, St. Jeor S, Rizvanov AA, Lombardi VC (2015) Multiplex analysis of serum cytokines in humans with hantavirus pulmonary syndrome. Front Immunol 6:432. doi:10.3389/fimmu.2015.00432
74. Muyangwa M, Martynova EV, Khaiboullina SF, Morzunov SP, Rizvanov AA (2015) Hantaviral proteins: structure, functions, and role in hantavirus infection. Front Microbiol 6:1–10. doi:10.3389/fmicb.2015.01326
75. Nelson R, Canate R, Pascale JM, Dragoo JW, Armien B, Armien AG, Koster F (2010) Confirmation of Choclo virus as the cause of hantavirus cardiopulmonary syndrome and high serum antibody prevalence in Panama. J Med Virol 82:1586–1593
76. Nichol ST, Spiropoulou CF, Morzunov S, Rollin PE, Ksiazek TG, Feldmann H, Sanchez A, Childs J, Zaki S, Peters CJ (1993) Genetic identification of a hantavirus associated with an outbreak of acute respiratory illness. Science 262:914–917

77. Padula PJ, Colavecchia SB, Martínez VP, Gonzalez Della Valle MO, Edelstein A, Miguel SD, Russi J, Riquelme JM, Colucci N, Almirón M, Rabinovich RD, Padula PJ, Colavecchia SB, Martínez VP, Gonzalez Della Valle MO, Edelstein A, Miguel SD, Russi J, Riquelme JRR (2000) Genetic diversity, distribution, and serological features of hantavirus infection in five countries in South America. J Clin Microbiol 38:3029–3035
78. Padula PJ, Edelstein A, Miguel SD, Lopez NM, Rossi CM, Rabinovich RD (1998) Hantavirus pulmonary syndrome outbreak in Argentina: molecular evidence for person-to-person transmission of Andes virus. Virology 241:323–330
79. Padula PJ, Rossi CM, Della Valle MO, Martinez PV, Colavecchia SB, Edelstein A, Miguel SDL, Rabinovich RD, Segura EL (2000) Development and evaluation of a solid-phase enzyme immunoassay based on Andes hantavirus recombinant nucleoprotein. J Med Microbiol 49:149–155
80. Pinna DM, Martinez VP, Bellomo CM, Lopez C, Padula P (2004) New epidemiologic and molecular evidence of person to person transmission of hantavirus Andes South. Medicina (B Aires) 64:43–46
81. Plyusnin A (2002) Genetics of hantaviruses: implications to taxonomy. Arch Virol 147:665–682. doi:10.1007/s007050200017
82. Plyusnin A, Morzunov SP (2001) Virus evolution and genetic diversity of hantaviruses and their rodent hosts. Curr Top Microbiol Immunol 256:47–75
83. Plyusnin A, Sironen T (2014) Evolution of hantaviruses: co-speciation with reservoir hosts for more than 100 MYR. Virus Res 187:22–26. doi:10.1016/j.virusres.2014.01.008
84. Prescott J, Safronetz D, Haddock E, Robertson S, Scott D, Feldmann H (2013) The adaptive immune response does not influence hantavirus disease or persistence in the Syrian hamster. Immunology 140:168–178. doi:10.1111/imm.12116
85. Raboni SM, de Borba L, Hoffmann FG, de Noronha L, Azevedo ML, Carstensen S, Mazzarotto GA, Bordignon J, dos Santos CN (2009) Evidence of circulation of Laguna Negra-like hantavirus in the central west of Brazil: case report. J Clin Virol 45:153–156
86. Raboni SM, Delfraro A, de Borba L, Teixeira BR, Stella V, de Araujo M, Carstensen S, Rubio G, Maron A, Lemos ERS, D'Andrea PS, Duarte Dos Santos CN (2012) Hantavirus infection prevalence in wild rodents and human anti-hantavirus serological profiles from different geographic areas of South Brazil. Am J Trop Med Hyg 87:371–378. doi:10.4269/ajtmh.2012.11-0762
87. Raboni SM, Levis S, Rosa EST, Bisordi I, Delfraro A, Lemos E, Correia DC, Duarte dos Santos CN (2007) Hantavirus infection in Brazil: development and evaluation of an enzyme immunoassay and immunoblotting based on N recombinant protein. Diagn Microbiol Infect Dis 58:89–97
88. Raboni SM, Probst CM, Bordignon J, Zeferino A, dos Santos CN (2005) Hantaviruses in Central South America: phylogenetic analysis of the S segment from HPS cases in Parana, Brazil. J Med Virol 76:553–562
89. Ramsden C, Holmes EC, Charleston MA (2009) Hantavirus evolution in relation to its rodent and insectivore hosts: no evidence for codivergence. Mol Biol Evol 26:143–153. doi:10.1093/molbev/msn234
90. Ramsden C, Melo FL, Figueiredo LM, Holmes EC, Zanotto PMA, Moreli ML, Moro De Sousa RL, Borges AA, Garcia De Figueiredo G, Bisordi I, Nagasse-Sugahara TK, Suzuki A, Pereira LE, Pereira De Souza R, Madia De Souza LT, Torres Braconi C, Araujo J (2008) High rates of molecular evolution in hantaviruses. Mol Biol Evol 25:1488–1492. doi:10.1093/molbev/msn093
91. Rasmuson J, Pourazar J, Mohamed N, Lejon K, Evander M, Blomberg A (2016) Cytotoxic immune responses in the lungs correlate to disease severity in patients with hantavirus infection. Eur J Clin Microbiol Infect Dis 35:713–721. doi:10.1007/s10096–016–2592-1
92. Razzauti M, Plyusnina A, Henttonen H, Plyusnin A (2008) Accumulation of point mutations and reassortment of genomic RNA segments are involved in the microevolution of Puumala hantavirus in a bank vole (*Myodes glareolus*) population. J Gen Virol 89:1649–1660. doi:10.1099/vir.0.2008/001248-0

93. Reusken C, Heyman P (2013) Factors driving hantavirus emergence in Europe. Curr Opin Virol 3:92–99. doi:10.1016/j.coviro.2013.01.002
94. Richter MH, Hanson JD, Cajimat MN, Milazzo ML, Fulhorst CF (2010) Geographical range of Rio Mamoré virus (family Bunyaviridae, genus hantavirus) in association with the small-eared pygmy rice rat *(Oligoryzomys microtis)*. Vector Borne Zoonotic Dis 10:613–620. doi:10.1089/vbz.2009.0115
95. Rosa ES, Mills JN, Padula PJ, Elkhoury MR, Ksiazek TG, Mendes WS, Santos ED, Araujo GC, Martinez VP, Rosa JF, Edelstein A, Vasconcelos PF (2005) Newly recognized hantaviruses associated with hantavirus pulmonary syndrome in northern Brazil: partial genetic characterization of viruses and serologic implication of likely reservoirs. Vector Borne Zoonotic Dis 5:11–19
96. da Rosa EST, Medeiros DBA, Nunes MRT, Simith DB, de Souza PA, Elkhoury MR, Lavocat M, Marques AAR, Via AV, D'Andrea P, Bonvicino CR, Lemos ERS, Vasconcelos PFC (2011) Pygmy rice rat as potential host of Castelo dos Sonhos hantavirus. Emerg Infect Dis 17:1527–1530. doi:10.3201/eid1708.101547
97. Safronetz D, Haddock E, Feldmann F, Ebihara H, Feldmann H (2011) In vitro and in vivo activity of ribavirin against Andes virus infection. PLoS One 6:e23560. doi:10.1371/journal.pone.0023560
98. Schmaljohn CS, Hooper JW (2001) Bunyaviridae: the viruses and their replication. In: Knipe B, Howley D (eds) Fields virology. Lippincott-Williams & Wilkins, Philadelphia, pp 1581–1633
99. Song JW, Gu SH, Bennett SN, Arai S, Puorger M, Hilbe M, Yanagihara R (2007) Seewis virus, a genetically distinct hantavirus in the Eurasian common shrew (*Sorex araneus*). Virol J 4:114
100. Song JY, Woo HJ, Cheong HJ, Noh JY, Baek LJ, Kim WJ (2016) Long-term immunogenicity and safety of inactivated Hantaan virus vaccine (Hantavax™) in healthy adults. Vaccine 34:1289–1295. doi:10.1016/j.vaccine.2016.01.031
101. Steppan S, Adkins R, Anderson J (2004) Phylogeny and divergence-date estimates of rapid radiations in Muroid rodents based on multiple nuclear genes. Syst Biol 53:533–553
102. Sumibcay L, Kadjo B, Gu SH, Kang HJ, Lim BK, Cook JA, Song J-W, Yanagihara R (2012) Divergent lineage of a novel hantavirus in the banana pipistrelle (*Neoromicia nanus*) in Côte d'Ivoire. Virol J 9:34. doi:10.1186/1743-422X-9-34
103. Suzan G, Giermakowski JT, Marce E, Suzan-Azpiri H, Armien B, Yates TL (2006) Modeling hantavirus reservoir species dominance in high seroprevalence areas on the Azuero peninsula of Panama. Am J Trop Med Hyg 74:1103–1110
104. Suzan G, Marce E, Giermakowski JT, Armien B, Pascale J, Mills J, Ceballos G, Gomez A, Aguirre AA, Salazar-Bravo J, Armien A, Parmenter R, Yates T (2008) The effect of habitat fragmentation and species diversity loss on hantavirus prevalence in Panama. Ann N Y Acad Sci 1149:80–83
105. Tang K, Zhang C, Zhang Y, Zhang Y, Zhuang R, Jin B, Ma Y (2015) Elevated plasma soluble CD14 levels correlate with the monocyte response status during Hantaan virus infection in humans. Viral Immunol 28:442–447. doi:10.1089/vim.2015.0040
106. Toro J, Vega JD, Khan AS, Mills JN, Padula P, Terry W, Yadón Z, Valderrama R, Ellis BA, Pavletic C, Cerda R, Zaki S, Wun-Ju S, Meyer R, Tapia M, Mansilla C, Baro M, Vergara JA, Concha M, Calderon G, Enria D, Peters CJ, Ksiazek TG (1998) An outbreak of hantavirus pulmonary syndrome, Chile, 1997. Emerg Infect Dis 4:687–694. doi:10.3201/eid0404.980425
107. Travassos da Rosa ES, Sampaio de Lemos ER, de Almeida Medeiros DB, Simith DB, de Souza PA, Elkhoury MR, Mendes WS, Vidigal JRB, de Oliveira RC, D'Andrea PS, Bonvicino CR, Cruz ACR, Nunes MRT, da Costa Vasconcelos PF (2010) Hantaviruses and hantavirus pulmonary syndrome, Maranhao, Brazil. Emerg Infect Dis 16:1952–1955. doi:10.3201/eid1612.100418
108. Vaheri A, Strandin T, Hepojoki J, Sironen T, Henttonen H, Mäkelä S, Mustonen J (2013) Uncovering the mysteries of hantavirus infections. Nat Rev Microbiol 11:539–550. doi:10.1038/nrmicro3066

109. Vaheri A, Vapalahti O, Plyusnin A (2008) How to diagnose hantavirus infections and detect them in rodents and insectivores. Rev Med Virol 18:277–288. doi:10.1002/rmv.581
110. Vasconcelos P, Lima V, Iversson L, Rosa M, Travassos da Rosa APA, Pereira L, Nassar E, Katz G, Matida L, Zaparoli M, Ferreira J, Peters C (1997) Hantavirus pulmonary syndrome in the rural area of Juquitiba, Sao Paulo metropolitan area, Brazil. Rev Inst Med Trop Sao Paulo 39:237–238
111. Vial PA, Valdivieso F, Ferres M, Riquelme R, Rioseco ML, Calvo M, Castillo C, Diaz R, Scholz L, Cuiza A, Belmar E, Hernandez C, Martinez J, Lee S-J, Mertz GJ, Abarca J, Tomicic V, Aracena ME, Rehbein AM, Velasquez S, Lavin V, Garrido F, Godoy P, Martinez C, Chamorro JC, Contreras J, Hernandez J, Pino M, Villegas P, Zapata V, Leon M, Vega I, Otarola I, Ortega C, Daube E, Huecha D, Neira A, Ruiz I, Nunez MA, Monsalve L, Chabouty H, Riquelme L, Palma S, Bustos R, Miranda R, Mardones J, Hernandez N, Betancur Y, Sanhueza L, Inostroza J, Donoso S, Navarrete M, Acuna L, Manriquez P, Castillo F, Unzueta P, Aguilera T, Osorio C, Yobanolo V, Mardones J, Aranda S, Carvajal S, Sandoval M, Daza S, Vargas F, Diaz V, Riquelme M, Munoz M, Carriel A, Lanino P, Hernandez S, Schumacher P, Yanez L, Marco C, Ehrenfeld M, Delgado I, Rios S, Vial C, Bedrick E (2013) High-dose intravenous methylprednisolone for hantavirus cardiopulmonary syndrome in Chile: a double-blind, randomized controlled clinical trial. Clin Infect Dis 57:943–951. doi:10.1093/cid/cit394
112. Vincent MJ, Quiroz E, Gracia F, Sanchez AJ, Ksiazek TG, Kitsutani PT, Ruedas LA, Tinnin DS, Caceres L, Garcia A, Rollin PE, Mills JN, Peters CJ, Nichol ST (2000) Hantavirus pulmonary syndrome in Panama: identification of novel hantaviruses and their likely reservoirs. Virology 277:14–19
113. Weiss S, Witkowski PT, Auste B, Nowa K, Weber N, Fah J, Mombouli JV, Wolfe ND, Drexler JF, Drosten C, Klempa B, Leendertz FH, Kruger DH (2012) Hantavirus in bat, Sierra Leone. Emerg Infect Dis 18:159–161. doi:10.3201/eid1801.111026
114. Witkowski PT, Drexler JF, Kallies R, Ličková M, Bokorová S, Mananga GD, Szemes T, Leroy EM, Krüger DH, Drosten C, Klempa B (2016) Phylogenetic analysis of a newfound bat-borne hantavirus supports a laurasiatherian host association for ancestral mammalian hantaviruses. Infect Genet Evol 41:113–119. doi:10.1016/j.meegid.2016.03.036
115. Zhang Y, Zhang C, Zhuang R, Ma Y, Zhang Y, Yi J, Yang A, Jin B (2015) IL-33/ST2 correlates with severity of haemorrhagic fever with renal syndrome and regulates the inflammatory response in Hantaan virus-infected endothelial cells. PLoS Negl Trop Dis 9:e0003514. doi:10.1371/journal.pntd.0003514
116. Zhang Y-Z, Holmes EC (2014) What is the time-scale of hantavirus evolution? Infect Genet Evol 25:144–145
117. Zou Y, Hu J, Wang ZX, Wang DM, Yu C, Zhou JZ, Fu ZF, Zhang YZ (2008) Genetic characterization of hantaviruses isolated from Guizhou, China: evidence for spillover and reassortment in nature. J Med Virol 80:1033–1041. doi:10.1002/jmv.21149

Chapter 12
Human Respiratory Syncytial Virus: Biology, Epidemiology, and Control

Edison Luiz Durigon, Viviane Fongaro Botosso, and Danielle Bruna Leal de Oliveira

1 Introduction

Acute respiratory infections (ARIs) are the most frequent infectious disease in humans, and the great majority of respiratory infections observed in medical practice around the world are of viral etiology [3, 34, 80]. During the period of 2000 to 2003, an estimate of 10,600,000 children under the age of 5 years died every year, and ARIs were responsible for nearly 19% of these deaths. Most of these fatalities were caused by bronchitis and pneumonia associated with viral infections [14, 57, 106, 119]. Viral respiratory infections are also associated with high morbidity in this age group worldwide. For example, 35% of hospitalized children in Brazil, 35% in Belgium, 22% in Italy, and 59% in the UK attending pediatric services were the result of viral respiratory infections [78, 84, 85, 87, 98].

Currently, the following viruses shall be considered causes of acute respiratory illness in children: human respiratory syncytial virus (HRSV); parainfluenza virus types 1, 2, 3, and 4 (PIV1, PIV2, PIV3, PIV4); influenza virus types A, B, C (IA, IB, IC); adenoviruses (ADV); coronaviruses HCoV-OC43, HCoV-229E, HCoV-HKU1, and HCoV-NL64; human rhinovirus (HRV); some subtypes of enterovirus (HEV-68); human metapneumovirus (hMPV); human bocavirus (HBoV); and WU and KI polyomavirus (Fig. 12.1). However, some viruses present high rates of co-detection,

E.L. Durigon (✉) • D.B.L. de Oliveira
Laboratory of Molecular and Clinical Virology, Department of Microbiology, Institute of Biomedical Sciences, University of São Paulo, São Paulo, Brazil
e-mail: eldurigo@usp.br; danibruna@gmail.com

V.F. Botosso
Scientific Development Division, Virology Branch, Institute Butantan, São Paulo, Brazil
e-mail: viviane.botosso@butantan.gov.br

J.E. Ludert et al. (eds.), *Human Virology in Latin America*,
DOI 10.1007/978-3-319-54567-7_12

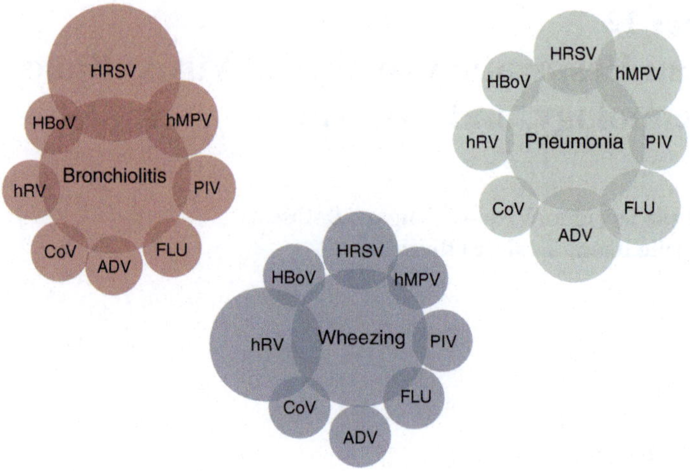

Fig. 12.1 Clinical association with major symptoms of disease and respiratory viruses. In South America, studies conducted in different regions of the country indicate the importance of viruses as etiological agents of low respiratory infection (LRI). These studies revealed the presence of different respiratory viruses in children and adults, such as human respiratory syncytial virus (HRSV), influenza A and B, HPIV, HAdV, HRV, hMPV, hBoV, and HCoV; the percentage of cases among children for some type of respiratory viruses ranged between 28.75% and 75%, whereas positivity in adults was 61.8% for at least one of the viruses studied [11, 15, 35, 39, 40, 85, 105, 111, 116]

as is the case of rhinovirus, enterovirus, human coronavirus and bocavirus, and polyomavirus, being questioned for its significance in the etiology of these infections [10, 45, 49, 64, 89, 91, 96, 114].

The seasonality of respiratory viruses is described in several studies, with some viral infections taking place throughout the year, such as influenza virus, with a predominance in the winter months [96], and others occurring chiefly in the late fall, winter, or early spring, such as HPIV, hMPV, HCoV, and HRSV [85, 99, 105, 114]. Adenoviruses are found worldwide and can circulate sporadically, endemically, or epidemically in the winter, spring, and early summer [103, 105].

Despite the great number of viral agents involved in respiratory infections and their importance, the HRSV is the leading cause of acute respiratory infections and one of the leading causes of hospitalization and death among children under 5 years of age worldwide. Each year, respiratory syncytial virus (RSV) infections lead to 2,100,000 outpatient attendances and 57,257 hospitalizations of children less than 5 years of age in the U.S. Additionally, RSV is responsible for 177,000 hospitalizations with 14,000 deaths among adults over 65 years of age [24].

Newborns, premature infants, and those with chronic lung disease are at greater risk of developing severe disease by infection with HRSV [48]. Despite their importance, there is no vaccine prophylaxis against HRSV infection or effective antiviral therapy available. Currently, in Latin America, only palivizumab (Pz) (Synagis; MedImmune, Gaithersburg, MD, USA) is being used in the prophylaxis and therapy of these infections [68].

2 History of the HRSV

In 1955, an outbreak of respiratory illness characterized by coughing, sneezing, and mucopurulent discharge was described in a colony of 20 chimpanzees at the Walter Reed Army Institute of Research (WRAIR) (Washington, DC, USA). During that episode, the RSV was isolated for the first time from a swab from the throat of a female chimpanzee and then called the chimpanzee coryza agent (CCA) [82]. Viral isolation was performed in liver cells, later being inoculated in various laboratory animals, mice, hamsters, rabbits, rats, and chimpanzees, the latter being the only ones to develop specific symptoms.

One of the attendants of the chimpanzees became sick and developed symptoms similar to those of the animals. Although the attempt at isolation of human respiratory syncytial virus was unsuccessful, an increase in antibody titer by complement fixation against CCA was detected. Parallel seroprevalence studies conducted in a human general population revealed the presence of antibodies to a new CCA agent in teenagers and adults.

The following year, Chanock and colleagues isolated a virus similar to the CCA of a child with pneumonia and another with croup, in Baltimore [25, 26]. The agent was named human respiratory syncytial virus, HRSV, to reflect its ability to form syncytia in cell culture and its tropism for the human respiratory tract.

Serological studies carried out at the time indicated that the majority of children in Baltimore had been infected with HRSV before 4 years of age. Similar investigations in diverse parts of the world indicated that the HRSV was associated with diseases of the lower respiratory tract [33]. Currently, HRSV is recognized as the viral agent more frequently related to cases of bronchiolitis and pneumonia during infancy and preschool age. About 95% of the children have the first HRSV infection in the first 2 years of life, and the peak incidence occurs in the first few months [3]. Approximately 40% of children develop symptoms of lower respiratory tract involvement during the first infection. Although reinfections are common during a lifetime, the clinical symptoms in older children and adults are less severe [55].

Some groups of patients are at risk of developing serious illness resulting from the lower respiratory tract infection by HRSV; these include children younger than 6 months of age, premature infants, immunodeficient children, and children with chronic lung disease or congenital heart disease [33]. There are also studies relating the HRSV to severe infections in the elderly [44, 117].

3 Classification

The human respiratory syncytial virus (HRSV) is a member of the order *Mononegavirales* (mono, from Greek, meaning "single, simple"; nega, from Latin, meaning "RNA negative polarity"; virales, from Latin, meaning virus), classified within the *Pneumoviridae* family and the genus *Orthopneumovirus*. Other members of the *Orthopneumovirus* genus are the bovine respiratory syncytial virus (BRSV) and the pneumonia virus of mice (murine pneumonia virus) [125].

4 Structure

The virion is pleomorphic with a diameter of 150–300 nm and is composed of an internal nucleocapsid of helical symmetry and an envelope derived from the host cytoplasmic membrane; viral glycoproteins that protrude from the envelope as 11- to 20-nm projections, separated by intervals of 6–10 nm, are involved in the processes of adherence and penetration of the virus. The viral genome is composed of a single-stranded RNA molecule, not segmented, and with negative polarity. Each infectious particle contains only one functional copy of the genome (Fig. 12.2) [33].

5 Genomic Organization

The virus contains a single-stranded negative-sense RNA genome with 15,222 nucleotides (nt), with molecular weight of 5×10^6 Da, which serves as a template for transcription of messenger RNAs (mRNAs), encoding for 11 proteins. The genome transcription takes place in the $5' \rightarrow 3'$ direction. The 3′-region of the genomic RNA consists of a region of 44 nucleotides that presumably contains the viral promoter [81]. The first 30 nt in this region are highly susceptible to inactivation by the insertion or deletion of nucleotides. This region is followed by 10 genes that encode 11 proteins, in the following order: *NS1*, *NS2*, *N*, *P*, *M*, *SH*, *G*, *F*, *M2*, *L*. The last gene, *L*, is followed by a region that is more tolerant to the insertion or deletion of nucleotides [81].

The first nine genes are separated by inter-gene regions ranging from 1 to 52 nt in size, which apparently do not have an important role in the modulation of gene expression and show little conservation among isolates [31, 71]. The beginning of each gene contains a conserved signal (gene start signal) composed of nine nucleotides, 3′-CCCCGUUUA,

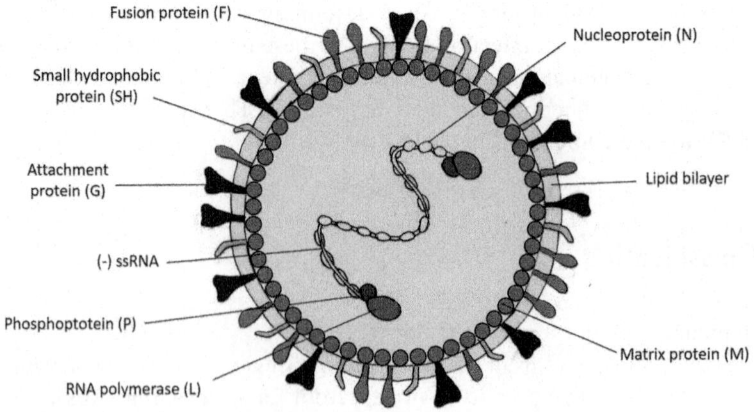

Fig. 12.2 Human respiratory syncytial virus (HRSV) virion structure

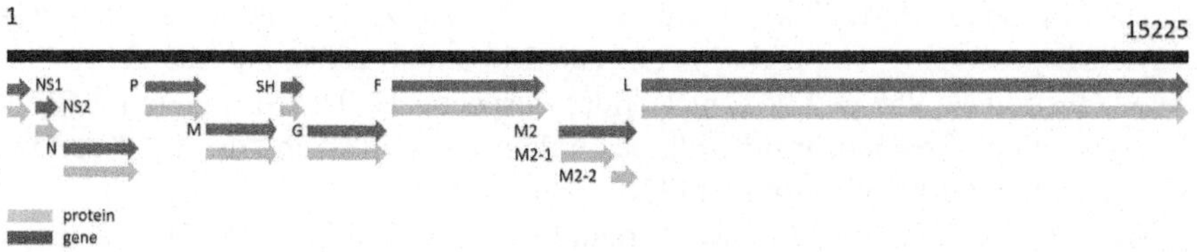

Fig. 12.3 Organization of the gene in the genome of HRSV. The genome is 15,225 nucleotides long, a single-stranded RNA with negative polarity. It has 10 genes encoding 11 proteins. The *M2* gene has two products: a nucleocapsid-associated transcription factor (M2-1) and another protein involved in genome replication (M2-2)

except for the *L* gene, which presents the following differences as underlined: 3′-*CCCUGUUUUA*. The genes end with a semi-conserved sign (gene end signal) composed of 12 or 13 nucleotides whose sequence is 3′-UCAAUNAAAUUU, which drives the end of transcription and polyadenylation. The last two genes, *M2* and *L*, have in common 68 nt. Consequently, the gene *L* has the initiation of the transcript inside the M2 gene [33]. The M2 has two ORFs (open reading frames), which give rise to proteins M2-1 and M2-2. The organization of the gene of HRSV is schematized in Fig. 12.3.

6 Proteins

In cells infected with HRSV, 11 proteins have been identified. Of these, 2 are nonstructural proteins, NS1 and NS2, present in abundance in the cells, but in small amounts in the virion. The others are structural proteins, M (matrix) and M2-1(transcriptiol elongation factor) proteins, N (nucleoprotein), P (phosphoprotein), L (large polymerase), and M22 nucleocapsid viral proteins, and 3 are surface glycoproteins G (attachment), F (fusion), and SH (small hydrophobic) [33]. The glycoproteins F and G are highly accessible to neutralizing antibodies, resulting in numerous changes in response to the host immune pressure [34] and therefore are the most studied.

The NS1 (molecular weight PM, 15.5 kDa) and NS2 (PM, 27 kDa) proteins have, respectively, 139 and 124 amino acids, and the genes that encode these have 532 and 503 nucleotides, respectively. Their functions are not well understood, but it is presumed that they are related to the structural regulation of RNA synthesis, the morphogenesis of the virion, or the interaction with the host cells [33].

The proteins P, L, and N are associated with genomic RNA and nucleocapsid, forming the ribonucleoprotein complex, considered as the minimum unit necessary for transcription and replication of the virus. The P protein is highly phosphorylated and acidic and has a key role in the regulation of the transcription and replication process. It has 241 amino acids and a molecular weight of 35 kDa, and the gene that encodes it has 914 nucleotides. The nucleoprotein N has 391 amino acids and a molecular weight of 43.4 kDa, and the gene that encodes it has 1203 nucleotides and is the main structural protein of the nucleocapsid, closely associated with the genomic RNA. The L protein, consisting of 2165 amino acids with a molecular weight of 250 kDa, is the largest viral protein. The gene that encodes the L protein has 6578 nucleotides [33].

The M proteins (PM, 27 kDa) and M2-1 (PM, 22 kDa) are internal and not glycosylated, possessing, respectively, 194 and 256 amino acids, and the genes that encode them have 958 and 961 nucleotides, respectively. The M protein mediates the association of nucleocapsid with the viral envelope [33], and the M2-1 acts on the elongation during transcription [32].

The SH protein is a small molecule (amino acids and 64 PM, 7.5 kDa), which is inserted in the cytoplasmic membrane of the host cell via a hydrophobic sequence, ranging from 14 to 41 amino acids. The function of this protein has not yet been clarified; however, because it is integrated in the membrane, it is assumed to be involved in adsorption, penetration, and denudation of the virus [33].

The glycoprotein F has 574 amino acids with a molecular weight of 70 kDa, and the gene that encodes it has 1903 nucleotides. Identified as a fusion protein, it is responsible for the attachment of the viral envelope with the plasma membrane of the host cell, releasing nucleocapsid directly within the cytoplasm. Also, it is responsible for the fusion of the cell infected with neighbor cells, favoring the formation of the syncytium [33].

The F protein is synthesized as an inactive precursor called F0, which consists of two domains, F2 (1–130 amino acids) and F1 (137–574 amino acids), and also has a cleaved peptide (131–136 amino acids). The F1 subunit is anchored to the membrane. The F1 subunit is relatively well preserved and is greatly affected in its function by deletions or substitutions of amino acids [33].

The glycoprotein G is a type II protein, which is anchored to the membrane next to its amino-terminal portion by a hydrophobic domain, non-cleaved, signal anchor type, that extends from residues 38 to 66 [73]. The G protein is 289 to 342 amino acids in length, depending on the viral strain. The G gene is composed of 918 to 1062 (group A) or 921 to 981 (group B) nucleotides [43, 102, 108].

The glycoprotein G is the viral attachment engaged in the adsorption of virus, and it has been shown that antibodies against the G glycoprotein inhibited the binding of virus to the cells [72]. The glycoprotein G is of special interest for showing the largest variability between the viral isolates [4, 52, 67] and can support large deletions or multiple amino acid substitutions without loss of function [43, 102]. This variability among strains of HRSV is a signature feature that can alter the pathogenicity and adaptation of the virus and contribute to the ability of the virus to cause repeated infections and outbreaks by escaping the immune system. The glycoproteins F and G are the most important proteins involved in a protective immune response [8, 66], and antibodies against them show strong neutralization activity in vitro [2, 123].

7 Replication

The cell receptor specific for the glycoprotein G was first identified by Krusat and Streckert [70], who showed that preincubation of the virus with heparin inhibited the infection in cell culture and that the G protein binds heparin. These results suggest that heparin or other glycosaminoglycans (GAGs) similar to heparin, present on the cell surface, are involved in the binding of the virus to the cell. The binding site of the glycoprotein G to the heparin (or another GAG) was mapped between the 184 and 198 amino acids of the protein G for group A and among the 183 to 197 amino acids for group B. Martinez et al. [77] confirmed that the presence of these receptors is critical for the binding of the virus.

The virus enters into the cell through fusion with the cell membrane. After penetration, the viral envelope remains as part of the cell membrane. The nucleocapsid is released into the cytoplasm and begins the process of transcription of the viral genome by the viral polymerase. The genes are transcribed in sense $3' \rightarrow 5'$ with a sign promoting to the 3′-side [33]. The peak of the synthesis of mRNAs occurs 16 h after infection, and the peak of proteins occurs at 18–20 h [6, 33]. In addition to the transcription and translation of proteins, another important step is the replication of the viral genome, which produces an intermediate positive (+ ssRNA), which will serve as a template to generate more copies of the viral genome (ssRNA). All the replication process takes place in the cytoplasm [33].

The maturation of the virus occurs in the first instance, with the combination of proteins N and P to the genomic RNA and subsequent addition of other auxiliary proteins to the nucleocapsid. The surface glycoproteins are inserted into the cytoplasmic membrane of the host cell. In the next step the matrix protein interacts by noncovalent forces to the cytoplasmic tails of the surface glycoprotein. The assembled internal structures of the virus interact with this surface and drive the budding, with the release of the virus, when the virus acquires the lipoprotein envelope [69].

8 Genetic Variability

The variability of the G protein is concentrated in the extracellular domain, where two variable regions have a high content of serine and threonine, between 69 to 164 and 207 to 298 amino acids, with approximately 56% divergence between groups A and B [66, 67]. Interspersing this region of high variability, there is a conserved region with a small segment of 13 amino acids (164–176) and four cysteine residues (C173, C176, C182, C186), which are well preserved in all samples of HRSV [97, 118], suggesting that this region is responsible for binding the virus to a cell receptor. However, data about the region for genotypes that emerged after 2010 are currently lacking.

The genotyping of HRSV-A and HRSV-B is based on the variability of the G-protein gene. For HRSV A, 11 genotypes were reported and designated as GA1, GA2, GA3, GA4, GA5, GA6, and GA7 [92, 93], SAA1 (South Africa, A1) [115], and more recently, NA1, NA2, NA3, and NA4 [102]. For HRSV-B, 17 genotypes have been described and designated as GB1, GB2, GB3, and GB4 [93], SAB1, SAB3 [115], BA1–BA6 (Argentina) [109], BA7–BA10 (Japan) [38], and B11 (Korea) [7]. Interestingly, strains belonging to genotype BA of HRSV-B exhibited duplication of 60 nucleotides (nt) in the second variable region protein gene *G*, but were not associated with more severe clinical manifestations [38, 108]. In Brazil, the only genotypes circulating currently from HRSV-A are NA2, NA3, and ON1 and BA genotyping from HRSV-B.

In 2012, Eshaghi et al. [43] detected in group A one repetition of 72 nucleotides (GTCAAGAGGAAACCCTCCACTCAACCACCTCCGAAG GCTATCTAAGCCCA TCACAAGTCTATACAACATCCG) in the C-terminal portion of the gene (G), being the largest duplication described in this group. This new genotype was called ON1 and was found in 10% of HRSV isolates. In 2013 this ON1 genotype was found in 75% of all isolates in Brazil [42, 80], and in 2015 the ON1 genotype had attained natural dominance and become the predominant genotype circulating in different areas of the world [107]. This area is specifically targeted for neutralizing antibodies, and these types of changes of structure can lead to changes in immunogenicity and pathogenicity of the virus. However, additional studies are still required to explore the pathogenicity, transmissibility, and replication of this new variant.

9 Epidemiology

In the 1990s several studies of molecular epidemiology were conducted based on partial sequences of genes *G* and *SH* and a restriction map of the *N* gene, enabling reaching some important conclusions about HRSV circulation:

1. The existence of several genotypes circulating concurrently in a single outbreak, with a predominance of one or two genotypes which tend to decrease in subsequent outbreaks until its disappearance [17, 19, 20, 27, 28, 30, 65, 75, 92, 93].
2. The genotypes of HRSV have worldwide distribution, and strains isolated in distinct communities and in different years may be more related to strains isolated in the same locality in two consecutive days, demonstrating a pattern of temporal and not necessarily geographic circulation [18, 50].
3. Within each strain (genotype) occurs a progressive buildup of amino acid changes [21].
4. Antigenic changes detected with a panel of anti-G monoclonal antibodies can be correlated with the position of the viruses in the phylogenetic trees [50].
5. The synonymous nucleotide substitutions have a uniform distribution over the *G* gene, and non-synonymous substitutions are accumulated in the two variable regions of the gene *G* [21, 50].

However, there are studies in which a minimal temporal variation in the gene encoding the G protein has been reported. A study performed in Cuba revealed the movement of extremely homogeneous samples during the 1994–1995 outbreaks, with a difference of just five nucleotides when compared to the sample long since isolated in 1956 [113].

The significance of the antigenic variation of HRSV groups in epidemiology is not yet clear. The antigenic dimorphism, although at modest rates, seems to contribute to the high incidence of reinfections during the first years of life. However, several reinfections in children involving viruses of the same group have been reported [60, 83]. In addition, there is no indication that reinfection with a heterologous group induces more serious clinical signs than reinfection with homologous samples [110].

The two groups (A and B) have been circulating concurrently in many epidemics for more than 20 years [12, 65], in diverse regions of the world, and with incidences that vary from year to year. Studies conducted in El Salvador, Santa Fe, and Buenos Aires in Argentina revealed the presence of both groups during outbreaks with prevalence of group A [23, 63, 121].

In some localities, such as Rochester and Boston in the U.S., Sapporo in Japan, and Rio de Janeiro, Porto Alegre, and Ribeirão Preto in Brazil, in addition to the co-circulation of the groups, the prevalence of groups A and B may switch over the years or show a balance of the frequencies of both groups [29, 56, 58, 104, 112].

Differences in pathogenicity between the two groups are not clear. Hall et al. [56] and Imaz et al. [63] verified increased severity in children infected with group A, although Zelaya et al. [121] found greater severity in children infected with group B. Other authors did not observe significant differences in pathogenicity between the groups [29, 104].

In a study carried out in Bogota, Colombia, a total of 13,488 samples of children hospitalized with a diagnosis of respiratory infection were tested for RSV during 5 years and 4,559 (33.8%) were found positive. The average age of patients analyzed in the study was 9.2 ± 8.5 months, and 71.7% of cases of HRSV infection occurred in the period from March to May, whereas 50% of the bronchiolitis cases were diagnosed from April to June during the years of the study [47].

In Chile, HRSV are detected as a single pathogen at 74/124 (58.7%) samples of nasopharyngeal aspirate of patients, and 28/124 (22.6%) samples were co-detected with HRV. Hospitalization was necessary in 77% of positive cases of HRSV (57/74), and 44.6% of these cases were considered serious; 53.6% (15/28) of cases coinfected by both viruses were hospitalized, too, but this coinfection does not increase the severity of illness [74].

In Brazil, many studies have already been carried out to investigate the etiology of acute respiratory diseases [12]. During the period of 2003–2009, nasopharyngeal aspirates were examined in more than 2000 children less than 5 years old, and HRSV were found in at least 42% of positivity between respiratory viruses identified in children hospitalized with acute respiratory disease [85, 105].

In countries in southern Latin America such as Argentina and Uruguay, outbreaks of HRSV occur predominantly during the winter months [22, 61]. In tropical and subtropical climates, the outbreaks are not always well defined, although in Ceara, located in the northeast of Brazil, HRSV caused yearly seasonal epidemics, generally from February until July (Moura et al. 2013). In Brazil, in the cities of Rio de Janeiro and São Paulo, HRSV outbreaks start in autumn (ranging from March to April) and extend until winter (July–August), with peak incidence occurring usually in May (Table 12.1) [85, 105].

Fortunately, fatalities from infection by HRSV are uncommon, and estimates indicate that the number of deaths is around 200–500 a year, 80% of which are of children under 1 year of age. However, mortality may increase significantly in children who present some background that predisposes to more serious diseases, such as congenital heart diseases and lung diseases, and premature infants, in which mortality by HRSV infection is around 10%, 5.5%, and 4.6%, respectively [41, 100]. High mortality rates may also be observed in individuals with immunodeficiency,

Table 12.1 Occurrence of positivity for human respiratory syncytial virus (HRSV) in different cities of Latin America at different times of the year

Country	City	Seasonality	Average of HRSV positive (%)
Argentina	Buenos Aires	Autumn – winter	23–26
Brazil	Tropical	Summer – winter	21–4
	Subtropical	Autumn – winter	42
Chile	Santiago	Winter	22.6–52.7
Colombia	Cali	Throughout the year	33.8
	Medellín	Summer	41.7
Costa Rica	San Jose	Autumn	15–20
Mexico	Mexico City	Winter – spring	36–55
	San Luis Potosí	Autumn – winter	24.8–46.7
Venezuela	Caracas	Throughout the year	31.6–66
Uruguay	Montevideo	Winter	56
Guatemala	Santa Rosa and Quetzaltenango	Autumn	24

congenital or induced by chemotherapy against cancer [54] or from organ transplants, especially in the first 20 days after the transplant [94]. Among bone marrow transplant recipients, the mortality of those who become infected with HRSV can reach 45% [14, 57].

A study conducted in the U.S. revealed the occurrence of 14,000 to 62,000 annual hospitalizations of the elderly with pneumonia associated with HRSV, at a cost of approximately 150,000,000–680,000,000 dollars to the health system and causing about 1,500–6,700 deaths per year (5–20 deaths/100,000) [124].

10 Laboratory Diagnosis

The laboratory diagnosis of HRSV can be carried out by the direct detection of viruses, viral antigens, or the viral genome or, indirectly, based on the detection of specific antibodies. For the routine clinical laboratory diagnosis using respiratory secretions as biological samples, the procedures may include viral isolation in cell culture, antigen detection by immunofluorescence or enzyme-linked immunosorbent assay, and viral RNA detection by reverse transcriptase (RT)-polymerase chain reaction (PCR). The best samples are those obtained by aspiration or washing nasopharyngeal secretions [76, 79]. The viral particle present in the secretions is highly labile, and the samples should be kept refrigerated during transportation to the laboratory and processing before inoculation in cell cultures.

The isolation in cell culture, regarded as the gold standard, can be carried out in a wide variety of human and animal cell lines, but HEp-2 and HeLa cells are the most used [110; Perini et al. 2007]. The cytopathic effect usually appears within 3–7 days after inoculation and is characterized by the presence of large syncytia resulting from cell fusion (Fig. 12.4). Nevertheless, as viral isolation in cell culture is difficult, the

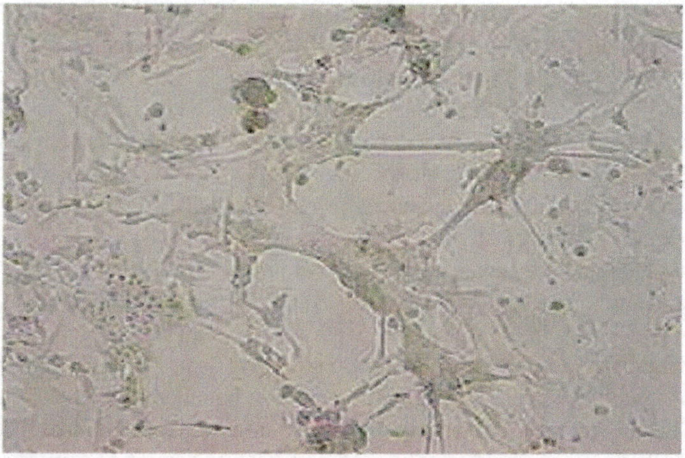

Fig. 12.4 Cytopathic effect of HRSV in HEp-2 cell line shows a large syncytium resulting from a fusion of the cells

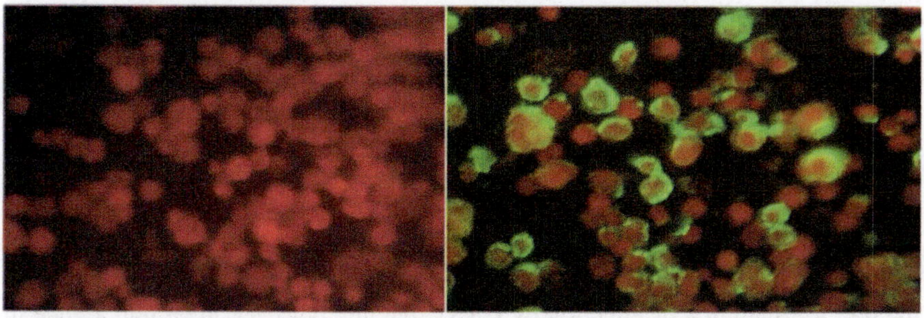

Fig. 12.5 Results of the indirect immunofluorescence (IFI) assay for detection of HRSV in nasopharyngeal aspirates. *Red*, negative results; *green*, positive detection of HRSV in the cytoplasm and membrane of the cell

diagnosis of infection is most often accomplished by detection of HRSV antigens in nasopharyngeal epithelial cells by immunofluorescence (Fig. 12.5) or enzyme-linked immunosorbent assay, faster methods that do not require the presence of infectious viral particles. These last two methods require the adequate preparation of the specimens by removing excess mucus. Finally, success of the immunofluorescence technique, aside from well-trained personnel and well-prepared samples, requires a minimum number of infected cells to enable a correct diagnosis [110].

Hughes et al. [62] compared the three diagnostic techniques for HRSV: isolation in cell culture, direct and indirect immunofluorescence (IFA), and enzyme-linked immunosorbent assay (ELISA). Both immunofluorescence-based methods detected more positive samples (showed higher sensitivity) than viral isolation. However, 15% of the samples found positive by viral isolation were negative by immunofluorescence, demonstrating the need for the use of at least two diagnostic methods.

The RT-PCR technique has been used both for the diagnosis [51, 59, 117] and for typing a sample to group A or B [53, 122]. It is considered a highly sensitive technique, especially useful in the diagnosis of infections, in which both the sample amount and the viral load in the sample are small, as is the case of samples taken from the elderly [117]. In the past decade, the molecular methods were considered as a gold standard, because of their specificity and ability of simultaneous detection of different viruses [90]. The advances in real time-RT-PCR (quantitative (q)RT-PCR) specificity and sensitivity for the detection of HRSV in clinical samples became more suitable for diagnosis in clinical laboratories [46].

The rapid antigen detection tests (RADTs) are dipstick-based immunoassays that allow for the rapid, qualitative detection of RSV antigen (viral fusion protein) directly from nasopharyngeal swab, nasopharyngeal aspirate, or nasal/nasopharyngeal wash specimens from symptomatic pediatric patients. The RADTs provide a result in 15 min, compared to approximately 90 min for a conventional IFA test and 2–3 h for ELISA [80]. Rapid tests may also be used as a point-of-care assay. These methods, although effective, may present several drawbacks, including price and skilled personnel. All these issues pose a challenge to hospitals and pediatric clinics to apply the best medical management for monitoring or treatment of children with suspected infection.

Serological diagnoses can be made through neutralization assays, complement fixation, or determination of class-specific immunoglobulins (IgG, IgM) by ELISA or immunofluorescence techniques. The diagnosis is based on the increase in antibody titer between acute and convalescent titers, performed in serum or saliva [110, 120]. The serology offers limited value in the diagnosis of primary infection in children less than 6 months of age because 40% of these cases present no increase in antibody titer. However, in infants and adults, the serology is regarded as a good indicator of reinfections [55]. The serology, therefore, is not the most appropriate method for diagnosis of infection by HRSV, having, however, great importance in clinical and epidemiological studies [36].

11 Treatment and Prevention

Ribavirin (1-β-D-ribofuranosyl-1,2,4-triazole-3-carboxamide), a nucleoside analogue of the guanosine, licensed since 1986, is the treatment of choice for RSV. Its use is indicated in the form of aerosol for the treatment of serious diseases caused by HRSV. Although several studies have demonstrated the effectiveness of ribavirin in inhibiting replication of the virus and the improvement of clinical conditions, resulting in a decrease in the need for supplemental oxygen and mechanical ventilation in children with lower respiratory tract infection, chronic lung disease, and infection in immunocompromised individuals, lately there has been controversy about the benefits of its use. Since 1989, several studies have appeared indicating that the use of ribavirin has minimal effect on disease outcome caused by HRSV, not showing evidence of decreased duration of hospitalization or the need of supporting therapy, in addition to the high cost and extended treatment (12 h or more of inhalation) [13, 37].

Several drug candidates have been studied in the past decades, including several inhibitors, targeting different HRSV proteins. Despite these efforts, until the present time there has been no antiviral drug approved for treatment (Heylen et al. 2017).

Development of an RSV vaccine has been hampered by the incidence of enhanced respiratory disease (ERD) following vaccination with formalin-inactivated RSV in the 1960s. Since its failure, multiple live virus vaccines have been developed, as well as other vaccine platforms, including virus-like particles, peptide-based vaccines, protein subunit vaccines, and plasmid DNA-based vaccines. Many of these vaccines have been evaluated in animals, and a few have been studied in humans. None, however, has shown sufficient promise to move toward licensure. It is clear that a better understanding of virus and host factors that contribute to both disease and protective immunity is still necessary to develop safe and effective RSV vaccines.

Alternative approaches to identify vaccine-relevant epitopes include the identification of neutralizing RSV protein epitopes to which a protective immune response can be safely generated and the development of modern pre- and post-RSV fusion (F) protein subunits. One obstacle to developing an RSV vaccine has been the difficulty in inducing long-term protective immunity, as evidenced by the repeated infections throughout life and the incomplete protection afforded to recipients of immune prophylaxis. In addition, an immunogenic approach targeted to a single neutralizing epitope mapped to the site A region may generate a focused immune response against RSV F, but in general, the polyclonal response generated by site A-based vaccines has been characterized by poor binding to intact RSV F protein, modest in vitro neutralization, and no evidence of protection to RSV challenges in vivo.

Palivizumab (Pz) (Synagis; MedImmune) is a humanized IgG monoclonal antibody that neutralizes HRSV through interaction with the HRSV F glycoprotein. Pz is the only FDA-approved prophylaxis against HRSV infection [5, 101]. Five monthly Pz injections spanning the annual HRSV epidemic period have been shown to reduce hospitalizations among high-risk children in the U.S. However, the quasispecies nature of RNA viruses allows rapid emergence of escape mutants to the immune pressure. The increasing use of Pz in high-risk children and immunocompromised patients provides opportunities for Pz-resistant mutants to arise and persist among humans [1, 9, 86, 88, 123]. However, little is known of these mutations in patients who did not use Pz.

References

1. Adams O, Bonzel L, Kovacevic A, Mayatepek E, Hoehn T, Vogel M (2010) Palivizumab-resistant human respiratory syncytial virus infection in infancy. Clin Infect Dis 51(2):185–188
2. Anderson LJ, Bingham P, Hierholzer JC (1988) Neutralization of respiratory syncytial virus by individual and mixtures of F and G protein monoclonal antibodies. J Virol 62(11):4232–4238
3. Anderson LJ, Parker RA, Strikas RL (1990) Association between respiratory syncytial virus outbreaks and lower respiratory tract deaths of infants and young children. J Infect Dis 161(4):640–646

4. Anderson LJ, Hierholzer JC, Tsou C, Hendry RM, Fernie BF, Stone Y, Mcintosh K (1985) Antigenic characterization of respiratory syncytial virus strains with monoclonal antibodies. J Infect Dis 151:626–633
5. Anon (1998) Palivizumab, a humanized respiratory syncytial virus monoclonal antibody, reduces hospitalization from respiratory syncytial virus infection in high-risk in infants. Pediatrics 102:531–537
6. Arslanagic E, Matsumoto M, Suzuki K, Nerome K, Tsutsumi H, Hung T (1996) Maturation of respiratory syncytial virus within HEp-2 cell cytoplasm. Acta Virol 40(4):209–214
7. Baek YH, Choi EH, Song MS, Pascua PN, Kwon HI, Park SJ, Lee JH, Woo SI, Ahn BH, Han HS, Hahn YS, Shin KS, Jang HL, Kim SY, Choi YK (2012) Prevalence and genetic characterization of respiratory syncytial virus (RSV) in hospitalized children in Korea. Arch Virol 157(6):1039–1050
8. Beeler JA, Coelingh van Wyke K (1989) Neutralization epitopes of the F glycoprotein of respiratory syncytial virus: effect of mutation upon fusion function. J Virol 63:2941–2950
9. Boivin G, Caouette G, Frenette L, Carbonneau J, Ouakki M, Serres G (2008) Human respiratory syncytial virus and other viral infections in infants receiving palivizumab. J Clin Virol 42(1):52–57
10. Boivin G, De Serres G, Côte S, Gilca R, Abed Y, Rochette L, Bergeron MG, Dery EP (2003) Human metapneumovírus infection in hospitalized children. Emerg Infect Dis 9(6):634–640
11. Bonfim CM, Nogueira ML, Simas PV, Gardinassi LG, Durigon EL, Rahal P, Souza FP (2011) Frequent respiratory pathogens of respiratory tract infections in children attending daycare centers. J Pediatr (Rio) 87(5):439–444
12. Botosso VF, Zanotto PM, Ueda M, Arruda E, Gilio AE, Vieira SE, Stewien KE, Peret TC, Jamal LF, Pardini MI, Pinho JR, Massad E, Sant'anna OA, Holmes EC, Durigon EL, VGDN Consortium (2009) Positive selection results in frequent reversible amino acid replacements in the G protein gene of human respiratory syncytial virus. PLoS Pathog 5(1):e1000254
13. Bricks LF (2000) Prevention of respiratory syncytial virus infections. Rev Hosp Clin Fac Med SP 56:79–90
14. Bryce J, Boschi-Pinto C, Shibuya K, Black RE (2005) Who estimates the causes of death in children. Lancet 365(23):1147–1152
15. Bueno IA, Riccetto AG, Morcillo AM, Arns CW, Baracat EC (2012) Respiratory syncytial vírus, infants and intensive therapy. Braz J Infect Dis 16(1):86–89
16. Cane PA (2001) Molecular epidemiology of respiratory syncytial virus. Rev Med Virol 11:103–116
17. Cane PA, Matheus DA, Pringle CR (1991) Identification of variable domains of the attachment (G) protein of subgroup A respiratory syncytial virus. J Gen Virol 72:2091–2096
18. Cane PA, Matthews DA, Pringle CR (1992) Analysis of relatedness of subgroup A respiratory syncytial viruses isolated worldwide. Virus Res 25:15–22
19. Cane PA, Matthews DA, Pringle CR (1994) Analysis of respiratory syncytial virus strain variation in successive epidemics in one city. J Clin Microbiol 32:1–4
20. Cane PA, Pringle CR (1991) Respiratory syncytial virus heterogeneity during an epidemic: analysis by limited nucleotide sequencing (SH gene) and restriction mapping (N gene). J Gen Virol 72:349–357
21. Cane PA, Prince CR (1995) Evolution of subgroup a respiratory syncytial virus: Evidence for progressive accumulation of amino acid changes in the attachment protein. J Virol 69:2918–2925
22. Carballal G, Videla CM, Espino MA, Savy V, Uez O, Sequeira MD, Knez V, Requeijo PV, Posse CR, Miceli I (2001) Multicentered study of viral acute lower respiratory infections in children from four cities of Argentina, 1993–1994. J Med Virol 64:164–174
23. Carballal G, Videla CM, Sequeira MD, Mistchenko A, Requeijo PV, Arbiza J (2000) Respiratory syncytial virus: changes in prevalence of subgroups A and B among Argentinian children, 1990–1996. J Med Virol 61:275–279

24. CDC-Centers for Disease Control and Prevention (2014) Respiratory syncytial virus circulation in the United States, July 2012–June 2014. MMWR 62:141–144
25. Chanock RM, Roizman B, Myers R (1957) Recovery from infants with respiratory illness of virus related to chimpanzee coryza agent. I. isolation, properties and characterization. Am J Hyg 66:281–290
26. Chanock RM, Finberg L (1957) Recovery from infants with respiratory illness of a virus related to chimpanzee coryza agent (CCA). II. Epidemiological aspects of infection in infants and young children. Am J Hyg 66:291–300
27. Choi E, Lee H (2000) Genetic diversity and molecular epidemiology of the G protein of subgroups A and B of respiratory syncytial virus isolated over 9 consecutive epidemics in Korea. J Infect Dis 181:1547–1556
28. Christensen LS, Larsen LB, Johansen J, Andersen EA, Wejse C, Klug B, Hornsleth A (1999) The fluctuating pattern of various genome types of respiratory syncytial virus in copenhagen and some other locations in Denmark. APMIS 107:843–850
29. Cintra OA, Owa MA, Machado AA, Cervi MC, Figueiredo LT, Rocha GM et al (2001) Occurrence and severity of infections caused by subgroup A and B respiratory syncytial virus in children in southeast Brazil. J Med Virol 65(2):408–412
30. Wb C, Ej L, Wm S (1998) Genetic variability among group A and group B respiratory syncytial virus in a children's hospital. J Clin Microbiol 36:35552–35557
31. Collins PL, Dickens LE, Buckler-White A, Olmsted RA, Spriggs MK, Camargo E et al (1986) Nucleotide sequences for the gene junctions of human respiratory syncytial virus reveal distinctive features of intergenic structure and gene order. Proc Natl Acad Sci U S A 83(13):4594–4598
32. Collins PL, Hill MG, Camargo E, Grosfeld H, Chanock RM, Murphy BR (1995) Production of infectious human respiratory syncytial virus from cloned cDNA confirms an essential role for the transcription elongation factor from the 5′ proximal open reading frame of the M2 mRNA in gene expression and provides a capability for vaccine development. Proc Natl Acad Sci U S A 92(25):11563–11567
33. Collins PL, McIntosh K, Chanock RM (1996) Respiratory syncytial virus. In: Fields BN, Knipe DM, Howley PM (eds) Fields virology, 3rd edn. Lippincott-Raven, Philadelphia, pp 1313–1351
34. Collins PL, Chanock RM, Murphy FBR (2001) Respiratory syncytial vírus. In: Fields BN (ed) Fields virology, 4th edn. Lippincott-Raven, Philadelphia/New York
35. Costa LF, Yokosawa J, Mantese OC, Oliveira TF, Silveira HL, Nepomuceno LL, Moreira LS, Dyonisio G, Rossi LM, Oliveira RC, Ribeiro LZ, Queiróz DA (2006) Respiratory vírus in children younger than five years old with acute respiratory disease from 2001 to 2004 in Uberlândia, MG, Brazil. Mem Inst Oswaldo Cruz 101(3):301–306
36. Cox MJ, Azevedo RS, Cane PA, Massad E, Medley GF (1998) Seroepidemiological study of respiratory syncytial virus in São Paulo state, Brazil. Med Virol 55:234–239
37. Crowe JE (2002) Respiratory syncytial virus vaccine development. Vaccine 20:S32–S37
38. Dapat IC, Shobugawa Y, Sano Y, Saito R, Sasaki A, Suzuki Y, Kumaki A, Zaraket H, Dapat C, Oguma T, Yamaguchi M, Suzuki H (2010) New genotypes within respiratory syncytial virus group B, genotype BA in Niigata, Japan. J Clin Microbiol 48(9):3423–3427
39. Debur MC, Vidal LR, Stroparo E, Nogueira MB, Almeida SM, Takahashi GA, Rotta I, Pereira LA, Silveira CS, Delfraro A, Nakatani SM, Skraba I, Raboni SM (2010 Dec) Impact of human metapneumovirus infection on in and outpatients for the years 2006-2008 in Southern Brazil. Mem Inst Oswaldo Cruz 105(8):1010–1018
40. Durigon GS, Oliveira DB, Vollet SB, Storni JG, Felício MC, Finelli C, Piera J, Magalhães M, Caldeira RN, Barbosa ML, Durigon EL, Berezin EN (2010) Hospital-acquired human bocavirus in infants. J Hosp Infect 76(2):171–173
41. Durigon GS, Oliveira DB, Felicio MC, Finelli C, Pereira MF, Storni JG et al (2015) Poor outcome of acute respiratory infection in young children with underlying health condition in Brazil. Int J Infect Dis 34:3–7

42. Colmanetti TC, Oliveira DBL, Botosso VF, Thomazelli LM, Pinez CMN, Crema D, Mesquita FS, Vieira SE, Gilio AE, Martinez MB, Durigon EL (2015) Naturally dominance of RSV, a N276S mutation in protein palivizumab binding site in new genotypes oni-1 and na1-2-3. Presented at Clinical Virological Symposium.2015
43. Eshaghi A, Duvvuri VR, Lai R, Nadarajah JT, Li A, Patel SN, Low DE, Gubbay JB (2012) Genetic variability of human respiratory syncytial virus A strains circulating in Ontario: a novel genotype with a 72 nucleotide G gene duplication. PLoS One 7(3):e32807
44. Falsey AR (2005) Respiratory syncytial virus infection in elderly and high-risk adults. Exp Lung Res 31(1):77
45. Forster J, Ihorst G, Rieger CH, Stephan V, Frank HD, Gurth H, Berner R, Rohwedder A, Werchau H, Schumacher M, Tsai T, Petersen G (2004) Prospective population-based study of viral lower respiratory tract infection in children under 3 years of age. Eur J Pediatr 163:709–716
46. Fry AM, Chittaganpitch M, Baggett HC, Peret TC, Dare RK, Sawatwong P et al (2010) The burden of hospitalized lower respiratory tract infection due to respiratory syncytial virus in rural Thailand. PLoS One 5(11):e15098
47. Gamba-Sanchez N, Rodriguez-Martinez CE, Sossa-Briceno MP (2016) Epidemic activity of respiratory syncytial virus is related to temperature and rainfall in equatorial tropical countries. Epidemiol Infect 144(10):2057–2063
48. Garcia CG, Bhore R, Soriano-Fallas A, Trost M, Chanson R, Ramilo O, Mejias A (2010) A Risk factors in children hospitalized with RSV bronchiolitis versus non-RSV bronchiolitis. Pediatrics 126(6):e1453–e1460
49. Garcia MLG, Gabin MO, Rey CC, Alvarez MIG, Ruiz JA, Sierra AA, Breña PP (2001) Infecciones virales de vias respiratórias inferiors en lactentes hospitalizados: etiología, características clínicas y factores de riesgo. An Esp Pediatr 55:101–107
50. Garcia O, Martin M, Dopazo J, Arbiza J, Frabasile S, Russsi J, Hortal M, Perez-Brena P, Martinez I, Garcia-Barreno B, Melero JA (1994) Evolutionary pattern of human respiratory syncytial virus (subgroup A): cocirculating lineages and correlation of genetic and antigenic changes in the G glicoprotein. J Virol 68:5448–5459
51. Gilbert LL, Dakhama A, Bone BM, Thomas EE, Hegele RG (1996) Diagnosis of viral respiratory tract infections in children by using a reverse transcription-PCR panel. J Clin Microbiol 34:140–143
52. Gimenez HB, Hardman N, Keir HM, Cash P (1986) Antigenic variation between human respiratory syncytial virus isolates. J Gen Virol 67(pt 5):863–870
53. Gottschalk J, Zbinden R, Kaempf L, Heinzer I (1996) Discrimination of respiratory syncytial virus subgroups A and B by reverse transcription-PCR. J Clin Microbiol 34:41–43
54. Hall CB, Keith RP, Macdonald NE, Gala CL, Menegus ME, Suffin SC, Cohen HJ (1986) Respiratory syncytial viral infection in children with compromised immune function. N Engl J Med 315:77–81
55. Hall CB, Walsh EE, Long CE, Schnabel KC (1991) Immunity to and frequency of reinfection with respiratory syncytial virus. J Infect Dis 163:693–698
56. Hall CB, Walsh EE, Schnabel KC, Long CE, Mcconnochie KM, Hildreth SW, Anderson LJ (1990) Occurrence of groups A and B of respiratory syncytial virus over 15 years: associated epidemiologic and clinical characteristics in hospitalized and ambulatory children. J Infect Dis 162:1283–1290
57. Harrington RD, Hackman EC, Storch GA, Osborne B, Gleaves CA, Benson A, Meyers JD (1992) An outbreak of respiratory syncytial virus in a bone marrow transplant center. J Infect Dis 153:291–297
58. Hendry RM, Pierik LT, Mcintosh K (1989) Prevalence of respiratory syncytial virus subgroups over six consecutive outbreaks: 1981–1987. J Infect Dis. 160:185–190
59. Henkel JH, Aberle SW, Kundi M, Popow-Kraupp T (1997) Improved detection of respiratory syncytial virus in nasal aspirates by seminested RT-PCR. J Med Virol 53:366–371
60. Hierholzer JC, Tannock GA, Hierholzer CM, Coombs RA, Kennett ML, Phillips PA, Gust ID (1994) Subgrouping of respiratory syncytial virus strains from Australia and Papua New Guinea by biological and antigenic characteristics. Arch Virol 136:133–147

61. Hortal M, Suarez A, Deleon C, Estevan M, Mogdasy MC, Russi JC, Contera M, Meny M (1994) Etiology and severity of community acquired pneumonia in children from Uruguay: a 4-year study. Rev Inst Med Trop São Paulo 36:255–264
62. Hughes JH, Mann DR, Hamparian VV (1988) Detection of virus respiratory syncytial virus in clinical specimens by viral culture, direct and indirect immunofluorescence, and enzyme immunoassay. J Clin Microbiol 26:588–591
63. Imaz MS, Sequeira MD, Videla C, Veronessi I, Cociglio R, Zerbini E, Carballal G (2000) Clinical and epidemiological characteristics of respiratory syncytial virus subgroups A and B infections in Santa Fe, Argentina. J Med Virol 61:76–80
64. Iwane MK, et al., New Vaccine Surveillance Network (2004) Population-based surveillance for hospitalizations associated with respiratory syncytial vírus, influenza vírus e parainfluenza víruses among young children. Pediatrics 113:1758–1764
65. Johansen J, Christensens LS, Hornsleth A, Klug B, Hansn KS, Nir M (1997) Restriction pattern variability of respiratory syncytial virus during three consecutive epidemics in Denmark. APMIS 105:303–308
66. Johnson PR Jr, Olmsted RA, Prince GA, Murphy BR, Alling DW, Walsh EE, Collins PL (1987) Antigenic relatedness between glycoproteins of human respiratory syncytial virus subgroups A and B: evaluation of the contributions of F and G glycoproteins to immunity. J Virol 61:3163–3166
67. Johnson PR, Spriggs MK, Olmsted RA, Collins PL (1987) The G glycoprotein of human respiratory syncytial viruses of subgroups A and B: extensive sequence divergence between antigenically related proteins. Proc Natl Acad Sci U S A 84(16):5625–5629
68. Kfouri RA, Wagner NH (2011) Infecção pelo virus Sincicial Respiratório. In: Neto VA (ed) Imunizações: atualizações, orientações e segestões. Segmento Farma, São Paulo
69. Kingsbury DW (1990) Paramyxoviridae and their replication. In: Fields BN (ed) Fields virology. Lippincott-Raven, Philadelphia, pp 945–962
70. Krusat T, Streckert HJ (1997) Heparin-dependent attachment of respiratory syncytial virus (RSV) to host cells. Arch Virol 142(6):1247–1254
71. Kuo L, Fearns R, Collins PL (1996) The structurally diverse intergenic regions of respiratory syncytial virus do not modulate sequential transcription by a dicistronic minigenome. J Virol 70(9):6143–6150
72. Levine SR, Klaiber-Franco R, Paradiso PR (1987) Demonstration that glycoprotein G is the attachment protein of respiratory syncytial virus. J Gen Virol 68:2521–2524
73. Lichtenstein DL, Roberts SR, Wertz GW, Ball LA (1996) Definition and functional analysis of the signal/anchor domain of the human respiratory syncytial virus glycoprotein. J Gen Virol 77(pt 1):109–118
74. Luchsinger V, Ampuero S, Palomino MA, Chnaiderman J, Levican J, Gaggero A, Larrañaga CE (2014) Comparison of virological profiles of respiratory syncytial virus and rhinovirus in acute lower tract respiratory infections in very young Chilean infants, according to their clinical outcome. J Clin Virol 61(1):138–144
75. Lukic-Grlic A, Cane PA, Bace A, Pringle CR, Mlinaric-Galinovic G, Popow-Kraupp T (1998) Antigenic and genomic diversity of central European respiratory syncytial virus strains. Arch Virol 143(7):1441–1447
76. Mackie PLK, Madge PJ, Getty S, Paton JY (1991) Rapid diagnosis of respiratory syncytial virus by using pernasal swabs. J Clin Microbiol 29:2653–2655
77. Martinez I, Valdes O, Delfraro A, Arbiza J, Russi J, Melero JA (1999) Evolutionary pattern of the G glycoprotein of human respiratory syncytial virus from antigenic group B: the use of alternative codons and lineage diversification. J Gen Virol 80:125–130
78. Massin MM, Montesanti J, Gerard P, Lepage (2006) Spectrum and frequency of illness presenting to a pediatric emergency department. Acta Clin Belg 61:161–165
79. Masters HB, Weber KO, Groothuis JR, Wren CG, Lauer BA (1987) Comparison of nasopharyngeal washing and swab specimens for diagnosis of respiratory syncytial virus by EIA, FAT, and cell culture. Diagn Microbiol Infect Dis 8:101–105

80. Mesquita FS, Oliveira DBL, Crema D, Botosso VF, Pinez CMN, Colmanetti TC, Thomazelli, LM, Gilio AE, Martinez MB, Durigon EL (2017) Rapid antigen detection test for respiratory syncytial virus (RSV) diagnosis as a diagnostic tool for use in pediatrics. J Pediatr (Rio J). 2016 Nov 25. pii: S0021-7557(16)30287-X. doi: 10.1016/j.jped.2016.06.013. [Epub ahead of print]
81. Mink MA, Stec DS, Collins PL (1991) Nucleotide sequences of the 3′ leader and 5′ trailer regions of human respiratory syncytial virus genomic RNA. Virology 186:615–624
82. Morris JA Jr, Blount RE, Savage RE (1956) Recovery of cytopathogenic agent from chimpanzees with coryza. Proc Soc Exp Med 92:544–550
83. Mufson MA, Belshe RB, Örvell C, Norrby E (1987) Subgroups characteristics of respiratory syncytial virus strains recovered from children with two consecutive infections. J Clin Microbiol 25:1535–1539
84. Nicholson KG, McNally T, Silverman M, Simons P, Stockton JD, Zambon MC (2006) Rates of hospitalisation for influenza, respiratory syncytial vírus and human metapneumovírus among infants and young children. Vaccine 24:102–108
85. Oliveira DBL, Durigon EL, Carvalho ACL, Leal AL, Souza TS, Thomazelli LM, Moraes CTP, Vieira SE, Al G, Stewien KE (2009) Epidemiology and genetic variability of human metapneumovírus during a 4-year-long study in Southeastern Brazil. J Med Virol 81:915–921
86. Oliveira DBL, Iwane MK, Prill MM, Weinberg GA, Williams JV, Griffind MR, Szilagyi PG, Edwards KM, Staat MA, Hall CB, Durigon EL, Erdman DD (2015) Molecular characterization of respiratory syncytial viruses infecting children reported to have received palivizumab immunoprophylaxis. J Clin Virol 65:26–31
87. Olszewska W, Zambon M, Openshaw PJM (2002) Development of vaccines against common colds. Br Med Bull 62:99–111
88. Papenburg J, Carbonneau J, Hamelin ME, Isabel S, Bouhy X, Ohoumanne N, Déry P, Paes BA, Corbeil J, Bergeron MG, De Serres G, Boivin G (2012) Molecular evolution of respiratory syncytial virus fusion gene Canada, 2006–2010. Emerg Infect Dis 18:120–124
89. Pavia AT (2011) Viral infections of the lower respiratory tract: old viruses, new viruses, and the role of diagnosis. Clin Infect Dis 52(suppl 4):S284–S289
90. Pecchini R, Rogério P, Berezin EN, Calahani Felício MC, Passos SD, de Souza MCO et al (2008) Incidence and clinical characteristics of the infection by the respiratory syncytial virus in children admitted in Santa Casa de São Paulo Hospital. Braz J Infect Dis 12(6):476–479
91. Peng D, Zhao D, Liu J, Wang X, Yang K, Xicheng H et al (2009) Multipathogen infections in hospitalized children with acute respiratory infections. Virol J 6(1):155–161
92. Peret TC, Hall CB, Hammond GW, Piedra PA, Storch GA, Sullender WM et al (2000) Circulation patterns of group A and B human respiratory syncytial virus genotypes in 5 communities in North America. J Infect Dis. 181(6):1891–1896
93. Peret TC, Hall CB, Schnabel KC, Golub JA, Anderson LJ (1998) Circulation patterns of genetically distinct group A and B strains of human respiratory syncytial virus in a community. J Gen Virol 79:2221–2229
94. Pohl C, Green M, Wald ER, Ledesma-Medina J (1992) Respiratory syncytial virus infections in pediatric liver transplant recipients. J Infect Dis 165:166–169
95. Pontoriero AV, Baumeister EG, Campos AM, Savy VL, Lin YP, Hay A (2003) Antigenic and genomic relation between human influenza víruses that circulated in Argentina in the period 1995–1999 and the corresponding vaccine components. J Clin Virol 28(2):130–140
96. Robinson JL, Lee BE, Bastien N, Li Y (2005) Seasonality and clinical features of human metapneumovírus infection in children in northern Alberta. J Med Virol 76:98–105
97. Sanz MC, Kew OM, Anderson LJ (1994) Genetic heterogeneity of the attachment glycoprotein G among group A respiratory syncytial viruses. Virus Res 33(3):203–217
98. Sauro A, Barone F, Blasio G, Russo L, Santillo L (2006) Do influenza and acute respiratory infective diseases weigh heavily on general practitioners' daily practice? Eur J Gen Pract 12:34–36
99. Shao X, Guo X, Esper F, Weibel C, Kahn JS (2007) Seroepidemiology of group I human coronaviruses in children. J Clin Virol 40(3):207–213

100. Shay D, Holman RC, Roosevelt GE, Clarke MJ, Anderson LJ (2001) Bronchiolitis-associated mortality and estimates of respiratory syncytial virus -associated deaths among US children, 1979–1997. J Infect Dis 183:16–22
101. Shadman KA, Wald ER (2011) A review of palivizumab and emerging therapies for respiratory syncytial virus. Expert Opin Biol Ther 11:1455–1467
102. Shobugawa Y, Saito R, Sano Y, Zaraket H, Suzuki Y, Kumaki A, Dapat I, Oguma T, Yamaguchi M, Suzuki H (2009) Emerging genotypes of human respiratory syncytial virus subgroup A among patients in Japan. J Clin Microbiol 47(8):2475–2482
103. Spencer MJ, Cherry JD (1987) Adenoviral infections. In: Feigin RD, Cherry JD (eds) Textbook of pediatric infectious diseases, 2nd edn. Elsevier, Philadelphia, pp 1688–1708
104. Straliotto SM, Nestor SM, Siqueira MM (2001) Respiratory syncytial virus groups A and B in Porto Alegre, Brazil, from 1990 to 1995 and 1998. Mem Inst Oswaldo Cruz 96(2):155–158
105. Thomazelli LM, Vieira SE, Leal AL, Silva TS, Oliveira DB, Golono MA, Gillio AE, Stewien KE, Erdman DD, Durigon EL (2007) Surveillance of eight respiratory víruses in clinical samples of pediatric patients in southeast Brazil. J Pediatr 83(5):422–428
106. Tregoning JS, Schwarze J (2010) Respiratory viral infections in infants: causes, clinical symptoms, virology, and immunology. Clin Microbiol Rev 23(1):74–98
107. Trento A, Abrego L, Rodriguez-Fernandes R, Gonzales-Sanches MI, Gonzales-Martines F, Delfraro A, Pascale JM, Arbiza J, Melero JA (2015) Conservation of G protein epitopes in respiratory syncytial virus (group a) despite broad genetic diversity: is antibody selection involved in virus evolution?. J Virol. piiJVI.00467–15
108. Trento A, Galiano M, Videla C, Carballal G, García-Barreno B, Melero JA, Palomo C (2003) Major changes in the G protein of human respiratory syncytial virus isolates introduced by a duplication of 60 nucleotides. J Gen Virol 84(pt 11):3115–3120
109. Trento A, Viegas M, Galiano M, Videla C, Carballal G, Mistchenko AS, Melero JA (2006) Natural history of human respiratory syncytial virus inferred from phylogenetic analysis of the attachment (G) glycoprotein with a 60-nucleotide duplication. J Virol 80(2):975–984
110. Tristram DA, Welliver RC (1996) Respiratory syncytial virus. In: Diagnostic procedure for viral, rickettsial and chlamydial infections, 7th edn. American Public Health Association, Washington, DC
111. Tsuchiya LRRV, Costa LMD, Raboni SM, Nogueira MB, Pereira LA, Rotta I, Takahashi GRA, Coelho M, Siqueira MM (2005) Viral respiratory infection in Curitiba, Southern Brazil. J Infect 51:401–407
112. Tsutsumi H, Onuma M, Suga K, Honjo T, Chiba Y, Chiba S, Ogra PL (1988) Occurrence of respiratory syncytial virus subgroup A and B strains in Japan, 1980 to 1987. J Clin Microbiol 26:1171–1174
113. Valdez O, Martinez I, Valdivia A, Cancio R, Savón C, Goyenechea A, Melero J (1998) Unusual antigenic and genetic characteristics of human respiratory syncytial virus isolated in Cuba. J Virol 72:7589–7592
114. van den Hoogen BG, van Doornum GJ, Fockens JC, Cornelissen JJ, Beyer WE, de Groot R, Osterhaus AD, Fouchier RA (2003) Prevalence e clinical symptoms of human metapneumovírus infection in hospitalized patients. J Infect Dis 188:1571–1577
115. Venter M, Madhi SA, Tiemessen CT, Schoub BD (2001) Genetic diversity and molecular epidemiology of respiratory syncytial virus over four consecutive seasons in South Africa: identification of new subgroup A and B genotypes. J Gen Virol. 82(pt 9):2117–2124
116. Se V, Stewien KE, Queiróz DAO, Durigon EL, Torok TJ, Anderson LJ, Miyao CR, Hein N, Botosso VF, Pahl MM, Gilio AE, Ejzenberg B, Okay Y (2001) Clinical patterns and seasonal trends in respiratory syncytial vírus hospitalizations in São Paulo, Brazil. Rev Inst Med Trop São Paulo 43:125–131
117. Walsh EE, Falsey AR, Swinburne IA, Formica MA (2001) Reverse transcription polymerase chain reaction (RT–PCR) for diagnosis of respiratory infection in adults: use of single-tube "hanging droplet" nested PCR. J Med Virol 63:259–263
118. Wertz GW, Collins PL, Huang Y, Gruber C, Levine S, Ball LA (1985) Nucleotide sequence of the G protein gene of human respiratory syncytial virus reveals an unusual type of viral membrane protein. Proc Natl Acad Sci U S A 82(12):4075–4079

119. Williams BG, Gouws E, Boschi-Pinto C, Bryce J, Dye C (2002) Estimates of world-wide distribution of child deaths from acute respiratory infections. Lancet Infect Dis 2:25–32
120. Wilson SD, Roberts K, Hammond K, Ayres JG, Cane PC (2000) Estimation of incidence of respiratory syncytial virus infection in schoolchildren using salivary antibodies. J Med Virol 61:81–84
121. Zelaya EAC, Petterson CA, Forsgren M, Orvell C, Strannegard O (1994) Respiratory syncytial virus infection in hospitalized patients and healthy children in El Salvador. Am J Trop Med Hyg 51:577–584
122. Zheng H, Peret TC, Randoph VB, Crowley JC, Anderson LJ (1996) Strain-specific reverse transcriptase PCR assay: means to distinguish candidate vaccine from wilde-type strains of respiratory syncytial virus. J Clin Microbiol 34:334–337
123. Zhu Q, McAuliffe JM, Patel NK, Palmer-Hill FJ, Yang CF, Liang B et al (2011) Analysis of respiratory syncytial virus preclinical and clinical variants resistant to neutralization by monoclonal antibodies palivizumab and/or motavizumab. J Infect Dis 203:674–682

Web Sites

124. CDC- Center for Disease Control and Prevention (2014) Respiratory syncytial virus circulation in the United States, July 2012-Jun 2014. MMWR 62:141–144. Accessed [Feb 2016]
125. ICTV. International Committee on Taxonomy of Viruses. Available from: http://www.ictvonline.org/virusTaxonomy.asp. [Mar 2015]. Available from: http://www.ictvonline.org/virusTaxonomy.asp. [3 Mar 2016]

Chapter 13
Influenza Viruses, Biology, Epidemiology, and Control

Elsa G. Baumeister and Andrea V. Pontoriero

1 Introduction

Influenza viruses are RNA viruses of negative polarity and segmented, belonging to the family *Orthomyxoviridae*. This family includes six genera: *Influenzavirus A*; *Influenzavirus B*; *Influenzavirus C*; *Isavirus; Thogotovirus*, which includes the *Thogoto virus* and *Dhori virus* including *Infectious salmon anemia virus*; and *Quaranjavirus* [17]. Influenza A viruses are classified into subtypes based on the antigenicity of the surface molecules hemagglutinin (H) and neuraminidase (N). Influenza A viruses infect a wide variety of birds and mammals. These viruses continually mutate, exhibiting well-studied patterns, such as antigenic drift and reassortment of genomic segments. In the past century, influenza A viruses have caused three pandemics [14], as they are known to be epidemics that extend over more than one continent. Influenza B viruses infect humans, and there is only one subtype; they have a low potential to cause pandemics, although they can cause severe respiratory illnesses. Influenza C viruses infect humans and pigs and in general cause mild respiratory illness.

Wild aquatic birds, such as ducks and gulls among others, are the most important natural reservoir of these viruses [33]. In this type of bird circulate all H and N subtypes, and it is thought that wild birds are the source of the transmission of the virus to other animal species, including poultry. Many of the subtypes of influenza A viruses

E.G. Baumeister (✉)
Laboratorio Nacional de Referencia de Influenza y Otros Virus Respiratorios, Centro Nacional de Influenza de OPS/OMS, Buenos Aires, Argentina
e-mail: ebaumeister@anlis.gov.ar

A.V. Pontoriero
Laboratorio Nacional de Referencia de Sarampión y Rubéola OPS/OMS, Departamento Virología, INEI-ANLIS "Carlos G. Malbrán", Buenos Aires, Argentina
e-mail: aponto@anlis.gov.ar

J.E. Ludert et al. (eds.), *Human Virology in Latin America*,
DOI 10.1007/978-3-319-54567-7_13

infect birds asymptomatically, that is, without causing disease or causing mild symptoms; however, some virus infections, for example, H5 and H7 subtypes, can cause severe illness and death in some species of wild and domestic birds, such as chickens and turkeys [34]. Pigs are also natural reservoirs of influenza A viruses, from which a limited number of subtypes have been isolated.

Every year, around 500 million people worldwide fall ill because of influenza virus A infection, of which between 3 and 5 million become severe cases leading to about 300,000 deaths. These cases of influenza occur regularly during the cold or rainy seasons each year and are known as epidemics or outbreaks of seasonal influenza.

Since 1977, in the human population two influenza A virus subtypes, known as H1N1 and H3N2, circulate seasonally, taking into account the characteristics of the proteins present on the surface, along with influenza virus B. The frequency of these three groups of viruses varies temporarily and geographically. In 2009, a new subtype H1N1 emerged with great antigenic/genomic changes that allowed them to reach the pandemic state. This new virus emerged from a virus generated by reassortment between a human virus, a porcine virus from North America, a porcine virus from Eurasian, and an avian virus [24].

2 Nomenclature

Influenza viruses belong to the *Orthomyxoviridae* family and consist of six genera:

1. *Influenzavirus A*. Type species, *Influenza A virus*
2. *Influenzavirus B*. Type species, *Influenza B virus*
3. *Influenzavirus C*. Type species, *Influenza C virus*
4. *Isavirus*. Type species, *Infectious salmon anemia virus*
5. *Quaranjavirus*. Species, *Johnston Atoll virus*, and type species, *Quaranfil virus*
6. *Thogotovirus*. Species, *Dhori virus*, and type species, *Thogoto virus*

The *Influenzavirus A* and *Influenzavirus B* genera cause epidemics every year. The emergence of a new and different virus in the human population can cause an influenza pandemic. Influenza C virus infections cause mild respiratory disease and are considered not to cause epidemics.

Influenza A viruses are divided into subtypes based on the two surface proteins of the virus: the hemagglutinin (H) and neuraminidase (N). In nature, there are at least 18 different subtypes of H and 11 different subtypes of N [31, 32]. In turn, within each subtype, influenza A viruses can be divided into different strains with different antigenic characteristics. Current influenza A virus subtypes detected in the human population are A(H1N1) and A(H3N2). In the spring of 2009, a new influenza virus A(H1N1) emerged and began to cause illness in people. This virus has major changes in comparison with seasonal influenza A(H1N1) virus and has caused an influenza pandemic after 40 years. The virus, known as A(H1N1)pdm09, has largely replaced the previously circulating H1N1 virus among humans. The influenza B viruses are not divided into subtypes, but different antigenic strain characteristics can be distinguished.

The Centers for Disease Control and Prevention (CDC), located in Atlanta, GA (USA), follows an internationally accepted nomenclature convention for influenza virus. This convention was recommended by the World Health Organization (WHO) in 1979. This nomenclature establishes the use of the following components:

1. The antigenic type (e.g., A, B, C).
2. The host of origin, for example, pig, horse, chicken, or in the case of environmental samples. It is not specified in the case of human viruses.
   a. Geographic origin (e.g., Denver, Taiwan).
3. Number assigned by the laboratory of origin to the sample from which the virus was isolated (e.g., 15, 7).
4. Year of isolation (e.g., 1957, 2009).
5. For influenza virus A, the description of H and N antigens appears in brackets [e.g., (H1N1), (H5N1)].

For example:

- A/duck/Czechoslovakia/1956 (H4N6) for a virus originated in ducks
- A/Sydney/5/97 (H3N2) for a virus originated in human

3 Structure and Biology

Under the electron microscope, these viruses have a pleomorphic appearance, with an average diameter of 100 nm (one ten-thousandth of a millimeter). The viral particle has an envelope composed of a lipid bilayer in which the H glycoproteins and N and lesser amounts of M2 transmembrane protein are inserted (see Fig. 13.1). Lining the inside of

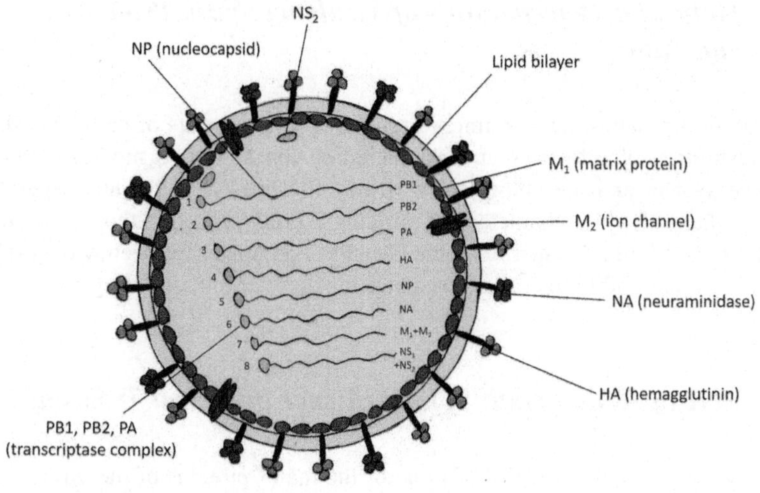

Fig. 13.1 Influenza A virion structure

the lipid membrane is a layer formed by the matrix protein M1. Inside the sheath is the viral genome. The genomic segments are covered with nucleoprotein NP and are also associated with viral RNA polymerase, which is a protein complex consisting of two basic subunits (PB1, PB2) and an acidic subunit (PA). The viral genome consists of eight segments of single-stranded RNA of negative polarity, ranging from 890 to 2,350 nucleotides in size, with some variation depending on the virus strain. In total, the genome has approximately 13,600 nucleotides and encodes 11 viral proteins. All segments encode a protein, with the exception of the *PB1* gene, which in some strains also encodes the PB1-F2 protein; the gene protein matrix, which encodes two proteins, M1 and M2; and the smallest gene, NS, which codes for proteins NS1 and NS2.

3.1 Viral Tropism

In humans, the virus usually enters the body through the nose or mouth and infects the cells lining the respiratory tract, joining sialic acid (SA) molecules on the surface of the cells to start the infection. The SA is an abundant molecule in all cells, a part of sugar chains bound to proteins or lipids, and defines the influenza virus tropism. This binding occurs because of the specificity of different virus strains for different types of links of SA with sugar, the upstream carbohydrate, which is generally galactose. Thus, human viruses recognize the conformation of SA with galactose in the alpha 2–6 position, whereas avian viruses join SA with union 2.3 [30]. The affinity for SA partly explains the restriction host of influenza viruses. In the swine tracheal epithelial cells, both types of SA links exist, which allow the pig to be naturally infected by swine, avian, or human viruses. This mechanism enables genetic rearrangements among animal species originating influenza viruses with combinations of gene segments from different origins.

3.2 Molecular Determinants of Virulence: Viral Proteins and Pathogenesis

Several viral proteins have an important role in some aspects of pathogenesis and host restriction of influenza viruses, including the ability to modulate the host immune system and the ability to replicate efficiently at different temperatures, among others. The best characterized proteins in order of their pathogenic potential are H, PB1, and PB2. Furthermore, the N and M2 proteins have been widely studied because of their ability to confer resistance to approved antiviral drugs.

3.3 Role of Hemagglutinin in Virulence and Viral Tropism

This glycoprotein, along with N, is one of the major proteins of the viral particle. The most neutralizing antibodies are produced against this main antigen, and these are capable of neutralizing virus infectivity. The H, as its name suggests, is capable of binding erythrocytes, a property that has been used in techniques for the classification of different subtypes. The H has a very important role in virus entry into the host cell, mediating the viral adsorption through its interaction with the SA. H is synthesized as a precursor protein called H0, which is cleaved at a specific site, resulting in the generation of the subunits HA1 and HA2. The cleavage site is characterized by being composed of basic amino acids. Proteolytic cleavage is essential for infectivity because it exposes a hydrophobic peptide at the amino terminus of HA2, which is responsible for mediating the fusion of the viral membrane and endocytic vesicle formed when the particle is internalized into the cell.

With the purpose of describing the spatial dissemination dynamics of influenza A(H3N2) within South America, 316 HA1 sequences of A(H3N2) viruses from Argentina, Brazil, Chile, Paraguay, Uruguay, Bolivia, Colombia, French-Guyana, Peru, and Venezuela collected between 1999 and 2012 were analyzed together with 153 available contemporary sequences from Australia, Hong Kong, the UK, and the U.S. Phylogenetic analyses revealed that influenza A(H3N2) sequences from South America were highly intermixed with sequences from other geographic regions, although a clear geographic virus population structure was detected globally. Fourteen clades mostly (≥80%) composed of influenza sequences from South American countries were identified. Bayesian phylogeographic analyses of those clades support a significant role of both temperate and tropical regions in the introduction and dissemination of new influenza A(H3N2) strains within South America and identify an intensive bidirectional viral exchange between different geographic areas. These findings indicated that seasonal influenza A(H3N2) epidemics in South America are seeded by both the continuous importation of viral variants from other geographic regions and the short-term persistence of local lineages. These results support a complex meta-population model of influenza A(H3N2) dissemination in South America, with no preferential direction in viral movement between temperate and tropical regions [7].

3.4 *Viral Replication and Protein PB2*

Protein PB1 and PA form the viral replication complex. PB2 has been associated with the transmissibility of the virus through the air and also with the restriction host.

3.5 *PB1-F2: A Cell Death Inductor Factor*

Protein PB1-F2 is a small protein encoded by the gene that also encodes the subunit viral polymerase PB1. PB1-F2 is not expressed by all influenza strains. It has been found that this protein interacts with the mitochondria membrane, causing its permeation and cytochrome C release, which induces cell death. It has also been reported that PB1-F2 exacerbates inflammatory response during viral infection in mice and increases the frequency and severity of secondary bacterial pneumonias.

3.6 NS1 and Control of the Innate Immune Response

During viral replication, transcriptional activation factors that stimulate the production of interferon are triggered; this is a strategy to prevent cell viral infections. The NS1 protein is a viral protein that antagonizes the interferon response of the cell. It was found that highly virulent strains, such as H5N1 avian, not only confer resistance to antiviral effects of interferon but also induce an exacerbated pro-inflammatory cytokine response. During pregnancy, immunological and hormonal alterations place women at increased risk for influenza-related severe illnesses, including hospitalization and death. Although A(H1N1) pdm09 infection resulted in increased disease severity in pregnant women, the precise mechanisms responsible for this risk have yet to be established. The role of host chemokines and cytokine profiles in A(H1N1)pdm09 infection regarding disease severity in pregnant women with confirmed influenza A(H1N1)pdm09 infection was investigated. Results revealed that pregnancy-related reductions in interferon (IFN)-β and transforming (TGF)-β expression levels and elevated levels of pro-inflammatory cytokines could explain the increased severity of infection and death of pregnant women [26].

3.7 NS2

This protein is also encoded by the viral RNA segment 8, and it has been detected associated with M1 protein in the virion. It is possible to find it in the nucleus and cytoplasm. It possesses a nuclear export signal that would function to export the RNPs from the nucleus.

3.8 M2: A Viral Ion Channel

The M2 protein is the less abundant viral coat. This protein functions as an ion channel that allows the entry of protons into the viral particle, facilitating the release of the viral genome into the cytoplasm. Adamantanes selectively block the channel formed by M2, which inhibits the release and therefore the replication of the viral genome. These compounds have been used against influenza outbreaks for many years; however, it has been found that adamantane resistance appears rapidly and frequently in influenza wild strains. Most human and swine A(H3N2) strains currently circulating are resistant to these antivirals; moreover, the new A(H1N1) pdm09 viruses are also resistant to these drugs.

3.9 Neuraminidase and Viral Dissemination

This glycoprotein, neuraminidase, is a sialidase whose function is to remove not only the SA of the viral glycoproteins of newly synthesized virus but also the SA present on the cell surface, allowing efficient release of virus from the infected cell to infect new cells. Inhibition of this activity produces the de novo synthesized virions that remain attached to the cell surface, inhibiting their spread to other cells. The N is the target of oseltamivir and zanamivir antivirals. These molecules are specific inhibitors of the N-sialidase activity.

4 Genetic Drift and Shift

Influenza viruses are constantly changing, and they can change in two different ways. One way is called "antigenic drift," which are small changes in the genes of influenza viruses that happen continually over time as the virus replicates. These small genetic changes usually produce viruses that are quite closely related to one another, which can be illustrated by their close location on a phylogenetic tree. Viruses that are closely related to each other usually share the same antigenic properties, and an immune system exposed to a similar virus will usually recognize it and respond. However, these small genetic changes can accumulate over time and result in viruses that are antigenically different (further away on the phylogenetic tree). When this happens, the body's immune system may not recognize those viruses. Genetic changes that result in a virus with different antigenic properties are the main reason why people can be infected by influenza virus more than once. This is also why the influenza vaccine composition must be reviewed each year and updated as needed to keep pace with evolving viruses [9]. In Mexico, whole-genome sequencing studies of human A(H3N2) isolates from 2003 to 2012 showed that different viral lineages co-circulate within the same season and can also persist locally in between different influenza seasons, increasing the chance for genetic reassortment events. A novel minor cluster was also identified, named here as Korea, that circulated worldwide during 2003 [11].

The other type of change is called "antigenic shift." Antigenic shift is an abrupt, major change in the influenza A viruses, resulting in new H or new H and N proteins in influenza viruses that infect humans. Shift results in a new influenza A subtype or a virus with a H or a H and N combination that has emerged from an animal population which is so different from the same subtype in humans that most people do not have immunity to the new (i.e., novel) virus. Such a "shift" occurred in the spring of 2009, when an H1N1 virus with a new combination of genes emerged to infect people and quickly spread, causing a pandemic. When this shift happens, most people have little or no protection against the new virus [9]. A(H1N1)pdm09 was originated from a quadruple viral reassortment: viral genes from North American swine, Eurasian swine, and avian and human populations.

On June 11, 2009, in acknowledgment of sustained global human-to-human transmission, the WHO declared that the novel influenza A(H1N1)pdm09 virus was pandemic. Although this virus has not been reported as more virulent than seasonal influenza A(H1N1), its high transmissibility in an immunologically naïve population implied the potential for substantial morbidity and mortality and mandated close surveillance for evolution toward increased virulence. Although worldwide the pandemic influenza A(H1N1) case fatality rate (CFR) was 0.4%, that rate in Argentina was 4.5%. After performing full-genome sequencing of strains isolated from mild and severe cases, no evidence was observed that the high CFR can be attributed directly to viral changes. Further, no evidence of reassortment with human or animal locally circulating strains, mutations associated with resistance to antiviral drugs, or genetic drift that might contribute to virulence was documented [3].

Analysis of 13 samples of 2009 H1N1 pandemic virus circulating in Paraguay in 2009 showed that these viruses clustered in a single genetic group. Neither the mutation related to exacerbation of disease (D239G in H) nor that related to antiviral resistance (H275Y in neuraminidase), both detected in neighboring countries, was found [12].

To generate information about the disease burden caused by 2009 influenza A(H1N1)pdm09 in children in Argentina, a retrospective case series study was conducted. The results obtained indicated that hospitalization rates were double in comparison with those for seasonal influenza in 2008. The overall rate of death was 1.1 per 100,000 children, as compared with 0.1 per 100,000 children for seasonal influenza in 2007. This investigation concluded that influenza A(H1N1)pdm09 was associated with pediatric death rates that were ten times the rates registered for seasonal influenza in previous years in Argentina [18].

Molecular characterization of circulating influenza A viruses in all regions of the world is essential to detect mutations potentially involved in increased virulence, antiviral resistance, and immune escape. To gain insight into these matters, a phylogenetic analysis of the N gene of 146 pandemic H1N1 (H1N1pdm) influenza A virus strains isolated in Argentina, Brazil, Chile, Paraguay, Peru, and Uruguay from 2009 to 2013 was performed. The comparison of vaccine strain A/California/7/2009 was included in the influenza vaccine recommended for the Southern Hemisphere from 2010 through 2016. Strains differ from vaccine in two predicted B-cell epitope regions present at positions 102–103 and 351–352 of the NA protein. Moreover, vaccine and strains isolated in Paraguay differ also in an epitope present at position 229. The analysis of the N gene of 2009 to 2013 H1N1 South American strains revealed several genetic and antigenic differences in the N of influenza A(H1N1)pdm09 among vaccine and strains circulating in South America [10].

Influenza viruses are changing by antigenic drift all the time, but antigenic shift happens only occasionally. Type A viruses undergo both kinds of changes; influenza type B viruses change only by the more gradual process of antigenic drift.

In the past century, three antigenic shifts occurred in the influenza A virus circulating in humans that were responsible for pandemics: in 1918, with the emergence of an A(H1N1) virus; in 1957, when the A(H1N1) virus was replaced by a virus subtype A(H2N2); and in 1968, when an A(H3N2) virus replaced the A(H2N2) viral subtype. In 1977, the A(H1N1) subtype was reintroduced in humans and did not replace the circulating subtype H3N2. The 1977 (H1N1) strain was replaced by the A(H1N1)pdm09 in 2009 and co-circulates together with A(H3N2) viral subtype up to the present worldwide, and they are responsible for the seasonal outbreaks that occur each year.

5 Viral Entry, Release, and Transmission

The main targets of the influenza virus are the columnar epithelial cells of the respiratory tract. These cells may be susceptible to infection if the viral receptor is present and functional. Thus, viral receptors are determinants of tropism. Viruses replicate in the respiratory epithelium, causing a localized infection. The new virions are overturned in the inner portion of the respiratory tract, transported through secretions, and disseminated out of the body when the individual sneezes, coughs, talks, or laughs.

The disease caused by the influenza virus is highly contagious. Transmission occurs by air in most cases, by coughing, talking, or sneezing, but can also spread through contact with surfaces contaminated with the respiratory secretions of sick individuals. Adults eliminate the virus from 1 day before the onset of symptoms until 5–10 days later; in children, elimination can start several days before and continue for 10 days or more after the onset of symptoms. Immunocompromised people can spread the virus for weeks.

6 Epidemiology and Local and Global Geographic Distribution

Human influenza viruses, including the new A(H1N1)pdm09 virus that has affected the human population since 2009, are distributed worldwide. This virus reaches peak prevalence in winter and rainy seasons. In Argentina particularly, although it is possible to detect viral circulation throughout the year, an increase of viral activity is observed between May and July, depending on the season. The number of positive cases detected and the types and viral subtypes circulating each year also vary by season. In 2012, the transmission pattern of influenza viruses was different in the four countries of the Southern Cone in South America. Surprisingly, peak activities were detected very late in this region. Increased viral circulation in the area was first noted at the end of April to early May in Chile and Paraguay; Argentina and Uruguay begun to report active transmission in early June. Particularly in Argentina, the peak was registered in August, 10 weeks later than in the past 9 years [6].

7 Ecology

Influenza A viruses can be found in various animals such as ducks, chickens, pigs, whales, horses, seals, and dogs. Influenza B viruses circulate only among humans. All known subtypes of influenza A virus were detected in birds, with the exception of H18N11 and H17N10 subtypes, which were only found in bats. Wild birds are the primary natural reservoir for all subtypes of influenza A viruses and are thought to be the source of influenza A viruses in all other animals. Argentina had been searched for evidences of influenza A infection in Antarctica through antibody and genome detection in different migratory bird species. Between December 2001and May 2004, sera and cloacal swabs were collected in different locations. Antibody against the A(H3N2), A(H1N1), A(H9N2), A(H5N1), and A(H7N2) subtypes and sequences of the M gene was detected in migratory birds in Antarctica, showing that these species had been in contact with influenza A virus during their lifetime. These findings were obtained also in young birds, born in Antarctica, and would suggest the local acquisition of the infection [2, 4].

As part of the ongoing efforts on animal influenza surveillance in Argentina, Xu et al. [35] described the isolation of an H9N2 virus from a wild aquatic bird (*Netta peposaca*), named as A/rosy-billed pochard/Argentina/CIP051–559/2007 (H9N2). Phylogenetic analysis of the HA gene revealed that the 559/H9N2 virus maintained an independent evolutionary pathway and shared a sister-group relationship with North American viruses, suggesting a common ancestor. The rest of the genome segments are clustered with viruses from South America. Experimental inoculation of the 559/H9N2 in chickens and quail revealed efficient replication and transmission only in quail. These viruses could easily jump to other bird species, thus highlighting the potential threat posed to local poultry. This study increases local understanding of H9N2 viruses in nature [35].

When a highly pathogenic virus (HPAI) appears, it produces severe symptoms and almost 100% mortality within 2 days. Since 2002, outbreaks of HPAI have occurred in the Americas: in Chile A(H7N3) 2002, U.S. (Texas) A(H5N2) 2004, 21 U.S. states A(H5) 2014–2015, and Canada (H7N3) 2004 were identified. In each of these outbreaks, a precursor virus of low pathogenicity mutated to become highly pathogenic after circulating in poultry [29].

The understanding of the global ecology of avian influenza A viruses is impeded by historically low levels of viral surveillance in Latin America. Through sampling and whole-genome sequencing of 31 avian influenza viruses from wild birds in Peru, ten HA subtypes (H1–H4, H6–H7, H10–H13) and eight N subtypes (N1–N3, N5–N9) were identified. The majority of those Peruvian avian influenza viruses were closely related to avian influenza viruses found in North America. However, unusual reassortants, including a H13 virus containing a PA segment related to extremely divergent Argentinean viruses, suggest that substantial avian diversity circulates undetected throughout South America [23].

In recent years, there has been extensive surveillance of the virus in aquatic birds in the Northern Hemisphere; however, in contrast, only a few studies have been attempted to detect avian influenza viruses in wild birds in South America. There are major flyways connecting South America to Central and North America, whereas avian migration routes between South America and the remaining continents are uncommon. As a result, it has been hypothesized that South American avian influenza strains would be more closely related to the strains from North America than to those from other regions in the world. The full genome of three avian influenza subtype H11N9 isolates obtained from ruddy turnstones (*Arenaria interpres*) on the Amazon coast of Brazil was studied. For all gene segments, all three strains consistently clustered together within evolutionary lineages of avian influenza viruses that had been previously described from aquatic birds in North America. In particular, the H11N9 isolates were remarkably closely related to avian influenza strains from shorebirds sampled at the Delaware Bay region, on the northeastern coast of the U.S. There was also evidence of genetic similarity to avian influenza strains from ducks and teals from the interior U.S. and Canada [16].

In 2002, the Chilean poultry industry was afflicted with a highly pathogenic avian influenza strain, which created economic loss and triggered the establishment of a surveillance program in wild birds. This effort consisted of periodic samplings of sick or suspicious animals found along the coast and analyses with standardized techniques for the detection of influenza A virus. The detection of three avian influenza strains (H13N2, H5N9, H13N9) in gulls from Chile between 2007 and 2009 showed highest similarities to viruses detected in wild birds from North America. These results suggest a dissemination route for influenza viruses along the coasts of the Americas [20].

Swine influenza virus (SIV) is enzootic in most regions with dense porcine populations. This disease is common in North and South America, Europe, Asia, and also in Africa. Although the viral subtypes found in the U.S. and Europe are the same, they are actually different strains. In Argentina, swine origin viruses have been identified in producing farm populations. An Argentinean study demonstrated the circulation of influenza A(H3N2) viruses in 19 pig farms between 2000 and 2002; the seroprevalence rate was 16.5% [5]. In sera collected in 2002, Piñeyro et al. [27] demonstrated the seroprevalence was 38.46% to 100% against H1N1 and 7.69% to 100% for H3N2 in 13 Argentinean swine.

In November 2008, an outbreak of respiratory disease in pigs consistent with SIV infection was detected in Argentina. Phylogenetic analysis revealed that the virus isolated shared nucleotide identities of 96–98% with A(H3N2) viruses that circulated in humans from 2000 to 2003. Sera collected from experimental inoculated animals mainly cross-reacted with noncontemporary human-origin H3N2 influenza viruses [8].

In 2009, A(H1N1)pdm09 transmission from human to pig was confirmed. In Argentina, seroepidemiological analyses performed in 17 pig farms showed that $\approx$41% of pigs had antibodies against A(H1N1) and A(H3N2) subtypes. Vaccines against swine influenza viruses were not licensed in Argentina in that period. In June–July 2009, an outbreak caused by influenza A(H1N1)pdm09 virus occurred on a pig farm. The virus was genetically related to the pandemic strain isolated in humans, and no evidence of further reassortment was confirmed [25]. Equine influenza occurs in almost all countries with a significant number of horses.

In July 2006, horses from various regions of Chile presented with fever, serious nasal discharge, dry cough, anorexia, and depression. The virus was identified as equine influenza virus H3N8. After performing sequencing of the H, N, and NP genes, important differences with the Santiago/85 isolate was observed, with a closer relationship to North American isolates, especially to the Florida lineage, and to Argentinean isolates from the 1990s [22].

There is no evidence that canine influenza virus is currently circulating outside the U.S. However, occasional infections with equine influenza virus are seen among dogs in other regions. In the UK, an equine virus H3N8 was responsible for outbreaks of respiratory disease in dog kennels in 2002 and 2003, and also in Australia in 2007. H3N2 viruses have been reported in dogs only in Korea.

8 Prophylaxis

In uncomplicated cases, bed rest with adequate hydration is the treatment of choice for most adolescents and young adult patients. Salicylates must be avoided in children of 18 years or younger because of the association to Reyes syndrome. Antibiotic treatment should be reserved for the treatment of secondary bacterial pneumonia. More severe cases or infections at high risk of complications can be treated with antiviral drugs. Two classes of antiviral drugs are available for the prevention and treatment of influenza: (a) the neuraminidase inhibitors (NAIs) zanamivir, oseltamivir, peramivir, and laninamivir, which are active against both influenza A and influenza B, and (b) the adamantanes, amantadine, and rimantadine, which are only active against influenza A. Before licensing of the NAIs in 1999, the adamantanes were the only drugs used for the treatment and prevention of influenza. These drugs target the virus M2 ion channel protein, involved in virus uncoating in the endosome. However, because of central nervous system complications in the elderly and lack of efficacy against influenza B, these agents were not widely employed. Additionally, since 2000, many viruses have acquired the substitutions L26F, L/V27A, A30T, S31 N, or G34E in the M2 gene conferring resistance, including the current human A(H3N2) and A(H1N1)pdm09 viruses and avian influenza A(H5N1) and A(H7N9) viruses, which have caused sporadic human infections. Resistance initially emerged in China, possibly related to the ready availability of the adamantanes in over-the-counter medications and use in poultry feed.

The second group of antivirals described here is called "NA inhibitor" (NAI) because these antiviral drugs bind to NA influenza virus and inhibit the enzymatic activity of this protein [21].

Antiviral agents may be prescribed as treatment to potentially shorten the duration and decrease the severity of influenza infection. Antivirals may also be prescribed for chemoprophylaxis to prevent/attenuate a potential influenza infection following contact with an infected individual or in vulnerable individuals during a community outbreak (e.g., nursing homes). When used as treatment, initiation of antiviral agents should not be delayed and ideally should be started within 48 h of the onset of symptoms.

When NA proteins change, the NAI can lose its ability to bind to and inhibit the function of the viral NA proteins, resulting in NAI resistance (non-susceptibility). A particular genetic change named the "H275Y" mutation is known to confer oseltamivir resistance in 2009 A(H1N1)pdm09 viruses. (The H275Y mutation is a substitution of histidine for tyrosine at position 275 in the NA.) This substitution prevents oseltamivir from inhibiting NA activity and allows the mutated virus to spread to healthy cells, which results in the drug not working as well. Although instances of antiviral drug resistance among influenza viruses have occurred during antiviral therapy, resistant influenza strains have also spread widely in the absence of such drug pressure. Viruses resistant to oseltamivir could not disseminate properly in the human population, which could explain the low frequency of resistant virus detected in the entire world. In a study performed in Central and South America in 2005–2008, the M2 and NA genes were sequenced, and resistance was inferred by comparison with published sequences and known resistant mutations. The results obtained indicated resistance to adamantanes in the majority of the A(H3N2) isolates, but in only one isolate of the influenza A(H1N1) viruses, and resistance to NAIs began to be detected in 2008 A(H1N1) isolates. Also, none of the influenza B viruses analyzed was resistant to NAIs. [13]. To spread in community settings, H275Y mutants must contain additional mutations. Oseltamivir-resistant A(H1N1)pdm09 strains with compensatory mutations were identified in different states of Brazil and in different countries [19]. Most of the influenza viruses tested during past seasons (2014, 2015) continued to be susceptible to the antiviral drugs recommended, although resistance to the adamantanes class of antiviral drugs among A/H3N2 and A/H1N1 viruses remains widespread [15]. In Argentina, in 2009, the first A(H1N1)pdm09 virus resistant to oseltamivir was detected in a 3-year-old child under treatment who had received an unrelated bone marrow transplantation and who had developed an upper respiratory tract infection with a long period of viral excretion. Between 2011 and 2015, in Argentina, using two rapid genotypic screenings, the H275Y substitution was found in 25 of 2216 influenza A(H1N1)pdm09 viruses tested and the E119V change in 1 of 1515 A(H3N2) viruses studied. Most of the viruses carrying the substitutions were collected from patients at risk without oseltamivir therapy. The resistant A(H1N1)pdm09 viruses showed the compensatory changes V241I and N369K. Oseltamivir appears to increase the chance of survival in people infected with H5N1 Asian lineage. These viruses are resistant to adamantanes and rarely to oseltamivir and zanamivir.

9 Prevention

Currently, the influenza vaccine, annually revised, includes three influenza virus strains: influenza A(H1N1)pdm09, A(H3N2), and influenza B. The vaccine includes seasonal strains and does not protect against the virus influenza C. According to the Centers for Disease Control and Prevention (CDC), influenza vaccine is the best way to protect people against influenza and prevent its spread. This vaccine can also reduce the severity of the disease if a person contracts a strain of the influenza that is not perfectly antigenically matched with the strain included in the vaccine formula.

About 2 weeks following vaccination are needed for antibodies to protect against the virus. Influenza vaccine will not protect against infections and diseases caused by other viruses that can also cause symptoms similar to those of influenza. This vaccine is available in the Southern Hemisphere countries during the autumn months (March and April), before influenza season starts. It contains the H and N viral surface proteins that have been previously grown in eggs, and it is administered subcutaneously. There are two different types of inactivated influenza vaccines, trivalent and quadruple. Trivalent vaccines protect against two influenza A viruses (H1N1 and H3N2) and an influenza B virus. Quadruple vaccines protect against two influenza A viruses and two influenza B viruses (Yamagata and Victoria lineages).

Influenza vaccine has to be applied every season because the body's immune response from vaccination declines over time, so an annual vaccine is needed for optimal protection and also because influenza viruses are constantly changing. The vaccine formula is reviewed each year and sometimes updated to keep up with changing influenza viruses. The selection of the strains included in the vaccine is based on the analysis of the strains circulating during the previous winter season in the countries of the Northern Hemisphere and the Southern Hemisphere carried out by an expert group coordinated by WHO.

Until 1998, for many years, WHO had held a consultation meeting once a year, in mid-February, to formulate a recommendation for the composition of inactivated influenza vaccines intended for the following winter. Epidemics of influenza occur at different times of the year in different parts of the world. Consequently, it was appropriate for WHO to review the recommendation twice a year. Since October 1998, a recommendation every February continues, which relates to the composition of vaccines intended for use for the following winter in the Northern Hemisphere (November to April), and a second recommendation made each September was implemented that relates to vaccines which will be used for the following winter in the Southern Hemisphere (May to October) [1]. To determine if the locally circulating strains are antigenically closely related to the vaccine strains administered, Argentinean strains were compared with the reference influenza viruses isolated from May 1994 to December 1997. Nasopharyngeal aspirates and nasopharyngeal swabs collected from hospitalized children and adults, respectively, with acute lower respiratory tract infection were tested. In this study, it has been shown that influenza A(H3N2) circulating in Argentina during the past 4 years matched partially with the antigens present in the vaccines administered during the 1994–1997 period. These antigenic variants sometimes circulated late in the year (October 1994 and 1997), initiating during the following influenza season and becoming prevalent: they were present in the vaccine formula administered in the Southern Hemisphere 2 years later [28]. Influenza vaccines can be administered from 6 months of age and also to people considered to belong to at-risk groups, such as pregnant women in the second or third trimester and people suffering from chronic respiratory diseases; or those with cardiac, metabolic, and immune compromise; or those who are morbidly obese. Healthcare workers who provide community services should consider vaccination to minimize disruption of essential activities during influenza outbreaks.

In addition to the trivalent and quadruple inactivated vaccine, other vaccines are currently available. In 2003, a live attenuated influenza virus vaccine was approved for use in the U.S. This vaccine is approved only for healthy individuals from 5 to 49 years old. In 2007, it was also approved for healthy children aged 24 to 59 months. It is administered nasally and provides mucosal, humoral, and cell-mediated immunity. The strain included, which is cold adapted, can grow in the upper respiratory tract where the temperature is lower than in the lower tract. It is attenuated by multiple changes in different segments of the genome. Because it is a live virus vaccine, when administered intranasally (as an aerosol) it generates a response in IgA and IgM/IgG.

Because the virus is grown in eggs, vaccine application is contraindicated in persons with hypersensitivity to any component thereof.

References

1. Anonymous (1998) Recommendation for the composition of influenza virus vaccines for use in 1999. Wkly Epidemiol Rec 73:305–308
2. Baumeister ELG, Pontoriero A et al (2004) Serological evidences of influenza A virus infection in Antarctica migratory birds. In: Options for the control of influenza. International congress series V. Elsevier, San Diego, pp 737–740
3. Baumeister E, Palacios G, Cisterna D et al (2010) Molecular characterization of severe and mild cases of influenza A (H1N1) 2009 strain from Argentina. Medicina (B Aires) 70:518–523
4. Baumeister EG, Salvador R, Vigo G et al (2005) Molecular evidences of the influenza A virus presence in migratory Antarctic birds. In: VII Congreso Argentino de Virología Libro de resúmenes. Sociedad Argentina de Virología, Ciudad de Buenos Aires
5. Baumeister EG, Perfumo C et al (2007) Influenza subtype H3 circulation in fattening pigs of Argentina. In: Proceedings, options for the control of influenza VI. International Medical Press 2008., London, p 333. ISBN 978-1-901-769-15-6 https://isirv.org/site/images/stories/publication/isirv_Options_VI_Proceedings_Book.pdf
6. Benedetti E, Daniels RS, Pontoriero A et al (2016) Influenza virus surveillance in Argentina during the 2012 season: antigenic characterization, genetic analysis and antiviral susceptibility. Epidemiol Infect 144:751–767
7. Born PS, Siqueira MM, Faria NR et al (2016) Phylodynamics of influenza A(H3N2) in South America, 1999–2012. Infect Genet Evol 43:312–320
8. Cappuccio JA, Pena L, Dibarbora M et al (2011) Outbreak of swine influenza in Argentina reveals a non-contemporary human H3N2 virus highly transmissible among pigs. J Gen Virol 92:2871–2878
9. CDC (2014) How the flu virus can change: "drift" and "shift". http://www.cdc.gov/flu/about/viruses/change.htm
10. Comas V, Moratorio G, Sonora M et al (2015) Phylogenetic analysis of the neuraminidase gene of pandemic H1N1 influenza A virus circulating in the South American region. Virus Res 197:1–7
11. Escalera-Zamudio M, Nelson MI, Cobian Guemes AG et al (2014) Molecular epidemiology of influenza A/H3N2 viruses circulating in Mexico from 2003 to 2012. PLoS One 9:e102453
12. Espinola EE, Amarilla AA, Martinez M et al (2014) Influenza A H1N1pdm 2009 virus in Paraguay: nucleotide point mutations in hemagglutinin and neuraminidase genes are not associated with drug resistance. Open Virol J 8:9–13
13. Garcia J, Sovero M, Torres AL et al (2009) Antiviral resistance in influenza viruses circulating in Central and South America based on the detection of established genetic markers. Influenza Other Respir Viruses 3:69–74
14. Guan Y, Vijaykrishna D, Bahl J et al (2010) The emergence of pandemic influenza viruses. Protein Cell 1:9–13
15. Hurt AC, Besselaar TG, Daniels RS et al (2016) Global update on the susceptibility of human influenza viruses to neuraminidase inhibitors, 2014–2015. Antivir Res 132:178–185
16. Hurtado R, Fabrizio T, Vanstreels RE et al (2015) Molecular characterization of subtype H11N9 avian influenza virus isolated from shorebirds in Brazil. PLoS One 10:e0145627
17. King Amq AM, Carstens EB, Lefkowitz EJ (2012) Virus taxonomy. Ninth report of the International Committee on taxonomy of viruses. Elsevier Academic Press, San Diego. www.trevorwilliams.info/ictv_iridoviridae_2012.pdf
18. Libster R, Bugna J, Coviello S et al (2010) Pediatric hospitalizations associated with 2009 pandemic influenza A (H1N1) in Argentina. N Engl J Med 362:45–55
19. Lopes E, Souza TM, Fintelman-Rodrigues N, Resende PC et al (2015) Oseltamivir-resistant influenza A(H1N1)pdm2009 strains found in Brazil are endowed with permissive mutations, which compensate the loss of fitness imposed by antiviral resistance. Mem Inst Oswaldo Cruz 110:101–105

20. Mathieu C, Moreno V, Pedersen J et al (2015) Avian influenza in wild birds from Chile, 2007–2009. Virus Res 199:42–45
21. Mckimm-Breschkin JL, Fry AM (2016) Meeting report: 4th ISIRV antiviral group conference: novel antiviral therapies for influenza and other respiratory viruses. Antivir Res 129:21–38
22. Muller I, Pinto E, Santibanez MC et al (2009) Isolation and characterization of the equine influenza virus causing the 2006 outbreak in Chile. Vet Microbiol 137:172–177
23. Nelson MI, Pollett S, Ghersi B et al (2016) The genetic diversity of influenza A viruses in wild birds in Peru. PLoS One 11:e0146059
24. Neumann G, Kawaoka Y (2011) The first influenza pandemic of the new millennium. Influenza Other Respir Viruses 5:157–166
25. Pereda A, Cappuccio J, Quiroga MA et al (2010) Pandemic (H1N1) 2009 outbreak on pig farm, Argentina. Emerg Infect Dis 16:304–307
26. Periolo N, Avaro M, Czech A et al (2015) Pregnant women infected with pandemic influenza A(H1N1)pdm09 virus showed differential immune response correlated with disease severity. J Clin Virol 64:52–58
27. Pineyro PE, Baumeister E, Cappuccio JA et al (2010) Seroprevalence of the swine influenza virus in fattening pigs in Argentina in the 2002 season: evaluation by hemagglutination-inhibition and ELISA tests. Rev Argent Microbiol 42:98–101
28. Savy VL, Baumeister EG, Pontoriero AV (1999) Antigenic relationship between influenza A (H3N2) strains circulating in Argentina and vaccine strains. Medicina (B Aires) 59:225–230
29. Senne DA, Suarez DL, Stallnecht DE et al (2006) Ecology and epidemiology of avian influenza in North and South America. Dev Biol (Basel) 124:37–44
30. Shinya K, Ebina M, Yamada S et al (2006) Avian flu: influenza virus receptors in the human airway. Nature (Lond) 440:435–436
31. Tong S, Li Y, Rivailler P et al (2012) A distinct lineage of influenza A virus from bats. Proc Natl Acad Sci U S A 109:4269–4274
32. Tong S, Zhu X, Li Y et al (2013) New world bats harbor diverse influenza A viruses. PLoS Pathog 9:e1003657
33. Webster RG, Bean WJ, Gorman OT et al (1992) Evolution and ecology of influenza A viruses. Microbiol Rev 56:152–179
34. Webster RG, Govorkova EA (2014) Continuing challenges in influenza. Ann N Y Acad Sci 1323:115–139
35. Xu K, Ferreri L, Rimondi A et al (2012) Isolation and characterization of an H9N2 influenza virus isolated in Argentina. Virus Res 168:41–47

Chapter 14
Adenoviruses: Biology and Epidemiology

Adriana E. Kajon and Ramón A. Gonzalez

1 Introduction

A few studies have been conducted by Latin American virologists on the role of adenoviral oncogenes in the biology of the virus [1], on the use of adenoviruses (AdV) as vectors [2], or on oncolytic AdV [3], and also in areas such as epidemiology of viral gastroenteritis and conjunctivitis. Without a doubt, the largest body of work representing AdV research in Latin American countries has focused on clinical and epidemiological studies of acute respiratory infection/acute respiratory disease (ARI/ARD) or influenza-like illness (ILI), and on the description of viral strains associated with severe disease, thus supporting the better understanding of the contribution of AdV infections to the global burden of respiratory disease. ARIs are still the leading cause of pediatric morbidity and mortality worldwide, with a significantly higher impact in low-income countries where they are associated with a high proportion of all pediatric deaths [4].

2 Adenovirus Structure, Genome Organization, and Replication

AdVs are non-enveloped icosahedral particles with a linear dsDNA genome of approximately 36 kb. The capsid is composed of 240 capsomers of hexon trimers, 12 pentameric penton capsomers, and 12 trimers of the fiber protein that project

A.E. Kajon (✉)
Lovelace Respiratory Research Institute, Albuquerque, NM, USA
e-mail: akajon@lrri.org

R.A. Gonzalez
Universidad Autónoma del Estado de Morelos, Cuernavaca, Mexico
e-mail: rgonzalez@uaem.mx

J.E. Ludert et al. (eds.), *Human Virology in Latin America*,
DOI 10.1007/978-3-319-54567-7_14

from the pentons. The AdV genome is organized in early (E1A, E1B, E2, E3, E4), intermediate (IX, IVa2), and late (major late, ML) transcription units. As with other DNA viruses that replicate in the nucleus of the host cell, the AdV replication cycle is divided into an early and a late phase, separated by the initiation of viral DNA synthesis. The early viral genes encode proteins whose functions establish conditions in the infected cell that are favorable for the synthesis of large quantities of viral macromolecules, whereas the late genes encode the viral structural proteins and other proteins that participate in the selective expression of viral genes, resulting in the production of progeny virions [5].

3 Human Adenoviruses: Classification and Association with Disease

Adenoviruses are classified in the family *Adenoviridae* and infect all major classes of vertebrates, from fish to mammals. The genus *Mastadenovirus* includes all human adenoviruses (HAdVs), which are classified into seven species designated human mastadenovirus A through G (HAdV-A–HAdV-G), based on biophysical, biochemical, and genetic criteria. Fifty-one serotypes and more than 70 human adenovirus genotypes have been recognized to date [5, 6]. Complete genomic sequences for these HAdVs are available in GenBank.

HAdVs circulate worldwide with no clear seasonality. They are transmitted through the oral–fecal route, by aerosols, and via fomites. Depending on the tropism of the virus, infection can result in respiratory disease (species HAdV-B, HAdV-C, HAdV-E, and occasionally HAdV-D), gastroenteritis (species HAdV-F and HAdv-G), or conjunctivitis (species HAdV-B, HAdV-D, and HAdV-E), accounting for an estimated 8% of all clinically relevant viral disease. It is estimated that worldwide AdV infections are responsible for 5–10% and 1–7% of all respiratory infections in children and adults, respectively. Common symptoms are fever, pharyngitis, tonsillitis, cough, and sore throat. Although uncommon in healthy adults, lower respiratory tract infections leading to pneumonia occur in up to 20% of newborns and infants. Severe infections in immunocompromised adults and children have been reported to result in fatalities exceeding 50% of cases [7, 8].

4 Adenovirus Acute Respiratory Infections in Latin America: Epidemiology and Impact

Many of the studies yielding information on the incidence and molecular epidemiology of HAdV respiratory infections in the past two decades resulted from sentinel surveillance efforts originally designed to determine the contribution of respiratory viruses to the burden of ARD, and more specifically ILI, and were funded by the local Ministries of Health, by the local research councils, and in

Fig. 14.1 Data available for Latin American countries regarding incidence and molecular epidemiology of adenovirus acute respiratory infections

several cases by the U.S. Department of Defense Global Emerging Infections Surveillance (GEIS) program. Although the majority of the publications resulting from these studies have a focus on epidemiology, relatively few of them include virus typing data. Unfortunately a considerable proportion of the work conducted in Latin America including data on HAdV respiratory infections is published in Spanish or Portuguese in journals with limited distribution. In the following sections, we present an overview of the activities and findings by the leading research groups operating in different countries clustered by region.

Figure 14.1 provides a summary of the type of information available for individual countries.

4.1 The Southern Cone

Building on a strong foundation of clinical reports and documenting the association of HAdV infection with pediatric severe ARD requiring hospitalization and resulting in pulmonary sequelae or death [9–17], numerous studies conducted since the 1980s by investigators in Argentina, Chile, and Uruguay have described the burden of ARD associated with HAdV infection during the past three decades and have identified the most prevalent types and genomic variants circulating in the region. The analysis of typing data reported by various studies suggests that the HAdV strains associated with pediatric ARD may be common to the region.

4.1.1 Argentina

The group of investigators led by Mercedes Weissenbacher at the Department of Microbiology, School of Medicine, University of Buenos Aires (UBA), conducted the very first study on clinical, etiological, and epidemiological features of acute lower respiratory tract infection (ALRI)) in children in Argentina. In 1990, the group published data gathered during a 40-month period from a cohort of 1003 children under 5 years of age (805 inpatients and 198 outpatients) [18, 19]. Viral etiology investigation performed on nasopharyngeal aspirates (NPAs) by indirect immunofluorescence (IF) and isolation in cell culture showed HAdV to be the second most frequently detected virus after respiratory syncytial virus (RSV).

The identification of the most prevalent types and genomic variants circulating in Argentina was possible through various collaborative studies initiated by a member of this group, Adriana Kajon, while she was a graduate student at BIO-SIDUS. These studies demonstrated the high prevalence of infections by species HAdV-B among pediatric cases of ARI requiring hospitalization in the late 1980s and early 1990s and in particular by variants of type HAdV-B7. As a result of these efforts, a new genome type of HAdV-B7 designated 7h was first identified in association with many of the severe and fatal cases of respiratory disease identified in the cohort just described [20–24]. This genomic variant was later found to be an intertypic recombinant strain with a HAdV-B7-like hexon and a HAdV-B3-like fiber [25]. Genome type 7h was found in a high proportion of specimens obtained from children hospitalized for ARI during the 1990s [26, 27] and was also found to circulate in Chile, Uruguay, and Brazil in the same period, as described next. The genome of this virus was completely sequenced for the Argentine strain 87–922 in 2011 (GenBank accession number JN860676). The bioinformatics analysis of this genome resulted in the assignment of a new type designation, HAdV-B66 [28].

The group of virologists working at the Department of Virology, Instituto Nacional de Enfermedades Infecciosas, ANLIS, "Carlos G. Malbrán," and led by Vilma Savy and Elsa Baumeister also contributed to the better understanding of the role of certain adenovirus infections in the etiology of pediatric ARD requiring hospitalization with the isolation and identification of various genomic variants of HAdV-B7 [29, 30] and as collaborators to some of the studies already mentioned.

The research group based at Centro de Educación Médica e Investigaciones Clínicas (CEMIC) Norberto Quirno and led by Guadalupe Carballal and Marcela Echavarria has conducted studies that have investigated the viral etiology of ARI in children in Buenos Aires since 1990. In 2001, the group reported data from the first multicenter study of viral etiology in pediatric cases of ALRI in Argentina [31]. The study enrolled 1278 children under 5 years of age, hospitalized in primary care centers in Buenos Aires, Córdoba, Santa Fé, and Mar del Plata cities, during a 2-year period (1993–1994). Cases of HAdV infection were identified in the four cities and represented 2.5% of all virus-positive cases. Work from this group using a polymerase chain reaction (PCR)-based approach to improve the sensitivity in the diagnosis of HAdV infections demonstrated a high incidence of HAdV coinfections and the presence of species HAdV-F in respiratory specimens of patients with ILI symptoms [32]. In collaboration with investigators at Facultad de Ciencias Exactas y Naturales, UBA, the CEMIC group developed a real-time PCR-based protocol to evaluate the susceptibility of HAdV to antiviral drugs [33]. The most recent study conducted by the CEMIC group enrolled 620 patients to investigate the viral etiology of ARI in hospitalized and outpatient children between 2008 and 2010 in Buenos Aires. HAdV infections were detected in 15 patients (2.4%). HAdV infections were significantly more frequent in hospitalized patients. In the emergency room, the incidence of HAdV was found to be 73 per 1000 children-years [34, 35].

Collaborative efforts between Adriana Kajon and Marcela Echavarria to type HAdV strains isolated from pediatric cases of ARI identified two novel intertypic recombinants of species HAdV-B [36, 37].

In 2004, Mariana Viegas and colleagues from the group led by Alicia Mistchenko at the Virology Laboratory, Hospital de Niños Ricardo Gutierrez reported data from a 5-year retrospective study evaluating the seasonality of respiratory viral infections in children under 5 years of age in Buenos Aires [38]. In the examined collection of 6083 virus-positive specimens, HAdV was found to be present in 473 (7.8%) and to circulate throughout the year.

The most comprehensive study reporting molecular typing data for Argentina in the past decade was conducted by Paola Barrero and colleagues from this group [39]. This study provided a thorough description of HAdV types detected in association with pediatric ALRI in Buenos Aires between 1999 and 2010. From a total of 743 HAdV-positive specimens, 654 (88%) represented cases of single infections and 89 (12%) were coinfections. From the 654 single HAdV infections, species HAdV-B was detected in 492 cases (75.23%), species HAdV-C in 138 cases (21.1%), species HAdV-E in 19 cases (2.91%), and species HAdV-D in 5 cases (0.76%). HAdV-B7 and HAdV-B3 were the most prevalent types ($n = 308$, 36.54%; $n = 230$, 27.28%, respectively), and HAdV-C1, HAdV-C2, HAdV-E4, HAdV-C5, HAdV-C6, HAdV-D8, HAdV-B11, HAdV-B14, and HAdV-B21 were also detected.

The group at Hospital de Niños Ricardo Gutierrez made other important contributions to the understanding of viral pathogenesis, including a seroepidemiology study [40], the characterization of cytokine responses in cases of severe adenovirus-associated disease [41], the description of a new genomic variant associated with severe disease [42], and the analysis of host cell gene expression during HAdV-B7h infection [43].

4.1.2 Chile

As in Argentina, the earliest contributions to the knowledge about the etiology of lower respiratory tract infections in infants in Chile date back to the late 1980s with a focus on patients admitted to the hospital [44, 45].

Manuela Vicente Suarez at the Instituto de Salud Pública de Chile (ISP) supported many of the early efforts of rapid viral diagnosis of adenovirus infections and of virus isolation in the country [46] that establish HAdV infections as the second most common cause of viral ALRI requiring hospitalization in Chile. A collection of 69 HAdV strains isolated by Manuela Vicente Suarez's laboratory at ISP from pharyngeal secretions of young children hospitalized with severe ALRI in Santiago, Chile, between 1984 and 1986 provided the first data on the types and genomic variants circulating in the country [20].

Surveillance of HAdV infection in children under 2 years of age hospitalized for ALRI in Chile was embraced by the group led by Luis Fidel Avendaño and Carmen Larrañaga from Instituto de Ciencias Biomédicas, Facultad de Medicina, Universidad de Chile, Santiago, in the late 1980s. Work from this group on a cohort of 3097 cases enrolled between 1988 and 1996 showed the HAdV isolation rate to be 12.6% and the most common admission diagnoses among HAdV-positive cases to be pneumonia and wheezing bronchitis (69.8%). The evaluation of the genetic variability of HAdV strains associated with infantile cases of ARD requiring hospitalization was conducted by restriction enzyme analysis (REA) in collaboration with Adriana Kajon [22, 23, 47, 48]. Genome typing of 221 HAdV isolates identified 87 as species HAdV-C and 134 as species HAdV-B. Genomic variant 7h was identified in 123 cases. Children infected with HAdV-B7h had longer hospital stays ($p < 0.01$), a higher frequency of rectal temperatures above 39°C ($p < 0.01$), and greater need for additional oxygen ($p < 0.02$) than those children infected with species HAdV-C. A paper published by this group in 2000 contributed to the better understanding of the clinical characteristics of respiratory infections by HAdV-B7h among Chilean children [49]. Using universal, species-specific, and 7h-specific PCR assays, Larrañaga and colleagues further established the contribution of HAdV-B7h infections to the etiology of severe ALRI requiring hospitalization [50].

Marisol Bruzzone from Eugenio Spencer's group at Departamento de Ciencias Biológicas, Facultad de Química y Biología, Universidad de Santiago de Chile developed a PCR-based assay for the specific detection of HAdV-B and described the genetic variability of Chilean isolates of genome type 7h [51, 52].

The laboratory of Marcela Ferrés at Pontificia Universidad Católica de Chile (PUCC) contributed to the clinical and epidemiological characterization of respiratory virus infections among adults hospitalized during the 2004 influenza season [53]. Viral etiology was determined in 87 cases of adult ILI hospitalized at the PUCC hospital between May and July 2004. HAdV infection was diagnosed in 1.2% of these cases by direct immunofluorescence assay (DFA).

4.1.3 Uruguay

Data on the incidence and impact of HAdV respiratory infections for Uruguay are significantly scarcer. The group led by Maria Hortal at Departamento de Laboratorios del Ministerio de Salud Pública in Montevideo had a leading role in the design and execution of the first studies conducted in the country to identify viral agents associated with ARI in children less than 5 years old [54–56]. Some of the HAdV isolates collected through these efforts were characterized by Kajon and colleagues to describe regional HAdV activity between 1991 and 1994 [22, 23, 57]. A small collection of 23 isolates from NPAs of children hospitalized for ARI in Uruguay between 1994 and 1998 was studied by the laboratory of Juan Ramon Arbiza at Facultad de Ciencias, Universidad de la República in Montevideo [58]. Fourteen of the examined isolates belonged to species HAdV-C and 9 to the species HAdV-B. Eight isolates were identified as corresponding to genome type 7h.

4.1.4 Brazil

A sentinel surveillance system for influenza and other respiratory viral infections was implemented by the Brazilian Ministry of Health in 2000 and now constitutes a network of outpatient clinics and emergency care departments in 26 of the 27 states. Respiratory specimens are tested in public health laboratories in each state for influenza, parainfluenza, RSV, and HAdVs. All positive and inconclusive samples, and 10% of the negative samples, are forwarded to one of the three reference laboratories for respiratory viruses [Instituto Evandro Chagas (IEC) in Belem, Instituto Adolfo Lutz in Sao Paulo, and Fundaçao Oswaldo Cruz (FIOCRUZ) in Rio de Janeiro]. Data gathered by the Ministry of Health between 2000 and 2010 showed that of 6421 samples with positive results, 742 were positive for HAdV (12%) [59].

Several studies conducted at FIOCRUZ in the 1980s and 1990s provided instrumental foundational data and developed technical resources for determining viral etiology in pediatric ARD in Brazil. In 1983 Sutmoller and colleagues identified HAdV in 47% of the virus-positive specimens obtained from the child population of a shanty town in Rio de Janeiro [60]. In 1991 Jussara Pereira do Nascimento reported data from the investigation of viral etiology in children less than 5 years old from 1982 to 1985 [61]. In her study, HAdV was found as the second most frequent virus isolated from nasopharyngeal secretions of symptomatic patients and serotypes 1, 2, and 7 as the most predominant. Nascimento and colleagues also described genomic variants identified among local strains of HAdV-B3 and HAdV-B7 [62, 63].

In collaboration with her colleagues at FIOCRUZ and IEC, Adriana Kajon identified genomic variants for a collection of HAdV strains of serotypes 1, 2, 3, 5, and 7 collected between 1976 and 1995 in Belem and Rio de Janeiro. [64].

In 2003, Maria de Albuquerque from Universidade Federal do Rio de Janeiro reported the first data for Brazil on the impact of ARI in the county's military population. The presence of adenovirus was detected by isolation in cell culture in 3 of 221 pharyngeal swabs obtained from military recruits in 2000 [65]. Only one of these 3 patients had symptoms of ARI.

In 2004 Selir Straliotto from Seção de Virologia do Laboratório Central de Saúde Pública in Porto Alegre, Rio Grande do Sul, in collaboration with Marilda Siqueira from FIOCRUZ, reported data for the South Region of Brazil [66]. In her study, the prevalence of viral respiratory infections was investigated among a cohort of 261 children less than 7 years of age admitted to the pediatric intensive care unit from June to December 1996 with pneumonia or bronchiolitis. HAdV infections were diagnosed by indirect IF of NPAs in 20 of these patients (7.7%). The case fatality rate of HAdV infection was 25% (5/20). Viruses were not typed in this study.

In collaboration with Marilda Siqueira's group at FIOCRUZ, Fernanda Moura and colleagues [67] characterized the HAdV strains isolated from 11 cases of infection diagnosed by indirect IF and cell culture between January and December 1998 at a pediatric clinic in Salvador, Northeast Brazil. HAdV infections represented 7.14% of all viral infections detected at the clinic that year. The majority (7 of 11) were upper respiratory tract infections (URI). Of these 7 cases, HAdV-C1 was identified by sero-neutralization in 4 cases, HAdV-C2 was identified in 2 cases, HAdV-C5 was identified in 1 case, and HAdV-B7 was identified in 4 cases. Later on, as the leader of her own group at Universidade Federal do Ceará, Moura conducted studies investigating the contribution of respiratory viral infections to childhood morbidity and mortality in the equatorial city of Fortaleza, Ceará. In a paper by Alonso and colleagues [68], this team of investigators reported data from the examination of cases of ARI seeking attention at a pediatric hospital between January 2001 and December 2008. In this cohort of patients, HAdV infection was detected in 5.5% of all virus-positive samples. In the recent publication by Pereira and colleagues [69], the group reported the results of a larger study conducted in Fortaleza between 2001 and 2013 and specifically describing the clinical and epidemiological profile of HAdV-associated pediatric ARI. HAdV infection was detected in 290 of the virus-positive samples and in only 15.86% of the children who required hospitalization. HAdV isolates were typed at Charlotte Harsi's laboratory at University of Sao Paulo (USP) by amplification and sequencing of the hexon hypervariable regions 1–6 following established protocols [70]. Species and type were identified in 189 strains: the predominant types were HAdV-B3 and HAdV-B7, present in 128 (67.72%) and 34 (18.00%) samples, respectively, followed by HAdV-C1 and HAdV-C,2 identified in 8 samples (4.24%) each. HAdV-E4 was detected in 6 (3.17%) samples.

Charlotte Harsi's group at USP characterized HAdV strains detected in NPAs of children hospitalized at USP's Pediatric Clinic in 1995 and 2000 [71]. Viruses were serotyped by neutralization assays and subsequently genome typed by REA. Nineteen isolates were characterized as HAdV-B, genome types 3a, 7h, and 7h1, and 11 as HAdV-C, genome types 1D10, 2D25, 5D2, and 6D3. Species HAdV-B showed epidemic infection patterns, with shifts in the predominant genome type. Genome types 7h, or the variant 7h1, dominated over genome type 7b among the 1995 isolates. In 2000, the variant 7h1 became predominant, and the emergence of type 3a was observed. The group also developed a PCR-based assay for the detection of HAdV in NPAS, and in collaboration with Dean Erdman and his team at the Centers for Disease Control and Prevention (CDC) [72] reported the identification of a novel genomic variant of HAdV-B7 with a unique deletion in the E3 region of the genome. The variant designated 7m was recovered from six epidemiologically unrelated cases of ARD with wheezing between September 1999 and March 2000 [73].

The USP group led by Edison Durigon reported the detection of HAdV in 6.8% of the children less than 5 years of age hospitalized for ALRI from January 1 to December 30, 2003 at the USP hospital [74].

In 2015 Giuliana Durigon and colleagues at USP reported the outcomes of pediatric ARD resulting from viral infection in a cohort of children under 2 years of age admitted to an urban tertiary hospital in Sao Paulo from March 2008 through February 2010 [75]. Using PCR-based assays, her study identified HAdV infections in 102 of 372 virus-positive specimens (27%). Descriptive analysis revealed a positive association between detection of HAdVs and the need of intensive care (20.6% HAdV positive vs. 13.1% HAdV negative; $\chi^2 = 4.048$; $p = 0.044$).

Sonia Raboni's group at Universidade Federal do Paraná in Curitiba conducted a systematic screening for respiratory viruses in pediatric outpatients from an emergency department in southern Brazil during two consecutive influenza seasons (2010 and 2011). A total of 492 children aged 24 to 59 months who presented with ARD symptoms and fever were enrolled in the study. HAdV infections were diagnosed in 18 cases in 2010 and in 19 cases in 2011 [76].

In 2013 the group led by Nancy Bellei at Departamento de Medicina, Universidade Federal de São Paulo, São Paulo compared the rates of detection of HAdV among immunocompetent and immunocompromised patients presenting with acute respiratory infection using two different methods, DFA and nested PCR [77]. HAdV was detected in 13.2% of the 643 patients tested by DFA or PCR: 6 of 139 (4.3%) adult emergency room patients, 7 of 205 (3.4%) healthcare workers, 4 of 69 (5.8%) renal transplant outpatients, and 68 of 230 (29.5%) hematopoietic stem cell transplant recipients. In 2012 Proenca-Modena and his colleagues from the School of Medicine of Ribeirao Preto of USP reported the results of their investigation of the role of respiratory viruses in chronic adenotonsillar diseases using TaqMan real-time PCR (qPCR) in nasopharyngeal secretions, tonsillar tissues, and peripheral blood from 121 children with chronic tonsillar diseases, without symptoms of ARD. HAdV was the most frequently detected virus, identified in 47.1% of the patients [78].

4.1.5 Paraguay

Available data for Paraguay are the result of a recent study conducted by Emilio E. Espínola and colleagues from Instituto de Investigaciones en Ciencias de la Salud, Universidad Nacional de Asunción. HAdV was detected in 9 of 50 (18%) NPAs obtained from children under 5 years of age with symptoms of severe ARI [79]. Viruses were typed by amplification and sequencing of a fragment of the hexon gene, allowing for the identification of HAdV-B16 in 1 specimen, HAdV-C1 in 2 specimens, HAdV-C2 in 1 specimen, HAdV-C5 in 2 specimens, HAdV-C6 in 2 specimens, and HAdV-D15 in 1 specimen.

4.2 *The Andean Region*

4.2.1 Peru

Available data on the incidence and molecular epidemiology of adenovirus infections in Peru became available as a result of two studies. The first one was led by Julia Ampuero from the U.S. Naval Medical Research Unit 6 (NAMRU-6) in Lima, and the other was conducted as a collaboration with Dirección Nacional de Salúd de Piura and Dirección General de Epidemiología, Ministry of Health, Lima. HAdV infections were detected in 2.5% of 26,375 participants visiting 38 health centers with ILI or severe ARI between 2000 and 2010 [80]. Typing by amplification and sequencing of a portion of the hexon gene [70] was carried out in 226 randomly selected HAdV isolates (41% of all) obtained from 2006 to 2010. Species HAdV-C was identified in the majority of the isolates (75.7%), followed by species HAdV-B (17.3%) and species HAdV-E (7.1%).

An earlier study by Josefina Garcia and colleagues from the U.S. Naval Medical Research Center Detachment in Lima [81] reported molecular typing data for 153 Peruvian strains isolated from respiratory specimens collected between 2006 and 2008. Species HAdV-B was identified in 32 specimens, species HAdV-C was identified in 114 specimens, and species HAdV-E was identified in 7 specimens.

The RESPIRA-Peru Project led by investigators at Vanderbilt University School of Medicine in Tennessee (USA) conducted active household-based surveillance of ARI among children under 3 years of age in the rural highland communities of San Marcos, Cajamarca, between May 2009 and September 2011 [82–84]. In this high-altitude rural setting with low population density, HAdV circulated year round with an incidence rate of 73 per 100 child-years.

4.2.2 Ecuador

NAMRU-6 established a collaborative network to determine the viral causes of ILI in Guayaquil and Quito. The study included 1702 participants, and 35% of oropharyngeal swabs were positive for CPE and IF for at least one virus. The most common viral agent was influenza A virus (21.6%). HAdV was isolated from 2.1% of specimens, and influenza B and parainfluenza were isolated from 6.4% and 2.1%, respectively. RSV and enterovirus were positive in only 0.9% of samples [85].

4.2.3 Colombia

The Colombian National Institute of Health initiated laboratory-based surveillance for respiratory viruses in 1997 to identify the viral agents associated with pediatric ARD throughout the country. In a study led by Diana Herrera-Rodriguez [86], 1743 children presenting with symptoms of ARD were examined from January 1997 to December 2003 at participating sentinel hospitals in Bogota and Manizales. A total of 610 patients (35%) were found to be positive for respiratory viruses. HAdV infection was diagnosed in 47 patients (2.6%). Most of the cases of HAdVs infection required hospitalization and oxygen support. The fatality/case ratio was 7%. A total of eight isolates were recovered and submitted to the CDC for molecular typing. Four isolates were identified as HAdV-B7, two as HAdV-C2, one as HAdV-C1, and one as HAdV-B3. In a similar study conducted by a group led by the same investigator [87] on a group of 138 children under 10 years of age seeking attention between March 2000 and September 2001 at the Central Military Hospital in Bogota, HAdV infection was diagnosed in only 4 patients under 3 years of age.

A large study was conducted by Arango and colleagues from Universidad de Antioquia and NAMRU-6 on 2039 participants reporting to one general hospital in

Medellin between 2007 and 2012 [88]. HAdV infections represented 14.9% of all viral infections and were more frequently detected among participants under the age of 4. HAdV infections were diagnosed exclusively by virus isolation and confirmed by DFA. HAdV isolates were not typed in any of these studies.

As part of an initiative to expand the knowledge of the molecular epidemiology of HAdV respiratory infections, a collaborative study between Universidad de Los Andes and Secretaria Distrital de Salud de Bogotá utilized a VA gene amplification and sequencing approach to identify strains detected in 48 NPAs collected between January and November 2008 from children under 5 years of age with ARD [89]. Species HAdV-B was identified in 41 respiratory specimens (85.42%), species HAdV-C was identified in 6 specimens (12.5%), and species HAdV-D was identified in 1 specimen (2.08%). HAdV-B3 was the best represented type in the sample ($n = 22$), followed by HAdV-B7 ($n = 19$).

A recent study conducted by investigators from the National University of Colombia in Bogota in collaboration with the District Health Department investigated the viral etiology of cases of severe ARI among adult patients admitted to seven sentinel surveillance institutions in Bogota in 2012 [90]. A PCR-based microarray diagnostic assay [91] was used to identify viruses present in NPAs or throat specimens. Viral identification was achieved for 69% of the 91 patients included in the analysis. HAdV infection was detected in 17 cases (18.7%), all identified during the rainy seasons (April to May and August to November). In 10 of these cases, HAdV was present as a coinfection with one or more respiratory viruses. Molecular typing was not carried out in this study.

4.2.4 Venezuela

Information about the epidemiology of ARI for Venezuela is still limited. A 4-year study (2006–2010) was conducted in two hospitals in Maracay as part of a collaborative sentinel surveillance of influenza-like illness effort between the Instituto de Investigaciones Biomedicas de la Universidad de Carabobo, the Hospital Jose Maria Carabaño Tosta, Hospital Nacional Central, and Dirección Nacional de Epidemiología, in Maracay, NAMRU-6, Lima, Peru, and the Naval Medical Research Center, Maryland, USA. HAdV was detected as the second most common viral pathogen (5.6%) after influenza A among symptomatic outpatients under 4 years of age using cell culture isolation and confirmation by DFA [92]. Virus isolates were not characterized in this study.

4.3 Central America and the Caribbean

4.3.1 Costa Rica

A study of 1017 NPAs from children hospitalized between 1997 and 1999 at the National Children Hospital in San José, Costa Rica, was carried out by Salas-Chavez and colleagues from the Instituto de Investigaciones en Salud (INISA),

Universidad de Costa Rica, and Hospital Nacional de Niños "Dr. Carlos Sáenz Herrera." IF was used to detect RSV, HAdV, influenza A and B, and parainfluenza 1, 2, and 3. HAdVs were detected in 9% of the cases, and RSV was present in 73% of the samples [93].

4.3.2 Guatemala

A survey for ARI in 3,964 hospitalized patients was conducted by the CDC International Emerging Infections Program, in collaboration with the Guatemala Ministry of Public Health and Welfare and the Universidad del Valle de Guatemala. Between November 2007 and December 2011, nasopharyngeal/oropharyngeal swabs were tested by PCR for the major respiratory viruses: *Chlamydia pneumoniae* and *Mycoplasma pneumoniae*. The study included all age groups. Viruses were detected in 71.8% of infants under 1 year old and in 52.6% of all cases. Using population denominators, the observed hospitalized ARI incidence was 128 cases per 100,000, with the highest rates seen among infants under 1 year old (1,703 per 100,000), followed by adults over 65 years old (292 per 100,000). HAdVs were found in 11.2% of all cases, second only to RSV (26.4%) [94].

4.3.3 El Salvador, Honduras, and Nicaragua

As part of a surveillance effort funded by the GEIS Program and designed to identify the viral agents associated with ILI in Nicaragua, Honduras, and El Salvador, Laguna-Torres from the U.S. Naval Medical Research Center Detachment in Lima and collaborators from the Ministries of Health of Nicaragua and El Salvador and the Instituto Hondureño de Seguridad Social [95] investigated the viral etiology in patients with symptoms of ILI, regardless of age and who sought attention or were hospitalized in participating health centers between August 2006 and April 2009. HAdV infection was diagnosed in 63 of the 1756 (3.6%) cases enrolled in the study. Among the 34 HAdVs isolates that were further processed for molecular typing, species HAdV-C clearly predominated (n = 31). The other 3 isolates belonged to species HAdV-B.

4.3.4 Cuba

The data available for Cuba are the result of studies carried out by investigators at Instituto de Medicina Tropical "Pedro Kourí" in Havana. In 2000 Tania Pumariega and colleagues reported the presence of HAdV in 8% of 128 nasopharyngeal swab specimens from infants and children under 5 years of age admitted with ARD in two pediatric hospitals in Havana between 1996 and 1997. Molecular typing by REA identified only species HAdV-C in these samples, with HAdV-1 and HAdV-6 each isolated from 4 samples and HAdV-2 from 1 [96]. In 2009 Belsy Acosta and colleagues in collaboration with the Centro Nacional de Microbiología-Instituto de Salud Carlos III in Spain published the results of a study designed to identify HAdV species and correlate clinical syndromes using nasopharyngeal swabs and pharyngeal washes collected from 512 patients with ARD between 2002 and 2006. DFA, isolation in cell culture, and PCR assays identified HAdV in 45 specimens (8%). The majority of the sampled patients (53%) were children under 5 years of age. Molecular typing conducted by amplification and sequencing of a portion of the hexon gene identified HAdV-D as the major species (59%), followed by HAdV-C (36%) and HAdV-B (4%). HAdV-C5 was the predominant type associated with bronchiolitis (71%), followed by HAdV-C6 (14%), and HAdV-D17 was the most common type detected in patients with URI (77%). All the respiratory specimens obtained from an atypical outbreak of acute febrile syndrome were positive for HAdV-D [97].

4.4 Mexico

The earlier studies designed to determine the contribution of HAdV to the etiology of ARI in Mexico were carried out by the Instituto Nacional de Enfermedades Respiratorias (INER). In 2003 Manjarrez and colleagues determined the prevalence of viruses in children with or without ARI. The study included 179 children with ARI and 179 with normal airway functions. Viruses were isolated from 49% of children with ARI and in 27% of children without ARI. The highest frequencies of respiratory viruses were found in infants under 1 year (56%) and in children from 2 to 3 years of age (54%). RSV was the most frequently isolated virus (38%), followed by HAdV (19%) [98]. Around the same period, a study conducted by the School of Medicine of the Universidad Autónoma de Nuevo León found viral infection in 30.7% of cases in 101 children hospitalized with ALRI by indirect IF. HAdV was detected in 6.5% of the cases [99]. In a different study by INER, nasopharyngeal exudates from 300 children were processed for viral isolation and IF. Viral infection was detected in 65% of the samples; 13% were positive for HAdV [100]. A later study in collaboration with Instituto Politécnico Nacional was designed to identify HAdV species in nasopharyngeal specimens of nonhospitalized children with ARI using PCR/restriction fragment length polymorphism (RFLP) assays. The study cohort included 100 children attending the same school. HAdV-C was the only species identified in a total of 23 samples [101]. A parallel effort by the Instituto de Diagnóstico y Referencia Epidemiológicos in collaboration with Instituto de Oftalmología FAP Conde de Valenciana in Mexico City performed IF and REA on 118 respiratory specimens from patients with ARI of all age groups in Mexico City and in the states of Nayarit, Michoacán, Hidalgo, and Tamaulipas. HAdVs were found in 22% of the samples. The most common type was HAdV-C5 (14%), followed by HAdV-C2 (5%) and HAdV-C1 (3.3%) [102].

As a result of the April 2009 outbreak of respiratory disease in Mexico City, the Ministry of Health of Mexico and the U.S. National Institute of Allergy and Infectious Diseases (NIAID) joined efforts to establish a Mexican Emerging Infectious Diseases Clinical Research Network (La Red). A large collaborative effort involving several institutions studied the etiology of ILI in Mexico City, enrolling 1065 patients from 2010 to 2011. HAdV was found in approximately 9% of the cases [103].

Through a collaboration between Hospital Infantil de México "Federico Gómez," Universidad Autónoma Metropolitana, Instituto Nacional de Enfermedades Respiratorias, Centro de Investigación y de Estudios Avanzados del Instituto Politécnico Nacional, Centro Médico Nacional Siglo XXI, and Instituto Mexicano del Seguro Social, the prevalence and genotype of HAdVs and coinfection with bocavirus were determined on a sample of 5185 nasopharyngeal swabs collected from 2005 to 2010 from both nonimmunosuppressed and immunosuppressed children diagnosed with pneumonia. Samples were processed by IF and PCR. A low incidence of HAdV (0.71%) was found. Typing was accomplished in only 12 of the 37 positive samples by amplification and sequencing of the hexon gene. HAdV-B3 was identified in 9 samples. HAdV-B55 (aka 11a) was identified in 1 sample, and HAdV-C2 and HAdV-C6 were identified in the remaining 2 samples, respectively [104].

In 2015 the group led by Carlos Arias from Instituto de Biotecnología, Universidad Nacional Autónoma de México, Cuernavaca, Morelos, and collaborators from Colegio de Pediatría del Estado de Veracruz [105] reported data from a study that evaluated the diversity of respiratory viruses present in nasal swabs of children seeking care in private pediatric practice settings in Veracruz using a multiplex PCR for 15 different viruses. The study included 525 children sampled from 2011 to 2012. HAdVs were the fourth most frequent virus detected (7.25%), after RSV (18.3%), rhinovirus (17.5%), and influenza A (9.1%).

5 Future Challenges

After more than three decades of surveillance efforts to describe the burden and etiology of ARD in Latin America, many issues remain to be addressed to gain a more panoramic understanding of the prevalence and impact of HAdV infections. Collaborative efforts using complementary expertise and resources should be encouraged. The use of standardized case definition criteria and diagnostic laboratory procedures will allow for the necessary comparison and integration of data sets. The wider implementation of PCR-based diagnostic methods should provide a higher sensitivity for the detection of HAdV in the context of disease and help fill the gap in knowledge regarding the identity of the most prevalent circulating types. Virus typing is critical to document evolving epidemiological patterns and for the detection of emerging HAdVs in the region. Larger surveillance networks are needed for more representative sampling in countries with climatic diversity that results from their great latitude coverage, such as Argentina and Chile.

References

1. Hidalgo P, Anzures L, Hernández-Mendoza A et al (2016) Morphological, biochemical, and functional study of viral replication compartments isolated from adenovirus-infected cells. J Virol 90(7):3411–3427
2. Mata-Espinosa DA, Mendoza-Rodríguez V, Aguilar-León D et al (2008) Therapeutic effect of recombinant adenovirus encoding interferon-gamma in a murine model of progressive pulmonary tuberculosis. Mol Ther 16(6):1065–1072
3. Weber HL, Gidekel M, Werbajh S et al (2015) A novel CDC25B promoter-based oncolytic adenovirus inhibited growth of orthotopic human pancreatic tumors in different preclinical models. Clin Cancer Res 21(7):1665–1674
4. Tregoning JS, Schwarze J (2010) Respiratory viral infections in infants: causes, clinical symptoms, virology, and immunology. Clin Microbiol Rev 23(1):74–98
5. Berk AJ (2013) Adenoviridae. In: Fields B, Howley P, Knipe D (eds) Fields virology, 6th edn. Lippincot Williams & Wilkins, Philadelphia, pp 1704–1731
6. Harrach B, Benkö M, Both GW et al (2011) Family *Adenoviridae*. In: King AMQ, Adams MJ, Carstens EB, Lefkowitz EJ (eds) Virus taxonomy: classification and nomenclature of viruses. Ninth report of the international committee on taxonomy of viruses. Elsevier, San Diego, pp 125–141
7. Lynch JP 3rd, Kajon AE (2016) Adenovirus: epidemiology, global spread of novel serotypes, and advances in treatment and prevention. Semin Respir Crit Care Med 37:586–602
8. Wold WSM, Ison MG (2013) Adenoviruses. In: Fields B, Howley P, Knipe D (eds) Fields virology, 6th edn. Lippincot Williams & Wilkins, Philadelphia, pp 1732–1767
9. Almarza MI, Wu E, Vicente M et al (1990) Hospital acute respiratory infections. Rev Chil Pediatr 61(4):185–188
10. Castro-Rodriguez JA, Daszenies C, Garcia M et al (2006) Adenovirus pneumonia in infants and factors for developing bronchiolitis obliterans: a 5-year follow-up. Pediatr Pulmonol 41(10):947–953
11. Colom AJ, Maffey A, Garcia Bournissen F et al (2015) Pulmonary function of a paediatric cohort of patients with postinfectious bronchiolitis obliterans. A long term follow-up. Thorax 70(2):169–174
12. Larrañaga C, Vicente M, Wu E et al (1988) Adenovirus in children with acute lower respiratory tract infections. Rev Chil Pediatr 59(5):312–317
13. Murtagh P, Kajon A (1997) Chronic pulmonary sequelae of adenovirus infection. Pediatr Pulmonol Suppl 16:150–151
14. Murtagh P, Giubergia V, Viale D et al (2009) Lower respiratory infections by adenovirus in children. Clinical features and risk factors for bronchiolitis obliterans and mortality. Pediatr Pulmonol 44(5):450–456
15. Siminovich M, Murtagh P (2011) Acute lower respiratory tract infections by adenovirus in children: histopathologic findings in 18 fatal cases. Pediatr Dev Pathol 14(3):214–217
16. Teper AM, Kofman CD, Maffey AF et al (1999) Lung function in infants with chronic pulmonary disease after severe adenoviral illness. J Pediatr 134(6):730–733
17. Wu E, Martínez V, Alvarez AM et al (1990) Fatal cases of adenovirus infection. Rev Chil Pediatr 61(4):177–184
18. Avila M, Salomón H, Carballal G et al (1990) Isolation and identification of viral agents in Argentinian children with acute lower respiratory tract infection. Rev Infect Dis Suppl 8:S974–S981
19. Weissenbacher M, Carballal G, Avila M et al (1990) Etiologic and clinical evaluation of acute lower respiratory tract infections in young Argentinian children: an overview. Rev Infect Dis 12(suppl 8):S889–S898
20. Kajon AE, Suarez MV (1990) Molecular epidemiology of adenoviruses isolated from hospitalized children with severe lower acute respiratory infection in Santiago, Chile. J Med Virol 30(4):294–297

21. Kajon AE, Wadell G (1992) Molecular epidemiology of adenoviruses associated with acute lower respiratory disease of children in Buenos Aires, Argentina (1984–1988). J Med Virol 36(4):292–297
22. Kajon AE, Suárez MV, Avendaño LF et al (1993) Genome type analysis of South American adenoviruses of subgenus C collected over a 7-year period. Arch Virol 132(1–2):29–35
23. Kajon AE, Wadell G (1994) Genome analysis of South American adenovirus strains of serotype 7 collected over a 7-year period. J Clin Microbiol 32(9):2321–2323
24. Murtagh P, Cerqueiro C, Halac A et al (1993) Adenovirus type 7h respiratory infections: a report of 29 cases of acute lower respiratory disease. Acta Paediatr 82(6-7):557–561
25. Kajon AE, Wadell G (1996) Sequence analysis of the E3 region and fiber gene of human adenovirus genome type 7 h. Virology 215(2):190–196
26. Videla C, Carballal G, Misirlian A, Aguilar M (1998) Acute lower respiratory infections due to respiratory syncytial virus and adenovirus among hospitalized children from Argentina. Clin Diagn Virol 10:17–23
27. Videla C, Carballal G, Kajon A (1999) Genomic analysis of adenovirus isolated from Argentinian children with acute lower respiratory infections. J Clin Virol 14(1):67–71
28. Dehghan S, Liu EB, Seto J et al (2012) Five genome sequences of subspecies B1 human adenoviruses associated with acute respiratory disease. J Virol 86(1):635–636
29. Baumeister EG, Fernández Cobo M, Savy V (1993) Detection of a new genomic variation of adenovirus 7 associated to severe infections of the lower respiratory tract. Medicina (B Aires) 53(3):193–196
30. Baumeister E, Fernández Cobo M, Pontoriero A, Savy V (1997) Characterization of isolated adenovirus associated with acute lower respiratory infection in pediatrics. Enferm Infecc Microbiol Clin 15(10):528–531
31. Carballal G, Videla CM, Espinosa MA et al (2001) Multicentered study of viral acute lower respiratory infections in children from four cities of Argentina, 1993–1994. J Med Virol 64(2):167–174
32. Echavarria M, Maldonado D, Elbert G et al (2006) Use of PCR to demonstrate presence of adenovirus species B, C, or F as well as coinfection with two adenovirus species in children with flu-like symptoms. J Clin Microbiol 44(2):625–627
33. Gainotti R, Ricarte C, Ebekian B et al (2010) Real time PCR for rapid determination of susceptibility of adenovirus to antiviral drugs. J Virol Methods 164(1-2):30–34
34. Marcone DN, Ellis A, Videla C et al (2013) Viral etiology of acute respiratory infections in hospitalized and outpatient children in Buenos Aires, Argentina. Padiatr Infect Dis J 32(3):e105–e110
35. Marcone DN, Durand LO, Azziz-Baumgartner E et al (2015) Incidence of viral respiratory infections in a prospective cohort of outpatient and hospitalized children aged ≤5 years and its associated cost in Buenos Aires, Argentina. BMC Infect Dis 15:447
36. Kajon AE, Dickson LM, Murtagh P et al (2010) Molecular characterization of an adenovirus 3-16 intertypic recombinant isolated in Argentina from an infant hospitalized with acute respiratory infection. J Clin Microbiol 48(4):1494–1496
37. Kajon AE, de Jong JC, Dickson LM et al (2013) Molecular and serological characterization of species B2 adenovirus strains isolated from children hospitalized with acute respiratory disease in Buenos Aires, Argentina. J Clin Virol 58(1):4–10
38. Viegas M, Barrero PR, Maffey AF et al (2004) Respiratory viruses seasonality in children under five years of age in Buenos Aires, Argentina: a five-year analysis. J Infect 49(3):222–228
39. Barrero PR, Valinotto LE, Tittarelli E et al (2012) Molecular typing of adenoviruses in pediatric respiratory infections in Buenos Aires, Argentina (1999–2010). J Clin Virol 53(2):145–150
40. Mistchenko AS, Biscotti E, Grinstein S (1988) Seroepidemiology of adenovirus in normal subjects from Buenos Aires (Argentina). Rev Epidemiol Sante Publique 36(1):68–69
41. Mistchenko AS, Diez RA, Mariani AL et al (1994) Cytokines in adenoviral disease in children: association of interleukin-6, interleukin-8, and tumor necrosis factor alpha levels with clinical outcome. J Pediatr 124(5):714–720

42. Mistchenko AS, Robaldo JF, Rosman FC et al (1998) Fatal adenovirus infection associated with new genome type. J Med Virol 54(3):233–236
43. Barrero PR, Yankilevich P, Diez RA, Mistchenko AS (2006) Bioarray analysis of differential cellular gene expression after adenovirus 7 h infection. Rev Argent Microbiol 38(1):32
44. Ceruti E, Díaz A, Vicente M et al (1991) Etiology of lower respiratory infections in hospitalized infants. Rev Chil Pediatr 62(3):155–166
45. Escobar AM, Martínez F, Ceruti E (1988) Etiology of acute lower respiratory tract infections in hospitalized infants: virologic studies. Rev Chil Pediatr 59(6):349–353
46. Torres G, Vicente M (1992) Detection of respiratory syncytial and adenovirus in nasopharyngeal aspirates: comparison of cellular cultures and immunofluorescence. Rev Med Chil 120(4):415–419
47. Kajon AE, Larrañaga C, Suárez MV et al (1994) Genome type analysis of Chilean adenovirus strains isolated in a children's hospital between 1988 and 1990. J Med Virol 42(1):16–21
48. Larrañaga C, Kajon A, Villagra E, Avendaño LF (2000) Adenovirus surveillance on children hospitalized for acute lower respiratory infections in Chile (1988–1996). J Med Virol 60(3):342–346
49. Palomino MA, Larrañaga C, Avendaño LF (2000) Hospital-acquired adenovirus 7h infantile respiratory infection in Chile. Pediatr Infect Dis J 19(6):527–531
50. Larrañaga C, Martínez HJ, Palomino MA et al (2007) Molecular characterization of hospital-acquired adenovirus infantile respiratory infection in Chile using species-specific PCR assays. J Clin Virol 39(3):175–181
51. Bruzzone MS, Fuentes L, Spencer E (2000) Specific subgroup B adenovirus diagnosis by PCR of the fibre gene. J Infect 40(2):154–159
52. Bruzzone M, Barro M, Spencer E (2001) Identification of adenovirus 7 h heterogeneity in the E3 region. Biol Res 34(2):75–82
53. Rabagliati BR, Serri VM, Perret PC et al (2006) Clinical and epidemiological characteristics of respiratory virus infections among adults hospitalized during 2004 influenza season. Rev Chil Infectol 23(2):111–117
54. Hortal M, Russi JC, Arbiza JR et al (1990) Identification of viruses in a study of acute respiratory tract infection in children from Uruguay. Rev Infect Dis 12(suppl 8):S915–S922
55. Hortal M, Contera M, Mogdasy C, Russi JC (1994) Acute respiratory infections in children from a deprived urban population from Uruguay. Rev Inst Med Trop Sao Paulo 36(1):51–57
56. Hortal M, Suarez A, Deleon C et al (1994) Etiology and severity of community acquired pneumonia in children from Uruguay: a 4-year study. Rev Inst Med Trop Sao Paulo 36(3):255–264
57. Kajon AE, Mistchenko AS, Videla C et al (1996) Molecular epidemiology of adenovirus acute lower respiratory infections of children in the south cone of South America (1991–1994). J Med Virol 48(2):151–156
58. Frabasile S, Vitureira N, Perez G et al (2005) Genotyping of Uruguayan human adenovirus isolates collected between 1994 and 1998. Acta Virol 49(2):129–132
59. Freitas FT (2013) Sentinel surveillance of influenza and other respiratory viruses, Brazil, 2000–2010. Braz J Infect Dis 17(1):62–68
60. Sutmoller F, Nascimento JP, Chaves JRS et al (1983) Viral etiology of acute respiratory diseases in Rio de Janeiro: first two years of a longitudinal study. Bull World Health Organ 61(5):845–852
61. Nascimento JP, Siqueira MM, Sutmoller F et al (1991) Longitudinal study of acute respiratory diseases in Rio de Janeiro: occurrence of respiratory viruses during four consecutive years. Rev Inst Med Trop Sao Paulo 33(4):287–296
62. Moraes MT, Leite JP, Siqueira MM et al (1997) Genomic characterization of adenovirus serotype 7 isolated in Brazil from acute respiratory disease patients during the period from 1980 to 1991. Rev Inst Med Trop Sao Paulo 39(4):185
63. Moraes MT, da Silva M, Leite JP, Nascimento JP (1998) Genetic an antigenic analysis of adenovirus type 3 strains showing intermediate behavior in standard seroneutralization test. Mem Inst Oswaldo Cruz 93(2):231–235

64. Kajon AE, Portes SA, de Mello WA et al (1999) Genome type analysis of Brazilian adenovirus strains of serotypes 1, 2, 3, 5, and 7 collected between 1976 and 1995. J Med Virol 58(4):408–412
65. de Albuquerque MC, da Silva FM, Soares CC et al (2003) Adenoviruses isolated from civilian and military personnel in the city of Rio de Janeiro, Brazil. Rev Inst Med Trop Sao Paulo 45(4):233–236
66. Straliotto SM, Siqueira MM, Machado V et al (2004) Respiratory viruses in the pediatric intensive care unit: prevalence and clinical aspects. Mem Inst Oswaldo Cruz 99(8):883–887
67. Moura FE, Mesquita JR, Portes SA et al (2007) Antigenic and genomic characterization of adenovirus associated to respiratory infections in children living in Northeast Brazil. Mem Inst Oswaldo Cruz 102(8):937–941
68. Alonso WJ, Laranjeira BJ, Pereira SA et al (2012) Comparative dynamics, morbidity and mortality burden of pediatric viral respiratory infections in an equatorial city. Pediatr Infect Dis J 31(1):e9–14
69. Pereira SA, Florêncio CM, Marinheiro JC et al (2016) Adenoviruses and acute respiratory infections in children living in an equatorial area of Brazil. Epidemiol Infect 144:355–362
70. Lu X, Erdman DD (2006) Molecular typing of human adenoviruses by PCR and sequencing of a partial region of the hexon gene. Arch Virol 151(8):1587–1602
71. Moura PO, Roberto AF, Hein N et al (2007) Molecular epidemiology of human adenovirus isolated from children hospitalized with acute respiratory infection in São Paulo, Brazil. J Med Virol 79(2):174–181
72. Marinheiro JC, Sanalios RB, Santos DC et al (2009) Duplex-PCR assay for the detection of adenovirus and respiratory syncytial virus in nasopharyngeal samples. Mem Inst Oswaldo Cruz 104(1):118–120
73. Marinheiro JC, Dos Santos TG, Siqueira-Silva J et al (2011) A naturally occurring human adenovirus type 7 variant with a 1743 bp deletion in the E3 cassette. J Gen Virol 92(10):2399–2404
74. Thomazelli LM, Vieira S, Leal AL et al (2007) Surveillance of eight respiratory viruses in clinical samples of pediatric patients in Southeast Brazil. J Pediatr 83(5):422–428
75. Durigon GS, Oliveira DB, Felicio MC et al (2015) Poor outcome of acute respiratory infection in young children with underlying health condition in Brazil. Int J Infect Dis 34:3–7
76. Giamberardin HI, Homsani S, Bricks LF et al (2016) Clinical and epidemiological features of respiratory virus infections in preschool children over two consecutive influenza seasons in southern Brazil. J Med Virol 88(8):1325–1333
77. Watanabe A, Carraro E, Camargo C (2013) Human adenovirus detection among immunocompetent and immunocompromised patients presenting acute respiratory infection. Rev Soc Bras Med Trop 46(2):161–165
78. Proenca-Modena JL, Pereira Valera FC, Jacob MG et al (2012) High rates of detection of respiratory viruses in tonsillar tissues from children with chronic adenotonsillar disease. PLoS One 7(8):e42136
79. Espínola EE, Russomando G, Basualdo W et al (2012) Genetic diversity of human adenovirus in hospitalized children with severe acute lower respiratory infections in Paraguay. J Clin Virol 53(4):367–369
80. Ampuero JS, Ocaña V, Gómez J et al (2012) Adenovirus respiratory tract infections in Peru. PLoS One 7(10):e46898
81. García J, Sovero M, Laguna-Torres VA et al (2009) Molecular characterization of adenovirus circulating in central and South America during the 2006–2008 period. Influenza Other Respir Viruses 3(6):327–330
82. Budge PJ, Griffin MR, Edwards KM, et al., RESPIRA-PERU Group (2014) A household-based study of acute viral respiratory illnesses in Andean children. Pediatr Infect Dis J 33(5):443–447

83. Grijalva CG, Griffin MR, Edwards KM et al (2014) Concordance between RT-PCR-based detection of respiratory viruses from nasal swabs collected for viral testing and nasopharyngeal swabs collected for bacterial testing. J Clin Virol 60(3):309–312
84. Howard LM, Johnson M, Williams JV et al (2015) Respiratory viral detections during symptomatic and asymptomatic periods in young Andean children. Pediatr Infect Dis J 34(10):1074–1080
85. Douce RW, Aleman W, Chicaiza-Ayala W et al (2011) Sentinel surveillance of influenza-like-illness in two cities of the tropical country of Ecuador: 2006–2010. PLoS One 6(8):e22206
86. Herrera-Rodríguez DH, de la Hoz F, Mariño C et al (2007) Adenovirus in children under five years of age. Circulation patterns and clinical and epidemiological characteristics in Colombia, 1997–2003. Rev Salud Publica (Bogota) 9(3):420–429
87. Herrera-Rodríguez DH, de la Hoz F, Mariño C et al (2007) Respiratory virus in children aged less than 10 years old suffering from respiratory infection in the Hospital Militar Central in Bogota from 2000-2001. Rev Salud Publica (Bogota) 9(4):576–586
88. Arango AE, Jaramillo S, Perez J et al (2015) Influenza-like illness sentinel surveillance in one hospital in Medellin, Colombia.2007–2012. Influenza Other Respir Viruses 9(1):1–13
89. Rojas LJ, Jaramillo CA, Mojica MF et al (2012) Molecular typing of adenovirus circulating in a Colombian paediatric population with acute respiratory infection. Epidemiol Infect 140(5):818–822
90. Remolina YA, Ulloa MM, Vargas H et al (2015) Viral infection in adults with severe acute respiratory infection in Colombia. PLoS One 10(11):e0143152
91. Frobert E, Escuret V, Javouhey E et al (2011) Respiratory viruses in children admitted to hospital intensive care units: evaluating the CLART® Pneumovir DNA array. J Med Virol 83(1):150–155
92. Comach G, Teneza-Mora N, Kochel TJ et al (2012) Sentinel surveillance of influenza-like illness in two hospitals in Maracay, Venezuela: 2006–2010. PLoS One 7(9):e44511
93. Salas-Chaves P, Alfaro-Bourrouet W (2005) Seasonal variation of viral acute respiratory infection in children. Rev Mexicana Pediatr 72(1):5–8
94. Verani JR, McCracken J, Arvelo W et al (2013) Surveillance for hospitalized acute respiratory infection in Guatemala. PLoS One 8(12):e83600
95. Laguna-Torres VA, Sánchez-Largaespada JF, Lorenzana I et al (2011) Influenza and other respiratory viruses in three central American countries. Influenza Other Respir Viruses 5(2):123–134
96. Pumariega T, Savón C, Muné M et al (2000) Isolation and identification of adenovirus in hospitalized children, under five years, with acute respiratory disease, in Havana, Cuba. Mem Inst Oswaldo Cruz 95(6):859–861
97. Acosta B, Valdes O, Piñon A et al (2009) Molecular characterization of adenoviral infections in Cuba: report of an unusual association of species D adenoviruses with different clinical syndromes. Arch Virol 154:619–627
98. Manjarrez ME, Rosete DP, Rincón M et al (2003) Comparative viral frequency in Mexican children under 5 years of age with and without upper respiratory symptoms. J Med Microbiol 52:579–583
99. Muraira GA, Villarreal CE, Quiroga GA et al (2002) Viral agents in children hospitalized by lower respiratory infections. Rev Mexicana Pediatr 69(5):185–189
100. Cabello C, Manjarrez ME, Olvera R et al (2006) Frequency of viruses associated with acute respiratory infections in children younger than five years of age at a locality of Mexico City. Mem Inst Oswaldo Cruz 101(1):21–24
101. Rosete DR, Manjarrez ME, Barrón BL (2008) Adenoviruses C in non-hospitalized Mexican children older than five years of age with acute respiratory infection. Mem Inst Oswaldo Cruz 103(2):195–200
102. Mejía-López H, López-Martínez I (2007) Adenovirus aislados de pacientes mexicanos con infección respiratoria aguda. Rev Biomed 18:146–152

103. Galindo-Fraga A, Ortiz-Hernandez AA, Ramırez-Venegas A et al (2013) Clinical characteristics and outcomes of influenza and other influenza-like illnesses in Mexico City. Int J Infect Dis 17:e510–e517
104. Uribe-Gutiérrez G, Hernández-Santos H, Manjarrez-Zavala ME (2014) Prevalence and genotypes of the adenovirus infection as well detection of co-infection with Bocavirus in Mexican immunosuppressed and non-immunosuppressed children with pneumonia. Clin Lab 60:1277–1285
105. Wong-Chew RM, Espinoza MA, Taboada B et al (2015) Prevalence of respiratory virus in symptomatic children in private physician office settings in five communities of the state of Veracruz, Mexico. BMC Res Notes 8:261

Chapter 15
Measles and Rubella in the Americas: The Path to Elimination

Marilda Mendonça Siqueira and David W.G. Brown

1 Introduction: Measles

Measles is one of the most infectious diseases known and was almost universal in the population before vaccination was introduced. The measles virus (MeV) spreads rapidly by airborne or droplet or by direct or indirect contact with the nasal and throat secretions of infected persons. The virus remains active and contagious in small airborne particles leading to direct aerosol transmission by inhalation and through indirect contamination of surfaces for up to 2 h. Once the virus is inhaled, the primary target cell is infected and systemic spread ensues. The incubation period ranges from 10 to 14 days before clinical signs appear. The prodromal phase is characterized by fever and malaise, associated with cough, coryza, and conjunctivitis, and the maculopapular rash usually appears about 14 days after exposure. During this stage, Koplik's spots can be observed on the buccal mucosa. Clinical complications of measles include pneumonia, gastroenteritis, otitis media, blindness, measles inclusion body encephalitis (MIBE), and subacute sclerosing panencephalitis (SSPE), a rare central nervous system disease. SSPE is characterized by a progressive functional degeneration that occurs several years after the primary measles infection. It is caused by a persistent infection and is estimated to occur in

M.M. Siqueira (✉)
Measles/Rubella National (MoH) and Regional (WHO) Reference Laboratory- Instituto Oswaldo Cruz- FIOCRUZ, Rio de Janeiro, Brazil
e-mail: mmsiq@ioc.fiocruz.br

D.W.G. Brown
Measles/Rubella National (MoH) and Regional (WHO) Reference Laboratory- Instituto Oswaldo Cruz- FIOCRUZ, Rio de Janeiro, Brazil

National Infection Service, Public Health England-London, London, UK
e-mail: david.brown@ioc.fiocruz.br

J.E. Ludert et al. (eds.), *Human Virology in Latin America*,
DOI 10.1007/978-3-319-54567-7_15

4 to 11 patients per 100,000 cases. The main risk factor is infection at an early age. Measles also causes immunosuppression. MeV-infected individuals are contagious from 4 days before to 4 days after rash onset [8, 16, 35, 65]. The high infectivity of MeV implies that a high level of population immunity is required to interrupt transmission. A live attenuated vaccine was developed in the 1960s, and most vaccines in current use derive from the prototype Edmonston strain (Moraten, Schwarz, and Edmonston-Zagreb). A few vaccines derive from other wild-type viruses (e.g., CAM-70). Measles vaccine is often delivered in combination with live attenuated vaccine for rubella (MR: measles-rubella vaccine) and mumps (MMR: measles, mumps, and rubella vaccine) [35, 65].

The MeV is a paramyxovirus, genus *Morbillivirus*. The virus particles are enveloped, 100–200 nm in diameter, with a single-stranded, nonsegmented negative-sense RNA genome that is 15,894 nucleotides in length. The genome contains six genes encoding nucleoprotein (N), phosphoprotein (P), matrix (M), fusion (F), hemagglutinin (H), and polymerase (L). The two viral membrane glycoprotein spikes (hemagglutinin tetramers and fusion protein trimmers) have key roles in viral entry and are the main targets of the virus-neutralizing antibodies. The H binds to the cellular receptor, and the F protein causes the fusion of virus and host cell membranes, viral penetration, and hemolysis. Although both the H and F proteins are the targets of neutralizing antibodies, the early humoral immune response is mainly directed against the N protein [25, 26, 35, 75].

2 Rubella

Rubella, also known as German measles, is less infectious than measles. It is a common mild self-limiting illness of children and young adults, and 50% of cases are asymptomatic or unrecognized. The illness is characterized by rash, lymphadenopathy, and low-grade fever. Conjunctivitis, coryza, sore throat, cough, and occasional headache and malaise are seen. The rash is pink or red, begins on the face, and then spreads downward to the rest of the body. The disease is transmitted via direct droplet contact from nasopharyngeal secretions. Symptoms appear 14 to 21 days after exposure. Up to 70% of adult women with rubella experience arthritis, but arthritis is rarely seen in children and men. Rubella is a teratogenic virus, and infection in pregnancy may have potentially devastating effects on the developing fetus and cause a number of anomalies. The manifestations of congenital rubella infection (CRI) vary depending upon the timing of maternal infection. Congenital rubella syndrome (CRS) refers to a collection of birth defects (hearing impairment, congenital heart defects, cataracts/congenital glaucoma, pigmentary retinopathy, splenomegaly, microcephaly) resulting from infection in pregnancy [6, 29, 39, 46, 48].

Rubella virus (RubV) was isolated in the early 1960s, and live attenuated rubella vaccines became available by the end of the same decade [59]; it is generally given in combination as MR or as MMR vaccine. RubV is the single member of the *Rubivirus* genus of the *Togaviridae* family. The virion is a small lipid-enveloped spherical particle 40–80 nm in size, containing a linear positive-sense single-stranded RNA genome approximately 9700 nucleotides in length. Five protein products are encoded by the genome: two nonstructural proteins (P90 and P150) and three virion proteins, two in the envelope (E1 and E2) and one in the core (capsid or C protein) [39].

3 Laboratory Diagnosis of Measles and Rubella

Case confirmation, which depends on epidemiological data, clinical presentation, and results of laboratory tests, is necessary for public health and outbreak control. Laboratory confirmation is based on the detection of MeV-specific IgM in a serum sample or the detection of MeV RNA by reverse transcription-polymerase chain reaction (RT-PCR) in throat/nasal swab, oral fluid, or urine samples. Enzyme immunoassay (EIA) is the most commonly used method for detecting IgM and IgG. Serum samples should be collected at the first contact with a suspected case. Oral fluid and dried capillary blood are used as an alternative to serum in some surveillance systems. RT-PCR is most sensitive for diagnosis if samples (throat or nasal swabs, oral fluid, urine, peripheral blood mononuclear cells) are collected as early as possible after the onset of rash [6, 65].

4 Molecular Epidemiology of Measles and Rubella

Genetic characterization of both measles and rubella is a valuable tool for laboratory-based surveillance and molecular epidemiological studies. The genetic information obtained can be used to define the pattern of MeV and RubV circulation in the population as endemic or imported. Genotyping results can confirm or disprove epidemiological linkages between cases. Virological surveillance provides a sensitive method to describe the transmission patterns of the viruses, which is required to document the interruption of transmission of endemic measles or rubella. The absence of endemic genotypes is one of the criteria for verifying measles elimination in a country or region [65, 81].

MeVs can be grouped according to the World Health Organization (WHO) protocol into eight clades, A–H, subdivided into 24 genotypes based on the nucleotide sequencing of the 450 nucleotides at the carboxyl-terminal of the N gene. The complete sequence of the H gene is also used for this purpose. A standardized methodology and nomenclature are used to define and describe measles sequence and genotypes to enable international sharing [82]. More recently, the use of whole-genome measles sequencing has been investigated to see if it provides additional discrimination within the same genotype; this can help to define if a single genotype identified during large or long-lasting outbreaks is the result of single or multiple introductions of virus [33].

For RubV, a 739 nucleotide genotyping region (nucleotides 8731–9469) has been designated by the WHO for molecular epidemiology studies. Analysis of this region describes a genetic variability of two clades, which include 13 distinct genotypes: 9 genotypes (1B, 1C, 1D, 1E, 1F, 1G, 2A, 2B, 2C) are recognized, and 4 genotypes (1a, 1h, 1i, 1j) are provisional [85]. Globally, only limited data have been collected for rubella.

5 Global Measles and Rubella Elimination

The measles and rubella vaccines are highly effective and have been available since 1963 and 1969, respectively. They have an excellent safety profile. The feasibility of measles eradication has been discussed for more than 30 years. Measles and rubella meet all the biological criteria for disease eradication: (1) humans are the sole pathogen reservoir; (2) an accurate diagnostic test exists; and (3) an effective, practical intervention (vaccine) is available at reasonable cost. Interruption of transmission in large geographic areas for prolonged periods further supports the feasibility of eradication [22, 50].

6 Measles

In 1980, before measles vaccine was used globally, an estimated 2.6 million deaths from measles occurred worldwide [72]. In 1989, the WHO resolved to reduce measles morbidity by 90% and measles mortality by 95% compared with disease burden in the pre-vaccination era [86]. In 1990, the World Summit for Children set a target of 90% for measles vaccine coverage by 2000 and for other vaccines used in the Expanded Program on Immunization (EPI) [76]. By the end of 1997, global measles morbidity and mortality had decreased 74% and 85%, respectively. At that time, three of six WHO regions had established measles elimination targets: the Americas by 2000, the European Region by 2007 and the Eastern Mediterranean Region by 2010. The WHO African, Southeast Asian, and Western Pacific regions retained the World Health Assembly morbidity and mortality reduction target [36]. Although there had been substantial progress in controlling measles worldwide, the WHO estimated that about 875,000 children died of measles in 1999. This estimate is 56% of all estimated deaths from vaccine-preventable diseases of childhood for that year, making measles the leading cause of vaccine-preventable child mortality, and in 2000 measles remained the fifth leading cause of childhood mortality, accounting for 5% of all deaths among children aged less than 5 years [36, 71]. In 2001, the WHO and the United Nations Children's Fund published a 5-year strategic plan to reduce measles mortality by half by 2005. Strategies include providing a second opportunity for measles immunization to all children through nationwide supplementary immunization activities, increasing routine vaccination coverage, and

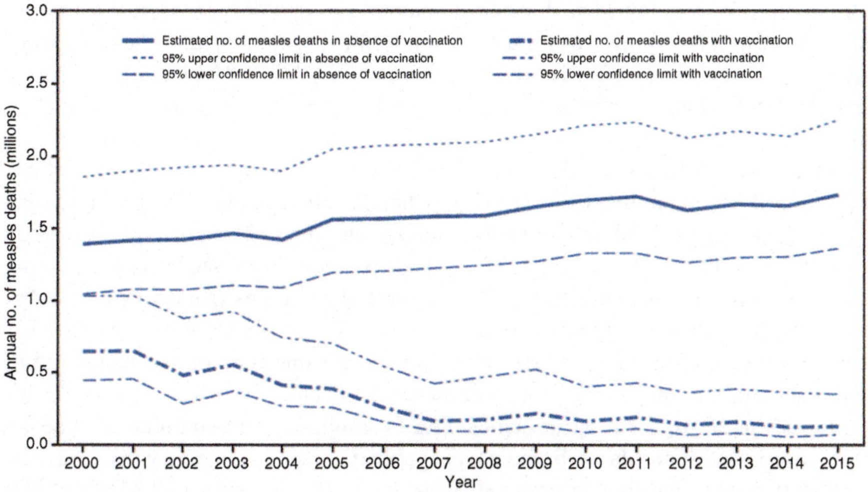

Fig. 15.1 Estimated annual number of measles deaths with and without vaccination, worldwide, 2000–2015. Compared with no measles vaccination, measles vaccination prevented an estimated cumulative total of 20.3 million deaths during 2000–2015. (From [58])

improving surveillance with laboratory confirmation of all suspected measles cases [87]. During the period 2000–2015, the annual reported measles incidence declined by 75% worldwide from 146 to 36 cases per million population, and annual estimated measles deaths declined by 79%, from 651,600 to 134,200. Figure 15.1 shows the estimated annual number of measles deaths with and without vaccination, worldwide, in 2000–2015 [58].

7 Measles Elimination in the Americas

In May 1985, after the successful eradication of smallpox in the Americas, the Pan American Health Organization (PAHO) proposed to eradicate polio from the region by 1990. This goal was achieved in 1991 and created a platform for building measles elimination strategies.

During 1971–1977, 28 countries in the Americas were reporting an average of 258,634 measles cases per year and inter-epidemic periods of 2–3 years. With the widespread use of measles vaccine, the intervals between epidemics increased, and a reduction in measles cases was observed in many countries, including Canada, parts of Brazil, Chile, Costa Rica, Mexico, and the U.S. [12, 55].

Cuba was the first country in the region to use a strategy called "the catch-up campaign" in 1986 [31]. In 1988, the Ministers of Health in the English-speaking Caribbean countries and territories declared their commitment to eliminate endemic measles by 1995, and a measles catch-up campaign in 1991 achieved 90% vaccine coverage. The Caribbean was the first area of the world to eliminate measles [41].

Brazil and Chile followed with catch-up campaigns in 1992. This strategy proved successful in interrupting viral transmission in these countries after the resurgence of measles in several countries in the Americas in 1988–1991, including Mexico and the USA, other countries in the region took up the challenge to eliminate measles [12, 66, 73].

In the early 1990s, several countries in the Americas adopted the measles elimination strategy recommended by PAHO, which has three components: (1) one-time mass vaccination of children and adolescents (catch-up), (2) routine immunization of successive birth cohorts (keep up), and (3) periodic mass vaccination of young children to prevent accumulation of susceptible populations (follow-up). In 2003, after the resolution to eliminate rubella and CRS from the Americas by 2010, a fourth strategy, referred to as "speed up," called for one-time mass vaccination of older adolescents and adults with combined MR vaccine [14, 24].

In addition to vaccination strategies, PAHO emphasized the importance of sensitive, case-based surveillance based on a capable diagnostic laboratory network. The vaccination and surveillance strategies were revised to include use of MMR or MR vaccines, to fully integrate measles and rubella surveillance, and to establish CRS surveillance [14]. A laboratory network was developed that comprises 1 global specialized (CDC/USA), 2 regional reference (NML/Canada and FIOCRUZ/Rio de Janeiro) laboratories, and 21 national and 141 subnational laboratories, with the objective of providing a timely diagnostic serological result, virus detection in clinical samples, and phylogenetic analysis of virus strain diversity for molecular epidemiology studies [77]. Figure 15.2 shows the impact of the measles and rubella control strategies used in Latin American countries (LAC) [54].

After the adoption of these elimination strategies, the number of confirmed measles cases declined from about 250,000 in 1990 to 2,109 in 1996 in the region of the Americas [37]. In 1997, there was a resurgence of MeV circulation, which began with a large urban outbreak in Sao Paulo and resulted in 52,284 confirmed cases in Brazil caused by measles genotype D6 [57]. The outbreak spread to Argentina and Bolivia, where a large number of measles cases occurred in the region during 1998 and 1999, respectively. Uruguay, Dominican Republic, Chile, and Haiti also

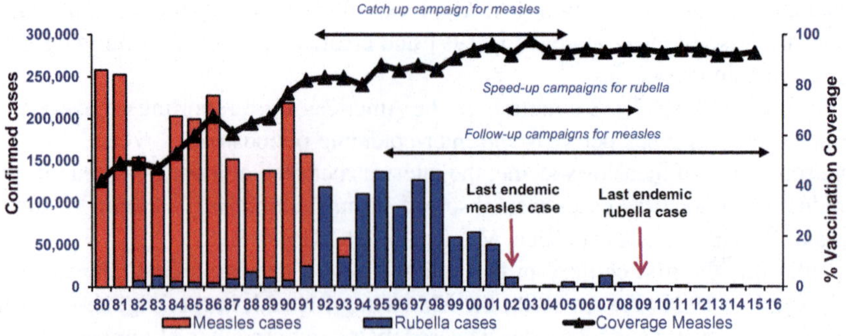

Fig. 15.2 Impact of measles and rubella elimination strategies: the Americas, 1980–2016. (From [54]. Reprinted with the permission of the Pan American Health Organization)

reported outbreaks caused by D6 strains, and this became the endemic genotype in the Americas at that time [7, 9, 23, 68].

In Argentina, 10,673 confirmed cases were reported during the years 1997–2000. The majority were in unvaccinated infants and preschool children. Cases declined after a follow-up campaign, which achieved 98% coverage, among children aged 1 to 4 years. The last cases were identified in February 2000 [23].

In 1999, to achieve the elimination goal, Brazil implemented a Supplementary Emergency Measles Action Plan, in which one measles surveillance technician was designated for each state. In 2000–2001, an MR campaign in women of childbearing age achieved 94% vaccine coverage. The number of cases declined, and only small outbreaks or single imported cases were detected; in 2001, only 1 of 5599 suspected cases was confirmed, and this was shown to be imported from Japan. These data suggested that indigenous circulation of measles in Brazil had been interrupted [60].

In 1994, a national catch-up campaign was conducted in Bolivia, which achieved 96% coverage. In 1998, after an importation of measles from Argentina, an epidemic started in the country that lasted until 2000, with 2567 cases reported, most of whom were unvaccinated. Fifty-five percent occurred in children less than 5 years old. A national house-to-house vaccination campaign for children aged 6 months to 4 years was implemented. The last endemic case occurred in October 2000 [61].

Costa Rica introduced the MMR vaccine in 1986, and this shifted susceptibility to older age groups, leading to the introduction of a second dose of MMR in 1992, administered at the age of 7 years. In 2000, the country set a goal of rubella and CRS elimination, and a national vaccination campaign targeting men and women 15 to 19 years of age was implemented; this, together with the strengthening of routine vaccination, led to measles elimination. The last measles endemic case was confirmed in 1999 and the last rubella and CRS case in 2001 [49].

In 1994, Venezuela launched a catch-up campaign and in 1998 a follow-up campaign that coincided with the introduction of MMR. In 2001, the country responded to an outbreak by vaccinating all children less than 15 years of age. With this strategy, endemic transmission of the MeV was interrupted in 2002. In February 2006, a measles outbreak began with a distinct clinical and epidemiological picture compared to classical outbreaks. The outbreak occurred in an older age group, and clinical disease was mild. Indigenous circulation of measles was interrupted in 2007, providing an opportunity to review the effectiveness of the classical control measures [67].

Mexico experienced a huge measles epidemic in 1989–1990, resulting in 5899 deaths. Following this epidemic, measles elimination efforts were strengthened with the introduction of two doses of MMR vaccine with high coverage. Indigenous measles transmission was interrupted in 1999. Since then, occasional reintroduction of the measles virus leading to small outbreaks or sporadic cases has been reported [63, 66].

Many activities were established in LAC to demonstrate the absence of measles circulation in the region. One of these activities was the periodic case finding; for example, in Brazil, from July 2000 to March 2001, 917 active case searches were conducted in 25 states, which involved review of 2,581,542 charts and 18,118 interviews conducted with health professionals. Of the 180 suspected measles cases and 1070 suspected rubella cases, 142 and 740 had been previously reported, respectively. No new cases were confirmed [60].

Another important strategy was the laboratory testing in each suspected measles or rubella case. It is recognized that when measles is well controlled, diagnosis based on clinical grounds alone is inaccurate and can be confused with other infectious exanthema. Studies in Brazil involving laboratory testing of febrile rash illnesses found that a variety of etiological agents were associated with clinical features similar to measles or rubella, and so cases suspected of measles or rubella should not be confirmed or discarded based only on clinical data [21, 52, 78].

Some challenges in case classification have been reported by the countries that have eliminated measles. Measles cases have been reported in vaccinated people, who presented with nonclassical signs and/or symptoms; their illness was of short duration. The virus was often detectible by RT-PCR, but the cases have low or undetectable IgM antibodies in the presence of high IgG levels. This picture is typical of a secondary immune response suggesting a reinfection in a vaccinee with low antibody levels. Detailed epidemiological linkage data are important to identify these cases [4, 38, 51, 70].

Sero-surveys can provide important information on population immunity, to identify risk groups in the population and to guide immunization strategy. Such studies are usually conducted by random population sampling using sera, and studies have been conducted in LAC [53, 74]. Another approach has used convenience samples collected for other purposes, including the use of residual sera collected for patient investigations, plasma from blood donors, or just serological data generated by regular testing during pregnancy or clinical follow-up. This approach represents a trade-off between the representativeness of the samples and the practicalities of collection and costs as illustrated by some Brazilian studies [3, 15].

The countries of the Americas successfully interrupted endemic measles transmission within 8 years of setting a regional elimination goal [2, 12]. Although the LAC has maintained measles elimination since 2002, measles outbreaks continue to occur globally, resulting in cases imported into the Americas with the risk of potential spread [64]. In 2003, 105 confirmed measles cases were reported in the Americas, the lowest number ever reported in the region. Of these, 1 occurred in Chile, 1 in Costa Rica, 2 in Brazil, 15 in Canada, 42 in the U.S., and 44 in Mexico [83]. During the period 2003–2010, low numbers of confirmed measles cases were reported in the Americas, with an annual mean of 155 confirmed cases, which represents a 99.9% decrease compared with the period 1987–1994. Three different genotypes were associated with imported cases in the LAC from 2007 to 2009 (D4 in Peru and Chile, D8 in Argentina, B3 in Venezuela) [14, 64]. In the period 2010–2016, a range of different genotypes was identified in sporadic cases or small outbreaks in various LAC; most were directly associated with imported cases. The most prolonged outbreak occurred in two states in Brazil (Pernambuco state from March 2013 to March 2014 and Ceara state from December 2013 to July 2015) with 1728 confirmed cases, and genotype D8 was identified throughout the outbreak [27, 45]. Surveillance data for measles, combined with these molecular epidemiology studies, indicate that the countries of the Americas are continually exposed to imported measles viruses from other regions of the world where measles continues to be endemic.

8 Rubella

The primary goal of rubella vaccination is to prevent congenital rubella infection, including CRS. The burden of CRS is typically underestimated in routine surveillance. Before the introduction of rubella vaccine, the incidence of CRS varied from 0.1 to 0.2 cases per 1000 live births during endemic periods and from 0.8 to 4 cases per 1000 live births during rubella epidemics [18, 84]. A study modeling the incidence of CRS in developing countries estimated that about 110,000 cases occurred in 1996, excluding the contribution of fetal loss from CRS [19].

Increasing numbers of countries and regions have developed rubella elimination goals, building on the progress of the measles elimination strategy (as described here), because rubella is one component of MMR or MR vaccines. The introduction of rubella vaccine is cost-effective and cost beneficial, but requires ongoing strengthening of routine immunization service and surveillance systems [62, 84].

Since 1996, when only 83 WHO Member States used rubella-containing vaccines (RCVs) in their national immunization schedules, there has been an important increase in the number of countries introducing rubella vaccine, which is delivered in most countries as MMR during the second year of life [84]. The level of introduction varies by WHO region, and Fig. 15.3 shows countries with and without rubella

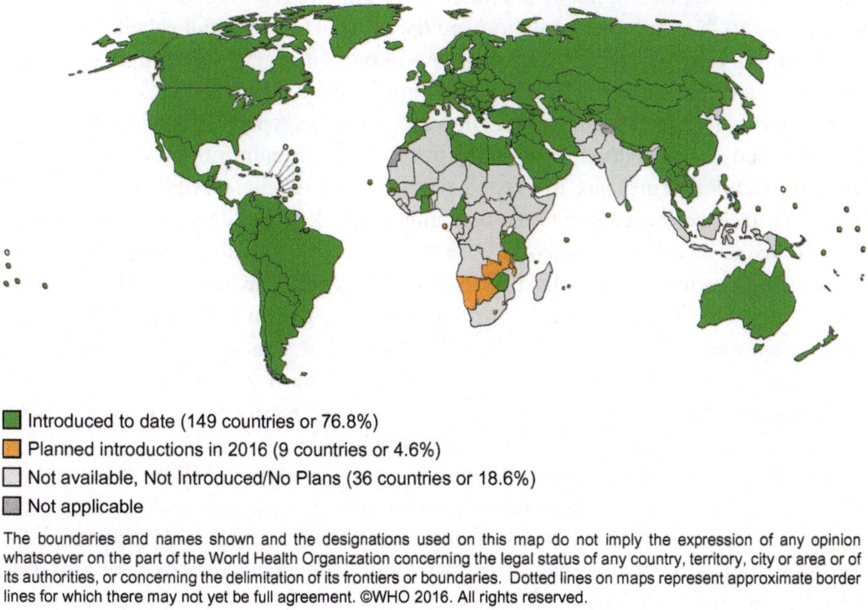

Fig. 15.3 Countries with rubella vaccine in their national immunization programs and countries planning to introduce rubella vaccination in 2016–2017. (From [80]). The boundaries and names shown and the designations used on this map do not imply the expression of any opinion whatsoever on the part of the World Health Organization concerning the legal status of any country, territory, city, or area or of its authorities, or concerning the delimitation of its frontiers or boundaries. *Dotted lines* on maps represent approximate border lines for which there may not yet be full agreement. (©WHO 2016. All rights reserved)

vaccine as part of their immunization programs [80]. The global burden of CRS has been estimated using seroprevalence and immunization coverage. This analysis indicated that CRS had decreased in the three WHO regions (America, Europe, Eastern Mediterranean) that had introduced widespread RCV by 2010, reaching fewer than 2 per 100,000 live births (the Americas and Europe) and 25 (95% CI, 4–61) per 100,000 live births (the Eastern Mediterranean). The estimated incidence in 2010 ranged from 90 (95% CI, 46–195) in the Western Pacific, excluding China, to 116 (95% CI, 56–235), and 121 (95% CI, 31–238) cases per 100,000 live births in Africa and Southeast (SE) Asia, respectively, where the highest number of cases was predicted. The burden of CRS is still significant [79].

The Region of the Americas eliminated rubella in 2009 [11], and the European and Western Pacific regions have established rubella elimination goals. The Southeast Asia Region has a goal to control CRS, and the Eastern Mediterranean and African Regions have not yet established goals for rubella control [34, 43].

9 Rubella Elimination in the Americas

Data from the regional measles surveillance system documented the widespread circulation of rubella virus in many countries in the Americas during the 1990s. In 1997, in response to the ongoing circulation and the potential for a major rubella epidemic in the region, the PAHO Technical Advisory Group on vaccine-preventable diseases recommended the implementation of a regional initiative to accelerate rubella and CRS control. To achieve this, PAHO developed a rubella and CRS control strategy that included the introduction of RCV into routine immunization, vaccination of women of childbearing age, integrated measles, and rubella surveillance including rubella virus detection and the implementation of a CRS surveillance system [10, 12]. Since 1998, all 35 countries and nine territories have reported rubella-suspected cases.

The implementation of robust CRS surveillance in LAC supported by timely clinical and laboratory diagnosis ensured the detection and classification of suspected cases as imported or endemic. This distinction was critical for confirming CRS elimination, as is well illustrated by the data from Costa Rica [49]. A complementary approach was to investigate, in the laboratory, suspect cases of CRS presenting with a single compatible manifestation [42].

The strategy of high coverage childhood immunization with MR proved successful in controlling disease and in reducing the number of measles and rubella cases. However, it led to an upward age shift in susceptible groups to adolescents and the adult population, as documented in Brazil, Chile, and Costa Rica [32, 49, 60]. For rubella, this shift posed a potential risk, because it could have lead to rubella outbreaks in young adult populations with an increased risk of CRS cases, as was documented in Brazil during a rubella outbreak in 2001 [44].

One important strategy used during the CRS elimination program was the vaccination of women of childbearing age and young adult men. Concern was raised about the use of this strategy during a mass vaccination campaign because it would lead to some women being vaccinated without knowing if they were pregnant or to women receiving the vaccine in the month before conception. At the time, there was no evidence that the vaccine could adversely affect fetal development and lead to congenital malformation, but only a few studies with limited numbers of cases had been published addressing the safety of rubella vaccine in pregnancy. These studies did not provide support for indiscriminate vaccination of pregnant women. Studies were conducted during the mass campaign of women of childbearing age in Argentina, Brazil, Costa Rica, Ecuador, El Salvador, and Paraguay, which provided evidence of rubella vaccine safety in this group [5, 13, 20, 69]. In addition, the vaccination of young adult men proved to be essential for rubella and CRS control. Initially, Brazil and Chile did not use this strategy. They immunized only women during their first MR adult vaccination campaign and needed to conduct another campaign including both men and women [32, 47]. In 2003, during a meeting of the 44th PAHO Directing Council, composed of the Ministers of Health of PAHO Members States, set the goal of eliminating rubella and CRS by 2010. In 2006, using these strategies, the incidence of rubella and CRS was substantially reduced across the Americas [11]. The inclusion of Guatemala in this initiative was delayed by the historically low investment in healthcare in the country. However, because of the country's resource mobilization and advocacy efforts (economic analysis, potential donors, disseminating information, among others), a national MR vaccination campaign was implemented in 2007, covering 99% of the target population. Data from epidemiological surveillance over subsequent years revealed that the country had been free of endemic RubV circulation [30].

Molecular epidemiological data are an important tool for classifying cases as imported or endemic and for documenting measles and rubella virus elimination. Limited rubella genotype information is available for many countries [1]. In LAC, the scale of rubella molecular data has improved over the years. The geographic distribution of wild-type rubella genotypes in the Americas, interpreted in the context of their global distribution, has contributed to the documentation of rubella elimination from some countries. During 1997–2007, genotype 1C was detected in Bolivia, Chile, Ecuador, El Salvador, Honduras, Mexico, Panama, Peru, and Venezuela. This genotype was restricted to the Americas; it was considered an endemic strain, and its disappearance was documented and provided evidence of the elimination of endemic rubella in these areas. Genotype 1E was reported in Bahamas, Guyana, and Suriname [40]. In Brazil, rubella virological surveillance before 2005 showed genotype 1B circulating in Rio de Janeiro state in 1997 and genotype 1G in 1996, 1997, 1999, 2000, and 2001, so this genotype was considered to be endemic in Brazil. A genotype 1J strain isolated in 2005 was considered an importation, although the index case was not identified. During 2006, a rubella outbreak that spread throughout the country until 2008 started in Brazil. All viruses detected from 14 states belonged to genotype 2B during this period. In addition, this genotype was found in outbreaks in Chile and Argentina and so was considered endemic in the Americas [1, 28, 40, 44]. A retrospective study from 1996 to 2009 in Sao Paulo, Brazil, identified four genotypes (1a, 1B, 1g, 2B) involved in congenital infections [17]. However, intense virological surveillance after 2006 provided evidence for the elimination of endemic 2B strain from the Americas [1]. During the period 2010–2016, imported RubV from different genotypes has been identified in LAC. As rubella continues to circulate in many regions of the world, controlling spread from imported rubella cases will be a challenge for the Americas.

10 Declaration of Elimination

In 2011, PAHO published a "Recommendation for documentation and verification of measles, rubella and CRS elimination in the Americas" to be used for the countries to document the elimination [56]. The documentation is based on the epidemiology of measles, rubella, and CRS, the quality of surveillance, molecular epidemiology, vaccinated population cohorts, and sustainability of measles, rubella, and CRS surveillance. At the same time, PAHO established an International Expert Committee (IEC) to verify the documents produced by the countries demonstrating elimination. Rubella was considered eliminated from the Americas in 2015 and measles in 2016.

11 Challenges and Perspectives

The demonstration by the Americas that measles and rubella elimination can be accomplished and maintained is a historic achievement and serves as an example for other regions. In the Americas, the challenge now is maintaining the high levels of vaccine coverage together with the comprehensive surveillance necessary to sustain elimination, in the face of competing priorities. Work is needed to produce a more heat-stable vaccine and easier modes of administration. Promising results have been achieved administering vaccines by aerosol and using micro-needles. Simplified diagnostic tests that can be used in the field or in the clinic are being evaluated and offer promise to enhance case-based surveillance.

References

1. Abernathy ES, Hubschen JM, Muller CP et al (2011) Status of global virologic surveillance for rubella viruses. J Infect Dis 204(suppl 1):S524–S532
2. Andrus JK, De Quadros CA, Solorzano CC et al (2011) Measles and rubella eradication in the Americas. Vaccine 29(suppl 4):D91–D96
3. Artimos De Oliveira S, Bastos Camacho LA, Uzeda Barreto MC et al (2011) Serologic status of women in an urban population in Brazil before and after rubella immunization campaign using routine screening data. J Infect Dis 204(suppl 2):S664–S668
4. Artimos De Oliveira S, Jin L, Siqueira MM et al (2000) Atypical measles in a patient twice vaccinated against measles: transmission from an unvaccinated household contact. Vaccine 19:1093–1096

5. Badilla X, Morice A, Avila-Aguero ML et al (2007) Fetal risk associated with rubella vaccination during pregnancy. Pediatr Infect Dis J 26:830–835
6. Banatvala JE, Brown DW (2004) Rubella. Lancet 363:1127–1137
7. Baumeister E, Siqueira MM, Savy V et al (2000) Genetic characterization of wild-type measles viruses isolated during the 1998 measles epidemic in Argentina. Acta Virol 44:169–174
8. Bellini WJ, Rota JS, Lowe LE et al (2005) Subacute sclerosing panencephalitis: more cases of this fatal disease are prevented by measles immunization than was previously recognized. J Infect Dis 192:1686–1693
9. Canepa E, Siqueira MM, Hortal M et al (2000) Recent measles viral activity in Uruguai: serological and genetic approaches. Acta Virol 44:35–39
10. Castillo-Solorzano C, Carrasco P, Tambini G et al (2003) New horizons in the control of rubella and prevention of congenital rubella syndrome in the Americas. J Infect Dis 187(suppl 1):S146–S152
11. Castillo-Solorzano C, Marsigli C, Bravo-Alcantara P et al (2011) Elimination of rubella and congenital rubella syndrome in the Americas. J Infect Dis 204(suppl 2):S571–S578
12. Castillo-Solorzano C, Marsigli C, Danovaro-Holliday MC et al (2011) Measles and rubella elimination initiatives in the Americas: lessons learned and best practices. J Infect Dis 204(Suppl 1):S279–S283
13. Castillo-Solorzano C, Reef SE, Morice A et al (2011) Rubella vaccination of unknowingly pregnant women during mass campaigns for rubella and congenital rubella syndrome elimination, the Americas 2001–2008. J Infect Dis 204(suppl 2):S713–S717
14. Castillo-Solorzano CC, Matus CR, Flannery B et al (2011) The Americas: paving the road toward global measles eradication. J Infect Dis 204(suppl 1):S270–S278
15. Castro-Silva R, Camacho LA, Amorim L et al (2003) Serological surveillance of measles in blood donors in Rio de Janeiro, Brazil. Rev Panam Salud Publica 14:334–340
16. Chen RT, Goldbaum GM, Wassilak SG et al (1989) An explosive point-source measles outbreak in a highly vaccinated population. Modes of transmission and risk factors for disease. Am J Epidemiol 129:173–182
17. Curti SP, Figueiredo CA, De Oliveira MI et al (2013) Molecular epidemiology of rubella viruses involved in congenital rubella infections in Sao Paulo, Brazil, between 1996 and 2009. J Med Virol 85:2034–2041
18. Cutts FT, Robertson SE, Diaz-Ortega JL et al (1997) Control of rubella and congenital rubella syndrome (CRS) in developing countries. Part 1: Burden of disease from CRS. Bull World Health Organ 75:55–68
19. Cutts FT, Vynnycky E (1999) Modelling the incidence of congenital rubella syndrome in developing countries. Int J Epidemiol 28:1176–1184
20. Da Silva E, Sa GR, Camacho LA, Stavola MS et al (2011) Pregnancy outcomes following rubella vaccination: a prospective study in the state of Rio de Janeiro, Brazil, 2001–2002. J Infect Dis 204(suppl 2):S722–S728
21. De Moraes JC, Toscano CM, De Barros EN et al (2011) Etiologies of rash and fever illnesses in Campinas, Brazil. J Infect Dis 204(suppl 2):S627–S636
22. De Quadros CA, Andrus JK, Danovaro-Holliday MC et al (2008) Feasibility of global measles eradication after interruption of transmission in the Americas. Expert Rev Vaccines 7:355–362
23. De Quadros CA, Izurieta H, Carrasco P et al (2003) Progress toward measles eradication in the region of the Americas. J Infect Dis 187(suppl 1):S102–S110
24. De Quadros CA, Olive JM, Hersh BS et al (1996) Measles elimination in the Americas. Evolving strategies. JAMA 275:224–229
25. De Swart RL, Yuksel S, Langerijs CN et al (2009) Depletion of measles virus glycoprotein-specific antibodies from human sera reveals genotype-specific neutralizing antibodies. J Gen Virol 90:2982–2989
26. De Swart RL, Yuksel S, Osterhaus AD (2005) Relative contributions of measles virus hemagglutinin- and fusion protein-specific serum antibodies to virus neutralization. J Virol 79:11547–11551

27. Dias Leite R, Naaman Berezin E (2015) Measles in Latin America: current situation. J Pediatric Infect Dis Soc 4:179–181
28. Donadio FF, Siqueira MM, Vyse A et al (2003) The genomic analysis of rubella virus detected from outbreak and sporadic cases in Rio de Janeiro state, Brazil. J Clin Virol 27:205–209
29. Enders G, Nickerl-Pacher U, Miller E et al (1988) Outcome of confirmed periconceptional maternal rubella. Lancet 1:1445–1447
30. Flores A, Villeda JA, Rodriguez-Fernandez R et al (2011) Advocacy and resource mobilization for rubella elimination in Guatemala. J Infect Dis 204(suppl 2):S598–S602
31. Galindo MA, Santin M, Resik S et al (1998) Eradication of measles in Cuba. Rev Panam Salud Publica 4:171–177
32. Gallegos D, Olea A, Sotomayor V et al (2011) Rubella outbreaks following virus importations: the experience of Chile. J Infect Dis 204(suppl 2):S669–S674
33. Gardy JL, Naus M, Amlani A et al (2015) Whole-genome sequencing of measles virus genotypes H1 and D8 during outbreaks of infection following the 2010 Olympic Winter Games reveals viral transmission routes. J Infect Dis 212:1574–1578
34. Grant GB, Reef SE, Dabbagh A et al (2015) Global progress toward rubella and congenital rubella syndrome control and elimination: 2000–2014. MMWR Morb Mortal Wkly Rep 64:1052–1055
35. Griffin DE (2013) Measles virus. In: Knipe DM, Howley PM (eds) Fields virology. Wolters Kluwer/Lippincott Williams & Wilkins Health, Philadelphia, pp 1042–1069
36. Henao-Restrepo AM, Strebel P, John Hoekstra E et al (2003) Experience in global measles control, 1990–2001. J Infect Dis 187(suppl 1):S15–S21
37. Hersh BS, Tambini G, Nogueira AC et al (2000) Review of regional measles surveillance data in the Americas, 1996-99. Lancet 355:1943–1948
38. Hickman CJ, Hyde TB, Sowers SB et al (2011) Laboratory characterization of measles virus infection in previously vaccinated and unvaccinated individuals. J Infect Dis 204(suppl 1): S549–S558
39. Hobman TC (2013) Rubella virus. In: Knipe DM, Howley PM (eds) Fields virology. Wolters Kluwer/Lippincott Williams & Wilkins Health, Philadelphia, pp 687–711
40. Icenogle JP, Siqueira MM, Abernathy ES et al (2011) Virologic surveillance for wild-type rubella viruses in the Americas. J Infect Dis 204(suppl 2):S647–S651
41. Irons B, Dobbins JG, Caribbean Vaccine M (2011) The Caribbean experience in maintaining high measles vaccine coverage. J Infect Dis 204(suppl 1):S284–S288
42. Jimenez G, Avila-Aguero ML, Morice A et al (2007) Estimating the burden of congenital rubella syndrome in Costa Rica, 1996–2001. Pediatr Infect Dis J 26:382–386
43. Lambert N, Strebel P, Orenstein W et al (2015) Rubella. Lancet 385:2297–2307
44. Lanzieri TM, Segatto TC, Siqueira MM et al (2003) Burden of congenital rubella syndrome after a community-wide rubella outbreak, Rio Branco, Acre, Brazil, 2000 to 2001. Pediatr Infect Dis J 22:323–329
45. Leite RD, Barreto JL, Sousa AQ (2015) Measles reemergence in Ceara, Northeast Brazil, 15 years after elimination. Emerg Infect Dis 21:1681–1683
46. Martinez-Quintana E, Castillo-Solorzano C, Torner N et al (2015) Congenital rubella syndrome: a matter of concern. Rev Panam Salud Publica 37:179–186
47. Menegolla IA, Bercini MA, Schermann MT et al (2011) Outbreak of rubella after mass vaccination of children and adult women: challenges for rubella elimination strategies. Rev Panam Salud Publica 29:243–251
48. Miller E, Cradock-Watson JE, Pollock TM (1982) Consequences of confirmed maternal rubella at successive stages of pregnancy. Lancet 2:781–784
49. Morice A, Avila-Aguero ML, Salas-Peraza D et al (2011) Approach to verify the status of measles, rubella, and congenital rubella syndrome elimination in Costa Rica. J Infect Dis 204(suppl 2):S690–S697
50. Moss WJ, Strebel P (2011) Biological feasibility of measles eradication. J Infect Dis 204(suppl 1):S47–S53

51. Oliveira SA, Siqueira MM, Camacho LA et al (2003) Use of RT-PCR on oral fluid samples to assist the identification of measles cases during an outbreak. Epidemiol Infect 130:101–106
52. Oliveira SA, Siqueira MM, Camacho LA et al (2001) The aetiology of maculopapular rash diseases in Niteroi, State of Rio de Janeiro, Brazil: implications for measles surveillance. Epidemiol Infect 127:509–516
53. Oliveira SA, Siqueira MM, Mann GF et al (1996) Measles antibody prevalence after mass immunization campaign in Niteroi, state of Rio de Janeiro, Brazil. Rev Inst Med Trop Sao Paulo 38:355–358
54. Pan American Health Organization (2016) Impact of measles and rubella elimination strategies: the Americas, 1980–2016. PAHO, oral presentation by Dr Gloria Rey. Reprinted with the permission of the Pan American Health Organization
55. Pan American Health Organization (1979) Measles in the region of the Americas, 1971–1978. EPI Newsl 1(3):4–5
56. Pan American Health Organization (2011) Plan of action: recommendation for documentation and verification of measles, rubella and congenital rubella syndrome elimination in the Americas. PAHO, Washington, DC
57. Pan American Health Organization (1998) Update: Sao Paulo measles outbreak. EPI Newsl 20:5–6
58. Patel MK, Gacic-Dobo M, Strebel P et al (2016) Progress towards regional measles elimination worldwide, 2000–2015. Morb Mortal Wkly Rep (MMWR) 65:1228–1233
59. Plotkin SA (1999) Rubella. In: Plotkin SA, Orenstein WA (eds) Vaccines. Saunders, Philadelphia, pp 409–440
60. Prevots DR, Parise MS, Segatto TC et al (2003) Interruption of measles transmission in Brazil, 2000–2001. J Infect Dis 187(suppl 1):S111–S120
61. Quiroga R, Barrezueta O, Venczel L et al (2003) Interruption of indigenous measles transmission in Bolivia since October 2000. J Infect Dis 187(suppl 1):S121–S126
62. Robertson SE, Featherstone DA, Gacic-Dobo M et al (2003) Rubella and congenital rubella syndrome: global update. Rev Panam Salud Publica 14:306–315
63. Rota J, Lowe L, Rota P et al (2006) Identical genotype B3 sequences from measles patients in 4 countries, 2005. Emerg Infect Dis 12:1779–1781
64. Rota PA, Brown K, Mankertz A et al (2011) Global distribution of measles genotypes and measles molecular epidemiology. J Infect Dis 204(suppl 1):S514–S523
65. Rota PA, Moss WJ, Takeda M et al (2016) Measles. Nat Rev Dis Primers 2:1–16
66. Santos JI, Nakamura MA, Godoy MV et al (2004) Measles in Mexico, 1941–2001: interruption of endemic transmission and lessons learned. J Infect Dis 189(Suppl 1):S243–S250
67. Sarmiento H, Cobo OB, Morice A et al (2011) Measles outbreak in Venezuela: a new challenge to postelimination surveillance and control? J Infect Dis 204(suppl 2):S675–S682
68. Siqueira MM, Castro-Silva R, Cruz C et al (2001) Genomic characterization of wild-type measles viruses that circulated in different states in Brazil during the 1997 measles epidemic. J Med Virol 63:299–304
69. Soares RC, Siqueira MM, Toscano CM et al (2011) Follow-up study of unknowingly pregnant women vaccinated against rubella in Brazil, 2001–2002. J Infect Dis 204(suppl 2):S729–S736
70. Sowers SB, Rota JS, Hickman CJ et al (2016) High concentrations of measles neutralizing antibodies and high-avidity measles IgG accurately identify measles reinfection cases. Clin Vaccine Immunol 23:707–716
71. Stein CE, Birmingham M, Kurian M et al (2003) The global burden of measles in the year 2000: a model that uses country-specific indicators. J Infect Dis 187(suppl 1):S8–14
72. Strebel PM, Cochi SL, Hoekstra E et al (2011) A world without measles. J Infect Dis 204(suppl 1): S1–S3
73. Strebel PM, Henao-Restrepo AM, Hoekstra E et al (2004) Global measles elimination efforts: the significance of measles elimination in the United States. J Infect Dis 189(suppl 1): S251–S257
74. Suarez-Ognio L, Adrianzen A, Ortiz A et al (2007) A rubella serosurvey in postpartum women in the three regions of Peru. Rev Panam Salud Publica 22:110–117

75. Tahara M, Burckert JP, Kanou K et al (2016) Measles virus hemagglutinin protein epitopes: the basis of antigenic stability. Viruses 8:216–230
76. United Nations Children's Fund (1990) Plan of action for implementing the world declaration on the survival, protection and development of children in the 1990s. UNICEF, New York. https://www.unicef.org/wsc/plan.htm
77. Venczel L, Rota J, Dietz V et al (2003) The measles laboratory network in the region of the Americas. J Infect Dis 187(suppl 1):S140–S145
78. Vianna RA, De Oliveira SA, Camacho LA et al (2008) Role of human herpesvirus 6 infection in young Brazilian children with rash illnesses. Pediatr Infect Dis J 27:533–537
79. Vynnycky E, Adams EJ, Cutts FT et al (2016) Using seroprevalence and immunisation coverage data to estimate the global burden of congenital rubella syndrome, 1996–2010: a systematic review. PLoS One 11:e0149160
80. World Health Organization (2016) Countries with rubella vaccine in the national immunization programme and planned introductions in 2016–2017. In: Vaccine introduction slides. http://www.who.int/immunization/monitoring_surveillance/burden/en/
81. World Health Organization (2007) Manual for the diagnosis of measles and rubella infection. WHO, Geneva
82. World Health Organization (2012) Measles virus nomenclature update: 2012. Wkly Epidemiol Rec 87:73–81
83. World Health Organization (2004) Progress towards measles elimination, Western Hemisphere, 2002–2003. Wkly Epidemiol Rec 79:149–151
84. World Health Organization (2011) Rubella vaccines: WHO position paper. Wkly Epidemiol Rec 86:301–316
85. World Health Organization (2013) Rubella virus nomenclature update: 2013. Wkly Epidemiol Rec 88:337–343
86. World Health Organization (1989) World Health Assembly (1989) Executive summary resolution WHA42.32. WHO, Geneva
87. World Health Organization and UNICEF (2002) WHO-UNICEF joint statement on strategies to reduce measles mortality worldwide. Wkly Epidemiol Rec 77:224–228

Part V
Sexually and Blood Borne Transmitted Diseases

Chapter 16
Hepatitis B Viruses

Selma A. de Gomes, Natalia M. Araujo, Diego Flichman, Rodolfo Campos, and Arturo Panduro

1 History

Epidemics of jaundice have been reported since before the Christian era [114]. However, the association of the clinical manifestation to a parenteral transmissible form of hepatitis did not occur until the late nineteenth century. Outbreaks of jaundice accompanied by symptoms of hepatitis were observed after vaccination during the years 1930–1940, when using vaccines stabilized with human lymph. Blood transfusions and repeated use of nonsterile needles were also identified as causes of hepatitis outbreaks [43, 138]. In 1965, Blumberg and colleagues [18] published what would become one of the most important landmarks in the field of viral hepatitis. During extensive work seeking to characterize inherited polymorphic genetic traits, they found, in a serum sample of a native Australian, an antigen that reacted specifically with antibodies from the serum of a U.S. hemophiliac patient. This "Australia antigen (Au)," which was relatively rare in North America and Western Europe but prevalent in African and Asian populations, was not readily associated

S.A. de Gomes (✉) • N.M. Araujo
Laboratório de Virologia Molecular, Instituto Oswaldo Cruz, Fiocruz, Av. Brasil 4365, 21040-36 Rio de Janeiro, Brazil
e-mail: selma@ioc.fiocruz.br; nmaraujo@ioc.fiocruz.br; https://ssl.gstatic.com/ui/v1/icons/mail/images/cleardot.gif

D. Flichman • R. Campos
Cátedra de Virología, Facultad de Farmacia y Bioquímica, Universidad de Buenos Aires, Junín 956 48 piso, C1113AAB, Ciudad Autónoma de Buenos Aires, Argentina
e-mail: dflichman@ffyb.uba.ar; rcampos@ffyb.uba.ar

A. Panduro
Department of Molecular Biology in Medicine, Civil Hospital of Guadalajara "Fray Antonio Alcalde" and University of Guadalajara, Hospital 278, Col Centro, Guadalajara, Jalisco 44280, Mexico
e-mail: apanduro@prodigy.net.mx

J.E. Ludert et al. (eds.), *Human Virology in Latin America*,
DOI 10.1007/978-3-319-54567-7_16

with serum hepatitis. Purification of the Au antigen, now known as the hepatitis B virus surface antigen (HBsAg) from the serum of patients with hepatitis, enabled the identification of the complete viral particle of hepatitis B virus (HBV) by electron microscopy [15, 19]. The availability of assays to detect the Au antigen allowed researchers to confirm the long-held suspicion that HBV was a blood-borne virus responsible for acute and chronic hepatitis in different parts of the world [20, 124].

2 Classification and Virion Structure

HBV is the prototype of the *Hepadnaviridae* family, which is composed of DNA viruses with tropism for liver cells that share similarities in genome organization, intracellular life cycle, and a unique mechanism of replication. They are the only DNA viruses of animals that replicate their genome by reverse transcription. The family is further split into two recognized genera: *Orthohepadnavirus* and *Avihepadnavirus*, the latter representing viruses that infect birds (ducks, herons, cranes, geese, parrots) [93], and the former including viruses that infect mammals (e.g., humans and nonhuman primates, squirrels, woodchucks, bats) [101].

HBV produces three types of viral particles, which may be identified by electron microscopy of infected serum preparations. The infectious viral particle, also called the Dane particle, has an outer diameter of approximately 42 nm. Its nucleocapsid has an icosahedral symmetry made up of 120 dimers of the core protein (hepatitis B core antigen, HBcAg), which encloses a single copy of the genome. Capsids are surrounded by an outer lipoprotein envelope embedded with surface viral glycoproteins (HBsAg), which are present in three different forms: large (L), middle (M), and small (S), and at different amounts in the viral envelope (rate of 1:1:4) [50]. These infectious particles have a density of 1.22 g/cm^3 in cesium chloride equilibrium gradients [29], and their concentration in the serum of infected individuals may be greater than 10^9 particles/ml. In addition, incomplete and noninfectious particles are found; these can have either a spherical or filamentous shape. These particles have an outer diameter of approximately 22 nm and a density of 1.18 g/cm^3 in cesium chloride equilibrium gradients. They are composed exclusively of HBsAg and some lipids derived from the host cells. They are found at high levels in the serum of infected individuals (concentration about 10^{13} particles/ml). Although not infectious, these particles are highly immunogenic and effectively induce a neutralizing anti-HBs antibody response [122].

The HBV genome consists of a molecule of a partially double-stranded circular DNA (where both strands are not covalently closed) of approximately 3200 nucleotides, one of the smallest genomes of the human viruses. Next to the 5′-ends of both DNA strands, there is a small sequence of 11 nucleotides that are directly repeated and so-called direct repeats (DR1 and DR2; Fig. 16.1). DR1 and DR2 are important sequences for the initiation of viral replication. The entire HBV genome is organized into four open reading frames (ORFs) known as pre-S/S, pre-C/C, P, and X [110]. All genes have regions of overlap with other genes and are encoded by the

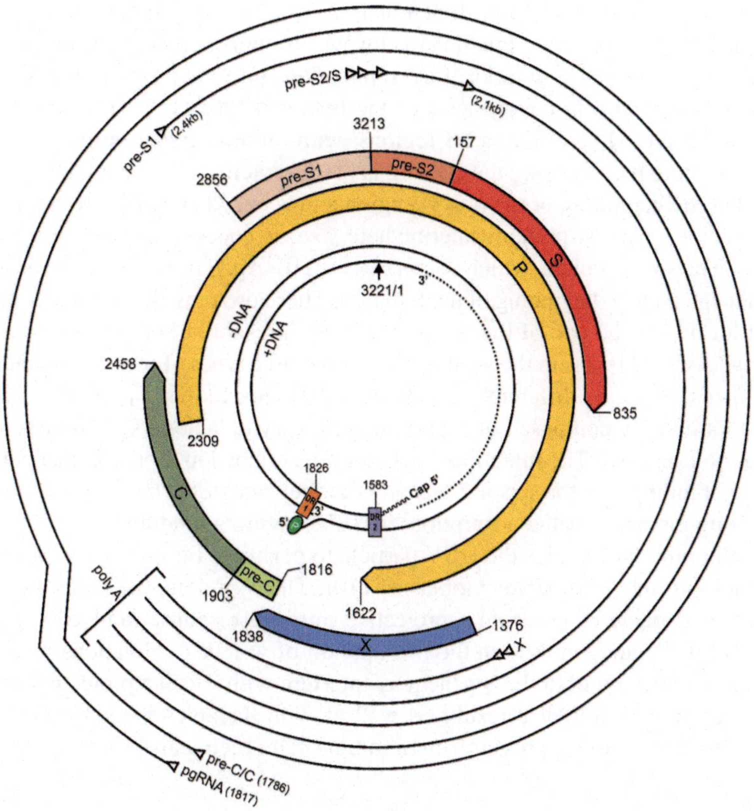

Fig. 16.1 Hepatitis B (HBV) genome organization. The circular viral DNA with direct repeats DR1 and DR2 is represented by the *inner circle*. The *dotted line* indicates the single-stranded region of the genome. The length of the genome is 3221 bp, the size of HBV/A. Numbering starting from a unique *Eco*RI site from most HBV/A isolates. The *outer rows* (in *color*) indicate the four open reading frames: pre-S/S, pre-C/C, P, and X. *External lines* indicate the viral RNAs with the location of the 5′-ends (triangles) and to the 3′-(polyadenylation site) common to all RNAs. (Scheme adapted from [42])

complete strand (see Fig. 16.1). The partially overlapping arrangement of the different ORFs allows the HBV to encode 50% more protein than would be expected by the size of its genome [42]. The ORF pre-S/S is responsible for the synthesis of the L, M, and S proteins which form HBsAg. The pre-C/C ORF is responsible for the synthesis of HBcAg and a secretable antigen found in sera of infected individuals, called HBeAg. The detection of this antigen is used in the clinical assessment of patients as it is indicative of active viral replication. The P ORF covers approximately three quarters of the genome and encodes the HBV DNA polymerase, a multifunctional protein with three functional domains: the terminal protein domain located at the N terminal portion of the protein, the reverse transcriptase domain, and an RNase H domain at the C-terminal portion of the protein. HBV polymerase is homologous to other reverse transcriptases [137]. The X ORF is responsible for the synthesis of HBxAg, a multifunctional viral regulatory protein that modulates the transcription process, signaling pathways, protein degradation, and cellular response to stresses. These tasks affect replication and viral proliferation; this protein is also involved in oncogenesis of the host cell [56, 111]. The pre-S/S gene includes the pre-S1, pre-S2, and S regions, with three in-frame initiation codons. The large protein composing the HBsAg is the L protein, whose initiation codon is located at the beginning of the pre-S1 region and is encoded by the pre-S1, pre-S2, and S regions. The M protein (intermediate size) is encoded by the pre-S2 and S regions. The small protein, which comprises the HBsAg, S protein, has the initiation codon located at the beginning of the S region. These proteins all have the same stop codon located at the end of the S region and are present in both glycosylated and non-glycosylated forms in the viral particle. The three types of protein are unevenly distributed among the different forms of viral particles. Subviral particles of 22 nm are predominantly composed of S protein, with varying amounts of M protein and little or no L protein. The infectious particles are enriched in L protein that contains the virus-binding site for specific receptors on hepatocytes. The M protein acts as connecting element for the adsorption of HBV, having a binding region to human serum albumin that enables the HBV particle to penetrate through albumin cellular receptors into the hepatocyte cytoplasm [110]. The S protein, the main protein of HBsAg, is capable of inducing a protective immune response (anti-HBs) against HBV and is the antigen used in the formulation of vaccines. Mutations in specific epitopes occurring within the S gene may interfere with vaccine protection and the interpretation of serological results as well as with therapies based on the use of specific antibodies to suppress infection in transplant recipients [106].

3 Replication

Hepatocytes are the primary site of HBV replication. In vitro HBV infection has been achieved only in human hepatocyte primary cell cultures [41]. Although HBV is not cultivable in cell lines, several human hepatoma-derived cell lines such as HepG2, HuH-6, and HuH-7 are able to support HBV replication by using an integrated or transfected HBV genome as a template [112, 129]. The unavailability of a reproducible system for HBV cultivation hampers studies of viral replication and life cycle. The use of animal models in experimental infections is limited because HBV infects only humans and some nonhuman primates.

The HBV replication strategy is unique among DNA animal viruses by proceeding through a step of reverse transcription of a pre-genomic RNA intermediate [80]. The initial stage of viral infection is the adsorption of the viral particle to susceptible hepatocytes (Fig. 16.2). Once in the cytoplasm of the hepatocyte, the virion loses its envelope, and the nucleocapsid is transported to the cell nucleus, where the relaxed circular DNA genome is released and then converted into a covalently closed circular DNA (cccDNA) form; this occurs after the repair of the incomplete positive strand of the viral genome by a cellular DNA polymerase. The cellular RNA polymerase II is responsible for transcribing cccDNA into genomic (longer than one unit of the

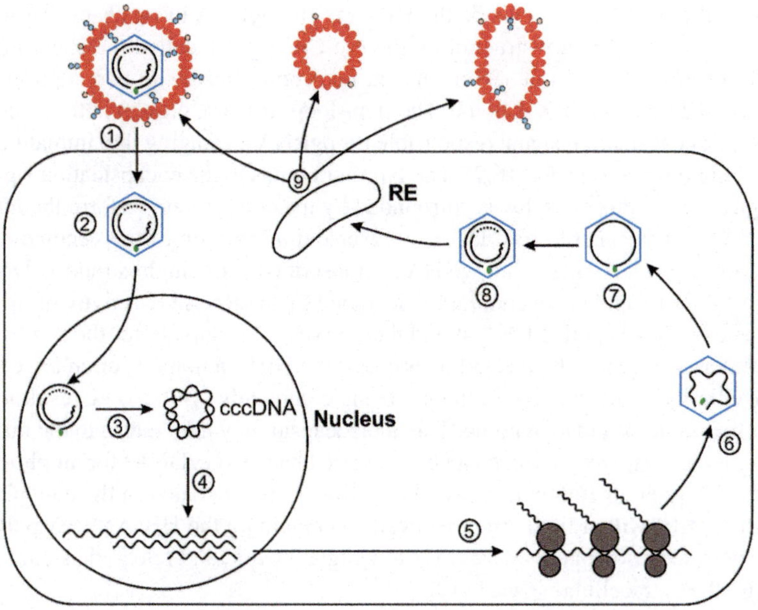

Fig. 16.2 Replicative cycle of HBV. (*1*) Penetration. (*2*) Uncoating. (*3*) The genomic DNA is converted to cccDNA from inside the nucleus. (*4*) Transcription of the viral RNAs. (*5*) Translation of the viral proteins. (*6*) Encapsidation of the pgRNA into core particles. (*7*) Synthesis of the negative strand of viral DNA. (*8*) Synthesis of the complementary strand and migration of the progeny core to the endoplasmic reticulum (ER), for the acquisition of envelope (HBsAg). (*9*) Release of enveloped virions and the other two forms of viral particles, spherical and filamentous. (Scheme adapted from [42])

genome of about 3.5 kb) and subgenomic viral RNAs (sizes between 0.9 and 2.4 kb), specialized in translating different gene products. All transcripts are capped at the 5′-end and share the same polyadenylation signal located in the pre-C/C gene (see Fig. 16.1). The four open reading frames of the HBV responsible for the synthesis of seven different products (HBeAg, HBcAg, P, HBx, and the envelope proteins L, M, and S) are controlled by four promoters (pre-S1, pre-S2/S, pre-C/C, and X). The subgenomic RNAs act exclusively as messenger RNAs for translation of the three envelope proteins (L, M, and S) and the X protein (see Fig. 16.2). The pre-S1 promoter controls transcription of a 2.4-kb subgenomic RNA that is the sole messenger for the L protein. The pre-S2/S promoter controls transcription of a family of 2.1-kb RNAs with microheterogeneity at the 5′-end; one of the transcripts starts immediately before the pre-S2 region initiation codon and the other after this start codon. Therefore, one of the 2.1-kb transcripts gives rise to the M protein and the remaining 2.1-kb RNAs originate the small S protein. The pre-C/C promoter controls transcription of several RNA molecules of about 3.5 kb, which have heterogeneity at the 5′-end. Among these RNAs, the pre-genomic RNA (pgRNA) is the template for the synthesis of genomic DNA using a complex reverse transcription process. The pgRNA also serves as the template for the synthesis of HBcAg protein and DNA polymerase protein. The 3.5-kb RNAs with 5′-ends located upstream of the start codon of the pre-C region encode the HBeAg. The pgRNA has a characteristic and well-conserved secondary structure in the pre-C/C region, called the encapsidation signal. A terminal region of 130 redundant nucleotides results in two copies of this structure at each end of the genome. The stem-loop structure present in the 5′-portion forms the encapsidation signal responsible for pgRNA packaging into immature capsids during replication (Fig. 16.2). The P protein binds to the encapsidation signal in the pgRNA, and this complex is surrounded by the core protein to form the nucleocapsid. The synthesis of HBV DNA by reverse transcription of the genome occurs inside the capsid. At first, the pgRNA is transcribed into single-stranded DNA of negative polarity and is concomitantly degraded by the RNase H activity of the viral polymerase. This negative DNA strand then serves as a template for the synthesis of the DNA positive strand. During this process, the HBV genome is circularized with the peculiarity of the positive strand not being completely synthesized, resulting in a partially double-stranded genome. The nucleocapsid may then return to the nucleus, releasing the viral DNA (which can be converted back to cccDNA) for amplification of cccDNA pool, or follow the assembly pathway, being coated in the endoplasmic reticulum (ER) with a lipid–protein envelope containing the HBsAg, transported to the protein secretion pathway from ER to Golgi network, and released as viral progeny into the extracellular space [110].

4 Transmission and Pathogenesis

Hepatitis B virus (HBV) transmission occurs primarily by parenteral exposure to blood, blood products, and organ tissue transplantation or through sexual and perinatal exposure [132]. HBV is mainly found in the blood of infected individuals, but can also be detected in bodily fluids such as urine, saliva, nasopharyngeal fluid, semen, and menstrual fluid [30].

HBV is highly resistant to environmental conditions; therefore, infection may occur through the use of dental instruments, syringes, needles, piercing tools used for acupuncture, tattooing instruments, and the sharing of needles and syringes among injectable drug users [1, 44]. Vertical transmission (mother to child) is often documented in populations living in highly endemic regions and is mostly caused by the exposure of the newborn to the mother's blood during delivery (particularly when mothers are HBeAg positive). However, infections also happen after birth from the close mother–child contact [75]. In the U.S., sexual transmission is the most common route of HBV transmission [27]. Those who are at higher risk of infection are healthcare providers and emergency responders, sexually active heterosexuals, men who have sex with men, individuals diagnosed with a sexually transmitted disease, illicit drug users (injecting, inhaling, snorting, pill popping), sexual contacts or close household members of an infected person, and kidney dialysis patients [39].

In Latin American (LA), the routes of HBV transmission are highly variable and are associated with heterogeneous disease distribution among countries. The Amazon basin is the most affected region in LA [117] with the highest prevalence reported in older age groups. In the Amazon horizontal transmission seems to be the most common route of infection, through either parenteral or sexual exposure. Cultural practices such as bloodletting, scarification, and tattooing contribute to an increased risk of horizontal transmission in this region. Vertical transmission and childhood horizontal transmission seem to be important routes of transmission in isolated Indian tribes with a high prevalence of HBV, such as in the Yanomami communities of the Upper Orinoco Basin where the majority of the HBsAg carriers (75%) were women of childbearing age [127].

The pathogenesis of HBV is not associated with a direct cytopathic effect of HBV on hepatocytes. Thus, it is assumed that the hepatic damage is related to an immune-mediated host response targeted against specific viral antigens in the infected hepatocytes. The elimination of HBV appears to be associated with the combined action of immune cells, the humoral response, and intracellular virus inactivation induced by the action of cytokines.

HBV may cause acute, fulminant, chronic hepatitis. During chronic infection, the disease may progress to liver cirrhosis or hepatocellular carcinoma (HCC). Most symptoms of acute hepatitis appear between 45 and 180 days after infection and include fatigue, anorexia, nausea, abdominal discomfort, vomiting, and mild hepatomegaly. During this period, there is an increase in serum transaminases and bilirubin levels, indicating liver cell damage with high levels of HBsAg and HBV DNA. In a few cases (<1%), acute infection may result in fulminant hepatic insufficiency, characterized by the development of hepatic encephalopathy a few weeks after the onset of symptoms of hepatitis, with consequent risk of death.

Chronic infection is defined as the persistence of HBsAg in the serum of an individual for 6 months or more. The risk of developing chronic infection varies inversely with age. A high percentage (>90%) of infants less than 1 year old exposed to HBV develop chronic hepatitis B infections because of the immaturity of their immune system. In adults, approximately 5% of infected individuals develop chronic hepatitis that may progress to a serious liver disease with risk of cirrhosis or HCC. The great majority of patients with chronic hepatitis B remain asymptomatic for many years. The severity of liver damage over the course of chronic hepatitis can lead to the development of cirrhosis (approximately 20% of cases) or hepatocellular carcinoma. The natural history of chronic hepatitis B consists of four phases, which are not necessarily sequential: immune tolerance, immunoreactive HBeAg positive, immunoreactive anti-HBe positive, and inactive carrier (see following). In some cases (20–30%), those in the inactive carrier stage may undergo spontaneous reactivation, usually in patients receiving immunosuppressive therapy.

5 Laboratory Diagnosis

Hepatitis B virus (HBV) diagnosis allows differentiating susceptible individuals and protected individuals (positive for anti-HBs antibodies) from those with an acute or a chronic infection, and between the latter, those who need to receive antiviral therapy.

Primary infection is characterized by serological detection of HBsAg and HBeAg (a serological marker of active viral replication), the presence of anti-HBc antibodies (total IgG and IgM), and high levels of HBV DNA ($>10^6$ IU/ml). In the blood of those individuals who resolve the infection, seroconversion to anti-HBs and anti-HBe antibody-positive status occurs, along with the maintenance of lifelong anti-HBc IgG antibodies. The presence of anti-HBs antibodies is pathognomonic of those who have resolved the infection or been vaccinated; those with a previous infection are distinguished by the presence of anti-HBc antibodies.

The persistence of HBsAg in the blood for more than 6 months is an indication of chronic infection. The immune tolerance phase of chronic infection is characterized by high levels of viral replication ($>10^6$ IU/ml) and positivity for HBsAg, HBeAg, anti-HBc IgG antibodies, and normal alanine aminotransferase (ALT) levels. The immune activation phase is characterized by an HBeAg-positive stage, with elevated HBV DNA ($>10^6$ IU/ml) and elevated ALT levels (with or without flares). The HBsAg and anti-HBc antibodies remain detectable, as in all phases of chronic infection. In addition, anti-HBc IgM may be detectable at lower levels than in acute infection and is usually associated with ALT flares. The following phase of chronic infection is characterized by the HBeAg seroconversion to anti-HBe (immunoreactive anti-HBe positive phase). In this phase, an abrupt reduction of HBV DNA (10^6–10^3 IU/ml) is observed. The last phase is the inactive carrier state, which is characterized by low or undetectable viral loads ($<2 \times 10^3$ IU/ml), detection of HBsAg, anti-HBc, anti-HBe, and normal ALT levels.

6 Treatment Prevention and Control

The treatment of chronic hepatitis B aims to reduce or suppress viral replication and prevent the progression of liver damage that may lead to cirrhosis and liver failure, as well as the development of hepatocellular carcinoma and subsequent death. Currently, two strategies are commonly used in the treatment of chronic hepatitis: (1) the use of alpha-interferon or pegylated alpha-interferon that have direct action against HBV and stimulate the immune response of the host and (2) nucleotide analogues (NAs). The NAs are oral antiviral agents that have a fast and powerful inhibitory effect on the reverse transcriptase activity of the HBV polymerase, are safe and effective in suppressing viral DNA, and promote the normalization of transaminase and histological improvement [136]. Interferon therapies have direct antiviral activity and immunostimulatory properties; their potential advantages compared to NAs include the lack of drug resistance, administration for a limited period of time, and a higher rate of HBeAg and HBsAg seroconversion. The disadvantages are mainly related to the parenteral route of administration, poor tolerance associated with important side effects, high cost, and its effectiveness limited to a small proportion of highly selected patients. Therefore, interferon therapy has currently been largely replaced by NAs.

The main goal of NA treatment is to control viral replication, to induce HBeAg to anti-HBeAg seroconversion, and to induce remission. It is important to note that current NA treatments usually do not allow seroconversion to anti-HBsAg because of maintenance of cccDNA. As Na treatment do not affect the cccDNA, NA therapy does not eradicate HBV from the liver. Hence, the ultimate goal is to achieve sustained suppression of viral replication, preventing progression to cirrhosis, liver failure, and terminal hepatocellular carcinoma. Accordingly, NA treatment is long term with the consequent risk of selection of resistant strains. It should be emphasized that individuals with chronic infections and evidence of liver injury (immune-activation phase, either HBeAg or anti-HBe positive) are the most suitable candidates for antiviral therapy [121].

Currently, five NA drugs are licensed for the treatment of HBV infection: lamivudine (LAM), adefovir (ADV), entecavir (ETV), telbivudine (LdT), and tenofovir (TDF). With the spontaneous variability of the viral genome, pharmacological pressure may select viral species that exhibit enhanced replication ability in the new environment created by the use of antivirals. LAM was the first NA approved in the mid-1990s. Although LAM is well tolerated and low in cost, resistance may occur in up to 80% of patients after 5 years of therapy. Resistant variants to most of the currently available NAs have now been observed but with significantly different frequencies. Mutations that confer resistance to NAs are located in the reverse transcriptase (rt) domain of the viral polymerase gene. LAM and TdT are no longer indicated as first-line treatments for HBV [40] because of the high rates of occurrence of resistance. ETV and TDF are currently considered the most potent antiviral agents and are low risk for inducing resistance.

On the other hand, an effective and efficient hepatitis B vaccine is currently available. The first vaccine prototype was obtained by heat inactivation of HBV from the plasma of chronically infected individuals; it was then replaced by a plasma-derived vaccine containing noninfectious HBV particles. Finally, in the mid-1980s, the current vaccine was developed, which consists of a yeast-derived recombinant HBsAg protein and is effective at producing protection in up to 95% of immunocompetent recipients. Currently, 184 of the 194 WHO member states (94.8%) have nationwide vaccination programs, and global coverage with three doses is estimated at 82% [26].

Despite the high efficacy of the HBV vaccine, breakthrough infections from vaccine escape mutations have been reported in vaccinated individuals, which accentuates the importance of the escape mutants. The neutralizing antibodies are primarily specific for the "a" determinant of the HBsAg. Mutations causing a conformational change within this epitope could affect HBsAg antigenicity and be responsible for the escape from vaccine-induced immunity. In addition, these variants may also provide false-negative results in serological tests, which are known as false occult hepatitis B infection (OBI) [106, 113].

The prevalence of chronic hepatitis B has declined dramatically worldwide since vaccination program implementation; however, the massive implementation of the vaccine has acted as a powerful selective force for vaccine escape mutants. For instance, HBV prevalence in Taiwan declined from 8.6% to 2.1% between 1984 and 1999, whereas vaccine escape mutants increased from 7.8% to 28.1% during the same time period [52]. By using mathematical models, it was estimated that, in the coming decades, vaccine escape mutants will become the dominant HBV quasi-species globally; if this occurs, reformulation of the vaccine should be considered [133].

7 Epidemiology

The World Health Organization estimated that, worldwide, about 2 billion people have been infected with HBV [132]. Among them, 240 million are chronic HBV carriers. Overall, almost half the global population lives in areas where HBV infection is highly endemic (HBsAg seroprevalence >8%). High endemicity is associated with densely populated locations, such as in Asia, and with poverty in regions lacking economic and hygienic resources including China, Indonesia, Nigeria, and many other regions in Asia and Africa. An intermediate level of HBV endemicity (HBsAg prevalence between 2% and 7%) is observed in southern Europe, the Middle East, South Asia, and in some LA countries including Guatemala, Belize, El Salvador, Honduras, Haiti, the Dominican Republic, Puerto Rico, Ecuador, Venezuela, Guyana, Surinam, and French Guyana. Many developed nations, including the U.S., fall into the low endemicity category (<2%) [90, 103, 113]. In LA, most Central and South American countries are now considered to be low prevalence, including most regions of Brazil, Mexico, Honduras, Nicaragua, Costa Rica, Panama, Cuba, Paraguay, Uruguay, Chile, Argentina, Peru, and northern Colombia. However, even in countries with low prevalence, it is possible to find isolated areas of high prevalence, such as the western Amazon basin, including Brazil and Peru, with observed HBsAg seroprevalence rates greater than 10%. A systematic review of studies reporting worldwide HBsAg seroprevalence, data collected over a 27-year period (1980–2007) determined that the prevalence of chronic HBV infection decreased in most regions: this was particularly evident in central sub-Saharan Africa, tropical and central LA, Southeast Asia, and Central Europe. The decline in HBV prevalence may be associated with better hygiene and quality of life as well as expanded immunization [90]. A recent study conducted in Brazil [117] showed a reduction in HBV prevalence countrywide, classifying Brazil as a whole as a low endemicity country. However, isolated regions with high prevalence persist, particularly the Amazon, as well as in specific groups such as homeless people in large cities and isolated Afro-descendant communities in the center of Brazil. The prevalence of anti-HBs antibodies alone (compatible with vaccine response) ranges from 50% to 90%. However, isolated and distant localities still have low coverage rates of vaccination. There is a need to intensify vaccination strategies for young people and adults in specific regions with persistently high HBV infection prevalence [117].

8 Genetic Variability

8.1 Genotypes and Geographic Distribution

Human HBV isolates have been classified into eight genotypes, denoted A (HBV/A) to H (HBV/H) and two tentative (HBV/I and HBV/J) genotypes based on a sequence divergence greater than 7.5% in the entire genome [11, 85–87, 118, 120, 128]. The significant diversity within some HBV genotypes (A–D, F, and I) has led to division into numerous sub-genotypes based on intergroup nucleotide divergence between 4% and 7.5% over the full-length genome [97, 115]. Figure 16.3 shows HBV genotypes and sub-genotypes. HBV genotypes have different genome lengths, ranging from 3182 nucleotides (nt) for HBV/D and HBV/J to 3248 nt for HBV/G. The

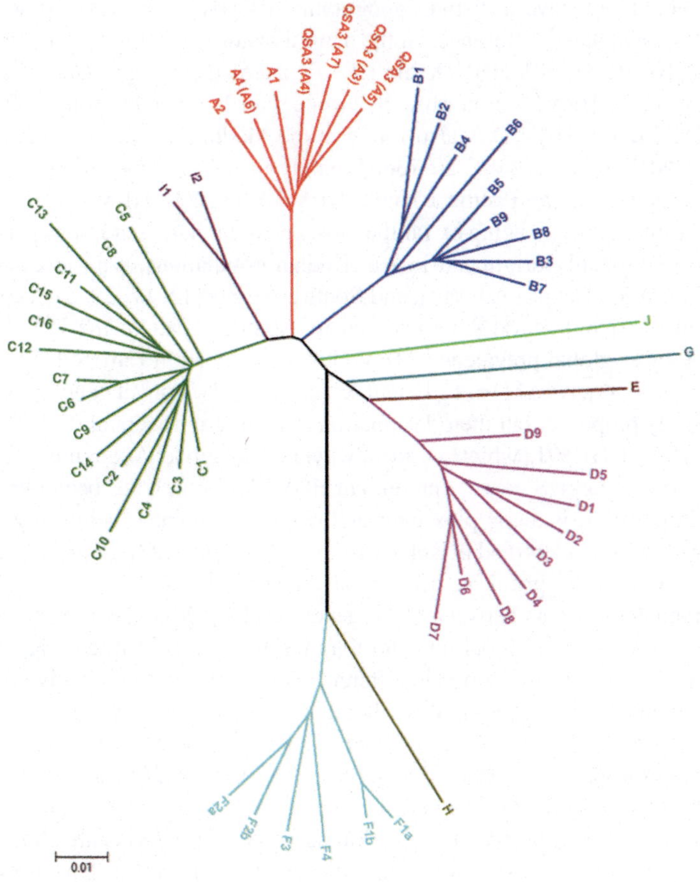

Fig. 16.3 Neighbor-joining phylogenetic tree based on HBV full-length genomes of all genotypes and sub-genotypes. HBV/A sub-genotypes before reclassification are shown in *parentheses*. After reclassification, HBV/A3 to HBV/A5 and HBV/A7 were grouped as quasi-sub-genotype A3 (QS-A3). HBV/A6 was renamed as HBV/A4. GenBank accession numbers of the sequences are listed here: A1, JN182318; A2, HE576989; A3, AB194951; A4, AY934764; A5, FJ692613; A6, GQ331047; A7, FN545833; B1, AB642091; B2, FJ899779; B3, GQ924617; B4, GQ924626; B5, GQ924640; B6, JN792893; B7, GQ358137; B8, GQ358147; B9, GQ358149; C1, AB697490; C2, GQ358158; C3, DQ089801; C4, HM011493; C5, EU410080; C6, EU670263; C7, GU721029; C8, AP011106; C9, AP011108; C10, AB540583; C11, AB554019; C12, AB554025; C13, AB644280; C14, AB644284; C15, AB644286; C16, AB644287; D1, GU456636; D2, GQ477452; D3, EU594434; D4, GQ922003; D5, GQ205377; D6, KF170740; D7, FJ904442; D8, FN594770; D9, JN664942; E, FN594748; F1a, AY090459; F1b, FJ709464; F2a, KC494405; F2b, DQ899146; F3, DQ899150; F4, AB166850; G, AB625342; H, AB516393; I1, FJ023659; I2, FJ023664; J, AB486012

genome size is 3212 nt for HBV/E; 3215 nt for genotypes B, C, F, H, and I; and 3221 nt for HBV/A. Compared to genotypes B, C, F, H, and I, HBV/A has 6 more nt in the polymerase gene, and HBV/D has a 33-nt deletion in the pre-S1 region. Both HBV/E and HBV/G have a deletion of 3 nt in the polymerase gene (codon 11). Finally, HBV/G has an insertion of 36 nt at the N-terminus of the core gene.

HBV genotypes have a distinct geographic distribution around the world and may be responsible for differences in the natural history and clinical outcome of the infection [10, 46, 58, 68]. HBV/A and HBV/D are distributed globally, with a high predominance of HBV/A in northwest Europe, North America, South Africa, and Brazil [58, 74, 83]. HBV/D is also found in the Mediterranean, India, and Russia [46, 134]. HBV/B and HBV/C are found mainly in Asia and the circumpolar north (HBV/B), as well as the Pacific Islands (HBV/C) [66, 83]. HBV/E is confined to West and Central Africa, being by far the most prevalent in these regions [4]. HBV/F and HBV/H probably originated in Amerindian populations, with localization of HBV/F in Alaska, Central America, and South America [12, 17, 32, 57, 126] and of HBV/H in Mexico, Central America, and the southern part of the U.S. [11, 91]. Despite a lower global prevalence, HBV/G is widespread in Europe [63, 131], the Americas [2, 8, 89], Asia [116, 123], and Africa [123]. Two novel genotypes, I and J, were recently proposed, but their designation remains controversial [62]. The single HBV isolate of HBV/J, which is closely related to gibbon/orangutan genotypes (polymerase and large S genes) and human HBV/C (C gene), has been identified in Japan [120]. HBV/I is a complex inter-genotypic recombinant between genotypes A, C, and G and was identified in isolates from Vietnam [48, 128], Laos [87], southern China [37, 119], and a primitive tribe in eastern India [6, 47].

Recombination events between HBV genotypes have been frequently described. Moreover, as a result of global human migration flows, a shift in the prevalence of HBV genotypes has been reported in different countries that will probably give rise to new recombinant forms. Among described HBV recombinants, about 60% are B/C and C/D hybrids [7]. Overall, recombinants display circulation patterns similar to those of their original genotypes. Recombinants of the globally widespread HBV/A and HBV/D are also found worldwide. On the other hand, hybrids of HBV genotypes B/C are limited to South and East Asia. Similarly, HBV/E hybrids are almost exclusively found in African countries [7]. As expected, HBV/F hybrids have been identified in countries from South America, with no reports outside this continent [8, 13, 55, 67]. Interestingly, A/C/G recombinants (HBV/I) have been identified in China, India, Laos, and Vietnam, although HBV/G has never been described in these countries.

8.2 Distribution of Genotypes and Sub-genotypes in Latin America

The distribution of HBV genotypes in LA can be traced from three main sources of ancestral populations: (1) Amerindians, who are mainly HBV/F carriers [25, 28] and may also have contributed HBV/H; (2) Europeans (colonizers and immigrants) responsible for the introduction of HBV/A (more specifically sub-genotype A2) and

HBV/D [16, 22, 45]; and (3) African slaves, who contributed to the introduction of HBV/A, sub-genotype A1 [59, 64]. HBV/E, restricted to West and Central Africa, has rarely been found in LA, suggesting that this genotype probably emerged only after the slave trade in the Americas [5]. The proportion of these three main ancestry populations (Africans, Amerindians, and Europeans) varies with the history of each country in LA. Thus, the Mexican population has a high degree of Amerindian ancestry that ranges from 38% to 76% depending on geographic region, followed by European ancestry (8.5–50%) and African ancestry (9–18%) [103]. Similarly, most of the population of Central and South America has the same sociodemographic pattern with regard to source of ancestry [103]. However, dissimilar distribution of these admixtures may be observed in some LA countries. In Argentina and Colombia, European ancestry is predominant (more than 70%), followed by Amerindian (about 20%), and African ancestry (3%). In Brazil, a similar proportion of European ancestry (about 70%) is found in the country; however, the African proportion may range from 30% in northeast to 10% in the northern region. On the other hand, the Amerindian proportion is highest in the north (19.4%) yet relatively similar in the other three geographic regions [92].

8.3 *The Unique Genotypes F and H in Latin America*

HBV/F, the main genotype of Amerindians, has been described as the predominant genotype in many countries of Central and South America. The overall prevalence of HBV/F in these countries depends on the degree of admixture of the population with Amerindians. Outside LA, HBV/F is also found in native Alaskan populations. HBV/H is mainly confined to Mexico and is the major genotype of both Amerindians and mestizos [33, 91]. HBV/F and HBV/H isolates display a close phylogenetic relationship and are the most divergent compared to the other HBV genotypes. HBV/F is highly divergent and has been classified into four sub-genotypes (F1–F4) [34, 71]. Recently, a new sub-genotype, F5, has been proposed in Panamanian blood donors [69]. Sub-genotype F1 was further subdivided into two clusters, F1a and F1b; similarly, F2 was subdivided in F2a and F2b. Three HBV isolates identified recently seems to form a separate cluster inside sub-genotype F4 [76]. The genetic distance among F sub-genotypes ranges from 3.91% to 7.41%. In contrast, within HBV/H isolates, an intra-genotypic divergence of only 0.032–3.82% has been found. Consequently, no HBV/H subgenotypes have been described [69, 103].

Recent studies regarding the origin, emergence, and distribution of HBV/F suggest that the ancestor was located in Central America or northern South America and subsequently spread in a north-to-south flow [69, 73, 126]. It has been proposed that the diversification and dissemination of HBV in the pre-Columbian Americas would be associated with two human migratory routes: one following the Pacific coastal route, leading to the spread of HBV/F1, and a separate, but contemporaneous, migration into the interior of South America leading to the diversification of sub-genotypes F2 to F4 [76, 103].

Fig. 16.4 Distribution of HBV/H and HBV/F sub-genotypes in Latin America (LA)

The present distribution of HBV/H and HBV/F sub-genotypes in LA is shown in Fig. 16.4. Among HBV/F1, cluster F1a is found in Costa Rica and El Salvador, while F1b is mainly found in Argentina and Chile and also has been detected in Brazil. Outside LA, F1b has been found in Alaska [11, 65, 94]. Sub-genotypes F2 and F3 both circulate in northern South America, with cluster F2a found mainly in Brazil and Venezuela and F2b restricted to Venezuela. Sub-genotype F3 is frequently found in Colombia, Venezuela, and Panama [32, 69, 73, 74, 79], while F4 is prevalent in Bolivia and Northern Argentina and also has been reported in southern and central Brazil [55, 73, 95]. HBV/F has been associated with a higher risk of HCC development [65]. Moreover, liver-related mortality with HBV/F is higher when compared to that of HBV/A and HBV/D [108]. Conversely, HBV/H clinical outcome among the native or mestizo Mexican population has been mainly associated with OBI (see following) and low viral load, being uncommon in the presence of HCC related to chronic HBV infection [105].

Finally, the geographic (sub)genotype distribution and the differential biological behaviors described for genotypes F and H justify deeper epidemiological studies of these HBV strains. These studies could contribute to the understanding of HBV features that may be relevant to clinical outcome, response to treatment, and immunization strategies in the populations of Mexico and Central and South America.

8.4 Tracking the Origin of Genotype A in LA

HBV/A is a ubiquitous genotype and has been previously classified into seven sub-genotypes (HBV/A1–HBV/A7). Recently, it has been proposed that sub-genotypes HBV/A1, HBV/A2, and HBV/A6 (now renamed as HBV/A4) fit the rules for HBV division into sub-genotypes, while the others form a single clade called "quasi-subgenotype-A3" (QS-A3) [97]. HBV/A1 is highly prevalent [97]. HBV/A1 is highly prevalent in Southeast Africa [21, 60] and predominant among people of African ancestry living in LA [3, 5, 70, 77] suggesting that this sub-genotype originated in African populations. HBV/A2 is the most frequent sub-genotype in northwest Europe and the U.S. [135] and has also been isolated in South Africa [21]. It has been suggested that HBV/A2 was introduced in Europe by Portuguese traders who traveled across southern Africa during the fifteenth century [49, 60]. HBV/A3 (now QS-A3) was originally isolated in Cameroon [61] and has been also identified in Gambia [49] and Mali [88]. Other QS-A3, previously named as HBV/A4 and HBV/A5 (QS-A3), were identified in Mali and Nigeria, respectively [49, 88]. HBV/A4 (previously named HBV/A6) was detected in Afro-Belgian patients [98]; HBV/A7 (renamed as QS-A3) has appeared in some individuals from Cameroon [54].

Until 1994, only three HBV strains from South America, two from Colombia [84] and one from Brazil [81], had been sequenced. All were classified as HBV/F. It was then assumed that HBV/F would be the most common genotype in Brazil, as it is typical of South Americans and the indigenous people of LA. However, further studies have demonstrated that HBV isolates from Brazil are mostly HBV/A and closely related to those from South Africa, suggesting an African origin for many Brazilian HBV/A isolates [9, 82].

To understand the diversification of HBV/A in LA, isolates from Argentina, Brazil, Costa Rica, Mexico, Nicaragua, Venezuela, and Uruguay were compared [72]. European HBV/A2 prevailed in most countries except in Brazil, where African HBV/A1 predominated [72]. In Brazil, HBV/A1 was found at a frequency about ten times higher than HBV/A2 [14, 64, 74, 78, 109]. This difference between Brazil and other LA countries may be understood by considering the colonization process of these countries. Brazil, which is surrounded by countries of Spanish colonization, is the only country in LA that was colonized by Portugal. Moreover, Brazil received about half the 12 million African slaves and was the last country to ban the Atlantic slave trade in the second half of the nineteenth century. The hypothesis that HBV/A1 was introduced into Brazil by the arrival of slaves [9, 64, 74] has been reinforced by the observations that HBV/A1 was almost the sole genotype found in semi-isolated

Afro-descendant communities [3, 70, 77]. In addition, people of African descent are more frequently infected with HBV/A1 than Caucasians, whereas the reverse is true for HBV/D [16]. Recently, using Bayesian analysis, the existence of an Asian-American clade within HBV/A1 has been suggested [64]. The close degree of relatedness of the Brazilian, Asian, and Somalian isolates suggests that the HBV/A1 strains predominant in Brazil did not originate from the 5 million slaves who were imported from Central and Western Africa from 1551 to 1840, but rather from the 300,000 to 400,000 captives forcibly removed from Southeast Africa at the middle of the nineteenth century [64]. In Haiti, where more than 90% of the population are descendants of African slaves, HBV/A is found in more than 70% of the population, and HBV/A1 is the predominant sub-genotype. Interestingly, about 20% of HBV/A isolates from this country belong to the rare sub-genotype HBV/A5 (QS-A3), which has been found only in the Bight of Benin, a former primary slave trading post [5].

9 Occult Hepatitis B Infection (OBI) in Latin America

OBI is defined as the detection of HBV DNA in serum or liver with a negative HBsAg result [99]. Other distinctive clinical features are viral loads lower than 200 IU/ml and the existence of two subtypes of OBI based on the seropositivity or the seronegativity of anti-HBc or anti-HB antibodies. This forthright definition is not free of debate because the accurate detection of OBI relies on the use of sensitive and specific diagnostic techniques for both HBsAg and HBV DNA. HBV DNA detection in liver biopsy or serum using highly sensitive polymerase chain reaction (PCR) techniques to amplify two or more genomic regions of the virus is a recommendation for OBI detection [99] that may be difficult to accomplish in many locations. Given the fact that most diagnostic techniques are standardized to cover the dynamic range of overt infection, the true prevalence of OBI may be underestimated by lack of sensitivity and overestimated if PCR cross-contamination is not controlled.

Despite these drawbacks, the prevalence of OBI tends to vary widely depending on the level of HBV endemicity in different geographic regions or clinical settings [100]. OBI is usually detected in high endemic regions or among several indigenous populations globally. In intermediate or low endemicity regions, it is observed in a number of high-risk groups, such as patients who are immunocompromised, have cancer, are coinfected with HCV or HIV, are organ transplant recipients, and are rheumatic or on hemodialysis. These high-risk patients have a common serological profile of apparently undetectable levels of HBsAg that may suddenly show up as flare episodes, that is, the sporadic and short-term appearance of HBsAg. Moreover, some may display a clinical feature denoted as “isolated anti-HBc” who are positive for anti-HBc antibody in the absence of HBsAg, in which serum HBV DNA may be positive in up to 50% of the cases [125]. Given these characteristics, it is recommended that patients with risk factors should receive follow-up testing for HBsAg, anti-HBc antibody, and viral load because OBI is also associated with hepatocellular carcinoma [53, 96].

On the other hand, transfusion-transmitted hepatitis from OBI is still of concern in many regions worldwide. Because HBsAg is the conventional serological marker for which blood banks screen, OBI in blood donors may be a risk factor [23]. The rate of OBI among blood donors in North America is estimated to be 1 in 350,000–610,000 [51]. These frequencies are unknown in many low- and middle-income countries of LA, which have fewer diagnostic resources available to supply secure blood units by using nucleic acid testing (NAT) technology. Further studies are needed to evaluate the risk of OBI in these settings and provide surrogate algorithms to avoid an iatrogenic transmission of HBV.

In regard to LA, large-scale epidemiological studies of HBV infection are not available in most countries, thus hindering the estimation of overt and OBI infection. Nonetheless, this region contains a promising opportunity to study the natural history of HBV infection in native and admixed human populations which have not been vaccinated or treated with antiviral therapy. Regardless of the overall endemicity, several studies carried out in indigenous populations in Argentina, Brazil, Colombia, Mexico, and Venezuela [24, 31, 35, 105, 130] have shown that HBV infections display low prevalence of HBsAg, high prevalence of anti-HBc antibody, and in some instances, a high prevalence of OBI. OBI is not restricted to one type of HBV; it prevails with nearly all genotypes reported worldwide. In this case, HBV/H in Mexico [36, 91, 105] and genotype F3 in Argentina [31] have been associated with this disease entity.

It has been proposed that one of the mechanisms of OBI may involve immunoregulatory factors that diminish the expression of HBsAg and genomic replication [100, 107]. In a recent study, a differential cytokine profile was detected between OBI with HBV/H compared to resolved infection in native Nahuas from Mexico [38]. On the other hand, from an evolutionary perspective, it has been proposed that OBI may be a biological adaptation among the indigenous groups in Mexico and perhaps in other groups in LA given the long-term and balanced host–virus relationship [103]. Interestingly, neither liver damage nor hepatocellular carcinoma related to the endemic genotypes H and F has been reported in these groups, although other environmental factors may be involved [102, 104]. Further studies are required to assess the differential response to HBV infection among the LA population that may have an impact on regional guidelines for the management of this disease.

References

1. Alter MJ (2003) Epidemiology and prevention of hepatitis B. Semin Liver Dis 23:39–46
2. Alvarado-Esquivel C, Sablon E, Conde-Gonzalez CJ et al (2006) Molecular analysis of hepatitis B virus isolates in Mexico: predominant circulation of hepatitis B virus genotype H. World J Gastroenterol 12:6540–6545
3. Alvarado-Mora MV, Romano CM, Gomes-Gouvea MS et al (2011) Molecular characterization of the hepatitis B virus genotypes in Colombia: a Bayesian inference on the genotype F. Infect Genet Evol 11:103–108
4. Andernach IE, Hubschen JM, Muller CP (2009a) Hepatitis B virus: the genotype E puzzle. Rev Med Virol 19:231–240

5. Andernach IE, Nolte C, Pape JW et al (2009b) Slave trade and hepatitis B virus genotypes and subgenotypes in Haiti and Africa. Emerg Infect Dis 15:1222–1228
6. Arankalle VA, Gandhe SS, Borkakoty BJ et al (2010) A novel HBV recombinant (genotype I) similar to Vietnam/Laos in a primitive tribe in eastern India. J Viral Hepatol 17:501–510
7. Araujo NM (2015) Hepatitis B virus intergenotypic recombinants worldwide: an overview. Infect Genet Evol 36:500–510
8. Araujo NM, Araujo OC, Silva EM et al (2013) Identification of novel recombinants of hepatitis B virus genotypes F and G in human immunodeficiency virus-positive patients from Argentina and Brazil. J Gen Virol 94:150–158
9. Araujo NM, Mello FC, Yoshida CF et al (2004) High proportion of subgroup A′ (genotype A) among Brazilian isolates of hepatitis B virus. Arch Virol 149:1383–1395
10. Araujo NM, Waizbort R, Kay A (2011) Hepatitis B virus infection from an evolutionary point of view: how viral, host, and environmental factors shape genotypes and subgenotypes. Infect Genet Evol 11:1199–1207
11. Arauz-Ruiz P, Norder H, Robertson BH et al (2002) Genotype H: a new Amerindian genotype of hepatitis B virus revealed in Central America. J Gen Virol 83:2059–2073
12. Arauz-Ruiz P, Norder H, Visona KA et al (1997) Genotype F prevails in HBV infected patients of hispanic origin in Central America and may carry the precore stop mutant. J Med Virol 51:305–312
13. Barbini L, Elizalde M, Torres C et al (2013) Molecular epidemiology and genetic diversity of hepatitis B virus in Mar del Plata city, Argentina. Infect Genet Evol 19:152–163
14. Barros LM, Gomes-Gouvea MS, Kramvis A et al (2014) High prevalence of hepatitis B virus subgenotypes A1 and D4 in Maranhao state, Northeast Brazil. Infect Genet Evol 24:68–75
15. Bayer ME, Blumberg BS, Werner B (1968) Particles associated with Australia antigen in the sera of patients with leukaemia, Down's syndrome and hepatitis. Nature (Lond) 218:1057–1059
16. Bertolini DA, Gomes-Gouvea MS, Carvalho-Mello IM et al (2012) Hepatitis B virus genotypes from European origin explains the high endemicity found in some areas from southern Brazil. Infect Genet Evol 12:1295–1304
17. Blitz L, Pujol FH, Swenson PD et al (1998) Antigenic diversity of hepatitis B virus strains of genotype F in Amerindians and other population groups from Venezuela. J Clin Microbiol 36:648–651
18. Blumberg BS, Alter HJ, Visnich S (1965) A "new" antigen in leukemia sera. JAMA 191:541–546
19. Blumberg BS, Gerstley BJ, Hungerford DA et al (1967) A serum antigen (Australia antigen) in Down's syndrome, leukemia, and hepatitis. Ann Intern Med 66:924–931
20. Blumberg BS, Larouze B, London WT et al (1975) The relation of infection with the hepatitis B agent to primary hepatic carcinoma. Am J Pathol 81:669–682
21. Bowyer SM, van Staden L, Kew MC et al (1997) A unique segment of the hepatitis B virus group A genotype identified in isolates from South Africa. J Gen Virol 78:1719–1729
22. Brichler S, Lagathu G, Chekaraou MA et al (2013) African, Amerindian and European hepatitis B virus strains circulate on the Caribbean Island of Martinique. J Gen Virol 94:2318–2329
23. Candotti D, Allain JP (2009) Transfusion-transmitted hepatitis B virus infection. J Hepatol 51:798–809
24. Cardona NE, Loureiro CL, Garzaro DJ et al (2011) Unusual presentation of hepatitis B serological markers in an Amerindian community of Venezuela with a majority of occult cases. Virol J 8:527
25. Casey JL, Niro GA, Engle RE et al (1996) Hepatitis B virus (HBV)/hepatitis D virus (HDV) coinfection in outbreaks of acute hepatitis in the Peruvian Amazon basin: the roles of HDV genotype III and HBV genotype F. J Infect Dis 174:920–926
26. Cassidy A, Mossman S, Olivieri A et al (2011) Hepatitis B vaccine effectiveness in the face of global HBV genotype diversity. Expert Rev Vaccines 10:1709–1715
27. CDC (2015) Sexual transmission and viral hepatitis. http://www.cdc.gov/hepatitis/populations/stds.htm. Accessed 19 Jun 2016

28. Crispim MA, Fraiji NA, Campello SC et al (2014) Molecular epidemiology of hepatitis B and hepatitis delta viruses circulating in the Western Amazon region, North Brazil. BMC Infect Dis 14:94
29. Dane DS, Cameron CH, Briggs M (1970) Virus-like particles in serum of patients with Australia-antigen-associated hepatitis. Lancet 1:695–698
30. Davison F, Alexander GJ, Trowbridge R et al (1987) Detection of hepatitis B virus DNA in spermatozoa, urine, saliva and leucocytes, of chronic HBsAg carriers. A lack of relationship with serum markers of replication. J Hepatol 4:37–44
31. Delfino CM, Berini C, Eirin ME et al (2012) New natural variants of hepatitis B virus among Amerindians from Argentina with mainly occult infections. J Clin Virol 54:174–179
32. Devesa M, Loureiro CL, Rivas Y et al (2008) Subgenotype diversity of hepatitis B virus American genotype F in Amerindians from Venezuela and the general population of Colombia. J Med Virol 80:20–26
33. Devesa M, Pujol FH (2007) Hepatitis B virus genetic diversity in Latin America. Virus Res 127:177–184
34. Devesa M, Rodriguez C, Leon G et al (2004) Clade analysis and surface antigen polymorphism of hepatitis B virus American genotypes. J Med Virol 72:377–384
35. di Filippo VD, Cortes-Mancera F, Payares E et al (2015) Hepatitis D virus and hepatitis B virus infection in Amerindian communities of the Amazonas state, Colombia. Virol J 12:172
36. Escobedo-Melendez G, Panduro A, Fierro NA et al (2014) High prevalence of occult hepatitis B virus genotype H infection among children with clinical hepatitis in west Mexico. Mem Inst Oswaldo Cruz 109:728–737
37. Fang ZL, Hue S, Sabin CA et al (2011) A complex hepatitis B virus (X/C) recombinant is common in Long An county, Guangxi and may have originated in southern China. J Gen Virol 92:402–411
38. Fierro NA, Roman S, Realpe M et al (2011) Multiple cytokine expression profiles reveal immune-based differences in occult hepatitis B genotype H-infected Mexican Nahua patients. Mem Inst Oswaldo Cruz 106:1007–1013
39. Francis DP, Favero MS, Maynard JE (1981) Transmission of hepatitis B virus. Semin Liver Dis 1:27–32
40. Gadano A, Daruich J, Cheinquer H et al (2011) Latin American guideline for the management of chronic hepatitis B. Acta Gastroenterol Latinoam 41:340–350
41. Galle PR, Hagelstein J, Kommerell B et al (1994) In vitro experimental infection of primary human hepatocytes with hepatitis B virus. Gastroenterology 106:664–673
42. Ganem D, Varmus HE (1987) The molecular biology of the hepatitis B viruses. Annu Rev Biochem 56:651–693
43. Gardner HT (1950) A note on the history of epidemic viral hepatitis in Germany. Am J Med 8:561–564
44. Goldstein ST, Alter MJ, Williams IT et al (2002) Incidence and risk factors for acute hepatitis B in the United States, 1982–1998: implications for vaccination programs. J Infect Dis 185:713–719
45. Gusatti CS, Costi C, Halon ML et al (2015) Hepatitis B virus genotype D isolates circulating in Chapeco, Southern Brazil, originate from Italy. PLoS One 10(8):e0135816
46. Hadziyannis SJ (2011) Natural history of chronic hepatitis B in Euro-Mediterranean and African countries. J Hepatol 55:183–191
47. Haldipur BP, Walimbe AM, Arankalle VA (2014) Circulation of genotype-I hepatitis B virus in the primitive tribes of Arunachal Pradesh in early sixties and molecular evolution of genotype-I. Infect Genet Evol 27:366–374
48. Hannoun C, Norder H, Lindh M (2000) An aberrant genotype revealed in recombinant hepatitis B virus strains from Vietnam. J Gen Virol 81:2267–2272
49. Hannoun C, Soderstrom A, Norkrans G et al (2005) Phylogeny of African complete genomes reveals a West African genotype A subtype of hepatitis B virus and relatedness between Somali and Asian A1 sequences. J Gen Virol 86:2163–2167
50. Heermann KH, Goldmann U, Schwartz W et al (1984) Large surface proteins of hepatitis B virus containing the pre-S sequence. J Virol 52:396–402

51. Hollinger FB, Sood G (2010) Occult hepatitis B virus infection: a covert operation. J Viral Hepatol 17:1–15
52. Hsu HY, Chang MH, Ni YH et al (2004) Survey of hepatitis B surface variant infection in children 15 years after a nationwide vaccination programme in Taiwan. Gut 53:1499–1503
53. Huang X, Hollinger FB (2014) Occult hepatitis B virus infection and hepatocellular carcinoma: a systematic review. J Viral Hepat 21:153–162
54. Hubschen JM, Mbah PO, Forbi JC et al (2011) Detection of a new subgenotype of hepatitis B virus genotype A in Cameroon but not in neighbouring Nigeria. Clin Microbiol Infect 17:88–94
55. Huy TT, Ushijima H, Sata T et al (2006) Genomic characterization of HBV genotype F in Bolivia: genotype F subgenotypes correlate with geographic distribution and T(1858) variant. Arch Virol 151:589–597
56. Kim CM, Koike K, Saito I et al (1991) HBx gene of hepatitis B virus induces liver cancer in transgenic mice. Nature (Lond) 351:317–320
57. Kowalec K, Minuk GY, Borresen ML et al (2013) Genetic diversity of hepatitis B virus genotypes B6, D and F among circumpolar indigenous individuals. J Viral Hepatol 20:122–130
58. Kramvis A, Kew M, Francois G (2005) Hepatitis B virus genotypes. Vaccine 23:2409–2423
59. Kramvis A, Kew MC (2007) Molecular characterization of subgenotype A1 (subgroup Aa) of hepatitis B virus. Hepatol Res 37:S27–S32
60. Kramvis A, Weitzmann L, Owiredu WK et al (2002) Analysis of the complete genome of subgroup A′ hepatitis B virus isolates from South Africa. J Gen Virol 83:835–839
61. Kurbanov F, Tanaka Y, Fujiwara K et al (2005) A new subtype (subgenotype) Ac (A3) of hepatitis B virus and recombination between genotypes A and E in Cameroon. J Gen Virol 86:2047–2056
62. Kurbanov F, Tanaka Y, Kramvis A et al (2008) When should "I" consider a new hepatitis B virus genotype? J Virol 82:8241–8242
63. Lacombe K, Massari V, Girard PM et al (2006) Major role of hepatitis B genotypes in liver fibrosis during coinfection with HIV. AIDS 20:419–427
64. Lago BV, Mello FC, Kramvis A et al (2014) Hepatitis B virus subgenotype A1: evolutionary relationships between Brazilian, African and Asian isolates. PLoS One 9(8):e105317
65. Livingston SE, Simonetti JP, McMahon BJ et al (2007) Hepatitis B virus genotypes in Alaska Native people with hepatocellular carcinoma: preponderance of genotype F. J Infect Dis 195:5–11
66. Locarnini S, Littlejohn M, Aziz MN et al (2013) Possible origins and evolution of the hepatitis B virus (HBV). Semin Cancer Biol 23:561–575
67. Lopez L, Flichman D, Mojsiejczuk L et al (2015) Genetic variability of hepatitis B virus in Uruguay: D/F, A/F genotype recombinants. Arch Virol 160:2209–2217
68. Marciano S, Galdame OA, Gadano AC (2013) HBV genotype F: natural history and treatment. Antivir Ther 18:485–488
69. Martinez AA, Zaldivar YY, CSS-NAT Group et al (2014) High diversity of hepatitis B virus genotypes in Panamanian blood donors: a molecular analysis of new variants. PLoS One 9(8):e103545
70. Matos MA, Reis NR, Kozlowski AG et al (2009) Epidemiological study of hepatitis A, B and C in the largest Afro-Brazilian isolated community. Trans R Soc Trop Med Hyg 103:899–905
71. Mbayed VA, Barbini L, Lopez JL et al (2001) Phylogenetic analysis of the hepatitis B virus (HBV) genotype F including Argentine isolates. Arch Virol 146:1803–1810
72. Mbayed VA, Pineiro y Leone FG et al (2009) Molecular characterization of hepatitis B virus genotype A from Argentina and Brazil. Arch Virol 154:525–529
73. Mello FC, Araujo OC, Lago BV et al (2013) Phylogeography and evolutionary history of hepatitis B virus genotype F in Brazil. Virol J 10:236
74. Mello FC, Souto FJ, Nabuco LC et al (2007) Hepatitis B virus genotypes circulating in Brazil: molecular characterization of genotype F isolates. BMC Microbiol 7:103
75. Merican I, Guan R, Amarapuka D et al (2000) Chronic hepatitis B virus infection in Asian countries. J Gastroenterol Hepatol 15:1356–1361

76. Mojsiejczuk LN, Torres C, Fainboin HA et al (2015) Identification of a new clade of hepatitis B virus genotype F. Infect Genet Evol 34:122–125
77. Motta-Castro AR, Martins RM, Araujo NM et al (2008) Molecular epidemiology of hepatitis B virus in an isolated Afro-Brazilian community. Arch Virol 153:2197–2205
78. Moura IF, Lopes EP, Alvarado-Mora MV et al (2013) Phylogenetic analysis and subgenotypic distribution of the hepatitis B virus in Recife, Brazil. Infect Genet Evol 14:195–199
79. Nakano T, Lu L, Hu X et al (2001) Characterization of hepatitis B virus genotypes among Yucpa Indians in Venezuela. J Gen Virol 82:359–365
80. Nassal M, Schaller H (1996) Hepatitis B virus replication: an update. J Viral Hepatol 3:217–226
81. Naumann H, Schaefer S, Yoshida CF et al (1993) Identification of a new hepatitis B virus (HBV) genotype from Brazil that expresses HBV surface antigen subtype adw4. J Gen Virol 74:1627–1632
82. Niel C, Moraes MT, Gaspar AM et al (1994) Genetic diversity of hepatitis B virus strains isolated in Rio de Janeiro, Brazil. J Med Virol 44:180–186
83. Norder H, Courouce AM, Coursaget P et al (2004) Genetic diversity of hepatitis B virus strains derived worldwide: genotypes, subgenotypes, and HBsAg subtypes. Intervirology 47:289–309
84. Norder H, Courouce AM, Magnius LO (1993) Complete nucleotide sequences of six hepatitis B viral genomes encoding the surface antigen subtypes ayw4, adw4q-, and adrq- and their phylogenetic classification. Arch Virol 8:189–199
85. Norder H, Courouce AM, Magnius LO (1994) Complete genomes, phylogenetic relatedness, and structural proteins of six strains of the hepatitis B virus, four of which represent two new genotypes. Virology 198:489–503
86. Okamoto H, Tsuda F, Sakugawa H et al (1988) Typing hepatitis B virus by homology in nucleotide sequence: comparison of surface antigen subtypes. J Gen Virol 69:2575–2583
87. Olinger CM, Jutavijittum P, Hubschen JM et al (2008) Possible new hepatitis B virus genotype, southeast Asia. Emerg Infect Dis 14:1777–1780
88. Olinger CM, Venard V, Njayou M et al (2006) Phylogenetic analysis of the precore/core gene of hepatitis B virus genotypes E and A in West Africa: new subtypes, mixed infections and recombinations. J Gen Virol 87:1163–1173
89. Osiowy C, Gordon D, Borlang J et al (2008) Hepatitis B virus genotype G epidemiology and co-infection with genotype A in Canada. J Gen Virol 89:3009–3015
90. Ott JJ, Stevens GA, Groeger J et al (2012) Global epidemiology of hepatitis B virus infection: new estimates of age-specific HBsAg seroprevalence and endemicity. Vaccine 30:2212–2219
91. Panduro A, Maldonado-Gonzalez M, Fierro NA et al (2013) Distribution of HBV genotypes F and H in Mexico and Central America. Antivir Ther 18:475–484
92. Pena SD, Di Pietro G, Fuchshuber-Moraes M et al (2011) The genomic ancestry of individuals from different geographical regions of Brazil is more uniform than expected. PLoS One 6(2):e17063
93. Piasecki T, Harkins GW, Chrzastek K et al (2013) Avihepadnavirus diversity in parrots is comparable to that found amongst all other avian species. Virology 438:98–105
94. Pineiro y Leone FG, Mbayed VA, Campos RH (2003) Evolutionary history of hepatitis B virus genotype F: an in-depth analysis of Argentine isolates. Virus Genes 27:103–110
95. Pineiro y Leone FG, Pezzano SC, Torres C et al (2008) Hepatitis B virus genetic diversity in Argentina: dissimilar genotype distribution in two different geographical regions; description of hepatitis B surface antigen variants. J Clin Virol 42:381–388
96. Pollicino T, Saitta C (2014) Occult hepatitis B virus and hepatocellular carcinoma. World J Gastroenterol 20:5951–5961
97. Pourkarim MR, Amini-Bavil-Olyaee S, Kurbanov F et al (2014) Molecular identification of hepatitis B virus genotypes/subgenotypes: revised classification hurdles and updated resolutions. World J Gastroenterol 20:7152–7168
98. Pourkarim MR, Lemey P, Amini-Bavil-Olyaee S et al (2010) Novel hepatitis B virus subgenotype A6 in African-Belgian patients. J Clin Virol 47:93–96
99. Raimondo G, Allain JP, Brunetto MR et al (2008) Statements from the Taormina expert meeting on occult hepatitis B virus infection. J Hepatol 49:652–657

100. Raimondo G, Caccamo G, Filomia R et al (2013) Occult HBV infection. Semin Immunopathol 35:39–52
101. Rasche A, Souza BF, Drexler JF (2016) Bat hepadnaviruses and the origins of primate hepatitis B viruses. Curr Opin Virol 16:86–94
102. Roman S, Fierro NA, Moreno-Luna LE et al (2013) Hepatitis B virus genotype H and environmental factors associated to the low prevalence of hepatocellular carcinoma in Mexico. J Cancer Ther 4:367–376
103. Roman S, Jose-Abrego A, Fierro NA et al (2014) Hepatitis B virus infection in Latin America: a genomic medicine approach. World J Gastroenterol 20:7181–7196
104. Roman S, Panduro A (2013) HBV endemicity in Mexico is associated with HBV genotypes H and G. World J Gastroenterol 19:5446–5453
105. Roman S, Tanaka Y, Khan A et al (2010) Occult hepatitis B in the genotype H-infected Nahuas and Huichol native Mexican population. J Med Virol 82:1527–1536
106. Romano L, Paladini S, Galli C et al (2015) Hepatitis B vaccination. Hum Vaccin Immunother 11:53–57
107. Samal J, Kandpal M, Vivekanandan P (2012) Molecular mechanisms underlying occult hepatitis B virus infection. Clin Microbiol Rev 25:142–163
108. Sanchez-Tapias JM, Costa J, Mas A et al (2002) Influence of hepatitis B virus genotype on the long-term outcome of chronic hepatitis B in western patients. Gastroenterology 123:1848–1856
109. Santos AO, Alvarado-Mora MV, Botelho L et al (2010) Characterization of hepatitis B virus (HBV) genotypes in patients from Rondonia, Brazil. Virol J 7:315
110. Seeger C, Zoulim F, Mason WS (2013) Hepadnaviruses. In: Knipe DM, Howley PM (eds) Virology, 6th edn. Lippincott Williams & Wilkins, Philadelphia, pp 2185–2221
111. Seifer M, Hohne M, Schaefer S et al (1991) In vitro tumorigenicity of hepatitis B virus DNA and HBx protein. J Hepatol 13:S61–S65
112. Sells MA, Zelent AZ, Shvartsman M et al (1988) Replicative intermediates of hepatitis B virus in HepG2 cells that produce infectious virions. J Virol 62:2836–2844
113. Shepard CW, Simard EP, Finelli L et al (2006) Hepatitis B virus infection: epidemiology and vaccination. Epidemiol Rev 28:112–125
114. Sherlock S (1987) The natural history of hepatitis B. Postgrad Med J 63:7–11
115. Shi W, Zhang Z, Ling C et al (2013) Hepatitis B virus subgenotyping: history, effects of recombination, misclassifications, and corrections. Infect Genet Evol 16:355–361
116. Shibayama T, Masuda G, Ajisawa A et al (2005) Characterization of seven genotypes (A to E, G and H) of hepatitis B virus recovered from Japanese patients infected with human immunodeficiency virus type 1. J Med Virol 76:24–32
117. Souto FJ (2016) Distribution of hepatitis B infection in Brazil: the epidemiological situation at the beginning of the 21st century. Rev Soc Bras Med Trop 49:11–23
118. Stuyver L, De Gendt S, Van Geyt C et al (2000) A new genotype of hepatitis B virus: complete genome and phylogenetic relatedness. J Gen Virol 81:67–74
119. Su H, Liu Y, Xu Z et al (2014) A novel complex A/C/G intergenotypic recombinant of hepatitis B virus isolated in southern China. PLoS One 9(1):e84005
120. Tatematsu K, Tanaka Y, Kurbanov F et al (2009) A genetic variant of hepatitis B virus divergent from known human and ape genotypes isolated from a Japanese patient and provisionally assigned to new genotype J. J Virol 83:10538–10547
121. Terrault NA, Bzowej NH, Chang KM et al (2016) AASLD guidelines for treatment of chronic hepatitis B. Hepatology 63:261–283
122. Tiollais P, Pourcel C, Dejean A (1985) The hepatitis B virus. Nature (Lond) 317:489–495
123. Toan NL, Song le H, Kremsner PG et al (2006) Impact of the hepatitis B virus genotype and genotype mixtures on the course of liver disease in Vietnam. Hepatology 43:1375–1384
124. Tong MJ, Sun SC, Schaeffer BT et al (1971) Hepatitis-associated antigen and hepatocellular carcinoma in Taiwan. Ann Intern Med 75:687–691
125. Torbenson M, Thomas DL (2002) Occult hepatitis B. Lancet Infect Dis 2:479–486

126. Torres C, Pineiro y Leone FG, Pezzano SC et al (2011) New perspectives on the evolutionary history of hepatitis B virus genotype F. Mol Phylogenet Evol 59:114–122
127. Torres JR, Mondolfi A (1991) Protracted outbreak of severe delta hepatitis: experience in an isolated Amerindian population of the Upper Orinoco basin. Rev Infect Dis 13:52–55
128. Tran TT, Trinh TN, Abe K (2008) New complex recombinant genotype of hepatitis B virus identified in Vietnam. J Virol 82:5657–5663
129. Tsurimoto T, Fujiyama A, Matsubara K (1987) Stable expression and replication of hepatitis B virus genome in an integrated state in a human hepatoma cell line transfected with the cloned viral DNA. Proc Natl Acad Sci U S A 84:444–448
130. Viana S, Parana R, Moreira RC et al (2005) High prevalence of hepatitis B virus and hepatitis D virus in the western Brazilian Amazon. Am J Trop Med Hyg 73:808–814
131. Vieth S, Manegold C, Drosten C et al (2002) Sequence and phylogenetic analysis of hepatitis B virus genotype G isolated in Germany. Virus Genes 24:153–156
132. WHO (2015) Hepatitis B. In: World Health Organization fact sheet no. 204. http://www.who.int/mediacentre/factsheets/fs204/en/. Accessed 19 Jun 2016
133. Wilson JN, Nokes DJ, Carman WF (1999) The predicted pattern of emergence of vaccine-resistant hepatitis B: a cause for concern? Vaccine 17:973–978
134. Yousif M, Kramvis A (2013) Genotype D of hepatitis B virus and its subgenotypes: an update. Hepatol Res 43:355–364
135. Zehender G, Svicher V, Gabanelli E et al (2015) Reliable timescale inference of HBV genotype A origin and phylodynamics. Infect Genet Evol 32:361–369
136. Zoulim F (2011) Hepatitis B virus resistance to antiviral drugs: where are we going? Liver Int 31:111–116
137. Zoulim F, Locarnini S (2009) Hepatitis B virus resistance to nucleos(t)ide analogues. Gastroenterology 137:1593–1608
138. Zuckerman AJ, Howard CR (1979) Hepatitis viruses of man. Academic Press, London

Chapter 17
Molecular Evolution of Hepatitis C Virus: From Epidemiology to Antiviral Therapy (Current Research in Latin America)

Natalia Echeverría, Pilar Moreno, and Juan Cristina

1 Introduction

Hepatitis C infection is globally widespread, and to date, there is an estimate of 130–150 million people worldwide living with chronic and progressive liver disease as a result of infection by hepatitis C virus (HCV) [196]. The development of cirrhosis and hepatocellular carcinoma (HCC)) are long-term complications in 15% to 30% of chronically infected patients, which is the reason why HCV infection is one of the most common indications for liver transplantation [81, 143].

Despite the urgent need to prevent dissemination of its etiological agent, there is currently no prophylactic vaccine against HCV [93]. However, significant advances have been made in relation to drug development in the past decade, leading to the approval of many direct antiviral agents (DAAs), which have greatly contributed to achieve higher sustained virological response (SVR) rates [67, 92, 129, 182]. Until 2010, the standard of care (SOC) therapy involved pegylated interferon-α (IFN-α-PEG) and ribavirin (RBV) [35, 110], administered for 24 or 48 weeks depending on the infecting viral genotype. Yet, since 2011 a new standard of care (NSOC) therapy has been approved for patients infected with HCV genotype 1, by including one or more DAAs in combination with IFN-α-PEG and/or RBV [67, 92, 129, 182]. Unfortunately, therapies including interferon present several drawbacks resulting from IFN inaccessibility in some countries, its tolerability, and patient response to treatment (some genotypes of HCV respond better than others) [36]. In addition, adverse effects such as rash have also been associated with the NSOC [171], but even more worrying is the high cost of the newly approved DAAs, particularly for Latin American countries [140, 177].

N. Echeverría • P. Moreno • J. Cristina (✉)
Laboratorio de Virología Molecular, Centro de Investigaciones Nucleares, Facultad de Ciencias, Universidad de la República, 11400 Montevideo, Uruguay
e-mail: necheverria@cin.edu.uy; pmoreno@cin.edu.uy; cristina@cin.edu.uy

J.E. Ludert et al. (eds.), *Human Virology in Latin America*,
DOI 10.1007/978-3-319-54567-7_17

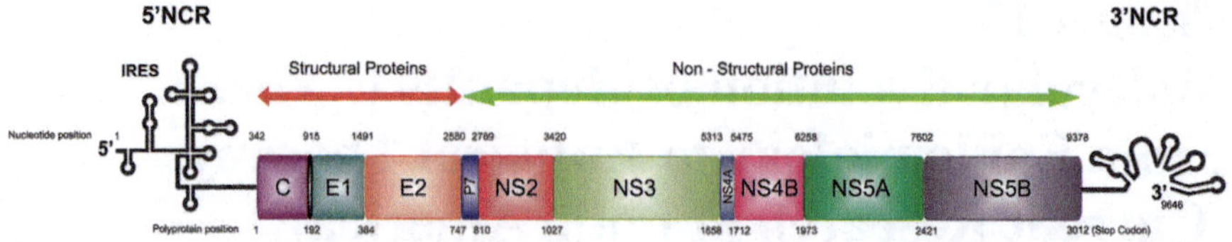

Fig. 17.1 Organization of hepatitis C virus (HCV) genome and hepatitis C virus polyprotein processing. Schematic representation of the 9.6-kb positive-stranded RNA genome. The 5′- and 3′-non-coding regions (NCRs) are shown as well as the internal ribosomal entry site (IRES)-driven translated polyprotein precursor, which is processed into the mature structural and nonstructural proteins. Nucleotide and amino acid positions are shown by *numbers* on the upper and lower part of the scheme, respectively. (Modified from Echeverría et al. 2015 [52])

As a blood-borne pathogen, the main route of HCV transmission is direct or indirect exposure to any source of contaminated blood: transfusion of unscreened blood or blood products, intravenous drug use, use of poorly sterilized surgical aids, organ transplants, and work-related accidents in healthcare centers, as well as vertical transmission from mother to child, although the latter is much less common [174, 195, 196].

HCV belongs to the family *Flaviviridae* and is classified as a member of the *Hepacivirus* genus, mainly because of specific genome features [174]. HCV is a single-stranded, positive-sense RNA virus with a genome approximately 9600 nucleotides in length. Most of the genome is composed of a single open reading frame (ORF) that encodes three structural (core, E1, E2) and seven nonstructural (p7, NS2, NS3, NS4A, NS4B, NS5A, NS5B) proteins (Fig. 17.1) [15, 52]. Functionally important short noncoding regions at each end of the genome (5′-NCR and 3′-NCR) are required for its translation and replication [62, 95]. Differing from the canonical eukaryotic mechanism of translation initiation (cap-dependent), this virus utilizes a different mechanism which is dependent on an internal ribosomal entry site (IRES) located within the 5′-untranslated region (5′-UTR), which interacts directly with the 40S ribosomal subunit [148].

Because of the self-encoded and error-prone RNA-dependent RNA polymerase and to the selective pressure exerted by the host immune system, HCV has diversified into seven major genetic lineages (genotypes 1–7) (Fig. 17.2) [52, 176] which differ in 31–33% of nucleotide sites along the complete genome [174]. Genotypes 1–6 of HCV contain a series of more closely related subtypes that typically differ from each other by at least 15% in nucleotide positions within the coding region [176]. Subtypes 1a, 1b, and 3a are widely distributed and account for most of the infections in Western countries, including in Latin America [195].

HCV genetic variability is not evenly distributed across the viral genome. The regions of the genome involved in translation and replication are the most conserved. The 5′-NCR and the core region exhibit 90% and 81–88% sequence identity between distant strains, respectively [23, 149]. The regions coding for the membrane glycoproteins E1 and E2 are the most variable [9]. The hypervariable regions 1 and 2 (HVR1 and HVR2) of the E2 gene are the least conserved, with a sequence

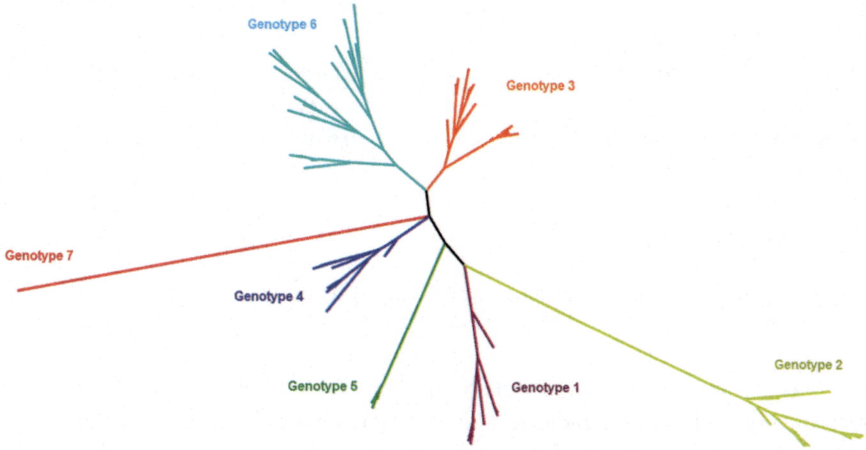

Fig. 17.2 Evolutionary tree of the seven genotypes and all known subtypes of the hepatitis C virus. The tree was constructed with the maximum-likelihood method using GTR + I + G based on a 307-nucleotide sequence from the NS5B-coding region. Sequences were extracted. (From Yusim et al. 2005 [204]; modified from Echeverría et al. 2015 [52])

homology of only 50% between different isolates [75]. As many other RNA viruses, HCV has large population sizes, short generation times, and high replication rates [48], all well-known factors that contribute to its high genetic variability.

2 Prevalence of HCV Infection in Latin America

HCV prevalence data among the different countries in Latin America are difficult to gather. As the prevalence describes the proportion of a particular population that is infected with this virus at a certain time point, it is therefore needed to have precise data on the infected patients in each country to correctly assess this matter. Many efforts have been made to address this issue, in particular in risk populations as well as blood bank donors; however, quantifying HCV infection prevalence among the general population has proven to be the most difficult task. Nevertheless, the World Health Organization has been able to collect comprehensive information regarding HCV seroprevalence as well as viremic HCV prevalence in different regions [135, 197] (Table 17.1); the former has been estimated around 1% and the latter around 0.8%, with subtle variations depending on the specific region considered.

Specific information for each country has relied upon varied studies, all of which have their limitations, particularly regarding the population being recruited for the study and the methods used for screening HCV infections. Regarding HCV prevalence in the general population, there have only been a few complete studies in Argentina [151], Brazil [145], and Mexico [76, 108] reporting an overall lower prevalence than for Latin American regions together (0.32% between years 2000

Table 17.1 Estimated prevalence of hepatitis C virus (HCV) infection by 2013 [74, 197] in different regions in Latin America

Region	Anti-HCV prevalence (CI)[a] (%)	Viremic HCV prevalence (CI)[b] (%)
Caribbean	0.8 (0.2–1.3)	0.6 (0.1–0.9)
Andean region	0.9 (0.4–1.3)	0.6 (0.3–0.9)
Central region	1.0 (0.8–1.4)	0.8 (0.6–1.1)
Southern region	1.2 (0.5–2.1)	0.9 (0.4–1.6)
Tropical region	1.2 (0.9–1.2)	1.0 (0.7–1.0)

CI confidence interval
[a]Exposure to HCV indicated by the presence of antibodies against the virus
[b]Chronic HCV infection indicated by the presence of viral RNA

and 2007, 1.38% between years 2005 and 2009, 0.27–0.35% in 2012 and 2010, respectively), which may indicate an overestimation of the figures by the World Health Organization or an underestimation by these studies due to the selected population. Some other works report the HCV prevalence in specific locations within different countries, but this cannot be considered as the prevalence for the whole country as the population studied was restricted to a particular area, city, or characteristics: 0.4% among the Belize Defence Force by 1993 [37], 0.8% in Amerindian populations in Venezuela in 2002 [120], 0.66–5.68% in different Colombian regions before 2011 [6], and to 5.8% in the Brazilian Eastern Amazonas Region in 2010 [108]. Unexpectedly, a much higher seroprevalence (11.7%) was found in Lima, Peru, in 1994 [165].

Some countries have also attempted to report incidence rates or number of cases for a particular time period; for example, Brazil reported that of a total of 151,056 hepatitis cases recorded between 2001 and 2012, 30.3% corresponded to hepatitis C infections [78]; in Uruguay, 904 new HCV infections were reported to the Ministry of Health between years 2009 and 2012 [10], accounting for 0.026% of the country population.

Exhaustive studies have been performed to address HCV seroprevalence in blood donor banks. These studies have shown lower prevalence rates in this group than in the general population [17, 30, 107, 127, 137, 165, 189]. The Pan-American Health Organization has, additionally, gathered data on anti-HCV screening in blood banks all over Latin America [135], and by 2011, the Latin American country with the highest HCV prevalence was Guatemala (0.65%) whereas the lowest, 0.03%, was registered in Chile.

In contrast, among high-risk populations such as multi-transfused patients, drug users, commercial sex workers, and HIV-coinfected patients, HCV prevalence rates tend to be higher (see Alonso et al. 2015 for a detailed review); 12.7% of multi-transfused patients in Uruguay [106] and more than 50% of hemophiliacs and patients undergoing hemodialysis in Peru [165] exhibited HCV antibodies; despite these intermediate prevalence rates, the lowest found in hemodialysis patients was reported in Argentina, 3.6% [112], and the highest in Chile, 83.9% [109]. Lower ranges in seroprevalence were reported for commercial sex workers, ranging from 0% to 8.2% in female sex workers from Panama [77] and Argentina [138], respectively. HCV prevalence among drug users seems to be higher in injection drug users (IDU) [5], as much as 95% in two cities in Mexico [63] and lower in non-injection drug users, being mostly in the range 0% to 14% [5]. For HIV-coinfected patients, most of the studies focus on genotyping HCV and analyzing response to treatment; however, one study in Venezuela reported a seemingly low prevalence of coinfection with HCV, 0.7% [87].

In summary, all these data suggest an estimate of 4.9 million persons infected with HCV in 2013 in all Latin America [197].

3 Molecular Epidemiology and Evolutionary History

Epidemiology is defined as the study of the distribution and determinants of health-related states and the application of this study to the control of diseases and other health problems. This field has been one of the most studied for hepatitis C infection in many countries in Latin America, and, therefore, there is a great deal of information concerning HCV genotype distribution. In addition, bioinformatics has allowed researchers to investigate the evolutionary history of this virus in our region, being able to identify different introductions in different countries, the dynamics of the population dispersal, and the approximate time of the most recent common ancestor (MRCA).

Historically, 5′-NCR and NS5B regions have been those most frequently used to genotype and subtype HCV isolates, respectively. Until the early 2000s, the preferential genotyping method involved partially amplifying the 5′-NCR by polymerase chain reaction (PCR) and subsequently performing restriction fragment-length polymorphism analysis (RFLP), allowing us to determine mainly the genotypes (and for genotype 1 also the subtypes). However, researchers in Uruguay, Argentina, and Colombia started reporting a mutation within that region (G107A) that seems to incorporate a new restriction site for the *Rsa*I enzyme and therefore prevents the correct subtyping of genotype 1 isolates bearing this change [31, 32, 58, 68, 70, 122]. This finding was regarded as a regional diversification of HCV, although this specific mutation has also been found in isolates from other parts of the world [68]. These findings, together with the advent of affordable sequencing technologies, have caused a shift from this RFLP methodology, to direct sequencing of the regions of interest followed by phylogenetic studies to genotype (5′-NCR) and subtype (NS5B) HCV isolates.

Genotype distribution is similar for some neighboring countries; however, subtle differences can be found in subtype distribution.

In Chile, the most widespread subtype is 1b, accounting for at least 72.71% of the infections, followed by 3a (16.53%) and 1a (7.87%). In lesser proportion, minor lineages have also been detected (genotype 2, 1.98%; genotype 4, 0.6%; genotype 5a, 0.28%; genotype 6, 0.06%) in a cohort of 1766 HCV-infected patients from different Chilean regions [191]. However, it is important to note that the genotyping method was PCR-RFLP of the 5′-NCR region. Another study, with fewer patients ($n = 57$) and genotyping by direct sequencing of the NS5B region, detected a higher proportion of genotype 1b (82.4%) [103]. Although the difference can be attributable to the method used, the sampling number may have also influenced these results. Subtype 1b also seems to be the most widespread in the Caribbean Island of Martinique in mono-infected patients (78.7%), followed by subtype 1a (18%) and genotype 2 (6.8%) [114]. Similar genotype distributions have been found in different studies in Colombia and Cuba, indicating once again that the most widespread subtype is 1b, followed by 1a, and to a lesser extent 2, 3a, and 4 [58, 121, 164, 200]. In Mexico, different authors have found conflicting results concerning the most prevalent subtype, 1b [108] or 1a [139]. Although detection of genotype 3 seems to be increasing over time, there is no doubt that genotype 1 is predominant in the country [128], as it is also in Peru [165].

In contrast, most Brazilian and Uruguayan reports indicate 1a as the most prevalent subtype, followed by 3a, and then by 1b, which suggests a different evolutionary history of HCV in these countries [27, 31, 99, 105, 119, 173]. Also, few cases of genotype 2, 4, and 5a have been reported in Brazil [99, 119, 159, 205].

A different situation is seen in Venezuela, where subtypes 1a, 1b, and genotype 2 are almost equally distributed [87, 154]; the rest of the genotypes are found in less than 5% of the infected patients.

Finally, Argentina reflects a distinct scenario, where the distribution of each genotype varies significantly between different regions and provinces, 2c being as prevalent as genotype 1 in central Argentina [102, 158], 1a the most widespread subtype in the touristic city of Mar del Plata (76%) [44], or similar prevalence rates between 1a, 1b, and 2c (25% each) when analyzing the general population from 12 provinces representing all Argentinean regions [151].

By means of phylogenetic analyses, the evolutionary history of HCV in different countries has been addressed. This approach has also allowed characterizing the variability of different strains and how related they are to each other. Integrating epidemiological to genetic models by employing coalescent methods has also been crucial to infer changes in population sizes (population dynamics) and to trace back the most recent common ancestor (MRCA). Not only the NS5B region has been widely used for evolutionary studies, but also the core, E2 and NS5A regions [27, 33, 43, 45, 69, 73, 98, 99, 103, 122, 132, 167, 178].

Bayesian coalescent studies have shown rather complex evolutionary histories of HCV in Uruguay [27], Brazil [99], and Argentina [44, 45, 71, 102], where multiple introductions of a given subtype have been inferred. The epidemic history in Uruguay and Argentina suggests multiple introductions of subtypes 1a, 1b, and 3a (only for Uruguay), with a few country-specific strains being disseminated locally [27, 45]. Similar evolution patterns were observed in Brazil [99], but with the dissemination of a major clade for subtype 1a, whereas for 1b and 3a, the evidence indicates concurrent dissemination of multiple lineages.

In reference to the time of the MRCA (t_{MRCA}) for subtypes 1b and 1a, these are surprisingly alike between different countries in our region. For Uruguay, Brazil, and Argentina, MRCA has been estimated around the 1920s for 1b and about 40 to 60 years later for 1a [27, 44, 45, 71, 99, 102]. It seems that the 1b population is in a steady state nowadays, while 1a is still under expansion. In Colombia and Cuba, subtype 1b appears to have been introduced at a later time point (1950s and 1970s, respectively); however, the rate of population growth reached a steady state soon after the 1990s [73, 121].

The demographic history of HCV in Chile seems a little different; despite multiple introductions of subtype 1b, the t_{MRCA} appears to be earlier than in the neighboring countries, dating between 1893 and 1901, indicating an earlier introduction of HCV in Chile. Venezuela appears to be one of the first countries in Latin America in which HCV seems to have spread, but opposed to the rest, the initial subtype was 2j and its t_{MRCA} was estimated around 1785, having been followed by 1b around a century later and by 1a around the 1920s [178].

It is clear that HCV spread has followed different pathways depending on the country, but it seems that neighboring countries tend to exhibit rather similar genotype distributions and epidemic histories. All these abundant data are undoubtedly relevant for epidemiological surveillance of HCV dispersal as well as for choosing therapeutic options, which, as is discussed in the next sections, is highly dependent upon the infecting genotype.

4 HCV Variability: Mutation, Quasispecies, and Recombination

HCV evolution is a dynamic process [38] driven by different mechanisms to generate genetic variation, among which nucleotide mutation seems to be the most important, particularly in RNA viruses. These mutations are primarily the result of an error-prone replication cycle as a consequence of using an RNA-dependent RNA polymerase that lacks proofreading activity [174]. HCV high replication rate, large number of progeny and mutation rate (10^{-4} substitutions/site/round of replication [13, 41]) give rise to a large number of different but genetically related viral variants during infection, which circulate in vivo as a complex population commonly known as a quasispecies cloud [20, 47, 113]. At a particular point of infection, the quasispecies distribution reflects the balance between the generation of new variants, the necessity to preserve crucial viral functions, and the positive selective pressure exerted by the environment. Quasispecies dynamics confers RNA viruses the ability to adapt easily to any changing environment. This fact represents one of the major complications for the control and prevention of RNA viral diseases [59] since, by means of generating and selecting fitter variants, viruses can escape control by antiviral drugs [141].

Although the analysis of quasispecies evolution is a complex task, several studies around the world have attempted to explore the genetic variability within quasispecies level as a means to predict response to antiviral therapy [4, 39, 40, 42, 85, 153, 187]. However, results seem conflicting on whether a higher diversity indeed predicts non-response (NR) and a lower level predicts SVR. Cristina et al. [39] showed that response to antiviral therapy was independent from genetic variability within quasispecies populations at the beginning of therapy. Opposite to these findings, Jardim et al. [85] reported that low nucleotide diversity pre-therapy was associated with viral clearance. Dissimilar results were also found along treatment [4, 187]. These conflicting results could be explained by the use of different methodological approaches as well as the distinct genomic regions that were used to carry out these studies. The inconsistency in the results highlights the need for a deep analysis and comparison of different methods for studying quasispecies variability.

HCV also resorts to recombination, as well as mutation, to generate variability. Therefore, recombination is regarded as a key mechanism for the production of new genomes with selective growth advantage [198]. RNA recombination involves replication of genomic RNA, in the middle of which the viral RNA-dependent RNA polymerase complex switches from one parental strand to another. This template strand exchange mechanism is known as "copy choice" [34]. Recombination in HCV has been reported between different genotypes (inter-genotypic), between different subtypes (intra-genotypic), and even between different variants of the same quasispecies (intra-quasispecies) in different geographic locations. However, it has been rarely reported in Latin America [33, 126]. The first evidence for the possible existence of HCV recombination in our continent came from the analysis of a few HCV strains from Honduras in which partial sequences from different regions of the viral genome resulted in an HCV-discordant genotype [203]. However, no case of inter-genotypic recombination has been reported in Latin America, despite the existence of several from other regions [19, 46, 50, 82, 90, 91, 96, 100, 101, 123, 124, 133, 180, 201]. Nevertheless, examples of intra-genotypic recombination (1a/1b) have been identified in Peru [33] as well as in Uruguay [126]. As for intra-quasispecies recombination, only three reports have been published worldwide [125, 136, 170], but none of them corresponds to HCV variants isolated from Latin American patients. However, given that one of the detections has been reported by Uruguayan authors [125], it is clear that Latin American researchers have the means to detect these recombination events within a host, and although it seems to be more difficult, it is an attainable aim provided that the variability of the region under study allows the differentiation of the parental genomes.

It is worth noting that recombination may impact how patients respond to antiviral therapies as well as vaccine development, which makes it clearly important to determine the extent to which this mechanism plays a role in HCV evolution [126].

5 Antiviral Therapy and Predictors of Response to Treatment

5.1 Standard of Care: Pegylated Interferon-α and Ribavirin

As was mentioned earlier, the standard of care for HCV treatment has been facing numerous changes in the past 6 years. Historically (and up to 2010), the only available treatment was dual therapy, which involves the coadministration of pegylated interferon-α (IFN-α-PEG) and RBV for 24 or 48 weeks, depending mainly on the infecting HCV genotype. Beginning in 1992, the first drug to be used was IFN-α in monotherapy, but the SVR rates were less than 20% in genotype 1 and 50% in genotypes 2 and 3 [152]. The conjugation of a polyethylene glycol (PEG) residue to IFN and the coadministration of RBV, a nucleoside analogue, increased those rates to up to 50% in genotype 1 and to 70–80% in genotype 2 or 3 [66]. The PEG residue enables a longer IFN half-life, and RBV contributes to viral eradication by introducing mutations to an unbearable level into the viral genome. Although the addition of PEG and the coadministration of RBV improved the SVR rates, this is still insufficient for genotype 1-infected patients. This treatment has several secondary effects (flu-like symptoms, hemolytic anemia, and depression) and is contraindicated in patients with autoimmune or depressive disorders and in those with cerebrovascular diseases [14]. Moreover, adherence to treatment is low as a consequence of both secondary effects as well as the parenteral route of IFN administration (injection).

5.2 New Standard of Care: Direct Antiviral Agents

The reasons outlined earlier prompted the development of new therapeutic agents, mainly aiming at inhibiting viral proteins, the first being approved in 2011 [66].

These new antiviral agents, known as DAAs (direct antiviral agents), target three different viral proteins vital to the viral cycle: the protease NS3-4A, the polymerase NS5B, and the protein NS5A (involved in many different steps throughout the replication cycle). Their antiviral effectiveness has proven to be much higher (80% to almost 100% depending on patient characteristics) than dual therapy when combined with IFN-α-PEG/RBV or even in IFN-free regimens, combining two or three different DAAs [142]. Furthermore, they exhibit fewer adverse effects, and because they are administered orally, compliance to treatment is higher [11]. Numerous DAAs have already been approved around the world [181]: telaprevir (TVR), boceprevir (BOC), simeprevir (SMV), asunaprevir (ASV), vaniprevir (VPV), and paritaprevir (PPV, PTV) (NS3-4A inhibitors); sofosbuvir (SOF) and dasabuvir (DBV, DSV) (NS5B inhibitors); and daclatasvir (DCV), ledipasvir (LDV), and ombitasvir (OBV) (NS5A inhibitors). Many of these are also available in Latin America.

5.3 Identification and Characterization of New Anti-HCV Drugs

Even though all these agents have been developed in industrialized countries, researchers in Latin America have done considerable work focused on the identification or characterization of new anti-HCV drugs [16, 57, 86, 97, 111, 150, 162, 185].

In reference to the synthesis and characterization of novel inhibitor drugs against the polymerase, many compounds have been described in Latin America, including indol-based compounds [97] and isoflavonoids [57, 111]. As for potential NS3-4A inhibitors, many products from the Amazon region were evaluated by molecular dynamics simulation studies [150]. The results of this work showed that some of those compounds seem to be promising as novel anti-HCV therapies. On the other hand, caffeine [16], natural compounds isolated from Brazilian plants [86], and acetylsalicylic acid [162, 185], also were shown to be efficient inhibitors of HCV replication and translation in cell culture-based assays.

5.4 Viral Genetic Factors Associated with Response to Treatment

Several factors have been associated with response to treatment in HCV-infected patients. Some have been essential to decide the length of dual therapy with IFN and RBV; others have become more important with the advent of direct antiviral drugs. These factors can be classified as virus- or host-related factors [160]. Among the pretreatment viral factors, it has been long recognized that the virus genotype is one of the strongest baseline predictors of dual therapy: genotype 1 has shown the lowest SVR rates (around 50%) [66]. This factor remains important with the new standard of care therapies with protease inhibitors (PI) because these were designed in a genotype-specific manner. Furthermore, triple therapies including a PI in genotype 1-infected patients have shown that SVR rates are slightly higher in patients infected with genotype 1b because resistance-associated variants (RAVs) are more frequent in genotype 1a [84, 94, 166]. Another well-known pretreatment factor associated with response to dual therapy is low baseline HCV load, which has been shown to be a predictor of SVR [18, 207].

Additional viral factors include quasispecies complexity as well as mutations/substitutions in different regions of the viral genome/proteins. The genetic heterogeneity of the ISDR domain of HCV NS5A (IFN sensitivity-determining region) showed a correlation to IFN response, meaning that in most cases a higher number of substitutions within the ISDR was observed in responding patients [55, 144, 194]. Latin American studies have also found similar results [21, 22, 168].

Concomitant with the introduction of DAAs to treat HCV infections, the detection of RAVs has become of utmost importance all around the world. In our region substantial work has been done in Brazil to address this issue [25, 28, 79, 80, 104, 131, 146, 206], but data on other countries are scarce, with only a few reports from elsewhere: Argentina [60, 88, 169], Venezuela [88], Mexico [60], and Uruguay [54].

The main role of the NS3-4A protease is to cleave the viral polyprotein to render individual proteins. Therefore, drugs that bind to its active site prevent the viral cycle from continuing by blocking replication by the lack of a cleaved polymerase. Because a more diverse choice of approved PIs has been available for a longer time than inhibitors of NS5A and NS5B, numerous data exist concerning RAVs in the

NS3 coding region. The vast majority of these RAVs map near the catalytic triad residues (His57, Asp81, Ser139) in the tertiary structure of the protease. Most substitutions exhibit cross-resistance to more than one PI, such as V36A/G/L/M, F43I/L/S/V, T54A/S, V55A/I, Q80K/L/R, S122R, R155G/K/S/T, A156S/T/V, D168A/G/V, and V/I170A/T [193, 199]. Others confer resistance to one drug only: Q41R (SMV), I132V (TVR), S138T (SMV), V158I (BOC), M175L (BOC) [193, 199].

In reference to RAV circulation in the Latin American region, the deep sequencing approach as well as the mismatch amplification mutation assay (MAMA) PCR method allowed detection of drug resistance variants present at very low frequencies in both Mexico and Argentina (1–6%) [60, 169]. On the other hand, Brazil reported HCV isolates with baseline RAVs in 3.2% to 18.9% of DAA treatment-naïve patients [25, 104, 146, 206]. Surprisingly, a higher frequency of baseline RAVs has been found in Uruguayan patients infected with HCV genotype 1 (25%) [54].

With respect to Q80K, a mutation associated with resistance to SMV, it has been found at a high prevalence among treatment-naïve HCV carriers in the U.S. (46%) and in Europe (4–16%) [131, 193], whereas in Brazil it is less than 6% [79, 146]. In Uruguay, although the number of samples analyzed was small ($n = 20$), Q80K/L was found in three patients [54].

NS5A and NS5B proteins seem to have a higher resistance barrier, and therefore fewer positions have been described as RAVs [53, 130]. The most important substitution conferring resistance to SOF (a nucleos(t)ide analogue, inhibitor of NS5B) is S282T, but others have been associated with treatment failure as well, such as L159F, C316N, and V321A. Substitutions in C316 are, together with S556G, implicated in resistance to DSV (a non-nucleoside inhibitor of the polymerase). As for RAVs to NS5A inhibitors, the primary mutations are M28T, Q30E/H/R, L31F/M/V, P32L, and Y93C/H/N.

Only three studies have been undertaken in our region to gain insight into resistance profiles against NS5A and NS5B inhibitors. The only study addressing NS5A RAVs documented several mutations in the consensus sequences, including M28T, Y93H, and L31M in at least 7% of the 107 treatment-naïve patients recruited [147]. NS5B S282T substitution was not found in Argentina [169], Brazil [28], or Venezuela [88]; however, C316N seems to be rather frequent in both Brazil and Venezuela, being present in 24% and 18% of genotype 1b-infected patients, respectively.

Other types of drugs are also under development, the most promising being miravirsen, a novel therapeutic agent targeting a host factor. Miravirsen is currently in phase 2 clinical trials [186], and it is aimed at inhibiting HCV in an indirect way by targeting miR-122, a liver-specific micro-RNA highly expressed in hepatocytes that is essential for HCV replication cycle [89]. Although results are promising given its broad antiviral activity and relatively high resistance barrier compared to DAAs [134], there are safety concerns because critical cellular functions are attributed to miR-122: maintenance of liver homeostasis as well as tumor-suppressor activity [83]. In light of these findings and bearing in mind the variety of RAVs already described, other alternatives to target HCV directly are being studied in Latin

America, such as RNA interference targeting five different regions of the genome [24] or the use of catalytically active DNA molecules capable of cleaving RNA upon binding to it [163].

As can be concluded from the studies already mentioned, there are as many interesting ongoing studies as there is still much to be done in the remaining Latin American countries to provide useful information for physicians before assigning a particular treatment to a given patient or even developing newer therapeutic strategies.

5.5 Host Factors Associated with Response to Treatment

So far, we have addressed only viral genetic factors, but several host factors have also been shown to be involved in the development of HCV infection and the response to therapy [160], apart from miR-122, discussed earlier. African-American race and advanced fibrosis or cirrhosis are associated with lower response rates, when compared to other races and less liver damage. Furthermore, since 2009, several single-nucleotide polymorphisms (SNP) near the interleukin-28B gene (IL28B), also known as interferon-lambda 3 (IFN-λ3), have been reported to influence response to dual antiviral therapy (IFN-α-PEG plus RBV) [65, 155, 179, 183, 188, 202]. The SNPs identified through genome-wide association studies (GWAS) were mainly three: rs12979860, rs8099917, and rs12980275. For every SNP, risk or unfavorable alleles have been determined: rs12979860, T; rs8099917, G; and rs12980275, G. When combined in homozygosis (TT, GG, or GG) or heterozygosis with the major allele (TC, GT, or GA), these genotypes are associated with non-response to treatment and to lower rates of viral clearance. As these results have been validated in numerous clinical studies, IL28B genotyping has become important to assist clinicians with respect to the best therapeutic regimen to adopt for each patient; in particular, to decide the length of IFN-containing therapies or the implementation of a DAA-containing regimen.

Only a few countries in Latin America have investigated IL28B SNPs allelic and genotypic frequencies: Chile [8, 192], Mexico [61, 72, 118, 175], Argentina [64, 161], Brazil [29, 56, 156, 157, 172, 190], and Uruguay (unpublished data). Despite some discordant results, most of these studies reflect an association between favorable IL28B genotypes and spontaneous or treatment-induced viral clearance. Therefore, they cannot be denied as important and useful factors to consider when deciding on a particular therapeutic strategy.

Another interesting finding is that the frequency of protective alleles seems to be significantly lower in Amerindian and admixed populations from Mexico [72] and Argentina [184]. Similarly, in Uruguay, preliminary studies indicate that the frequency distribution for rs12979860 and rs8099917 genotypes among general population resembles that of an admixed population, instead of that of European ancestry. As observed for most other countries, genotype distribution varies when comparing healthy individuals and infected patients, but these changes seem to be statistically significant only when analyzing rs12979860 SNP (unpublished data). In Brazil, similar observations were made with respect to African descendants, who show a higher frequency of unfavorable genotypes [29]. Despite some discordant results, most of the studies do indeed reflect an association between favorable IL28B genotypes and spontaneous or treatment-induced viral clearance. Therefore, these cannot be denied as important and useful factors to consider when deciding on a particular therapeutic strategy.

In fact, all the data mentioned throughout this section, if anything, highlight how imperative it is for the rest of the countries in our region to focus on characterizing and analyzing all the factors that influence a patient's response to anti-HCV therapy, as a way of evolving to a more personalized and effective medicine.

6 Vaccine Candidates

Despite worldwide efforts to develop a prophylactic vaccine against HCV, it remains elusive. Nevertheless, several Cuban researchers have been actively working on vaccine candidates for the past 15 years [1–3, 7, 12, 26, 49, 51, 115–117]. Most of the work focused on the main HCV antigens: the core protein and the E2 glycoprotein. The successful heterologous expression of these proteins and chimeric ones in yeast (*Pichia pastoris*) [116] or bacteria (*Escherichia coli*) [3] indicated that it is feasible to generate HCV vaccine candidates that can elicit both humoral and cell-mediated immune responses in mice, sheep, and monkeys [1, 2, 115]. Besides protein-based candidates, DNA-based formulations have also been investigated. Vaccination of mice with a plasmid containing a truncated variant of the core was able to induce a slow but potent immune response [51]. Furthermore, coadministration of a combination of a plasmid encoding the first 650 amino acids of the polyprotein and a protein comprising the first 120 amino acids of the core also showed potent immune responses in rats [12] and in humans [26], with no observable toxicity or adverse local or systemic alterations. These results are promising because vaccination with this combination seems to be safe and well tolerated; however, further studies are needed to evaluate these features in a larger cohort of individuals.

7 Concluding Remarks and Future Challenges

As can be appreciated, a great amount of work has been pursued in Latin America on HCV molecular evolution, although there are still numerous countries for which data on HCV infection are scarce or not available, or even where studies are lacking. Additionally, mainly because of budget constraints, many studies have been hampered or their development has been slow when compared to industrialized countries. These facts highlight the main problem that Latin America faces when coping with epidemics: the difficulty in keeping up with the most up-to-date findings and techniques. Nevertheless, the work described in this chapter proves that the research in our region cannot be underestimated and that there are strong and well-prepared groups to face any necessary study to help in controlling and preventing HCV infection throughout Latin America.

Acknowledgments This work was supported by Agencia Nacional de Investigación e Innovación (ANII) through project FMV_1_2014_1_104171, Comisión Académica de Posgrado, UdelaR, PEDECIBA and program Grupos I+D CSIC, Uruguay.

References

1. Acosta-Rivero N, Alvarez-Obregón JC, Musacchio A, Falcón V, Dueñas-Carrera S, Marante J, Menéndez I, Morales J (2002) In vitro self-assembled HCV core virus-like particles induce a strong antibody immune response in sheep. Biochem Biophys Res Commun 290:300–304. doi:10.1006/bbrc.2001.6177
2. Acosta-Rivero N, Poutou J, Álvarez-Lajonchere L, Guerra I, Aguilera Y, Musacchio A, Rodríguez A, Aguilar JC, Falcon V, Álvarez-Obregon JC, Soria Y, Torres D, Linares M, Pérez A, Morales-Grillo J, Dueñas-Carrera S (2009) Recombinant in vitro assembled hepatitis C virus core particles induce strong specific immunity enhanced by formulation with an oil-based adjuvant. Biol Res 42:41–56. doi:10.4067/S0716-97602009000100005
3. Aguilar-Noriega D, Alvarez-Lajonchere L, Brown E, Santana FL, Dubuisson J, Wychowski C, Guerra I, Martínez-Donato G, Pérez A, Amador-Cañizares Y, Dueñas-Carrera S (2014) A chimeric protein encompassing hepatitis C virus epitopes is able to elicit both humoral and cell-mediated immune responses in mice. Biotechnol Appl Biochem 61:627–636. doi:10.1002/bab.1223
4. Alfonso V, Flichman DM, Sookoian S, Mbayed VA, Campos RH (2004) Evolutionary study of HVR1 of E2 in chronic hepatitis C virus infection. J Gen Virol 85:39–46. doi:10.1099/vir.0.19403-0
5. Alonso M, Gutzman A, Mazin R, Pinzon CE, Reveiz L, Ghidinelli M (2015) Hepatitis C in key populations in Latin America and the Caribbean: systematic review and meta-analysis. Int J Public Health 60:789–798. doi:10.1007/s00038-015-0708-5
6. Alvarado-Mora MV, Fernandez MFG, Gomes-Gouvêa MS, de Azevedo Neto RS, Carrilho FJ, Pinho JRR (2011) Hepatitis B (HBV), hepatitis C (HCV) and hepatitis delta (HDV) viruses in the Colombian population-how is the epidemiological situation? PLoS One 6(4):e18888. doi:10.1371/journal.pone.0018888
7. Amador-Cañizares Y, Martínez-Donato G, Álvarez-Lajonchere L, Vasallo C, Dausá M, Aguilar-Noriega D, Valenzuela C, Raíces I, Dubuisson J, Wychowski C, Cinza-Estévez Z, Castellanos M, Núñez M, Armas A, González Y, Revé I, Guerra I, Pérez Aguiar Á, Dueñas-Carrera S (2014) HCV-specific immune responses induced by CIGB-230 in combination with IFN-α plus ribavirin. World J Gastroenterol 20:148–162. doi:10.3748/wjg.v20.i1.148
8. Angulo J, Pino K, Pavez C, Biel F, Labbé P, Miquel JF, Soza A, López-Lastra M (2013) Genetic variations in host IL28B links to the detection of peripheral blood mononuclear cells-associated hepatitis C virus RNA in chronically infected patients. J Viral Hepat 20:263–272. doi:10.1111/jvh.12076
9. Argentini C, Genovese D, Dettori S, Rapicetta M (2009) HCV genetic variability: from quasispecies evolution to genotype classification. Future Microbiol 4:359–373. doi:10.2217/fmb.09.8
10. Armas D, Moreale J, González T (2013) Nuevos antivirales en la hepatitis C: Telaprevir y Boceprevir. Biomedicina (B Aires) 8(2):22–35

11. Attar BM, Van Thiel DH (2016) Hepatitis C virus: a time for decisions. Who should be treated and when? World J Gastrointest Pharmacol Ther 6:33–40. doi:10.4292/wjgpt.v7.i1.33
12. Bacardí D, Amador-Cañizares Y, Cosme K, Urquiza D, Suárez J, Marante J, Viña A, Vázquez A, Concepción J, Pupo M, Aldana L, Soria Y, Romero J, Madrigal R, Martínez L, Hernández L, González I, Dueñas-Carrera S (2009) Toxicology and biodistribution study of CIGB-230, a DNA vaccine against hepatitis C virus. Hum Exp Toxicol 28:479–491. doi:10.1177/0960327109106438
13. Bartenschlager R, Lohmann V (2000) Replication of hepatitis C virus. J Gen Virol 81:1631–1648. doi:10.1099/0022-1317-81-7-1631
14. Bartenschlager R, Lohmann V, Penin F (2013) The molecular and structural basis of advanced antiviral therapy for hepatitis C virus infection. Nat Rev Microbiol 11:482–496. doi:10.1038/nrmicro3046
15. Bartenschlager R, Penin F, Lohmann V, André P (2011) Assembly of infectious hepatitis C virus particles. Trends Microbiol 19:95–103. doi:10.1016/j.tim.2010.11.005
16. Batista MN, Carneiro BM, Braga ACS, Rahal P (2014) Caffeine inhibits hepatitis C virus replication in vitro. Arch Virol 160:399–407. doi:10.1007/s00705-014-2302-1
17. Bedoya JAP, Cortés Márquez MM, Cardona Arias JA (2012) Seroprevalence of markers of transfusion transmissible infections in blood bank in Colombia. Rev Saude Publica 46:950–959
18. Berg T, Sarrazin C, Herrmann E, Hinrichsen H, Gerlach T, Zachoval R, Wiedenmann B, Hopf U, Zeuzem S (2003) Prediction of treatment outcome in patients with chronic hepatitis C: significance of baseline parameters and viral dynamics during therapy. Hepatology 37:600–609. doi:10.1053/jhep.2003.50106
19. Bhattacharya D, Accola MA, Ansari IH, Striker R, Rehrauer WM (2011) Naturally occurring genotype 2b/1a hepatitis C virus in the United States. Virol J 8:458. doi:10.1186/1743-422X-8-458
20. Biebricher CK, Eigen M (2005) The error threshold. Virus Res 107:117–127. doi:10.1016/j.virusres.2004.11.002
21. Bittar C, Jardim ACG, Yamasaki LHT, de Queiróz ATL, Carareto CMA, Pinho JRR, de Carvalho-Mello IMVG, Rahal P (2010) Genetic diversity of NS5A protein from hepatitis C virus genotype 3a and its relationship to therapy response. BMC Infect Dis 10:36. doi:10.1186/1471-2334-10-36
22. Bolcic F, Laufer N, Torres C, Cassino L, Reynoso R, Quarleri J (2012) Longitudinal analysis of the 5′UTR, E2-PePHD and NS5A-PKRBD genomic regions of hepatitis C virus genotype 1a in association with the response to PEGinterferon and ribavirin therapy in HIV-coinfected patients. Antivir Res 95:72–81. doi:10.1016/j.antiviral.2012.05.015
23. Bukh J, Purcell RH, Miller RH (1992) Sequence analysis of the 5′ noncoding region of hepatitis C virus. Proc Natl Acad Sci U S A 89:4942–4946. doi:10.1073/pnas.89.11.4942
24. Carneiro B, Braga ACS, Batista MN, Harris M, Rahal P (2015) Evaluation of canonical siRNA and Dicer substrate RNA for inhibition of hepatitis C virus genome replication: a comparative study. PLoS One 10:1–18. doi:10.1371/journal.pone.0117742
25. de Carvalho IMVG, Alves R, de Souza PAVM, da Silva EF, Mazo D, Carrilho FJ, Queiroz ATL, Pessoa MG (2014) Protease inhibitor resistance mutations in untreated Brazilian patients infected with HCV: novel insights about targeted genotyping approaches. J Med Virol 86:1714–1721. doi:10.1002/jmv.24015
26. Castellanos M, Cinza Z, Dorta Z, Veliz G, Vega H, Lorenzo I, Ojeda S, Dueñas-Carrera S, Alvarez-Lajonchere L, Martínez G, Ferrer E, Limonta M, Linares M, Ruiz O, Acevedo B, Torres D, Márquez G, Herrera L, Arús E (2010) Immunization with a DNA vaccine candidate in chronic hepatitis C patients is safe, well tolerated and does not impair immune response induction after anti-hepatitis B vaccination. J Gene Med 12:107–116. doi:10.1002/jgm.1407
27. Castells M, Bello G, Ifrán S, Pereyra S, Boschi S, Uriarte R, Cristina J, Colina R (2015) Epidemic history of major genotypes of hepatitis C virus in Uruguay. Infect Genet Evol 32:231–238. doi:10.1016/j.meegid.2015.03.021

28. Castilho MCB, Martins AN, Horbach IS, Perez RDM, Figueiredo FAF, Pinto PDTA, Nabuco LC, De Lima DB, Tanuri A, Porto LC, Ferreira Júnior ODC (2011) Association of hepatitis C virus NS5B variants with resistance to new antiviral drugs among untreated patients. Mem Inst Oswaldo Cruz 106:968–975
29. Cavalcante LN, Abe-Sandes K, Angelo ALD, Machado TMB, Lemaire DC, Mendes CMC, Pinho JR, Malta F, Lyra LGC, Lyra AC (2012) IL28B polymorphisms are markers of therapy response and are influenced by genetic ancestry in chronic hepatitis C patients from an admixed population. Liver Int 32:476–486. doi:10.1111/j.1478-3231.2011.02653.x
30. Césaire R, Martial J, Maier H, Kerob-Bauchet B, Bera O, Duchaud E, Brebion A, Pierre-Louis S (1999) Infection with GB virus C/hepatitis G virus among blood donors and hemophiliacs in Martinique, a Caribbean Island. J Med Virol 59:160–163. doi:10.1002/(SICI)1096-9071(199910)59:2<160::AID-JMV6>3.0.CO;2-Y
31. Colina LR, María D, Mogdasy C, Cristina J, Uriarte R (2002) Caracterización molecular del virus de la hepatitis C en Montevideo-Uruguay. Rev Méd Uruguay 18:76–82
32. Colina R, Azambuja C, Uriarte R, Mogdasy C, Cristina J (1999) Evidence of increasing diversification of hepatitis C viruses. J Gen Virol 80:1377–1382
33. Colina R, Casane D, Vasquez S, García-Aguirre L, Chunga A, Romero H, Khan B, Cristina J (2004) Evidence of intratypic recombination in natural populations of hepatitis C virus. J Gen Virol 85:31–37. doi:10.1099/vir.0.19472-0
34. Cooper P, Steiner-Pryor A, Scotti P, Delong D (1974) On the nature of poliovirus genetic recombinants. J Gen Virol 23:41–49. doi:10.1099/0022-1317-23-1-41
35. Cornberg M, Hardtke S, Port K, Manns MP, Wedemeyer H (2012) Hepatitis C standard of care. Short Guid Hepat C. doi:10.1111/j.1945-5100.2009.tb02000.x
36. Cornberg M, Wedemeyer H, Manns MP (2002) Treatment of chronic hepatitis C with PEGylated interferon and ribavirin. Curr Gastroenterol Rep 4:23–30. doi:10.1007/s11894-002-0034-y
37. Craig PG, Bryan JP, Miller RE, Reyes L, Hakre S, Jaramillo R, Krieg RE (1993) The prevalence of hepatitis A, B and C infection among different ethnic groups in Belize. Am J Trop Med Hyg 49:430–434
38. Cristina J (2005) Genetic diversity and evolution of hepatitis C virus in the Latin American region. J Clin Virol 34(suppl 2):S1–S7
39. Cristina J, del Pilar MM, Moratorio G (2007) Hepatitis C virus genetic variability in patients undergoing antiviral therapy. Virus Res 127:185–194. doi:10.1016/j.virusres.2007.02.023
40. Cuevas JM, Gonzalez M, Torres-Puente M, Jiménez-Hernández N, Bracho MA, García-Robles I, González-Candelas F, Moya A (2009) The role of positive selection in hepatitis C virus. Infect Genet Evol 9:860–866. doi:10.1016/j.meegid.2009.05.007
41. Cuevas JM, González-Candelas F, Moya A, Sanjuán R (2009) Effect of ribavirin on the mutation rate and spectrum of hepatitis C virus in vivo. J Virol 83:5760–5764. doi:10.1128/JVI.00201-09
42. Cuevas JM, Torres-Puente M, Jiménez-Hernández N, Bracho MA, García-Robles I, Carnicer F, Del Olmo J, Ortega E, González-Candelas F, Moya A (2009) Combined therapy of interferon plus ribavirin promotes multiple adaptive solutions in hepatitis C virus. J Med Virol 81:650–656. doi:10.1002/jmv.21460
43. Culasso ACA, Baré P, Aloisi N, Monzani MC, Corti M, Campos RH (2014) Intra-host evolution of multiple genotypes of hepatitis C virus in a chronically infected patient with HIV along a 13-year follow-up period. Virology 449:317–327. doi:10.1016/j.virol.2013.11.034
44. Culasso ACA, Elizalde M, Campos RH, Barbini L (2012) Molecular survey of hepatitis C virus in the touristic city of Mar Del Plata Argentina. PLoS One. doi:10.1371/journal.pone.0044757
45. Culasso ACA, Farías A, Di Lello FA, Golemba MD, Ré V, Barbini L, Campos R (2014) Spreading of hepatitis C virus subtypes 1a and 1b through the central region of Argentina. Infect Genet Evol 26:32–40. doi:10.1016/j.meegid.2014.05.008

46. Demetriou VL, Kyriakou E, Kostrikis LG (2011) Near-full genome characterisation of two natural intergenotypic 2 k/1b recombinant hepatitis c virus isolates. Adv Virol 2011:1–8. doi:10.1155/2011/710438
47. Domingo E, Baranowski E, Ruiz-Jarabo CM, Martín-Hernández AM, Sáiz JC, Escarmís C (1998) Quasispecies structure and persistence of RNA viruses. Emerg Infect Dis 4:521–527. doi:10.3201/eid0404.980402
48. Domingo E, Holland JJ (1997) RNA virus mutations and fitness for survival. Annu Rev Microbiol 51:151–178. doi:10.1146/annurev.micro.51.1.151
49. Dorta-Guridi Z, Dueñas-Carrera S, Arús-Soler ER, Castellanos-Fernández MI, Cinza-Estévez Z (2015) Therapeutic vaccines against the hepatitis C virus in the age of direct-acting antivirals. Biotecnol Apl 32:1121–1124. doi:10.1016/B978-0-323-08375-1.00329-2
50. Du H, Qi Y, Hao F, Huang Y, Mao L, Ji S, Huang M, Qin C, Yan R, Zhu X, Zhang C (2012) Complex patterns of HCV epidemic in Suzhou: evidence for dual infection and HCV recombination in East China. J Clin Virol 54:207–212. doi:10.1016/j.jcv.2012.04.017
51. Dueñas-Carrera S, Alvarez-Lajonchere L, Alvarez-Obregón JC, Herrera A, Lorenzo LJ, Pichardo D, Morales J (2000) A truncated variant of the hepatitis C virus core induces a slow but potent immune response in mice following DNA immunization. Vaccine 19:992–997. doi:10.1016/S0264-410X(00)00209-7
52. Echeverria N, Moratorio G, Cristina J, Moreno P (2015) Hepatitis C virus genetic variability and evolution. World J Hepatol 7:831–845. doi:10.4254/wjh.v7.i6.831
53. Eltahla AA, Luciani F, White PA, Lloyd AR, Bull RA (2015) Inhibitors of the hepatitis C virus polymerase; mode of action and resistance. Viruses 7:5206–5224. doi:10.3390/v7102868
54. Echeverría N, Betancour G, Gámbaro F, Hernández N, López P, Chiodi D, Sánchez A, Boschi S, Fajardo A, Sóñora M, Moratorio G, Cristina J, Moreno P (2016) Naturally occurring NS3 resistance-associated variants in hepatitis C virus genotype 1: their relevance for developing countries. Virus Res 223:140–146. doi:10.1016/j.virusres.2016.07.008
55. Enomoto N, Sakuma I, Asahina Y, Kurosaki M, Murakami T, Yamamoto C, Izumi N, Marumo F, Sato C (1995) Comparison of full-length sequences of interferon-sensitive and resistant hepatitis C virus 1b: sensitivity to interferon is conferred by amino acid substitutions in the NS5A region. J Clin Invest 96:224–230. doi:10.1172/JCI118025
56. Fabrício-Silva GM, Poschetzky BS, Perez RDM, Dos Santos RC, Cavalini LT, Porto LC (2015) Association of cytokine gene polymorphisms with hepatitis C virus infection in a population from Rio de Janeiro, Brazil. Hepatic Med Evid Res 7:71–79
57. Fernandes TA, Manvar D, JLO D, Basu A, Nichols DB, Kaushik-Basu N, PRR C (2016) 5-Carba-pterocarpens: a new scaffold with anti-HCV activity. Eur J Med Chem 112:33–38. doi:10.1016/j.ejmech.2016.02.010
58. di Filippo D, Cortes-Mancera F, Beltran M, Arbelaez MP, Jaramillo S, Restrepo JC, Correa G, Navas M-C (2012) Molecular characterization of hepatitis C virus in multi-transfused Colombian patients. Virol J 9:242. doi:10.1186/1743-422X-9-242
59. Flint SJ, Enquiest LW, Krug RM, Racaniello VR, Skalka AM (2000) Principles of virology, molecular biology, pathogenesis, and control. Lancet Infect Dis 9:471–472. doi:10.1016/S1473-3099(09)70195-1
60. Fonseca-Coronado S, Escobar-Gutiérrez A, Ruiz-Tovar K, Cruz-Rivera MY, Rivera-Osorio P, Vazquez-Pichardo M, Carpio-Pedroza JC, Ruíz-Pacheco JA, Cazares F, Vaughan G (2012) Specific detection of naturally occurring hepatitis C virus mutants with resistance to telaprevir and boceprevir (protease inhibitors) among treatment-naïve infected individuals. J Clin Microbiol 50:281–287. doi:10.1128/JCM.05842-11
61. Fonseca-Coronado S, Vaughan G, Cruz-Rivera MY, Carpio-Pedroza JC, Ruiz-Tovar K, Ruiz-Pacheco JA, Escobar-Gutiérrez A (2011) Interleukin-28B genotyping by melt-mismatch amplification mutation assay PCR analysis using single nucleotide polymorphisms rs12979860 and rs8099917, a useful tool for prediction of therapy response in hepatitis C patients. J Clin Microbiol 49:2706–2710. doi:10.1128/JCM.00877-11

62. Friebe P, Bartenschlager R (2009) Role of RNA structures in genome terminal sequences of the hepatitis C virus for replication and assembly. J Virol 83:11989–11995. doi:10.1128/JVI.01508-09
63. Frost SDW, Brouwer KC, Firestone Cruz MA, Ramos R, Ramos ME, Lozada RM, Magis-Rodriguez C, Strathdee SA (2006) Respondent-driven sampling of injection drug users in two U.S.-Mexico border cities: recruitment dynamics and impact on estimates of HIV and syphilis prevalence. J Urban Heal doi:10.1007/s11524-006-9104-z
64. Galvan CA, Elbarcha OC, Fernandez EJ, Beltramo DM, Soria NW (2012) Distribution of polymorphisms in cytochrome P450 2B6, histocompatibility complex P5, chemokine coreceptor 5, and interleukin 28B genes in inhabitants from the central area of Argentina. Genet Test Mol Biomarkers 16:130–133. doi:10.1089/gtmb.2011.0058
65. Ge D, Fellay J, Thompson AJ, Simon JS, Shianna KV, Urban TJ, Heinzen EL, Qiu P, Bertelsen AH, Muir AJ, Sulkowski M, McHutchison JG, Goldstein DB (2009) Genetic variation in IL28B predicts hepatitis C treatment-induced viral clearance. Nature (Lond) 461:399–401. doi:10.1038/nature08309
66. Ghany MG, Nelson DR, Strader DB, Thomas DL, Seeff LB (2011) An update on treatment of genotype 1 chronic hepatitis C virus infection: 2011 practice guideline by the American Association for the Study of Liver Diseases. Hepatology 54:1433–1444. doi:10.1002/hep.24641
67. Ghany MG, Strader DB, Thomas DL, Seeff LB (2009) Diagnosis, management, and treatment of hepatitis C: an update. Hepatology 49:1335–1374. doi:10.1002/hep.22759
68. Gismondi MI, Becker PD, Valva P, Guzmán CA, Preciado MV (2006) Phylogenetic analysis of previously nontypeable hepatitis C virus isolates from Argentina. J Clin Microbiol 44:2229–2232. doi:10.1128/JCM.02569-05
69. Gismondi MI, Díaz Carrasco JM, Valva P, Becker PD, Guzmán CA, Campos RH, Preciado MV (2013) Dynamic changes in viral population structure and compartmentalization during chronic hepatitis C virus infection in children. Virology 447:187–196. doi:10.1016/j.virol.2013.09.002
70. Gismondi MI, Staendner LH, Grinstein S, Guzmán CA, Preciado MV (2004) Hepatitis C virus isolates from Argentina disclose a novel genotype 1-associated restriction pattern. J Clin Microbiol 42:1298–1301. doi:10.1128/JCM.42.3.1298-1301.2004
71. Golemba MD, Culasso ACA, Villamil FG, Bare P, Gadano A, Ridruejo E, Martinez A, Di Lello FA, Campos RH (2013) Hepatitis C virus diversification in Argentina: comparative analysis between the large city of Buenos Aires and the small rural town of O'Brien. PLoS One. doi:10.1371/journal.pone.0084007
72. Gonzalez-Aldaco K, Rebello Pinho JR, Roman S, Gleyzer K, Fierro NA, Oyakawa L, Ramos-Lopez O, Ferraz Santana RA, Sitnik R, Panduro A (2016) Association with spontaneous hepatitis C viral clearance and genetic differentiation of IL28B/IFNL4 haplotypes in populations from Mexico. PLoS One 11:e0146258. doi:10.1371/journal.pone.0146258
73. González-Horta EE, Marante J, Amador-Cañizares Y, Álvarez-Lajonchere L, Guerra I, Martínez-Donato G, Dueñas-Carrera S (2011) Analysis of hepatitis C virus core encoding sequences in chronically infected patients reveals mutability, predominance, genetic history and potential impact on therapy of Cuban genotype 1b isolates. Eur Rev Med Pharmacol Sci 15:1320–1327
74. Gower E, Estes C, Blach S, Razavi-Shearer K, Razavi H (2014) Global epidemiology and genotype distribution of the hepatitis C virus infection. J Hepatol 61:S45–S57. doi:10.1016/j.jhep.2014.07.027
75. Le Guillou-Guillemette H, Vallet S, Gaudy-Graffin C, Payan C, Pivert A, Goudeau A, Lunel-Fabiani F (2007) Genetic diversity of the hepatitis C virus: impact and issues in the antiviral therapy. World J Gastroenterol 13:2416–2426
76. Gutiérrez JP, Sucilla-Pérez H, Conde-González CJ, Izazola JA, Romero-Martínez M, Hernández-Ávila M (2016) Disminución de la seroprevalencia de hepatitis C en México: resultados de la Ensanut 2012. Salud Publica Mex 58:25–32

77. Hakre S, Arteaga G, Núñez AE, Bautista CT, Bolen A, Villarroel M, Peel SA, Paz-Bailey G, Scott PT, Pascale JM (2013) Prevalence of HIV and other sexually transmitted infections and factors associated with syphilis among female sex workers in Panama. Sex Transm Infect 89:156–164. doi:10.1136/sextrans-2012-050557
78. Hanus JS, Ceretta LB (2015) Major article incidence of hepatitis C in Brazil. Rev Soc Bras Med Trop 48:665–673
79. Hoffmann L, Faffe DS, Lima JFC, Capitanio TA, Cabral BCA, Ürményi TP, Coelho HSM, Rondinelli E, Villela-Nogueira CA, Silva R (2015) No correspondence between resistance mutations in the HCV-NS3 protease at baseline and early telaprevir-based triple therapy. BBA Clin 3:146–151. doi:10.1016/j.bbacli.2015.01.004
80. Hoffmann L, Ramos JA, de Souza EV, de Araújo Ramos AL, Villela-Nogueira CA, Urményi TP, Tanuri A, Rondinelli E, Silva R (2013) Dynamics of resistance mutations to NS3 protease inhibitors in a cohort of Brazilian patients chronically infected with hepatitis C virus (genotype 1) treated with pegylated interferon and ribavirin: a prospective longitudinal study. Virol J 10:57. doi:10.1186/1743-422X-10-57
81. Hoofnagle JH (2002) Course and outcome of hepatitis C. Hepatology 36(5 suppl 1):S21–S29. doi:10.1053/jhep.2002.36227
82. Hoshino H, Hino K, Miyakawa H, Takahashi K, Akbar SMF, Mishiro S (2012) Inter-genotypic recombinant hepatitis C virus strains in Japan noted by discrepancies between immunoassay and sequencing. J Med Virol 84:1018–1024. doi:10.1002/jmv.23300
83. Hsu SH, Wang B, Kota J, Yu J, Costinean S, Kutay H, Yu L, Bai S, Perle KL, Chivukula RR, Mao H, Wei M, Clark KR, Mendell JR, Caligiuri MA, Jacob ST, Mendell JT, Ghoshal K (2012) Essential metabolic, anti-inflammatory, and anti-tumorigenic functions of miR-122 in liver. J Clin Invest 122:2871–2883. doi:10.1172/JCI63539
84. Jacobson IM, McHutchison JG, Dusheiko G, Di Bisceglie AM, Reddy KR, Bzowej NH, Marcellin P, Muir AJ, Ferenci P, Flisiak R, George J, Rizzetto M, Shouval D, Sola R, Terg RA, Yoshida EM, Adda N, Bengtsson L, Sankoh AJ, Kieffer TL, George S, Kauffman RS, Zeuzem S (2011) Telaprevir for previously untreated chronic hepatitis C virus infection. N Engl J Med 364:2405–2416. doi:10.1056/NEJMoa1012912
85. Jardim ACG, Bittar C, Matos RPA, Yamasaki LHT, Silva RA, Pinho JRR, Fachini RM, Carareto CMA, de Carvalho-Mello IMVG, Rahal P (2013) Analysis of HCV quasispecies dynamic under selective pressure of combined therapy. BMC Infect Dis 13:61. doi:10.1186/1471-2334-13-61
86. Jardim ACG, Igloi Z, Shimizu JF, Santos VAFFM, Felippe LG, Mazzeu BF, Amako Y, Furlan M, Harris M, Rahal P (2015) Natural compounds isolated from Brazilian plants are potent inhibitors of hepatitis C virus replication in vitro. Antivir Res 115:39–47. doi:10.1016/j.antiviral.2014.12.018
87. Jaspe RC, Sulbarán YF, Loureiro CL, Martínez N, Devesa M, Rodríguez Y, Torres JR, Rangel HR, Pujol FH (2014) Genetic diversity of hepatitis B virus and hepatitis C virus in human immunodeficiency virus type 1-co-infected patients from Venezuela. J Med Microbiol 63:1099–1104. doi:10.1099/jmm.0.067496-0
88. Jaspe RC, Sulbarán YF, Sulbarán MZ, Loureiro CL, Rangel HR, Pujol FH (2012) Prevalence of amino acid mutations in hepatitis C virus core and NS5B regions among Venezuelan viral isolates and comparison with worldwide isolates. Virol J 9:214. doi:10.1186/1743-422X-9-214
89. Jopling CL, Yi M, Lancaster AM, Lemon SM, Sarnow P (2005) Modulation of hepatitis C virus RNA abundance by a liver-specific microRNA. Science 309(5740):1577–1581. doi:10.1126/science.1113329
90. Kageyama S, Agdamag DM, Alesna ET, Leaño PS, Heredia AML, Abellanosa-Tac-An IP, Jereza LD, Tanimoto T, Yamamura JI, Ichimura H (2006) A natural inter-genotypic (2b/1b) recombinant of hepatitis C virus in the Philippines. J Med Virol 78:1423–1428. doi:10.1002/jmv.20714
91. Kalinina O, Norder H, Mukomolov S, Magnius LO (2002) A natural intergenotypic recombinant of hepatitis C virus identified in St. Petersburg. J Virol 76:4034–4043. doi:10.1128/JVI.76.8.4034-4043.2002

92. Kanda T, Yokosuka O, Omata M (2013) Treatment of hepatitis C virus infection in the future. Clin Transl Med 2:9. doi:10.1186/2001-1326-2-9
93. Keyvani H, Fazlalipour M, Monavari SHR, Mollaie HR (2012) Hepatitis C virus-proteins, diagnosis, treatment and new approaches for vaccine development. Asian Pac J Cancer Prev 13:5931–5949. doi:10.7314/APJCP.2012.13.12.5917
94. Kieffer TL, Sarrazin C, Miller JS, Welker MW, Forestier N, Reesink HW, Kwong AD, Zeuzem S (2007) Telaprevir and pegylated interferon-alpha-2a inhibit wild-type and resistant genotype 1 hepatitis C virus replication in patients. Hepatology 46:631–639. doi:10.1002/hep.21781
95. Kieft JS, Zhou K, Jubin R, Murray MG, Lau JYN, Doudna JA (1999) The hepatitis C virus internal ribosome entry site adopts an ion-dependent tertiary fold. J Mol Biol 292:513–529. doi:10.1006/jmbi.1999.3095
96. Kurbanov F, Tanaka Y, Avazova D, Khan A, Sugauchi F, Kan N, Kurbanova-khudayberganova D, Khikmatullaeva A, Musabaev E, Mizokami M (2008) Detection of hepatitis C virus natural recombinant RF1_2k/1b strain among intravenous drug users in Uzbekistan. Hepatol Res 38:457–464. doi:10.1111/j.1872-034X.2007.00293.x
97. De la Vega AP, Alarcón DA, Gómez-Jeria JS (2013) Quantum chemical study of the relationships between electronic structure and pharmacokinetic profile, inhibitory strength toward hepatitis C virus NS5B polymerase and HCV replicons of indole-based compounds. J Chil Chem Soc 4:2148–2157
98. Lampe E, Espirito-Santo MP, Martins RMB, Bello G (2010) Epidemic history of hepatitis C virus in Brazil. Infect Genet Evol 10:886–895. doi:10.1016/j.meegid.2010.05.010
99. Lampe E, Lewis-Ximenez L, Espírito-Santo MP, Delvaux NM, Pereira SA, Peres-Da-Silva A, Martins RMB, Soares MA, Santos AF, Vidal LL, Germano FN, De Martinez AMB, Basso R, Rebello Pinho JR, Malta FM, Gomes-Gouvêa M, Moliterno RA, Bertolini DA, Fujishima MAT, Bello G (2013) Genetic diversity of HCV in Brazil. Antivir Ther 18:435–444. doi:10.3851/IMP2606
100. Lee YM, Lin HJ, Chen YJ, Lee CM, Wang SF, Chang KY, Chen TL, Liu HF, Chen YMA (2010) Molecular epidemiology of HCV genotypes among injection drug users in Taiwan: full-length sequences of two new subtype 6w strains and a recombinant form-2b6w. J Med Virol 82:57–68. doi:10.1002/jmv.21658
101. Legrand-Abravanel F, Claudinon J, Nicot F, Dubois M, Chapuy-Regaud S, Sandres-Saune K, Pasquier C, Izopet J (2007) New natural intergenotypic (2/5) recombinant of hepatitis C virus. J Virol 81:4357–4362. doi:10.1128/JVI.02639-06
102. Di Lello FA, Farias AA, Culasso AC, Perez PS, Pisano MB, Contigiani MS, Campos RH, Re VE (2015) Changing epidemiology of hepatitis C virus genotypes in the central region of Argentina. Arch Virol 160:909–915. doi:10.1007/s00705-015-2390-6
103. Di Lello FA, Pineiro YLFG, Munoz G, Campos RH (2009) Diversity of hepatitis B and C viruses in Chile. J Med Virol 81:1887–1894. doi:10.1002/jmv.21607
104. Lisboa-Neto G, Noble CF, Rebello Pinho JR, Malta FM, Gomes-Gouvea MS, Alvarado-Mora MV, Da Silva MH, Leite AGB, Piccoli LZ, Rodrigues FK, Carrilho FJ, Mendes-Correa MC (2015) Resistance mutations are rare among protease inhibitor treatment-naive hepatitis C genotype-1 patients with or without HIV coinfection. Antivir Ther 20:281–287. doi:10.3851/IMP2873
105. Lopes CLR, Teles SA, Espírito-Santo MP, Lampe E, Rodrigues FP, Motta-Castro ARC, Marinho TA, Reis NR, Silva ÁMC, Martins RMB (2009) Prevalência, fatores de risco e genótipos da hepatite C entre usuários de drogas. Rev Saude Publica 43:43–50. doi:10.1590/S0034-89102009000800008
106. López L, López P, Arago A, Rodríguez I, López J, Lima E, Insagaray J, Bentancor N (2005) Risk factors for hepatitis B and C in multi-transfused patients in Uruguay. J Clin Virol 34(suppl 2):S69–S74
107. Lopez-Balderas N, Bravo E, Camara M, Hernandez-Romano P (2015) Seroprevalence of hepatitis viruses and risk factors in blood donors of Veracruz, Mexico. J Infect Dev Ctries 9:274–282. doi:10.3855/jidc.4812

108. Lopez-Colombo A, Melendez-Mena D, Sedeno-Monge V, Camacho-Hernandez JR, Vazquez-Cruz E, Morales-Hernandez ER, Sosa-Jurado F, Marquez-Dominguez L, Santos-Lopez G (2014) Hepatitis C virus infection in patients and family members attending two primary care clinics in Puebla, Mexico. Ann Hepatol 13:746–752
109. De los Ríos R, Miyahira J, Colichon A, Cieza J (1997) Prevalencia de anticuerpos anti-hepatitis C en pacientes en hemodiálisis crónica. Rev Med Hered 8:67–71
110. Manns MP, Wedemeyer H, Cornberg M (2006) Treating viral hepatitis C: efficacy, side effects, and complications. Gut 55:1350–1359. doi:10.1136/gut.2005.076646
111. Manvar D, Fernandes TA, Domingos JLO, Baljinnyam E, Basu A, Junior EFT, PRR C, Kaushik-Basu N (2015) Synthesis and biological evaluation of α-aryl-α-tetralone derivatives as hepatitis C virus inhibitors. Eur J Med Chem 93C:51–54. doi:10.1016/j.ejmech.2015.01.057
112. Marinovich S, Lavorato C, Bisigniano L, Soratti C, Krogh DH, Celia E, Fernández V, Tagliafichi V, Diez GR, Fayad A, Haber V (2014) Registro Argentino de Diálisis Crónica SAN-INCUCAI 2013. Buenos Aires, Argentina
113. Martell M, Esteban JI, Quer J, Genescà J, Weiner A, Esteban R, Guardia J, Gómez J (1992) Hepatitis C virus (HCV) circulates as a population of different but closely related genomes: quasispecies nature of HCV genome distribution. J Virol 66:3225–3229
114. Martial J, Morice Y, Abel S, Cabié A, Rat C, Lombard F, Edouard A, Pierre-Louis S, Garsaud P, Béra O, Chout R, Gordien E, Deny P, Césaire R (2004) Hepatitis C virus (HCV) genotypes in the Caribbean island of Martinique: evidence for a large radiation of HCV-2 and for a recent introduction from Europe of HCV-4. J Clin Microbiol 42:784–791. doi:10.1128/JCM.42.2.784-791.2004
115. Martinez-Donato G, Amador-Cañizares Y, Alvarez-Lajonchere L, Guerra I, Pérez A, Dubuisson J, Wychowsk C, Musacchio A, Aguilar D, Dueñas-Carrera S (2014) Neutralizing antibodies and broad, functional T cell immune response following immunization with hepatitis C virus proteins-based vaccine formulation. Vaccine 32:1720–1726. doi:10.1016/j.vaccine.2014.01.046
116. Martínez-Donato G, Capdesuñer Y, Acosta-Rivero N, Rodríguez A, Morales-Grillo J, Martínez E, González M, Alvarez-Obregon JC, Dueñas-Carrera S (2007) Multimeric HCV E2 protein obtained from *Pichia pastoris* cells induces a strong immune response in mice. Mol Biotechnol 35:225–235. doi:10.1007/s12033-007-9002-y
117. Martínez-Donato G, Musacchio A, Alvarez-Lajonchere L, Acosta-Rivero N, Amador Y, Guerra I, Peña D, Pérez A, Castro J, Puentes P, Soria Y, Cosme K, Sanchez J, Dueñas-Carrera S (2010) Ratio of HCV structural antigens in protein-based vaccine formulations is critical for functional immune response induction. Biotechnol Appl Biochem 56:111–118. doi:10.1042/BA20090216
118. Martínez-Gómez LE, Chávez-Tapia NC, Burguete-García AI, Aguilar-Olivos N, Madrid-Marina V, Román-Bahena M, Orbe-Orihuela C, Misael U, Méndez-Sánchez N (2012) IL28B polymorphisms predict the response to chronic hepatitis C virus infection treatment in a Mexican population. Ann Hepatol 11:876–881
119. Mendes-Correa MC, Cavalheiro NP, Mello C, Barone AA, Gianini RJ (2008) Genotypic distribution of hepatitis C among hepatitis C and HIV co-infected patients in Brazil. Int J STD AIDS 19:595–599. doi:10.1258/ijsa.2007.007183
120. Monsalve-Castillo F, Chacin-Bonilla L, Atencio RJ, Espinoza LP, Costa-Leon L, Echevarria JM (2007) Low prevalence of hepatitis C virus infection in Amerindians from Western Venezuela. Mem Inst Oswaldo Cruz 102:107–110
121. Mora MVA, Romano CM, Gomes-Gouvea MS, Gutierrez MF, Carrilho FJ, Pinho JRR (2010) Molecular characterization, distribution, and dynamics of hepatitis C virus genotypes in blood donors in Colombia. J Med Virol 82:1889–1898. doi:10.1002/jmv.21908
122. Moratorio G, Martínez M, Gutiérrez MF, González K, Colina R, López-Tort F, López L, Recarey R, Schijman AG, Moreno MP, García-Aguirre L, Manascero AR, Cristina J (2007) Evolution of naturally occurring 5'-non-coding region variants of hepatitis C virus in human populations of the South American region. Virol J 4:79. doi:10.1186/1743-422X-4-79

123. Moreau I, Hegarty S, Levis J, Sheehy P, Crosbie O, Kenny-Walsh E, Fanning LJ (2006) Serendipitous identification of natural intergenotypic recombinants of hepatitis C in Ireland. Virol J 3:95. doi:10.1186/1743-422X-3-95
124. Morel V, Descamps V, François C, Fournier C, Brochot E, Capron D, Duverlie G, Castelain S (2010) Emergence of a genomic variant of the recombinant 2 k/1b strain during a mixed hepatitis C infection: a case report. J Clin Virol 47:382–386. doi:10.1016/j.jcv.2010.01.011
125. Moreno MP, Casane D, López L, Cristina J (2006) Evidence of recombination in quasispecies populations of a hepatitis C Virus patient undergoing anti-viral therapy. Virol J 3:87. doi:10.1186/1743-422X-3-87
126. Moreno P, Alvarez M, López L, Moratorio G, Casane D, Castells M, Castro S, Cristina J, Colina R (2009) Evidence of recombination in hepatitis C virus populations infecting a hemophiliac patient. Virol J 6:203. doi:10.1186/1743-422X-6-203
127. Muñoz G, Velasco M, Thiers V, Hurtado C, Brahm J, Larrondo-Lillo M, Guglielmetti A, Smok G, Brechot C, Lamas E (1998) Prevalence and genotypes of hepatitis C virus in blood donors and in patients with chronic liver disease and hepatocarcinoma in a Chilean population. Rev Med Chil 126:1035–1042
128. Muñoz-Espinosa LE, Trujillo-Trujillo ME, Martínez-Macías RF, Panduro A, Rivas-Estilla AM, Fierro NA, Silvera-Linares AL, Torres-Valadez R, Cordero-Pérez P, González-Aldaco K, Chen-López CY, José-Abrego A, Zuñiga-Noriega JR, Gutiérrez-Ruiz MC, Roman S (2015) Increase of drug use and genotype 3 in HCV-infected patients from Central West and Northeast Mexico. Ann Hepatol 14:642–651
129. Mutimer D, Aghemo A, Diepolder H, Negro F, Robaeys G, Ryder S, Zoulim F, Peck M, Craxi A, Fried M, Zeuzem S (2014) EASL clinical practice guidelines: management of hepatitis C virus infection. J Hepatol 60:392–420. doi:10.1016/j.jhep.2011.02.023
130. Nakamoto S, Kanda T, Wu S, Shirasawa H, Yokosuka O (2014) Hepatitis C virus NS5A inhibitors and drug resistance mutations. World J Gastroenterol 20:2902–2912. doi:10.3748/wjg.v20.i11.2902
131. Nishiya AS, de Almeida-Neto C, Ferreira SC, Alencar CS, Di-Lorenzo Oliveira C, Levi JE, Salles NA, Mendrone A, Sabino EC (2014) HCV genotypes, characterization of mutations conferring drug resistance to protease inhibitors, and risk factors among blood donors in São Paulo, Brazil. PLoS One 9:e86413. doi:10.1371/journal.pone.0086413
132. Nishiya AS, de Almeida-Neto C, Romano CM, Alencar CS, Ferreira SC, Di-Lorenzo-Oliveira C, Levi JE, Salles NA, Mendrone-Junior A, Sabino EC (2015) Phylogenetic analysis of the emergence of main hepatitis C virus subtypes in São Paulo, Brazil. Braz J Infect Dis 19:473–478. doi:10.1016/j.bjid.2015.06.010
133. Noppornpanth S, Lien TX, Poovorawan Y, Smits SL, Osterhaus ADME, Haagmans BL (2006) Identification of a naturally occurring recombinant genotype 2/6 hepatitis C virus. J Virol 80:7569–7577. doi:10.1128/JVI.00312-06
134. Ottosen S, Parsley TB, Yang L, Zeh K, van Doorn L-J, van der Veer E, Raney AK, Hodges MR, Patick AK (2015) *In vitro* antiviral activity and preclinical and clinical resistance profile of Miravirsen, a novel anti-hepatitis C virus therapeutic targeting the human factor miR-122. Antimicrob Agents Chemother 59:599–608. doi:10.1128/AAC.04220-14
135. PAHO (2013) Supply of blood for transfusion in Latin American and Caribbean countries 2010 and 2011. Washington, DC
136. Palmer BA, Moreau I, Levis J, Harty C, Crosbie O, Kenny-Walsh E, Fanning LJ (2012) Insertion and recombination events at hypervariable region 1 over 9.6 years of hepatitis C virus chronic infection. J Gen Virol 93:2614–2624. doi:10.1099/vir.0.045344-0
137. Paltanin LF, Reiche EMV (2002) Seroprevalence of anti-hepatitis C virus antibodies among blood donors, Brazil. Rev Saude Publica 36:393–399
138. Pando MA, Berini C, Bibini M, Fernandez M, Reinaga E, Maulen S, Marone R, Biglione M, Montano SM, Bautista CT, Weissenbacher M, Sanchez JL, Avila MM (2006) Prevalence of HIV and other sexually transmitted infections among female commercial sex workers in Argentina. Am J Trop Med Hyg 74:233–238

139. Panduro A, Roman S, Khan A, Tanaka Y, Kurbanov F, Martinez-Lopez E, Campollo O, Hernandez-Nazara Z, Mizokami M (2010) Molecular epidemiology of hepatitis C virus genotypes in West Mexico. Virus Res 151:19–25. doi:10.1016/j.virusres.2010.03.009
140. Pantiri K, Westerhout KY, Obando CA (2015) Cost-effectiveness analysis of simeprevir and sofosbuvir combination therapy for the treatment of genotype 1 HCV patients in the Dominican Republic. Value Health 18:A872. doi:10.1016/j.jval.2015.09.021
141. Pawlotsky JM (2006) Hepatitis C virus population dynamics during infection. Curr Top Microbiol Immunol 299:261–284. doi:10.1007/3-540-26397-7_9
142. Pawlotsky JM (2014) New hepatitis C therapies: the toolbox, strategies, and challenges. Gastroenterology 146:1176–1192. doi:10.1053/j.gastro.2014.03.003
143. Pawlotsky J-M (2003) The nature of interferon-alpha resistance in hepatitis C virus infection. Curr Opin Infect Dis 16:587–592. doi:10.1097/01.qco.0000104300.87920.03
144. Pawlotsky JM, Germanidis G, Neumann AU, Pellerin M, Frainais PO, Dhumeaux D (1998) Interferon resistance of hepatitis C virus genotype 1b: relationship to nonstructural 5A gene quasispecies mutations. J Virol 72:2795–2805
145. Pereira LMMB, Martelli CMT, Moreira RC, Merchan-Hamman E, Stein AT, Cardoso MRA, Figueiredo GM, Montarroyos UR, Braga C, Turchi MD, Coral G, Crespo D, Lima MLC, Alencar LCA, Costa M, dos Santos AA, Ximenes RAA (2013) Prevalence and risk factors of hepatitis C virus infection in Brazil, 2005 through 2009: a cross-sectional study. BMC Infect Dis 13:60. doi:10.1186/1471-2334-13-60
146. Peres-da-Silva A, de Almeida AJ, Lampe E (2012) Genetic diversity of NS3 protease from Brazilian HCV isolates and possible implications for therapy with direct-acting antiviral drugs. Mem Inst Oswaldo Cruz 107:254–261
147. Peres-da-Silva A, De Almeida AJ, Lampe E (2015) NS5A inhibitor resistance-associated polymorphisms in Brazilian treatment-naive patients infected with genotype 1 hepatitis C virus. J Antimicrob Chemother 70:726–730. doi:10.1093/jac/dku462
148. Pestova TV, Shatsky IN, Fletcher SP, Jackson RJ, Hellen CUT (1998) A prokaryotic-like mode of cytoplasmic eukaryotic ribosome binding to the initiation codon during internal translation initation of hepatitis C and classical swine fever virus RNAs. Genes Dev 12:67–83. doi:10.1101/gad.12.1.67
149. Piñeiro D, Martinez-Salas E (2012) RNA structural elements of hepatitis C virus controlling viral RNA translation and the implications for viral pathogenesis. Viruses 4:2233–2250. doi:10.3390/v4102233
150. Pinheiro AS, Duarte JBC, Alves CN, de Molfetta FA (2015) Virtual screening and molecular dynamics simulations from a bank of molecules of the Amazon region against functional NS3-4A protease-helicase enzyme of hepatitis C virus. Appl Biochem Biotechnol 176(6):1709–1721. doi:10.1007/s12010-015-1672-5
151. del Pino N, Oubiña JR, Rodríguez-Frías F, Esteban JI, Buti M, Otero T, Gregori J, García-Cehic D, Camos S, Cubero M, Casillas R, Guárdia J, Esteban R, Quer J (2013) Molecular epidemiology and putative origin of hepatitis C virus in random volunteers from Argentina. World J Gastroenterol 19:5813–5827. doi:10.3748/wjg.v19.i35.5813
152. Poynard T, Marcellin P, Lee SS, Niederau C, Minuk GS, Ideo G, Bain V, Heathcote J, Zeuzem S, Trepo C, Albrecht J (1998) Randomised trial of interferon alphab1;2b plus ribavirin for 48 weeks or for 24 weeks versus interferon alphab1;2b plus placebo for 48 weeks for treatment of chronic infection with hepatitis C virus. Lancet 352:1426–1432. doi:10.1016/S0140-6736(98)07124-4
153. Puig-Basagoiti F, Forns X, Furčić I, Ampurdanés S, Giménez-Barcons M, Franco S, Sánchez-Tapias JM, Saiz JC (2005) Dynamics of hepatitis C virus NS5A quasispecies during interferon and ribavirin therapy in responder and non-responder patients with genotype 1b chronic hepatitis C. J Gen Virol 86:1067–1075. doi:10.1099/vir.0.80526-0
154. Pujol FH, Loureiro CL (2007) Replacement of hepatitis C virus genotype 1b by genotype 2 over a 10-year period in Venezuela. J Clin Gastroenterol 41:518–520
155. Rallón NI, Naggie S, Benito JM, Medrano J, Restrepo C, Goldstein D, Shianna KV, Vispo E, Thompson A, McHutchison J, Soriano V (2010) Association of a single nucleotide

polymorphism near the interleukin-28B gene with response to hepatitis C therapy in HIV/hepatitis C virus-coinfected patients. AIDS 24:F23–F29. doi:10.1097/QAD.0b013e3283391d6d
156. Ramos JA, de Ramos ALA, Hoffmann L, de Perez RM, HSM C, Ürményi TP, Silva R, Rondinelli E, Villela-Nogueira CA (2012) A single nucleotide polymorphism, rs129679860, in the IL28B locus is associated with the viral kinetics and a sustained virological response in a chronic, monoinfected hepatitis C virus genotype-1 Brazilian population treated with pegylated interferon-rib. Mem Inst Oswaldo Cruz 107:888–892. doi:10.1590/S0074-02762012000700008
157. Ramos JA, Silva R, Hoffmann L, Ramos AL, Cabello PH, Urmenyi TP, Villela-Nogueira CA, Lewis-Ximenez L, Rondinelli E (2012) Association of IL-10, IL-4, and IL-28B gene polymorphisms with spontaneous clearance of hepatitis C virus in a population from Rio de Janeiro. BMC Res Notes 5:508. doi:10.1186/1756-0500-5-508
158. Ré V, Contigiani M, Yoshida CFT, Lampe E (2007) Identification of hepatitis C virus subtype 2c by sequencing analysis in patients from Córdoba, Argentina. Mem Inst Oswaldo Cruz 102:995–998. doi:10.1590/S0074-02762007000800016
159. Ribeiro LC, Souto FJD, Do Espírito-Santo MP, G-Oliveira R, Lampe E (2009) An autochthonous case of hepatitis C virus genotype 5a in Brazil: phylogenetic analysis. Arch Virol 154:665–670. doi:10.1007/s00705-008-0279-3
160. Ridruejo E (2012) Predictors of response to chronic hepatitis C treatment. Futur Virol 11:1089–1101
161. Ridruejo E, Solano A, Delettieres D, Martínez A, Marciano S, Mandó OG, Galdame O, Gadano A, Silva MO, Adrover R, Cocozzella D (2011) Interleukin-28B predicts Svr in Latin American hepatitis C genotype 1 patients treated with Peg IFN and ribavirin. J Hepatol 54:S465. doi:10.1016/S0168-8278(11)61179-3
162. Ríos-Ibarra CP, Lozano-Sepulveda S, Muñoz-Espinosa L, Rincón-Sánchez AR, Cordova-Fletes C, Rivas-Estilla AMG (2014) Downregulation of inducible nitric oxide synthase (iNOS) expression is implicated in the antiviral activity of acetylsalicylic acid in HCV-expressing cells. Arch Virol 159:3321–3328. doi:10.1007/s00705-014-2201-5
163. Robaldo L, Berzal-Herranz A, Montserrat JM, Iribarren AM (2014) Activity of core-modified 10-23 DNAzymes against HCV. Chem Med Chem 9:2172–2177. doi:10.1002/cmdc.201402222
164. MCM V, Rodríguez LL d l A, MCM V, Corredor MB, Frómeta SS, Hernández JM, Carrera SD, Wong MS, Samada M, Núñez MB, Alonso LV, da Silva Filho HP, Hübschen JM, Reis MG (2012) HCV genotype determination in monoinfected and HIV co-infected patients in Cuba. Trans R Soc Trop Med Hyg 106:711–717. doi:10.1016/j.trstmh.2012.08.001
165. Sanchez JL, Sjogren MH, Callahan JD, Watts DM, Lucas C, Abdel-Hamid M, Constantine NT, Hyams KC, Hinostroza S, Figueroa-Barrios R, Cuthie JC (2000) Hepatitis C in Peru: risk factors for infection, potential iatrogenic transmission, and genotype distribution. Am J Trop Med Hyg 63:242–248
166. Sarrazin C, Kieffer TL, Bartels D, Hanzelka B, Müh U, Welker M, Wincheringer D, Zhou Y, Chu HM, Lin C, Weegink C, Reesink H, Zeuzem S, Kwong AD (2007) Dynamic hepatitis C virus genotypic and phenotypic changes in patients treated with the protease inhibitor telaprevir. Gastroenterology 132:1767–1777. doi:10.1053/j.gastro.2007.02.037
167. Sede M, Jones LR, Moretti F, Laufer N, Quarleri J (2014) Inter- and intra-host variability of hepatitis C virus genotype 1a hypervariable envelope coding domains followed for a 4-11 year of human immunodeficiency virus coinfection and highly active antiretroviral therapy. Virology 471:19–28. doi:10.1016/j.virol.2014.09.016
168. Sede M, Laufer N, Ojeda D, Gun A, Cahn P, Quarleri J (2013) Analysis of sequences of hepatitis C virus NS5A genotype 1 in HIV-coinfected patients with a null response to nitazoxanide or peg-interferon plus ribavirin. Arch Virol 158:1907–1915. doi:10.1007/s00705-013-1687-6
169. Sede MM, Laufer NL, Quarleri J (2015) Previous failure of interferon-based therapy does not alter the frequency of HCV NS3 protease or NS5B polymerase inhibitor resistance-associated variants: longitudinal analysis in HCV/HIV co-infected patients. Int J Antimicrob Agents 46:219–224. doi:10.1016/j.ijantimicag.2015.04.011

170. Sentandreu V, Jiménez-Hernández N, Torres-Puente M, Bracho MA, Valero A, Gosalbes MJ, Ortega E, Moya A, González-Candelas F (2008) Evidence of recombination in intrapatient populations of hepatitis C virus. PLoS One. doi:10.1371/journal.pone.0003239
171. Sherman KE (2012) Managing adverse effects and complications in completing treatment for hepatitis C virus infection. Top Antivir Med 20:125–128
172. Da Silva Conde SRS, Soares Monteiro JCM, Silva Dos Santos BT, Fonseca Filgueiras NK, De Almeida Lins PA, Bonfim Freitas F, Da Silva GE, Demachki S, Ferreira De Araújo MT, Ishak R, Rosário Vallinoto AC (2014) SNP rs8099917 in gene IL28B might be associated with risk of chronic infection by HCV but not with response to treatment. Biomed Res Int 748606. doi:10.1155/2014/748606
173. Silva CM, Costi C, Krug LP, Ramos AB, Grandi T, Gandolfi VL, Menezes ME, Ocampos M, Niel C, Rossetti ML (2007) High proportion of hepatitis C virus genotypes 1 and 3 in a large cohort of patients from Southern Brazil. Mem Inst Oswaldo Cruz 102:867–870. doi:10.1590/S0074-02762007005000122
174. Simmonds P (2004) Genetic diversity and evolution of hepatitis C virus – 15 years on. J Gen Virol 85:3173–3188. doi:10.1099/vir.0.80401-0
175. Sixtos-Alonso MS, Avalos-Martinez R, Sandoval-Salas R, Dehesa-Violante M, García-Juarez I, Chávez-Ayala A, Domínguez-López A, Vargas-Voráckovà F, Toapanta-Yanchapaxi L, Amezcua-Guerra LM, Uribe M, Sánchez-Ávila JF (2015) A genetic variant in the interleukin 28B gene as a major predictor for sustained virologic response in chronic hepatitis C virus infection. Arch Med Res 46:448–453. doi:10.1016/j.arcmed.2015.07.001
176. Smith DB, Bukh J, Kuiken C, Muerhoff AS, Rice CM, Stapleton JT, Simmonds P (2014) Expanded classification of hepatitis C virus into 7 genotypes and 67 subtypes: updated criteria and genotype assignment web resource. Hepatology 59:318–327. doi:10.1002/hep.26744
177. Sophie L. Woolston, Nina Kim M (2016) Cost and access to direct-acting antiviral agents cost of direct-acting antiviral agents cost-effectiveness of direct-acting antiviral agents. http://www.hepatitisc.uw.edu.
178. Sulbarán MZ, di Lello FA, Sulbarán Y, Cosson C, Loureiro CL, Rangel HR, Cantaloube JF, Campos RH, Moratorio G, Cristina J, Pujol FH (2010) Genetic history of hepatitis C virus in Venezuela: high diversity and long time of evolution of HCV genotype 2. PLoS One 13;5(12):e14315. doi:10.1371/journal.pone.0014315
179. Suppiah V, Moldovan M, Ahlenstiel G, Berg T, Weltman M, Abate ML, Bassendine M, Spengler U, Dore GJ, Powell E, Riordan S, Sheridan D, Smedile A, Fragomeli V, Müller T, Bahlo M, Stewart GJ, Booth DR, George J (2009) IL28B is associated with response to chronic hepatitis C interferon-alpha and ribavirin therapy. Nat Genet 41:1100–1104. doi:10.1038/ng.447
180. Tallo T, Norder H, Tefanova V, Krispin T, Schmidt J, Ilmoja M, Orgulas K, Pruunsild K, Priimägi L, Magnius LO (2007) Genetic characterization of hepatitis C virus strains in Estonia: fluctuations in the predominating subtype with time. J Med Virol 79:374–382. doi:10.1002/jmv.20828
181. Tamori A, Enomoto M, Kawada N (2016) Recent advances in antiviral therapy for chronic hepatitis C. Mediat Inflamm 2016:1–11. doi:10.1155/2016/6841628
182. Tanaka A (2015) JSH guidelines for the management of hepatitis C virus infection (version 3). Nihon Rinsho 73:221–227
183. Tanaka Y, Nishida N, Sugiyama M, Kurosaki M, Matsuura K, Sakamoto N, Nakagawa M, Korenaga M, Hino K, Hige S, Ito Y, Mita E, Tanaka E, Mochida S, Murawaki Y, Honda M, Sakai A, Hiasa Y, Nishiguchi S, Koike A, Sakaida I, Imamura M, Ito K, Yano K, Masaki N, Sugauchi F, Izumi N, Tokunaga K, Mizokami M (2009) Genome-wide association of IL28B with response to pegylated interferon-alpha and ribavirin therapy for chronic hepatitis C. Nat Genet 41:1105–1109. doi:10.1038/ng.449
184. Trinks J, Caputo M, Hulaniuk M, Burgos Pratx L, Re V, Fortuny L, Frias A, Torres O, Nunez F, Gadano A, Corach D, Flichman D (2013) Distribution of genetic polymorphisms associated to hepatitis C virus (HCV) antiviral response in a multiethnic and admixed population. J Viral Hepatol 20:20. doi:10.1038/tpj.2014.20

185. Trujillo-Murillo K, Rincón-Sánchez AR, Martínez-Rodríguez H, Bosques-Padilla F, Ramos-Jiménez J, Barrera-Saldaña HA, Rojkind M, Rivas-Estilla AM (2008) Acetylsalicylic acid inhibits hepatitis C virus RNA and protein expression through cyclooxygenase 2 signaling pathways. Hepatology 47:1462–1472. doi:10.1002/hep.22215
186. U.S. National Institutes of Health (2013) ClinicalTrials.gov. In: clinicaltrials.gov. http://clinicaltrials.gov/. Accessed 6 Apr 2016.
187. Ueda E, Enomoto N, Sakamoto N, Hamano K, Sato C, Izumi N, Watanabe M (2004) Changes of HCV quasispecies during combination therapy with interferon and ribavirin. Hepatol Res 29:89–96. doi:10.1016/j.hepres.2004.02.014
188. Urban TJ, Thompson AJ, Bradrick SS, Fellay J, Schuppan D, Cronin KD, Hong L, McKenzie A, Patel K, Shianna KV, McHutchison JG, Goldstein DB, Afdhal N (2010) IL28B genotype is associated with differential expression of intrahepatic interferon-stimulated genes in patients with chronic hepatitis C. Hepatology 52:1888–1896. doi:10.1002/hep.23912
189. Valdivia JP, Vidal J, Bussalleu A, Huerta J, Ramirez D, Valdivia J, De los Rios R (2003) Infección por el Virus de la Hepatitis C en donantes del banco de sangre en el Hospital Cayetano Heredia (1998–2002). Rev Gastroenterol 23:22–28
190. Vasconcelos LRS, Moura P, do Carmo RF, Pereira LB, Cavalcanti Mdo S, DCBL A, Dutra RA, Pereira LM (2014) Low IL10 serum levels as key factor for predicting the sustained virological response to IFN??/ribavirin in Brazilian patients with HCV carrying IL28B CT/TT genotype. Hum Immunol 75:895–900. doi:10.1016/j.humimm.2014.06.017
191. Venegas M, Torres C, Brahm J (2013) Distribución nacional y regional de genotipos del virus hepatitis C en Chile. Rev Chil Infectol 1:566–568
192. Venegas M, Villanueva RA, González K, Brahm J (2011) IL28B polymorphisms associated with therapy response in Chilean chronic hepatitis C patients. World J Gastroenterol 17:3636–3639. doi:10.3748/wjg.v17.i31.3636
193. Vidal LL, Soares MA, Santos AF (2016) NS3 protease polymorphisms and genetic barrier to drug resistance of distinct hepatitis C virus genotypes from worldwide treatment-naïve subjects. J Viral Hepatol. doi:10.1111/jvh.12503.
194. Witherell GW, Beineke P (2001) Statistical analysis of combined substitutions in nonstructural 5A region of hepatitis C virus and interferon response. J Med Virol 63:8–16
195. World Health Organization (2014) Guidelines for the screening, care and treatment of persons with hepatitis C infection. WHO Library Catalog Data. doi:10.1186/1471-2334-13-288
196. World Health Organization (2015) Hepatitis C, fact sheet no. 164. Hepat C 4. doi:http://dx.doi.org/10.1016/S0140-6736(14)62401-6
197. World Health Organization (WHO) (2016) Guidelines for the screening, care and treatment of persons with hepatitis c infection. Updated version, April 2016. WHO Library Catalog Data.
198. Worobey M, Holmes EC (1999) Evolutionary aspects of recombination in RNA viruses. J Gen Virol 80:2535–2543
199. Wyles DL (2012) Beyond telaprevir and boceprevir: resistance and new agents for hepatitis C virus infection. Top Antivir Med 20:139–145
200. Yepes A, Álvarez C, Restrepo JC, Correa G, Zapata JC, Arango YAE (2002) Genotipos virales en pacientes con infección por el virus de la hepatitis C (VHC) en medellín. Gastroenterol Hepatol 25:334–335
201. Yokoyama K, Takahashi M, Nishizawa T, Nagashima S, Jirintai S, Yotsumoto S, Okamoto H, Momoi MY (2011) Identification and characterization of a natural inter-genotypic (2b/1b) recombinant hepatitis C virus in Japan. Arch Virol 156:1591–1601. doi:10.1007/s00705-011-1038-4
202. Yuan H, Adams-Huet B, Petersen T, Attar N, Lee WM, Jain MK (2012) A single nucleotide polymorphism in IL28B affects viral evolution of hepatitis C quasispecies after pegylated interferon and ribavirin therapy. J Med Virol 84:1913–1919. doi:10.1002/jmv.23407
203. Yun Z, Lara C, Johansson B, Lorenzana De Rivera I, Sönnerborg A (1996) Discrepancy of hepatitis C virus genotypes as determined by phylogenetic analysis of partial NS5 and core sequences. J Med Virol 49:155–160. doi:10.1002/(SICI)1096-9071(199607)49:3<155::AID-JMV1>3.0.CO;2-3

204. Yusim K, Richardson R, Tao N, Dalwani A, Agrawal A, Szinger J, Funkhouser R, Korber B, Kuiken C (2005) Los Alamos hepatitis C immunology database. Appl Bioinformat 4:217–225. doi:10.2165/00822942-200504040-00002
205. Zarife MA, de Oliveira EC, Romeu JM, dos Reis MG (2006) Detection of genotype 4 of the hepatitis C virus in Salvador, BA. Rev Soc Bras Med Trop 39:567–569. doi:10.1590/S0037-86822006000600012
206. Zeminian LB, Padovani JL, Corvino SM, Silva GF, Pardini MIDMC, Grotto RMT (2013) Variability and resistance mutations in the hepatitis C virus NS3 protease in patients not treated with protease inhibitors. Mem Inst Oswaldo Cruz 108:13–17
207. Zeuzem S, Buti M, Ferenci P, Sperl J, Horsmans Y, Cianciara J, Ibranyi E, Weiland O, Noviello S, Brass C, Albrecht J (2006) Efficacy of 24 weeks treatment with peginterferon alfa-2b plus ribavirin in patients with chronic hepatitis C infected with genotype 1 and low pretreatment viremia. J Hepatol 44:97–103

Chapter 18
Hepatitis D Virus

Raymundo Paraná, Maria Isabel Schinoni, and Mauricio de Souza Campos

1 Introduction

The hepatitis D virus, also known as hepatitis delta virus (HDV), is the sole representative of the *Deltaviridae* family, genus *Deltavirus*. The HDV was described in 1977 by Rizzetto et al. [37], and it is considered the most pathogenic among all hepatotropic viruses [20, 33, 38]. This defective virus depends on coinfection with hepatitis B virus (HBV) for its transmission, but its replicative cycle is independent. Remarkably, the HDV suppress other viral agent HBV replication and probably do so with hepatitis C virus (HCV) when the patient is infected by HBV/HDV/HCV [39].

Although the viral antigen particles serve as housing and protection for HDV [16], thus allowing transmission, infectivity, and de novo penetration in hepatocytes, the HDV replication is completely independent of HBV replication in hepatocytes [24]. The HDV is still able to inhibit the synthesis of viral antigens during the superinfection in HBV carriers, particularly of the surface antigen (HBsAg) and core antigen (HBcAg) [26, 38].

The HDV genome sequence revealed a significant heterogeneity for this agent; the first three differentiated clones were identified and referred to as genotypes I and II (IIa, IIb) and III [18]. Later, other HDV genotypes had been described and named by Roman algorithms from I to VIII. Each HDV genotype presents a unique geographic distribution and different clinical course in terms of disease severity. Some data report that genotype I subtype Ia and genotype type III are the most

R. Paraná • M. de Souza Campos
Federal University of Bahia, Salvador, State of Bahia, Brazil
e-mail: unif@svn.com.br; mauriciodesouzacampos@hotmail.com

M.I. Schinoni (✉)
Institute of the Federal University of Bahia-Bra\l and researcher of Hepatology Center of the University Hospital Professor Edgar Santos, Federal University of Bahia,
Salvador, State of Bahia, Brazil
e-mail: mariaschinoni4@gmail.com

J.E. Ludert et al. (eds.), *Human Virology in Latin America*,
DOI 10.1007/978-3-319-54567-7_18

severe, and genotype II has been associated with less severe disease [24]. Nonetheless, in general, HDV causes severe forms of hepatitis, especially in hyper-endemic areas, as in Africa and Western Brazilian Amazon [20, 22, 23] where cases of acute HDV infection frequently evolve toward fulminant hepatitis [19, 22, 23].

2 The Viral Agent

The HDV is biologically presented as a single satellite agent and human subviral particle that necessarily depends on the helping function promoted by the HBV, through its envelope protein (HBsAg), to complete its life cycle. HDV replication occurs exclusively in hepatocytes, yet no receptors for the HDV have been identified in these cells.

The HDV virion consists of a small spherical enveloped particle, ranging in diameter from 35 to 37 nm [30, 38] (Fig. 18.1). This particle is externally coated by the HBsAg and presents an internal ribonucleoprotein complex formed by the hepatitis D antigen (HDAg) and the circular genomic RNA.

The HDV is classified as an HBV satellite subvirus based on three biological properties: (1) the HDV is composed of small-size, single-stranded, circular, and faulty RNA; (2) it is unable to promote infection in the absence of HBV; and (3) it is dependent on the functional aid of a DNA virus [30, 38].

The HDV antigen (HDVAg) is composed of two proteins: a short one denominated HDAg-S, with a molecular weight of 24 kDa (195 amino acids), responsible for replication of viral genetic material (RNA), and the other called long HDAg-L, 27 kDa (214 amino acids) used to inhibit RNA replication and to promote the RNA packaging of the virions [31, 38]. During viral replication, these two proteins are located in the nucleus of liver cells.

The HDV genome consists of a single RNA molecule with a length of 1.75 kb [30, 38]. After genomic denaturation, the RNA appears as a single, rolled, and

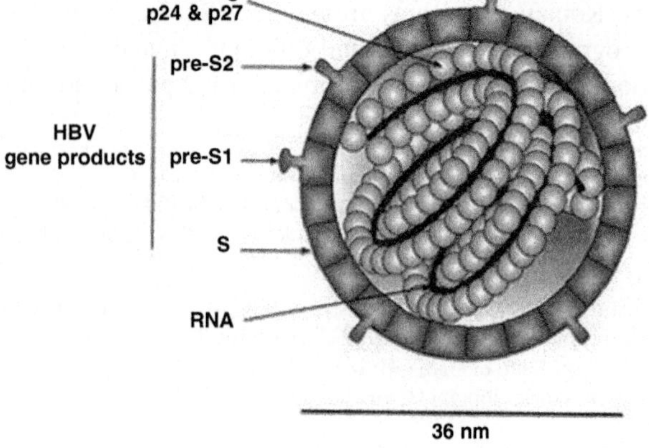

Fig. 18.1 Schematic model of the hepatitis D virus (HDV)

circular particle containing 1676 to 1683 nucleotides. During HDV infection, HBV provides particles of HBsAg, which will work as actual envelopes for HDV, thus guaranteeing its integrity and the ability to become infective to humans [30, 38].

The genetic sequence of the HDV genome revealed a high level of heterogeneity. Three clones of the HDV were originally identified, differentiated, and referred to as types I, II (IIa and IIb), and III [11, 44]. However, recent studies analyzing the genome of the virus in Africa suggest the existence of three other HDV genotypes [11, 12, 18, 35, 41].

Several studies performed in the 1980s showed data on HDV infection in South America, but until that point there were no studies on the viral dynamics of this virus. Alvarado-Mora et al. [1] conducted an evolutionary analysis of hepatitis delta genotype 3 (HDV/3) prevalent in South America. The aim was to estimate the nucleotide substitution rate, to determine the time of most recent ancestor (TMRCA), and to characterize the epidemic history and evolutionary dynamics. Furthermore, they characterized the presence of HBV/HDV infection in seven samples collected from patients who died of fulminant hepatitis from the Colombian Amazon region and included them in the evolutionary analysis. This was the first study reporting HBV and HDV sequences from the Amazon region of Colombia. Of the seven Colombian patients, five were positive for HBV DNA and HDV RNA. Of these, two samples were successfully sequenced for HBV (sub-genotypes F3 and F1b), and the five samples that were HDV positive were classified as HDV/3. By using all HDV/3 available reference sequences with sampling dates (n = 36), they estimated the HDV/3 substitution rate in 1.07×10^{-3} substitutions per site per year (s/s/y), which resulted in a time to the most recent common ancestor (TMRCA) of approximately 85 years. Also, it was determined that HDV/3 spread exponentially from the early 1950s to the 1970s in South America. This work discusses for the first time the viral dynamics for the HDV/3 circulating in South America. They suggest that the measures implemented to control HBV transmission resulted in the control of HDV/3 spreading in South America, especially after the important rise in this infection associated with a huge mortality during the 1950s up to the 1970s. The differences found among HDV/3 and the other HDV genotypes concerning its diversity raised evidence of a different origin and/or transmission route [1].

There are few data regarding the biomolecular aspects of HBV/HDV coinfection. A cross-sectional study with 92 patients HBsAg(+)/anti-HDV IgG(+) followed at the Hepatitis Referral Centers of Porto Velho (RO), Rio Branco, and Cruzeiro do Sul (AC), Brazil, from March 2006 to March 2007 showed that in this area, HBV sub-genotype F2 is the most prevalent (40.2%), followed by the sub-genotypes A1 (15.2%) and D3 (8.7%), 16.4% were other sub-genotypes or genotypes, 4.3% were discordant, and 15.2% were not amplifiable. HDV genotype 3 (HDV-3) was found in all the HBV/HDV-infected patients who could be genotyped for HDV, confirming that HDV-3 can associate with non-F HBV genotypes. However, HDV-3 mutant strains were found in 29.3% of patients and were more frequently associated with non-F HBV genotypes than were no mutant strains, suggesting that the mutation may facilitate association of HDV-3 with non-F HBV genotypes [4].

More recently, a study was conducted to determine the serological prevalence and molecular features of HDV within an Amerindian community from Argentina exhibiting positivity for HBsAg and/or anti-HBc total Ig. Forty-six plasma samples were tested for the detection of total anti-HDV antibodies by enzyme-linked immunosorbent assay (ELISA). Concomitantly, a partial RNA region coding for the delta antigen (HDAg) was amplified by RT-nested polymerase chain reaction (PCR) (RT-nPCR). In silico translation of DNA sequences into the amino acid (aa) sequence of HDAg-S (aa 110–195) and HDAg-L (aa 110–214) was performed. Of 46 HDV nonreactive samples by ELISA, three were HDV RNA positive by RT-nPCR. These samples were anti-HBc-only positive, two of them identified as cases of occult hepatitis B infection (OBI). The three cases were HBeAg negative and showed normal ALT/AST levels. All sequences were ascribed to HDV genotype 1, but exhibited nucleotide differences in HDAg-L coding region, among which mutations at codons 197 and 201, reportedly known to promote in vitro an unsuitable interaction with HBsAg, were observed [45].

3 Epidemiology

HDV infection occurs practically all over the world (Fig. 18.2), with highly endemic areas well documented in South America and sub-Saharan Africa [35].

The vaccination campaign against the hepatitis B virus (HBV) in the past 20 years has decreased the HDV circulation in developed countries, but it is reentering Europe through immigration from Western countries. Currently, Mongolia and Pakistan have areas with the highest prevalence of the disease [36].

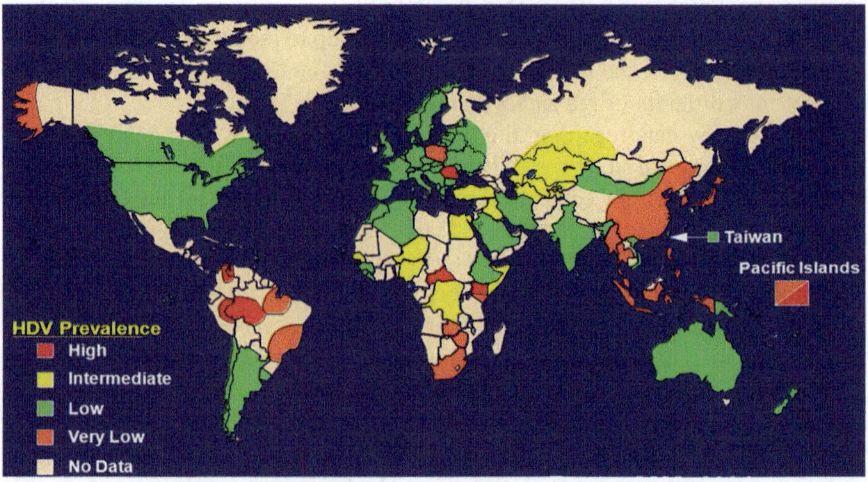

Fig. 18.2 Geographic distribution of infection by the hepatitis D virus (HDV)

4 Epidemiology in Latin America

The highest prevalence of HDV in Latin America is found in the Western Amazon Basin region, including Brazil, Peru, Ecuador, Venezuela, and Colombia [5].

4.1 HDV in Brazil

In Brazil, the Jurua, the Solimões, and the Purus Rivers in the state of Amazonas are considered highly endemic regions for hepatitis B and D infections, including cases of fulminant hepatitis [5].

A recent study in Manaus City recruited 222 AgHBs individuals [130 blood donors and 60 outpatient subjects from an outpatient clinic; most of the participants (150) lived in Manaus City and the rest (63) in remote isolated communities]. The results showed that the HBV/HDV coinfection rate was 8.5% in blood donors and 65% in outpatient subjects; of note, 47% of them were from the remote city Eirunepe, showing that the HDV is circulating in the remote cities of Amazonas State [13].

Most cases of hepatitis delta in Brazil are concentrated in the Western Amazon, a region highly impacted by this disease [12], affecting mainly children and young adults [35]. A study showed a prevalence of 20% HDV in VHB carriers in the Western Brazilian Amazon, and the infection spread to every population group, showing that this virus is disseminated throughout this region [25]. The HDV genotypes found in the Western Brazilian Amazon were I and III, but genotype III carriers were younger and showed more severe and symptomatic clinical presentations than the HDV genotype I carriers [32].

More recently, HDV infection has been also described among African-Brazilian isolated groups in northeastern Brazil [4]. A study evaluated the seroprevalence of HDV among HBsAg chronic carriers from Maranhão State, a region located in the northeast of Brazil. Among 133 patients, 5 had anti-HD, and 3 of them had also detectable HDV RNA. HDV genotypes were characterized using Bayesian phylogenetic analysis of nucleotide sequences from the HDAg coding region; HDV-3 was identified in 1 patient living in Maranhão State but born in the Amazonas State (Western Amazon Basin). Phylogenetic analysis shows that this HDV-3 sequence grouped with other HDV-3 sequences isolated in this state, which suggests that the patient probably contracted HDV infection there. Surprisingly, the other 2 patients were infected with HDV-8, an African genotype. These patients were born in and have always lived in Urbano Santos, a rural county of Maranhão State. Moreover, they had never gone to Africa and denied any contact with people from that continent. This is the first description of the HDV-8 in nonnative African populations. Presumably, this genotype may have been introduced into Brazil through slave trade from West Africa during the sixteenth to eighteenth centuries. This result indicates that it is necessary to perform clinical and epidemiological studies to investigate the presence of this infection in other areas in Brazil [4].

The existence of HDV superinfection in patients with chronic HBV infection shows that the Amazon basin offers environmental, social, cultural, and genetic characteristics that may contribute to HDV transmission [12]. In contrast, other regions of the country show very low HDV prevalence, either in risk groups or in the general population [24].

Severe cases of hepatitis caused by hepatitis B virus (HBV) or hepatitis D virus (HDV) are often seen in many Amazon countries, but there is a paucity of epidemiological studies on viral hepatitis in this area. A cross-sectional study to investigate the prevalence of markers for HBV and HDV was performed around 2004 in the Brazilian Amazon. Of the 2656 analyzed samples, 89 (3.3%) were positive for HBsAg, and 1628 (61.5%) were positive for IgG antibody to HBcAg. Markers for HDV were found in 47 cases (1.7%). Antibodies to HDV were associated with an Amerindian ethnic origin, a lower educational level, a history of acute viral hepatitis, a history of malaria, male gender, tattooing, and older age. The most frequent HBV genotypes were A and F. This study showed a high prevalence of HBV and HDV in the Western Brazilian Amazon, as well as the predominance of HBV genotypes A and F [27].

The HBV genotypes associated with HDV among Venezuelan Amerindian populations were genotypes I and III; only one HDV genotype I isolate was associated with HBV genotype D [42]. This coinfection triggered severe disease with acute liver failure and rapidly progressive chronic liver disease with a high mortality rate [37]. Although HBV and HDV share the same transmission routes, in certain northern areas of South America, such as the Brazilian and Venezuelan Amazon, HDV transmission seems to occur by inapparent exposure to probably related skin injuries [20, 34].

4.2 HDV in Colombia

In Colombia, cases of hepatitis D virus (HDV) infection have been officially described since 1985, mainly in the Amerindian population from Sierra Nevada de Santa Marta (North Caribbean Coast), Uraba (northwest), and Amazon (southeast) [15]. The incidence of HBV infection in Colombian Amazonas State in 2010 was 7.7 times higher than the average of the country, and in 2011 the circulation of HDV genotype III in the Amerindian population was reported [1].

A recent study of serological prevalence of HBV and HDV and molecular biology performed in 19 Amerindian communities with 861 individuals of Amazonas State in Colombia showed a prevalence of 43.5% of HDV infection with genotype III. A total of 23 of 861 individuals were positive for HBsAg detection by rapid test. Serological and/or molecular markers of HDV infection were demonstrated in 43.5% (10 of 23) of samples from Amerindians. The phylogenetic analysis demonstrated the exclusive circulation of HBV sub-genotype F1b of and HDV 3 in this population. Also a high frequency of HBV/HDV infection was found in the Amerindian (43.5%, 10 of 23). The circulation of HDV 3 and HBV sub-genotype F1b suggests a constant flow of these viral genotypes as a result of the interaction of the Amerindian populations from the Amazon basin [15].

4.3 HDV in Argentina

A study conducted to determine the serological prevalence and molecular features of HDV within an Amerindian community in Argentina exhibiting positivity for HBsAg and/or anti-HBc total Ig analyzed 46 plasma samples for the detection of total anti-HDV antibodies by ELISA. Concomitantly, a partial RNA region coding for the delta antigen (HDAg) was amplified by RT-nested PCR (RT-nPCR). This study showed that 3 of 46 HDV nonreactive samples by ELISA were HDV RNA positive by RT-nPCR. These samples were anti-HBc positive only, and 2 of them were identified as cases of occult hepatitis B infection. These results provide evidence of covert HDV infection even among occult hepatitis B infection, highlighting the need to reevaluate the currently applied guidelines for HDV diagnostic algorithms [14]. The same study also analyzed the prevalence of HBV and HDV infection in 56,983 blood donors in Buenos Aires City and Misiones Province (Northeast Region) and found a prevalence of 0.12% for HBsAg and 1.68 for anti-HBc antibodies in Buenos Aires and 0.73% and 8.55 in Misiones, respectively, with 1 case of HDV genotype I identified in Buenos Aires [14].

4.4 HDV in Peru

A cross-sectional study performed in 453 subjects (children and adults) living in three Apurimac region districts (Haquira, Challhuahuacho, Progreso) evaluated psychomotor development, intelligence coefficient, anxiety and depression levels, and the presence of communicable diseases (viral hepatitis B, C, and delta, syphilis, and HIV), heavy metals (lead in blood and cadmium, arsenic, and mercury in urine samples), and serum cholinesterase levels. The study aimed to determine the prevalence of communicable diseases, mental health, and environmental pollutant exposure in populations living near the Las Bambas mining project before exploitation phase. However, no cases of HIV, HCV, or HDV were found [3].

5 HDV Transmission

Hepatitis D virus (HDV) perinatal transmission depends on the infectivity of HBV carriers, occurring in HBV carrier mothers with a high viral load [21, 38]. Risk factors for acute HBV/HDV coinfection and HDV superinfection in HBV carriers may differ (Table 18.1).

In a patient who is coinfected with HBV and HDV, clinical illness is usually moderate, but it can become severe with acute liver failure. Fulminant hepatitis from coinfections is more likely than with infection with HBV alone. Clinical illness may be biphasic, with two aminotransferase peaks, first from HBV and then from HDV, although a monophasic illness with a single peak of enzyme levels may

Table 18.1 Risk factors for superinfection and coinfection of HBV/HDV

Risk factors for superinfection HDV acute in patients with HBsAg	Risk factors for acute HBV/HDV coinfection
People who use injectable drugs	People who have had blood transfusions
Prisoners (because of the vulnerability caused by sexual intercourse among interns or other practices, such as tattooing)	People who use injectable drugs
Hemophiliacs who have had blood transfusion (in case the blood is contaminated)	People who have tattoos
Hemodialysis patients (risk of contamination during the procedure)	People who have submitted to surgery in endemic areas
Institutionalized patients such as the elderly, or abandoned children who live in public housing	People who work in healthcare, such as doctors, nurses, pharmacists
People who work in healthcare, such as doctors, nurses, pharmacists	People who had engaged in sexual intercourse without male or female condoms
Homosexuals and prostitutes, because of the high number of sexual partners	Vertical transmission: the virus is transmitted from the mother to the child (always related to the infectivity of HBV)

also be observed. Coinfection is usually self-limited, and clearance of HBV results in clearance of HDV. Chronic HBV/HDV infection occurs in less than 5% of patients with coinfection.

The patient with previous chronic HBV infection provides a potential milieu for superinfection with HDV after exposure to someone infected with both HBV and HDV. This superinfection may be observed as an acute flare of hepatitis and sometimes leads to initial discovery of the underlying HBV infection, with misidentification of the illness as acute HBV infection. Measurement of the IgM anti-HBc titer can assist in differentiating chronic from acute HBV disease; circulating titers are typically low (or negative) in chronic HBV carriers but high in patients with acute HBV infection. Testing for HDV should be considered in any patient with HBV who has an acute flare of hepatitis and risk factors for HDV infection.

Superinfection with HDV can be self-limited and result in clearance of both viruses, although this outcome is uncommon. Most patients with superinfection develop a progressive form of chronic hepatitis. Superinfection is often seen as a worsening clinical illness in a previously stable chronic carrier of HBV. Clinical illness with superinfection can be rapidly progressive, leading to cirrhosis within 2 years in 10% to 15% of patients. The genotype of HBV may also have a function in the rapidity and severity of progressive disease. HDV will suppress HBV replication in simultaneously infected patients. Even HBV/HDV-infected patients with coexisting hepatitis C virus (HCV) infection will have reduced HCV replication.

When the sequence of the HDV genome was analyzed in several studies, some authors reported evidences of HDV transmission between couples who have intercourse. This finding suggests that sexual transmission of the HDV actually can occur. More recently, studies conducted in seven indigenous groups from the state of Amazonas suggest that sexual transmission is the most significant mechanism in the spread of HDV among this population [7]. Environmental and cultural factors related to human behavior could also be implicated in the HDV transmission. On the other hand, the existence of HDV animal reservoirs or insect-related transmission, although suspected, has never been well documented [6, 20, 22, 38].

6 Clinical Aspects

The HDV incubation period ranges from 28 to 180 days; infection can occur at the same time (coinfection) or after HBV infection (superinfection). In most coinfection cases, no progress to chronicity is observed, and it is postulated that less than 10% of patients become HDV chronic [2]. In contrast, HDV superinfection, involving previously symptomatic or asymptomatic HBsAg carriers, has a poor prognosis [2, 8].

The role of the HDV in hepatocellular carcinoma (HCC) has been suggested, but the association of the HDV with hepatocellular carcinoma does not seem to be frequent. However, superinfection with HDV in HBV carrier children can accelerate the progression to liver cirrhosis, and thus constitutes a risk for the development of hepatocellular carcinoma [7].

Peculiar fulminant HDV hepatitis occurred in tropical countries. The hallmarks of this disease in hepatic histology are mild to moderate eosinophilic necrosis, microvesicular steatosis (morula cells or spongiocytes), and the presence of HDAg at the core and cytoplasm of hepatocytes [18, 19].

Currently, it is suggested that the natural history of the disease and the prognosis of HDV infection are essentially related to the HBV genotypes [24].

7 Diagnosis

The laboratory diagnosis of HDV infection is complex for reasons of its own natural history (types of infection) and the use of several viral tests to identify both viruses. The clinical features alone have no specificity to define HDV diagnosis, so the serology and molecular biology are important tools to establish the diagnosis of HBV/HDV coinfection or superinfection.

The biochemical laboratory tests include the aminotransferases ALT and AST that announce the hepatic damage. The level of aminotransferases may be increased up to 25- to 100 fold; the bilirubin levels can also be high, and the prothrombin time can be prolonged in severe cases.

The serology is able to identify HBV markers such as HBsAg, anti-HBc, anti-HBc IgM, and HDV markers such as anti-HDV IgG and IgM. Table 18.2 shows the serological tests for hepatitis D and its clinical status.

The HDV is a pathogen that causes a severe and rapidly progressive disease of hepatocytes. The measurement of viral load in the peripheral blood of patients with HDV infections is important for diagnosis, treatment monitoring, and support for follow-up studies of viral replication during the course of the disease. Botelho et al. [10]

Table 18.2 Interpretation of serological results for hepatitis D

Forms	HBsAg	Anti-HBc total	Anti-HBc IgM	Anti-HDV total	Anti-HBs
Coinfection	(+)	(−)	(+)	(+)[a]	(−)
Superinfection	(+)	(+)	(−)	(+)[a]	(−)
Healing	(−)	(+)	(−)	(+)[b]	(+)

[a]IgM and IgG anti-HDV in high titers
[b]IgM and IgG anti-HDV in low titers

have developed an assay capable of detecting and quantifying the abundance of HDV particles in serum samples, based on reverse-transcription quantitative PCR (RT-qPCR). Two standards for calibration were produced for determining the viral load of HDV: a cDNA cloned into a linear plasmid and a transcribed RNA. For validating this assay, 140 clinical samples of sera were used, composed of 100 samples from patients who tested positive for anti-HDV and hepatitis B virus surface antigen (HBsAg) by ELISA, 30 samples from blood donors, 5 samples monoinfected with hepatitis B virus (HBV), and 5 samples monoinfected with hepatitis C virus (HCV). The HDV RT-qPCR assay performed better when calibrated using the standard based on HDV cDNA cloned into a linear plasmid, yielding an efficiency of 99.8% and a specificity of 100% in the in vitro assays. This study represents the first HDV RT-qPCR assay developed with clinical samples from Brazil and offers great potential for new clinical efficacy studies of antiviral therapeutics for use in patients with hepatitis delta in the Western Amazon region [10].

8 Treatment and Prevention

Treatment of chronic HDV infection has as an endpoint the early elimination of both viruses, interruption of viral replication, reduction of aminotransferase levels, and control of the chronic inflammatory process to reduce the chances of progression to cirrhosis and HCC. The treatment could include both viruses, HBV and HDV, depending on the viral load of each one of them.

In chronic HDV hepatitis, the only treatment so far available is interferon (INF)-α. Other antiviral drugs such as lamivudine, ribavirin, and adefovir dipivoxil have not exhibited a satisfactory therapeutic response [16, 29, 43]. In most studies, INF-2α at 9,000,000 IU or INF-2β at 10,000,000 IU subcutaneously, three times a week, was used, but without success. Although the therapeutic option available for patients with chronic HDV infection is limited to INF-α, with rare curative outcome, a new drug, Myrcludex B, has appeared as the first-in-class entry inhibitor, inactivating the hepatitis B virus (HBV) and hepatitis D virus (HDV) receptor sodium taurocholate co-transporting polypeptide [9]. The interim results of a pilot trial on chronically infected HDV patients treated with Myrcludex B, or pegylated interferon-alpha (PegIFNα-2a) or their combination, indicated that Myrcludex B was well tolerated and no serious adverse event occurred [9]. Although HBsAg levels remained unchanged, HDV RNA significantly declined at week 24 in all cohorts. HDV RNA became negative in two patients in the Myrcludex B and PegIFNα-2a cohorts and in five patients of the Myrcludex B + PegIFNα-2a cohort. ALT decreased significantly in the Myrcludex B cohort (six of eight patients), and HBV DNA was significantly reduced at week 24 in the Myrcludex B + PegIFNα-2a cohort. Virus kinetic modeling suggested a strong synergistic effect of Myrcludex B and PegIFNα-2a on both HDV and HBV [9]. Finally, in that study, Myrcludex B showed a strong effect on HDV RNA serum levels and induced ALT normalization under monotherapy. Synergistic antiviral effects on HDV RNA and HBV DNA in the Myr-IFN cohort indicated a benefit of the combination of entry inhibition with PegIFNα-2a to treat CHD patients [9].

More recently, PegIFN was evaluated alone or associated with nucleos(t)ide analogues. This study has analyzed data from the Hep-Net-International Delta Hepatitis Intervention Trial on 50 patients with compensated liver disease who tested positive for anti-HDV and HDV RNA. Subjects received pegylated INF-α 2a, with adefovir or placebo, or only adefovir, for 48 weeks. Twenty-four weeks after treatment was finished, 41 patients were evaluated for blood levels of HDV RNA and DNA, liver enzymes, and HBsAg; liver biopsy specimens were analyzed for fibrosis. Response to therapy was defined as end-of-treatment response or post-treatment week 24 virological response. In both cases, virological response was associated with undetectable HDV RNA levels. Patients with less than a 1 log decrease in HDV RNA at the end of treatment were considered null responders [28].

Based on univariate and multivariate analysis, the level of HDV RNA at week 24 of treatment was associated more strongly with response to therapy than other factors analyzed. The level of HBsAg at week 24 of treatment was associated with a response to therapy only in univariate analysis. Lack of HDV RNA at week 24 of treatment, or end of treatment, identified responders with positive predicted values of 71% and 100%, respectively. At 24 weeks after treatment, a decrease in HDV RNA level of less than 1 log, combined with no decrease in HBsAg level, identified null responders with a positive predictive value of 83%. A decrease in HDV RNA level of more than 2 log at week 24 of treatment identified null responders with a negative predictive value of 95%.

Based on an analysis of data from a large clinical trial, the level of HDV RNA at week 24 of treatment with pegylated interferon, with or without adefovir for 48 weeks, can identify patients who test negative for HDV RNA 24 weeks after the end of treatment. This information can be useful to help physicians to manage patients receiving therapy for chronic hepatitis D [40].

The treatment time should be as long as possible, not less than 12 months, even though the HDV eradication rates are below 20% [9]. It is necessary to follow up the patient to evaluate the therapeutic efficacy, which includes evaluation of clinical symptoms and measurement of aminotransferase levels and serum markers [7].

Active immunization (vaccine) against HBV is the best procedure to reduce the prevalence and incidence of infection HDV [7, 21, 38]. Among HBV chronic carriers living in endemic areas of HDV infection, or belonging to risk groups, prophylaxis of HDV superinfection still represents a great challenge. The risk groups (men who have sex with men, people who inject illicit drugs, hemodialysis patients) need close monitoring and advice about HDV transmission routes [20, 38]. In these areas, vaccination of all newborn children and children up to 9 years old has been suggested [13]. Passive immunization (anti-HBs immunoglobulin) against HBV has been used as a prophylactic action to prevent reinfection by HBV and hence reinfection with HDV in transplant patients because of chronic liver disease by D virus [17, 38].

9 Conclusions and Future Challenges

The mechanisms determining the persistence of HBV and HDV infection and long-term pathogenesis of HBV-associated liver disease appear to be multifactorial. Chronic HBV/HDV coinfection leads to the most severe form of chronic viral hepatitis, so it is important to elucidate the molecular mechanisms regulating virus–host interplay and pathogenesis.

New information on the molecular biology and life cycle of both viruses is allowing the development of more efficacious drugs to treat the coinfection, but it is necessary to await the results of the new drug trials being developed using larger cohorts to define the best treatment for these patients. Finally, recent studies have demonstrated that HDV infection continues to be an important health issue in the Brazilian Amazon and that the implementation of the HBV vaccination in rural Lábrea had little or no impact on the spread of HDV. Thus, HDV has not yet disappeared from HBV hyperendemic areas, reminding us that it is far from being a vanishing disease in the Amazon basin [31].

References

1. Alvarado-Mora MV, Romano CM, Gomes-Gouvêa MS, Gutierrez MF, Carrilho FJ, Pinho JR (2011) Dynamics of hepatitis D (delta) virus genotype 3 in the Amazon region of South America. Infect Genet Evol 11:1462–1468
2. Arboleda M, Castilho MC, Fonseca JCF, Albuquerque BC, Saboia RC, Yoshida CFT (1995) Epidemiological aspects of hepatitis B and D virus in the northern region of Amazonas, Brazil. Trans R Soc Trop Med Hyg 89:481–483
3. Astete J, Gastañaga Mdel C, Fiestas V, Oblitas T, Sabastizagal I, Lucero M, Abadíe Jdel M, Muñoz ME, Valverde A, Suarez M (2010) Communicable diseases, mental health and exposure to environmental pollutants in population living near Las Bambas mining project before exploitation phase. Rev Peru Med Exp Salud Publica 27:512–519
4. Barros A, Gomes-gouvêa B, Pinho B, Alvarado-mora BA, Dos Santos A, Mendes-corrêa AJM, Caldas AMT, Sousa MDC, Santos SP, Ferreira ASP (2011) Hepatitis delta virus genotype 8 infection in northeast Brazil: inheritance from african slaves. Virus Res 60:333–339
5. Bensag A (1983) Labrea hepatitis and other fulminant hepatitis in Serra Madureira Acre and Boca de Acre Amazonas Brasil. Rev Inst Med Trop 1404:182
6. Braga WSM, Brasil LM, Souza RAB, Castilho MCC, Fonseca JCF (2001) Ocorrência da infecção pelo vírus da hepatite B (VHB) e delta (HDV) em sete grupos indígenas do Estado do Amazonas. Rev Soc Bras Med Trop 34:349–355
7. Braga WS, Castilho Mda C, Borges FG, Leão JR, Martinho AC, Rodrigues IS, Azevedo EP, Barros Júnior GM, Paraná R (2012) Hepatitis D virus infection in the Western Brazilian Amazon—far from a vanishing disease. Rev Soc Bras Med Trop 45:691–695

8. Brasil. Ministerio da Saude (2002) Programa Nacional de Hepatites Virais: avaliacao da assistencia as hepatites virais no Brasil. Ministerio da Saude, Brasilia65
9. Bogomolov P, Alexandrov A, Voronkova N, Macievich M, Kokina K, Petrachenkova M, Lehr T, Lempp FA, Wedemeyer H, Haag M, Schwab M, Haefeli WE, Blank A, Urban S (2016) Treatment of chronic hepatitis D with the entry inhibitor myrcludex B: first results of a phase Ib/IIa study. J Hepatol 65:490–498
10. Botelho-Souza LF, dos Santos AO, Borzacov LM, Honda ER, Villalobos-Salcedo JM, Vieira DS (2014) Development of a reverse transcription quantitative real-time PCR-based system for rapid detection and quantitation of hepatitis delta virus in the western Amazon region of Brazil. J Virol Methods 197:19–24
11. Casey L, Brown TL, Colan EJ, Wignall F, Gerin JL (1993) A genotype of hepatitis D virus that occurs in northern America. Proc Natl Acad Sci U S A 90:9016–9020
12. Casey JL, Niro GA, Engle RE, Veja A, Gomez H, McCarthy WDM, Hyams KC, Gerin JL (1996) Hepatitis B virus (HBV)/hepatitis D virus (HDV) coinfection in outbreaks of acute hepatitis in the Peruvian Amazon basin: the roles of HDV genotype III and HBV genotype F. J Infect Dis 174:920–926
13. Crispim MA, Fraiji NA, Campello SC, Schriefer NA, Stefani MM, Kiesslich D (2014) Molecular epidemiology of hepatitis B and hepatitis delta viruses circulating in the Western Amazon region, North Brazil. BMC Infect Dis 14(94):1471–2334
14. Delfino CM, Eirin ME, Berini C, Malan R, Gentile E, Castillo A, Pedrozo W, Krupp R, Blejer J, Oubiña JR, Mathet VL, Biglione MM (2012) HDAg-L variants in covert hepatitis D and HBV occult infection among Amerindians of Argentina: new insights. J Clin Virol 54:223–228
15. Di Filippo VD, Cortes-Mancera F, Payares E, Montes N, de la Hoz F, Arbelaez MP, Correa G, Navas MC (2015) Hepatitis D virus and hepatitis B virus infection in Amerindian communities of the Amazonas state, Colombia. Virol J 12:172–175
16. Ferreira MS, Borges AS (2004) Hepatites por virus (A, B, D). In: Cinerman S, Cinerman B (eds) Condutas em infectologia. Atheneu, São Paulo, pp 249–258
17. Ferreira MS, Borges AS (2009) Hepatites viroticas. In: Rocha MOC, ERP P (eds) Fundamentos em infectologia. Rubio, Rio de Janerio, pp 351–406
18. Focaccia R (2005) Hepatites virais. In: Focaccia R (ed) Veronesi Tratado de Infectologia, 3rd edn. Atheneu, Rio de Janeiro, pp 427–538
19. Fonseca JCF, Ferreira LCL, Brasil LM, Castilho MC, Moss R, Barone M (1992) Fulminant Labrea hepatitis: role of HAV, HBV and HDV infection. Rev Inst Med Trop Sao Paulo 34:609–612
20. Fonseca JCF (1993) Hepatite delta. In: Fonseca JCF (ed) Hepatite delta. Imprensa Universitária, Manaus, pp 1–66
21. Fonseca JCF, Brasil LM, Castilho MC, Braga WSM, Souza RAB (1994) Hepatitis delta virus (HDV) infection in the Brazilian Amazon basin and its role in chronic liver disease. Hepatology 19:63I
22. Fonseca JCF, Brasil LM, Castilho MC, Botelho R, Braga WM, Ferreira LC (1999) Fulminant hepatitis and microvesicular steatosis (morula-like cells). A study in children and adolescents. Hepatology 30:284A
23. Fonseca JCF (2002) Hepatite D. Rev Soc Bras Med Trop 35:181–190
24. Hadler SC, Monzon M, Ponzetto A, Anzola R, Rivero D, Mandolfi A, Bracho A, Francis DP, Gerber MA, Thung S, Gerin J, Maynard JE, Popper H, Purcell RH (1983) Delta virus infection and severe hepatitis. An epidemic in the Yucpa Indians of Venezuela. Ann Intern Med 100:339–344
25. Jardi R, Rodriguez R, Buti M, Costa X, Cotrina M, Galimany R, Esteban R, Guardia J (2001) Role of hepatitis B, C, and D viruses in dual and triple infection: influence of viral genotypes and hepatitis B precore and basal core promoter mutations on viral replicative interference. Hepatology 34:404–410
26. Kay A, Melo da Silva E, Pedreira H, Negreiros S, Lobato C, Braga W, Muwonge R, Dény P, Reis M, Zoulim F, Trepo C, D'Oliveira A Jr, Salcedo JM, Schinoni MI, Parana R (2014) HBV/HDV co-infection in the Western Brazilian Amazonia: an intriguing mutation among HDV genotype 3 carriers. J Viral Hepatol 12:921–924

27. Keskin O, Wedemeyer H, Tüzün A, Zachou K, Deda X, Dalekos GN, Heidrich B, Pehlivan S, Zeuzem S, Yalçın K, Gürel S, Tabak F, Idilman R, Bozkaya H, Manns M, Yurdaydin C (2015) Association between level of hepatitis D virus RNA at week 24 of pegylated interferon therapy and outcome. Clin Gastroenterol Hepatol 13:2342–2349
28. Liang TJ (2009) Hepatitis B: the virus and disease. Hepatology 49:S13–S21
29. Negro F, Rizzetto M (1995) Diagnosis of hepatitis delta virus infection. J Hepatol 22:136–139
30. Oliviero S, D'adamio L, Chiarberge E, Brunetto R, Smedile A, Marchi M, Negro F, Bonino F (1991) Characterization of hepatitis Delta antigen gene of a highly pathogenic strain of hepatitis Delta virus. Prog Clin Biol Res 364:321–325
31. Paraná R, Kay A, Mollinet F, Viana S, Silva LK, Rios-Leite M, Tavares Neto J, Lobato C, Matione L, Trepo C (2006) HDV genotypes in the Western Brasilian Amazon region. An preliminary report. Am J Trop Med Hyg 75:475–479
32. Ponzetto A, Hoyer BH, Popper H, Engle R, Purcell RH, Gerin JL (1987) Tritation of the infectivity of hepatitis Delta virus in chimpanzees. J Infect Dis 155:72–78
33. Quintero A, Uzcátegui N, Loureiro CL, Villegas L, Illarramendi X, Guevara ME, Ludert JE, Blitz L, Liprandi F, Pujol FH (2001) Hepatitis delta virus genotypes I and III circulate associated with hepatitis B virus genotype F In Venezuela. J Med Virol 3:356–359
34. Radjef N, Ivannuishina V, Anais P, Trinchet J, Deny P (2001) Hepatitis D virus (HDV) genome analysis from Africa suggests the existence of more than three world-wide genotypes. J Hepatol 34(suppl 1):120
35. Rizzetto M (2016) The adventure of Delta. Liver Int 1:135–140
36. Rizzetto M, Macagno S, Chiaberge E, Negro F, Giacomo C, Doglia M, Macagno S, Verme G, Marinucci G, Alfani D, Milazzo F, Fassati LR (1987) Liver transplantation in hepatitis Delta virus disease. Lancet 2:469–471
37. Rizzeto M, Canese MG, Aricó S, Crivelli C, Trepo C, Bonino F, Verme G (1977) Immunofluorescence detection of a new antigen/antibody system (Delta/anti-Delta) associated with hepatitis B virus in liver and serum of HBsAg carriers. Gut 18:997–1003
38. Sagnelli E, Stroffolini T, Ascione A, Chiaramonte M, Craxi A, Giusti G, Piccinino F (1997) Decrease in HDV endemicity in Italy. J Hepatol 26:20–24
39. Saigal S, Smith HM, Heaton ND, Devlin J (1999) Suppression of hepatites B and hepatitis C viraemia by concurrent delta virus infection: its beneficial effect on the outcome of orthotopic liver transplantation. Hepatology 30:302A
40. Samuel D, Zignego AL, Reynes M, Feray C, Arulnaden JL, David MF, Gigou M, Bismuth A, Mathieu D, Gentilini P, Benhamou JP, Brechot C, Bismuth H (1995) Long-term clinical and virological outcome after liver transplantation for cirrhosis caused by chronic delta hepatitis. Hepatology 21:333–339
41. Shakil AO, Hadziyannis S, Hoofnagle JH, Di Bisceglie AM, Gerin JL, Casey JL (1997) Geographic distribution and genetic variability of hepatitis Delta virus genotype I. Virology 234:160–167
42. Viana S, Paraná R, Moreira RC, Compri AP, Macedo V (2005) High prevalence of hepatitis B virus and hepatitis D virus in the western Brazilian Amazon. Am J Trop Med Hyg 4:808–814
43. Wolters LMM, Van Nunen AB, Honkoop P, Vossen ACTM, Niesters HGM, Zonderuvan PE, Man RA (2000) Lamivudine-high dose interferon combination therapy for chronic hepatitis B patients co-infected with the hepatitis D virus. J Viral Hepatol 7:428–434
44. Wu JC, Chen CM, Sheen IJ, Lee SD, Tzeng HM, Choo KB (1995) Evidence of transmission of hepatitis Delta virus to spouses from sequence analysis of the viral genome. Hepatology 22:1656–1660
45. Wu JC, Choo KB, Chen CM, Chen TZ, Huo T, Lee SD (1995) Genotyping of hepatitis D virus by restriction-fragment length polymorphism and relation to outcome of hepatitis D. Lancet 346:939–941

Chapter 19
HIV Epidemiology in Latin America

Horacio Salomón and María de los Ángeles Pando

1 The Beginning

The acquired immunodeficiency syndrome (AIDS) was recognized in 1981 as a new clinical entity when several cases of a rare lung infection (*Pneumocystis carinii* pneumonia) and of an unusual cancer named Kaposi's sarcoma were reported among young, previously healthy, gay men from several cities of the United States of America (U.S.). Over the course of the following 2 years, similar cases were observed among injecting drug users, hemophiliacs, Haitians, female sex workers, and children born to women who presented with the same signs and symptoms. Then, it was evident that the infection was transmitted through unprotected sex with human immunodeficiency virus (HIV)-positive individuals, contact with infected blood, and from mother to child during pregnancy, at delivery, or through breast-feeding. The first studies revealed a progressive decline in CD4 T-lymphocyte counts in HIV-infected individuals, and as a result of this immune damage, these individuals developed diseases that were normally controlled by the immune system. In 1983, HIV was described as the causative agent of AIDS [1].

Since the identification of the etiological agent of AIDS, many efforts have been made around the world to prevent new transmissions and to improve the quality of life of those infected. A few years after the introduction of zidovudine (AZT) in 1987, several options of antiretroviral therapy started to change the clinical course of the disease, making the HIV infection a treatable chronic disease with the advent of the highly active antiretroviral therapy (HAART) in 1996. During the past years,

H. Salomón, PhD (✉) • M. de los Ángeles Pando, PhD
Instituto de Investigaciones Biomédicas en Retrovirus y SIDA (INBIRS), Facultad de Medicina, Universidad de Buenos Aires, Paraguay 2155, Piso 11, C1121ABG, Ciudad Autónoma de Buenos Aires, Argentina
e-mail: hsalomon@fmed.uba.ar; mpando@fmed.uba.ar

J.E. Ludert et al. (eds.), *Human Virology in Latin America*,
DOI 10.1007/978-3-319-54567-7_19

several improvements on antiretroviral treatment (i.e., new targets, less toxicity, and combination of antiretrovirals) have simplified and enhanced treatment adherence, making great progress on the survival of HIV-infected individuals, thus decreasing viral replication and slowing down the weakness of the immune system [2].

2 The Virus and the Replication Cycle

HIV is a retrovirus that belongs to the genus *Lentivirus*. Viral particles have a diameter of 120–150 nm and a lipid envelope with external glycoproteins (gp 120) and transmembrane proteins (gp 41). The matrix protein p17 is attached to the inside of the viral lipoprotein membrane. The capsid encloses the protein–nucleic acid complex: two copies of the RNA genome, the nucleoprotein, and the reverse transcriptase (RT). The viral particle contains, in addition, other enzymes necessary for replication: the integrase and the protease. The RNA genome contains three genes: *gag*, *pol*, and *env* flanked by two LTR (long terminal repeat) regions. The *gag* gene codes for the matrix (p17), the capsid (p24), and the nucleocapsid (p7); the *pol* gene for the reverse transcriptase (p66), the integrase (p32), and the protease (p11); and the *env* gene for the glycoproteins of the membrane (gp 120 and gp 41). In addition, the genome contains six regulatory genes (*vif*, *vpu*, *vpr*, *tat*, *rev*, and *nef*).

CD4 T lymphocytes are the main target of HIV, with CD4 and the chemokine receptors, CCR5 or CXCR4, being essential for the virus entry. Other cells bearing CD4 and chemokine receptors are also infected, including resting CD4 T cells, monocytes, macrophages, and dendritic cells. Following attachment and membrane fusion, the viral capsid is released into the cytoplasm. Then, conversion of viral RNA into proviral DNA occurs in the cytoplasm mediated by the viral enzyme RT. The replication of HIV is prone to error and characterized by a high spontaneous mutation rate. After transportation of the pre-integration complex into the nucleus, the integration of the proviral DNA into the host genome occurs with the viral enzyme integrase intervention. This integration of the proviral DNA into the host genome makes the infection exceedingly difficult to eradicate with the use of current therapies. With the intervention of host enzymes, HIV is transcribed, proteins are produced and cleaved, and mature virions are released. During the budding process, the lipid membranes of the new virus may incorporate host cell proteins, such as HLA [3, 4].

3 Transmission Routes

When the first AIDS cases were reported, transmission routes were clearly established. HIV can be spread in all the body secretions and fluids of an infected individual; thus unprotected sex with an infected individual, contact with blood or blood-contaminated materials, and vertical transmission (from infected mother to newborn) are considered the main transmission routes. Sexual contact is the most important HIV transmission route from the epidemiological point of view, accountable for most infections worldwide. Various factors have an influence on the transmission/infection risk regardless of the transmission route, such as the genetic background of the receptor, virus subtype, and levels of HIV viremia in the transmitter [3, 4].

Sexual contact as well as the presence of lesions in mucosa and specific intercourse practices can predispose for higher HIV transmission. Several studies worldwide demonstrated that individuals with ulcerative sexually transmitted infections have higher prevalence of HIV infection as the disruption of the mucosal/skin barrier can facilitate exposure to the virus. Regarding sexual practices, anal intercourse has been considered the main transmission route, particularly for the receptive partner, because of the anal mucosal trauma generated during the intercourse. Vaginal intercourse has a lower risk of HIV transmission, being higher for the female partner because of the major mucosal exposure. Oral sex has been described as a probable route of HIV acquisition if lesions are present in mucosa and the fluids exchanged contain blood; however, transmission is still controversial. The main prevention strategy against HIV transmission through sexual contact consists in the use of condoms for any kind of sexual contact. Nevertheless, over the past years, new prevention strategies have emerged. For example, male circumcision has been demonstrated to reduce the probability of HIV infection, and in consequence, it has been implemented as a prevention strategy in high prevalence settings. Multiple microbicide formulations (those providing a physical barrier or those based on antiretrovirals) and presentations (gels, creams, suppository) have been explored in clinical trials with promising results. However, an effective microbicide is not currently available for HIV prevention [5].

In relationship to vertical transmission (mother to child), it has been estimated that without intervention, approximately 40% of mothers transmit the virus to the newborns, either during pregnancy, at delivery, or during breastfeeding. Since antiretroviral therapy was implemented during pregnancy as well as elective cesarean, postexposure treatment of the newborn, and substitution of breastfeeding, vertical transmission has been almost eradicated in several countries and dramatically diminished in most of the world [6].

Contact with infected blood or blood products have a high probability of transmission, especially during transfusions. The implementation of screening at blood banks has largely reduced the transmission of HIV through transfusions at the global level. However, among injecting drug users who share paraphernalia, blood transmission continues being an important route. Needle exchange programs, implemented in several countries, demonstrated to be effective in reducing HIV and also other blood-borne transmitted infections such as hepatitis C virus. Exposure to blood through accidental injuries in health practitioners has been extensively explored, and in addition to the implementation of barrier prevention measures (gloves, goggles), postexposure prophylaxis (PEP) has been implemented, showing excellent results [7].

4 Global Epidemiology

The last UNAIDS report estimated that, by the end of 2014, a total of 36.9 million people were living with HIV worldwide, around 2 million people were infected, and 1.2 million people had died of AIDS-related illnesses during the same year. The number of newly infected individuals in 2014 is 35% lower than in 2000. However, even when new infections declined, the number of people living with HIV continues to increase, in part because approximately 15 million people globally are on antiretroviral treatment and AIDS-related diseases have decreased. Sub-Saharan Africa remains the region mostly affected by the epidemic with 25.8 million (range, 24 to 28.7 million) people living with HIV [8].

5 Latin American Epidemic

An estimated 1.7 million (1.4 to 2.0 million) people were living with HIV in Latin America by the end of 2014, including 87,000 (70,000–100,000) individuals newly infected during this year. Most of the cases are in adults, reaching 1.6 million (1.4–2.0). Approximately 60% of people living with HIV in the region were men, including heterosexual men and gay men and other men who have sex with men. Global HIV prevalence in adults in the region has been estimated in 0.4% (0.4–0.5). It has also been estimated that the number of new HIV infections in 2014 was 17% lower than in 2000 in Latin America. Also, AIDS-related deaths have decreased by 31% since 2000 [8, 9].

The number of new HIV infections in children aged 0 to 14 years declined by 73% between 2000 and 2014, that is, from 7400 to 2000 children recently infected with HIV. The rate of mother-to-child transmission (MTCT) of HIV in Latin America and the Caribbean decreased between 2010 and 2014, that is, from 14% to 7%. These data are in line with antiretroviral therapy (ART) coverage in HIV-positive pregnant women, which has increased significantly during the last decade. It was estimated that by 2014 96% of pregnant women from Latin America and the Caribbean received at least one prenatal care visit, 75% accessed HIV testing and counseling, and 81% of women who needed it received ART, a 43% increase over 5 years. However, coverage of antiretroviral therapy to prevent MTCT varies across countries, with less than 30% coverage in some countries such as Guatemala and the Bolivarian Republic of Venezuela. Ninety-eight percent of the countries in the region have national plans to eliminate HIV-MTCT, and in most of these countries the plans are integrated with the elimination of congenital syphilis. HIV-MTCT is less than 4% in 75% of the countries. On the basis of the progress made in the last years, and the pending gaps, it appears that the elimination of HIV-MTCT in the short term could be possible [10].

Latin America presents a heterogeneous epidemic. HIV prevalence varies considerably within and between countries in the region. The epidemic is mostly concentrated in urban settings. Even when the average HIV prevalence in most countries is much lower than in other regions, the numbers in some groups of the population are alarming. Most infections in Latin America are concentrated in the male-to-female transgender (MTFT) population, men who have sex with men (MSM), female sex workers (FSWs), and drug users (DUs). However, as the result of cultural, religious, regulatory, and socioeconomic differences, countries present several epidemiological differences [11]. Stigma and discrimination in some Latin American countries interfere with efforts to achieve universal access to HIV prevention, diagnosis, treatment, and care.

5.1 Men Who Have Sex with Men (MSM)

Gay, bisexual, and other men who have sex with men (MSM) remain one of the groups more affected by the HIV epidemic worldwide and also in Latin America. A global review found a pooled HIV prevalence of 14.9% (95% CI, 14.1–15.7) for Latin American countries [12]. However, HIV prevalence among MSM ranges from 7% in Nicaragua to 20% in Chile and Panama [9]. This high prevalence can be explained by the very high transmission efficiency of HIV across the rectal mucosa, estimated at 1.4% per act, being 18 fold higher than for vaginal intercourse [12, 13]. Data on HIV incidence are less available, but studies revealed consistently high rates of new infections among MSM (3.5% per 100 person-years in Peru) [13, 14] (Fig. 19.1).

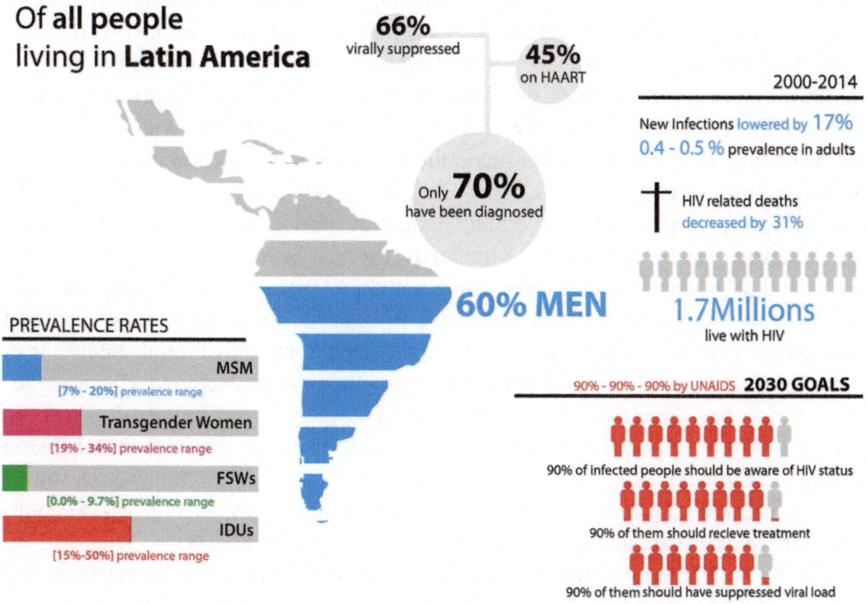

Fig. 19.1 State of the human immunodeficiency virus (HIV) epidemic in Latin America

Risk factor analyses revealed that unprotected anal intercourse (UAI) is the main transmission route for HIV in MSM in Latin America. Other studies also revealed a strong association between sexual role and HIV infection; those reporting a passive role are more likely to become HIV infected [15]. Non-injecting illegal drugs and alcohol use disorders are highly prevalent among MSM in some countries of the region and are associated with an increased risk of HIV infection and low ART adherence [16–19]. Another important risk factor for HIV infection is the presence of active sexually transmitted diseases (STD) [20] or a history of STD [21]. Transactional sex among MSM was also associated with a significant increase in HIV prevalence [22].

One international survey reported that criminalization of homosexuality through arrest and convictions was less common in Latin America than in other continents, which was reported by 9.7% among MSM in the region, as compared to 24% in sub-Saharan Africa. However, those suffering criminalization had lesser access to condoms, sexually transmitted infection (STI) treatment, and medical care [23]. In the same line, physical and psychological violence were also associated with higher risk exposure [24, 25]. Perception of stigma was also associated with lower testing frequency among MSM from El Salvador [26].

5.2 Transgender Women

Transgender women are individuals assigned male at birth but who identified themselves as female. They are exposed to stigma, discrimination, and familial/social exclusion from early childhood; in consequence, most of them are involved in sex work activities as one of the very few alternatives to earn a living and to pay for feminization procedures. Even when they represent a small number of individuals, a recent study revealed extremely elevated HIV prevalence rates with transgender women being the most affected group in the region. Seroprevalence studies revealed prevalence rates ranging from 19% in Uruguay to 34% in Argentina [27]. In most countries, sex work activity is the main economic activity for transgenders, exposing them to several risk factors, particularly drug use, sexual or other physical violence, and unprotected sexual intercourse with male partners, including clients and noncommercial (stable or casual) partners. Similar to MSM, HIV transmission in this group occurs mainly because of unprotected receptive anal intercourse. However, transgenders are more exposed to practicing unprotected sex because clients offer to pay more money or in other cases they are forced to have sex without condoms. Another possible risk factor in this group is the parental transmission through hormone or silicone injections outside the health system. It has been shown the high rates of substance use observed in transgender women from several countries, where data were available, such as Argentina and Brazil [28].

The high HIV prevalence and incidence reported among transgenders in the region call for an urgent need of prevention, treatment, and care services targeted for this group. Additionally, transgenders need to be considered a separate category on epidemiological studies. Until now, some countries often include transgenders as a subpopulation of MSM, hiding the HIV situation of this vulnerable group. During the last few years, most countries in the region have achieved innovative advances to guarantee the rights of lesbian, gay, bisexual, and transgender individuals, including national laws that allow same-sex marriage and the change of the individual's identity according to gender identification. However, at the beginning of 2016, approximately 73 countries in the world have laws criminalizing homosexuality, most of them located in Asia and Africa [29].

Because of the increasing number of countries in the region with regulations that allow genital surgery, HIV transmission through a neovagina should be considered in future studies. To date, there are no data on the likelihood of transgenders to become infected with HIV when they undergo these kinds of surgeries [30].

As well as in MSM, PrEP demonstrated to be effective in preventing HIV acquisition. However, some researchers have found more barriers to adherence [31].

5.3 Female Sex Workers

Commercial sex has been identified as one of the riskiest practices driving the HIV epidemic in several countries. Most sex workers worldwide are women; however, as previously mentioned, in some Latin American countries, transgender women constitute a growing portion of sex workers. HIV prevalence among FSWs in Latin America ranges from 0.0% in Chile to 9.7% in Honduras, with a pooled HIV prevalence of 6.1% [9, 32]. Data from Latin America evidence substantially higher levels of HIV infection among FSWs compared with women of reproductive age, although prevalence is much lower than that observed in sub-Saharan Africa, where HIV prevalence is higher than 35% [33]. The high HIV prevalence may be attributed to an increasing sexual exposure because of their activity. However, other factors such as poverty, discrimination, and gender disparity, as well as punitive laws, can remain barriers to the current prevention strategies. In several studies, FSWs reported the use of condoms more often with commercial partners than with noncommercial partners (either casual or steady) [34, 35].

Interventions to reduce HIV transmission in FSWs need to include behavioral individual interventions (e.g., condom use campaigns, FSW empowerment, and sexual health education) as well as political and legal changes that contribute to eliminating the penalization of sex work activity.

5.4 Drug Users

Individuals who use illegal drugs constitute a diversity group with different rates of risk for HIV transmission according to the route of administration. Even when injecting drug users (IDUs) have a higher risk for HIV acquisition if they share paraphernalia, the use of other drugs was associated with HIV transmission, even by the direct administration (e.g., sharing inhalator elements) or by the behavioral changes resulting from consumption (e.g., condom-less) [7].

The frequency of illegal drug use has been estimated in 4.8% in the region, with marijuana and cocaine as the most common drugs [36]. Regarding the administration route, IDU is not a common practice nowadays in most of the countries, and data are scarce [37, 38]. Studies performed in the region several years ago account for HIV prevalence near 50% in IDUs from Argentina and Brazil and lower than 15% in Peru and Paraguay [39]. However, some new studies revealed specific places and groups where IDU is still a problem. For example, 10.3% of female sex workers from several cities in Mexico reported injecting illegal drugs which were associated with client-perpetrated sexual violence. This association has significant implications for increasing FSW risk for HIV infection [40]. Another study performed in the same area reported a surprisingly high prevalence of the use of dead-space syringes among IDUs [41]. Efforts are needed to expand coverage of syringe exchange programs in these areas.

Studies among non-injecting drug users (NIDU) reveal that they have an increased sexual risk behavior, as can be seen in the high HIV prevalence. One review study from Peru revealed that drug use increases risky sexual behaviors, decreasing the use of condoms, and increasing the number of partners per year, the number of casual sex encounters, and the risk for sexually transmitted diseases (STDs) [42]. A high prevalence of HIV has been found among NIDU in Argentina, which was 6.3% [43]. In this study, the risk of being infected with HIV, HBV, and HCV was significantly associated with having had a sex partner who was either a IDU or who was known to be HIV positive. Because of the rising number of NIDU and the demonstrated risk for HIV and other STIs, it is essential to implement prevention strategies focused on this population. Studies from Brazil reported an HIV prevalence of 23.1% among drug users, being significantly associated to injecting drug use and syringe/needle sharing [44].

Illegal substance use has also been associated with lower adherence to ART among people living with HIV in Latin America [45].

6 Other Sexually Transmitted Infections (STIs)

Individuals most affected by HIV in Latin America are also exposed to other STIs. Syphilis is one of the most prevalent infections (more than 20%) among at-risk groups with critical hotspot cities such as Sao Paulo and Buenos Aires [46]. Syphilis infection was also found to predict HIV incidence, being associated with HIV acquisition in follow-up studies [47]. Other curable STIs, including syphilis, gonorrhea, HPV, and Chlamydia, were demonstrated to have high prevalence in several countries [48–52]. Studies also demonstrated that ulcerative STIs (including syphilis and herpes) could be important risk factors for HIV acquisition among MSM and transgenders [53, 54]. Previous hepatitis B virus infection is also common in older individuals in the at-risk group [55]. Because HBV vaccination has been implemented in most of the countries in the past decade, the decrease of this agent in the following years is expected.

7 HIV Molecular Epidemiology

Until now, two types of HIV have been described, HIV-1 and HIV-2. HIV-1 is responsible for the worldwide epidemic, and HIV-2 is much less common, less virulent, and mostly confined to West Africa. HIV-1 is characterized by extensive genetic diversity because of its high replication rate, the error-prone reverse transcriptase, and recombination events that may occur during virus replication. HIV-1 includes four groups, M (main), O (outlier), N (non-M, non-O), and P, which have different geographic distributions. The M group includes nine subtypes (A, B, C, D, F, G, H, I, J), as well as at least 79 circulating recombinant forms (CRFs) and multiple unique recombinant forms (URFs), which together account for more than 90% of the infections worldwide [56].

The HIV epidemic in Latin America is dominated by subtype B, which accounts for nearly 70% of the infections in the region. However, the epidemic is more diverse in the Southern cone where subtypes B, C, and F as well as B/C and B/F recombinants are the most prevalent. Inter-subtype recombinants were first reported as a CRF12_BF in Argentina and Uruguay in 2001 followed by several BF CRFs in Brazil, Uruguay, and Chile. The analysis of the temporal distribution of HIV subtypes and recombinants indicated an overall increase in the proportion of recombinants [57].

8 New Prevention Agenda

Among the new prevention strategies targeting at-risk groups, oral preexposure prophylaxis (PrEP) has been the best biomedical intervention to demonstrate efficacy in preventing HIV infection among MSM and transgenders, including studies performed in Latin American countries that suggest high acceptability [57, 58]. Studies in the region also explored the efficacy of male circumcision on HIV acquisition and demonstrated the efficacy of this intervention only for MSM who are primarily insertive [59–61]. Microbicide acceptability was also explored among MSM; however, even though acceptability is high, it varied by city, suggesting the importance of regional studies [62].

New testing technologies, such as rapid HIV testing, have been included in most countries in the last years; however, over-the-counter test kits are not available yet, even when research studies suggested that self-testing may have the potential to increase testing frequency of high-risk groups [63–65].

9 Ending the AIDS Epidemic by 2030: 90-90-90 UNAIDS Strategy

There is a strong global consensus that ending the AIDS epidemic is possible. This hope is based on scientific evidence and lessons learned during more than a decade of scaling up the AIDS response. HIV antiretroviral treatment can extend the life expectancy of people living with HIV and also prevent HIV transmission. Even when HIV infections may not disappear in the near future, the AIDS epidemic can be ended as a global health threat. UNAIDS has proposed that to achieve this by 2030, 90% of the people living with HIV should be aware of the HIV status, 90% of the people who know their status should receive treatment, and 90% of people on HIV treatment should have a suppressed viral load. The HIV cascade of care constitutes a useful epidemiological tool that allows monitoring HIV-positive individuals from diagnosis to viral suppression. Of the people estimated to be living with HIV in Latin America, only 70% have been diagnosed, 45% were on HAART, and an estimated 66% have become virally suppressed, achieving 28% of all the people living with HIV in the region. Latin America has a median of 80% retention on ART at 12 months after initiation.

However, some factors, such as psychological health problems or criminalization of homosexuality, can generate significantly lower access to prevention (condoms), STI treatment, and medical care [66, 67].

9.1 Pending Research

Most HIV prevalence studies focused on at-risk groups have taken place only in large metropolitan cities, obscuring the situation of small towns. One of the possible explanations about this limitation is the lower social tolerance of sexual diversity and/or sexual work in small cities that results in lower probability of sampling these groups for fear to being recognized. Future research studies and public health interventions need to implement new strategies to reach at-risk groups in all settings to diagnose and include all HIV-infected individuals in the healthcare system. Regarding methodology, only a few studies with probabilistic sampling methodologies have been performed in the region; the majority of the research studies were based on convenience sampling. In general, probabilistic methodologies are more difficult to implement and need more monetary resources [68]. To perform probabilistic methodologies, it is also necessary to know the number of individuals at each risk group. However, only small estimations have been performed in the region and mostly focused on specific cities [69, 70].

In Latin America, prevention efforts should continue to focus on reducing risk behavior among MTFTs, MSM, FSWs, and DUs as well as to ensure access to universal HIV testing and care. Even when governments from several countries have made many efforts to cover the cost of ART for all HIV patients, one of the challenges in the region is to focus more on allocating monetary resources on prevention aimed at those key populations with the highest vulnerabilities.

In summary, the HIV epidemic in Latin America is concentrated in transgender women and men who have sex with men, who have an estimated prevalence higher than 10%, whereas the prevalence in the general population does not reach 1%. Because these two groups account for many of the new infections in the region, it is crucial to continuously update the HIV epidemiological situation, including HIV prevalence, incidence studies, risk factor analyses, as well as continuum of care of those infected individuals. Prevention efforts should continue to focus on reducing risky behavior but need also to include new prevention strategies and access to anti-retroviral treatment to significantly reduce HIV transmission.

Acknowledgments We thank Carolina Gomez Carrillo for figure design.

References

1. Sharp PM, Hahn BH (2011) Origins of HIV and the AIDS pandemic. Cold Spring Harbor Perspect Med 1(1):a006841
2. Cihlar T, Fordyce M (2016) Current status and prospects of HIV treatment. Curr Opin Virol 26(18):50–56
3. Deeks SG, Overbaugh J, Phillips A, Buchbinder S (2015) HIV infection. Nat Rev Dis Primers 1:15035. doi:10.1038/nrdp.2015.35
4. Maartens G, Celum C, Lewin SR (2014) HIV infection: epidemiology, pathogenesis, treatment, and prevention. Lancet 19(384(9939):258–271
5. Padian NS, McCoy S, Karim SS et al (2011) HIV prevention transformed: the new prevention research agenda. Lancet 16(378(9787)):269–278
6. Stevens J, Lyall H (2014) Mother to child transmission of HIV: what works and how much is enough? J Infect 69(suppl 1):S56–S62
7. Patel P, Borkowf CB, Brooks JT et al (2014) Estimating per-act HIV transmission risk: a systematic review. AIDS 19(28(10)):1509–1519
8. UNAIDS (2015) AIDS by the numbers. Available at: http://www.unaids.org/sites/default/files/media_asset/AIDS_by_the_numbers_2015_en.pdf. http://www.unaids.org/sites/default/files/media_asset/AIDS_by_the_numbers_2015_en.pdf. Accessed June 2016
9. UNAIDS (2014) The gap report. 2014. Available at:http://files.unaids.org/en/media/unaids/contentassets/documents/unaidspublication/2014/UNAIDS_Gap_report_en.pdf. Accessed June 2016
10. PAHO (2015) Elimination of Mother-to-Child Transmission of HIV and Syphilis in the Americas. Update 2015. Available at: http://iris.paho.org/xmlui/bitstream/handle/123456789/18372/9789275118702_eng.pdf?sequence=3&isAllowed=y. Accessed June 2016
11. De Boni R, Veloso VG, Grinsztejn B (2014) Epidemiology of HIV in Latin America and the Caribbean. Curr Opin HIV AIDS 9(2):192–198
12. Beyrer C, Baral SD, van Griensven F et al (2012) Global epidemiology of HIV infection in men who have sex with men. Lancet 380(9839):367–377
13. Beyrer C, Sullivan P, Sanchez J et al (2013) The increase in global HIV epidemics in MSM. AIDS 27(17):2665–2678
14. Sanchez J, Lama JR, Peinado J et al (2009) High HIV and ulcerative sexually transmitted infection incidence estimates among men who have sex with men in Peru: awaiting for an effective preventive intervention. J Acquir Immune Defic Syndr 51(suppl 1):S47–S51

15. Meng X, Zou H, Fan S et al (2015) Relative risk for HIV infection among men who have sex with men engaging in different roles in anal sex: a systematic review and meta-analysis on global data. AIDS Behav 19(5):882–889
16. Ludford KT, Vagenas P, Lama JR et al (2013) Screening for drug and alcohol use disorders and their association with HIV-related sexual risk behaviors among men who have sex with men in Peru. PLoS One 8(8):e69966
17. Deiss RG, Clark JL, Konda KA et al (2013) Problem drinking is associated with increased prevalence of sexual risk behaviors among men who have sex with men (MSM) in Lima, Peru. Drug Alcohol Depend 132(1–2):134–139
18. Ferro EG, Weikum D, Vagenas P et al (2015) Alcohol use disorders negatively influence antiretroviral medication adherence among men who have sex with men in Peru. AIDS Care 27(1):93–104
19. Julio RS, Friedman RK, Cunha CB et al (2015) Unprotected sexual practices among men who have sex with women and men who have sex with men living with HIV/AIDS in Rio de Janeiro. Arch Sex Behav 44(2):357–365
20. Pitpitan EV, Goodman-Meza D, Burgos JL et al (2015) Prevalence and correlates of HIV among men who have sex with men in Tijuana, Mexico. J Int AIDS Soc 9(18):19304
21. Bautista CT, Sanchez JL, Montano SM et al (2004) Seroprevalence of and risk factors for HIV-1 infection among South American men who have sex with men. Sex Transm Infect 80(6):498–504
22. Oldenburg CE, Perez-Brumer AG, Reisner SL et al (2015) Transactional sex and the HIV epidemic among men who have sex with men (MSM): results from a systematic review and meta-analysis. AIDS Behav 19(12):2177–2183
23. Santos GM, Makofane K, Arreola S et al (2017) Reductions in access to HIV prevention and care services are associated with arrest and convictions in a global survey of men who have sex with men. Sex Transm Infect 93:62–64. pii: sextrans-2015-052386
24. Wheeler J, Anfinson K, Valvert D et al (2014) Is violence associated with increased risk behavior among MSM? Evidence from a population-based survey conducted across nine cities in Central America. Glob Health Action 7:24814
25. Cunha CB, De Boni RB, Guimarães MR et al (2014) Unprotected sex among men who have sex with men living with HIV in Brazil: a cross-sectional study in Rio de Janeiro. BMC Public Health 14:379
26. Andrinopoulos K, Hembling J, Guardado ME et al (2015) Evidence of the negative effect of sexual minority stigma on HIV testing among MSM and transgender women in San Salvador, El Salvador. AIDS Behav 19(1):60–71
27. Baral SD, Poteat T, Strömdahl S et al (2013) Worldwide burden of HIV in transgender women: a systematic review and meta-analysis. Lancet Infect Dis 13(3):214–222
28. Poteat T, Wirtz AL, Radix A et al (2015) HIV risk and preventive interventions in transgender women sex workers. Lancet 385(9964):274–286
29. Carroll, A (2016) State Sponsored Homophobia 2016: A world survey of sexual orientation laws: criminalisation, protection and recognition (Geneva; ILGA, May 2016). Available at: http://ilga.org/downloads/02_ILGA_State_Sponsored_Homophobia_2016_ENG_WEB_150516.pdf. Accessed June 2016
30. Poteat T, Reisner SL, Radix A (2014) HIV epidemics among transgender women. Curr Opin HIV AIDS 9(2):168–173
31. Deutsch MB, Glidden DV, Sevelius J et al (2015) HIV pre-exposure prophylaxis in transgender women: a subgroup analysis of the iPrEx trial. Lancet HIV 2(12):e512–e519
32. Baral S, Beyrer C, Muessig K et al (2012) Burden of HIV among female sex workers in low-income and middle-income countries: a systematic review and meta-analysis. Lancet Infect Dis 12(7):538–549
33. Beyrer C, Crago AL, Bekker LG et al (2015) An action agenda for HIV and sex workers. Lancet 385(9964):287–301
34. Prüss-Ustün A, Wolf J, Driscoll T et al (2013) HIV due to female sex work: regional and global estimates. PLoS One 8(5):e63476

35. Pando MA, Berini C, Bibini M et al (2006) Prevalence of HIV and other sexually transmitted infections among female commercial sex workers in Argentina. Am J Trop Med Hyg 74(2):233–238
36. García PJ, Bayer A, Cárcamo CP (2014) The changing face of HIV in Latin America and the Caribbean. Curr HIV/AIDS Rep 11(2):146–157
37. Rossi D, Pawlowicz MP, Rangugni V et al (2006) The HIV/AIDS epidemic and changes in injecting drug use in Buenos Aires, Argentina. Cad Saude Publica 22(4):741–750
38. Cohen J (2006) HIV/AIDS: Latin America & Caribbean. Argentina: up in smoke: epidemic changes course. Science 313(5786):487–488
39. Mathers BM, Degenhardt L, Phillips B et al (2008) 2007 Reference Group to the UN on HIV and Injecting Drug Use. Global epidemiology of injecting drug use and HIV among people who inject drugs: a systematic review. Lancet 372(9651):1733–1745
40. Semple SJ, Stockman JK, Pitpitan EV et al (2015) Prevalence and correlates of client-perpetrated violence against female sex workers in 13 Mexican cities. PLoS One 10(11):e0143317
41. Rafful C, Zule W, González-Zúñiga PE et al (2015) High dead-space syringe use among people who inject drugs in Tijuana, Mexico. Am J Drug Alcohol Abuse 41(3):220–225
42. Massa AA, Rosen MI (2012) The relationship between substance use and HIV transmission in Peru. J Addict Res Ther 3(3):128
43. Rossi D, Radulich G, Muzzio E et al (2008) Multiple infections and associated risk factors among non-injecting cocaine users in Argentina. Cad Saude Publica 24(5):965–974
44. Malta M, Magnanini MM, Mello MB et al (2010) HIV prevalence among female sex workers, drug users and men who have sex with men in Brazil: a systematic review and meta-analysis. BMC Public Health 10:317
45. De Boni RB, Shepherd BE, Grinsztejn B et al (2016) Substance use and adherence among people living with HIV/AIDS receiving cART in Latin America. AIDS Behav 20(11):2692
46. Zoni AC, González MA, Sjögren HW (2013) Syphilis in the most at-risk populations in Latin America and the Caribbean: a systematic review. Int J Infect Dis 17(2):e84–e92
47. Solomon MM, Mayer KH, Glidden DV et al (2014) Syphilis predicts HIV incidence among men and transgender women who have sex with men in a preexposure prophylaxis trial. Clin Infect Dis 59(7):1020–1026
48. Perez-Brumer AG, Konda KA, Salvatierra HJ et al (2013) Prevalence of HIV, STIs, and risk behaviors in a cross-sectional community- and clinic-based sample of men who have sex with men (MSM) in Lima, Peru. PLoS One 8(4):e59072
49. Soares CC, Georg I, Lampe E et al (2014) HIV-1, HBV, HCV, HTLV, HPV-16/18, and *Treponema pallidum* infections in a sample of Brazilian men who have sex with men. PLoS One 9(8):e102676
50. Pando MA, Balán IC, Marone R et al (2012) HIV and other sexually transmitted infections among men who have sex with men recruited by RDS in Buenos Aires, Argentina: high HIV and HPV infection. PLoS One 7(6):e39834
51. Dos Ramos Farías MS, Garcia MN, Reynaga E et al (2011) First report on sexually transmitted infections among trans (male to female transvestites, transsexuals, or transgender) and male sex workers in Argentina: high HIV, HPV, HBV, and syphilis prevalence. Int J Infect Dis 15(9):e635–e640
52. Creswell J, Guardado ME, Lee J et al (2012) HIV and STI control in El Salvador: results from an integrated behavioural survey among men who have sex with men. Sex Transm Infect 88(8):633–638
53. Sanchez J, Lama JR, Peinado J et al (2009) High HIV and ulcerative sexually transmitted infection incidence estimates among men who have sex with men in Peru: awaiting for an effective preventive intervention. J Acquir Immune Defic Syndr 51(suppl 1):S47–S51
54. Silva-Santisteban A, Raymond HF, Salazar X et al (2012) Understanding the HIV/AIDS epidemic in transgender women of Lima, Peru: results from a sero-epidemiologic study using respondent driven sampling. AIDS Behav 16(4):872–881

55. Lama JR, Agurto HS, Guanira JV et al (2010) Hepatitis B infection and association with other sexually transmitted infections among men who have sex with men in Peru. Am J Trop Med Hyg 83(1):194–200
56. Los Alamos sequence database. http://www.hiv.lanl.gov/. Last Access July 2016
57. Hemelaar J, Gouws E, Ghys PD, Osmanov S, WHO-UNAIDS Network for HIV Isolation and Characterisation (2011) Global trends in molecular epidemiology of HIV-1 during 2000-2007. AIDS 25(5):679–689. doi:10.1097/QAD.0b013e328342ff93
58. Grant RM, Lama JR, Anderson PL et al (2010) Preexposure chemoprophylaxis for HIV prevention in men who have sex with men. N Engl J Med 363(27):2587–2599
59. Peinado J, Lama JR, Galea JT et al (2013) Acceptability of oral versus rectal HIV preexposure prophylaxis among men who have sex with men and transgender women in Peru. J Int Assoc Provid AIDS Care 12(4):278–283
60. Sánchez J, Sal Y, Rosas VG, Hughes JP et al (2011) Male circumcision and risk of HIV acquisition among MSM. AIDS 25(4):519–523
61. Pando MA, Balan IC, Dolezal C et al (2013) Low frequency of male circumcision and unwillingness to be circumcised among MSM in Buenos Aires, Argentina: association with sexually transmitted infections. J Int AIDS Soc 16:18500
62. Kinsler JJ, Cunningham WE, Nureña CR et al (2012) Using conjoint analysis to measure the acceptability of rectal microbicides among men who have sex with men in four South American cities. AIDS Behav 16(6):1436–1447
63. Volk JE, Lippman SA, Grinsztejn B et al (2016) Acceptability and feasibility of HIV self-testing among men who have sex with men in Peru and Brazil. Int J STD AIDS 27(7):531–536
64. Lippman SA, Périssé AR, Veloso VG et al (2014) Acceptability of self-conducted home-based HIV testing among men who have sex with men in Brazil: data from an on-line survey. Cad Saude Publica 30(4):724–734
65. Bustamante MJ, Konda KA, Joseph Davey D et al (2016) HIV self-testing in Peru: questionable availability, high acceptability but potential low linkage to care among men who have sex with men and transgender women. Int J STD AIDS 28(2):133–137. Jan 29. pii: 0956462416630674. [Epub ahead of print]
66. Biello KB, Oldenburg CE, Safren SA et al (2016) Multiple syndemic psychosocial factors are associated with reduced engagement in HIV care among a multinational, online sample of HIV-infected MSM in Latin America. AIDS Care 28(suppl 1):84–91
67. Santos GM, Makofane K, Arreola S et al (2016) Reductions in access to HIV prevention and care services are associated with arrest and convictions in a global survey of men who have sex with men. Sex Transm Infect 93(1):62–64. pii: sextrans-2015-052386
68. Clark JL, Konda KA, Silva-Santisteban A et al (2014) Sampling methodologies for epidemiologic surveillance of men who have sex with men and transgender women in Latin America: an empiric comparison of convenience sampling, time space sampling, and respondent driven sampling. AIDS Behav 18(12):2338–2348
69. Cáceres C, Konda K, Pecheny M et al (2006) Estimating the number of men who have sex with men in low and middle income countries. Sex Transm Infect 82(suppl 3):iii3–iii9
70. Paz-Bailey G, Jacobson JO, Guardado ME et al (2011) How many men who have sex with men and female sex workers live in El Salvador? Using respondent-driven sampling and capture-recapture to estimate population sizes. Sex Transm Infect 87(4):279–282

Chapter 20
Human Papillomavirus Research in Latin America

María Alejandra Picconi and Luisa Lina Villa

1 Introduction

Papillomaviruses are 8000-base-pair (8000-bp), double-stranded, circular DNA viruses that can cause warty and neoplastic changes in epithelia from many host species. Their genome consists of double-stranded DNA and encodes sequences for six early (E1, E2, E4, E5, E6, E7) and two late (L1, L2) proteins involved in capsid formation [30, 79]. Viral types are defined as a viral genome with an L1 late gene sequence that is at least 10% different from that of any other type. Interestingly, differing from most other viruses, papillomavirus infection is determined by DNA detection and not viral isolation.

Of the nearly 200 human papillomavirus (HPV) types identified, approximately 40 can infect human mucosa, particularly the anogenital and aerodigestive tracts [8], albeit cervical HPV infections are best understood.

The International Agency for Research on Cancer (IARC) has classified 12 HPV types as group 1 carcinogens; they are called "high-risk HPVs" (hr-HPVs) and include the following types: 16, 18, 31, 33, 35, 39, 45, 51, 52, 56, 58, and 59 [12]. Among the hr-HPVs, HPV16 is by far the most carcinogenic in terms of numbers of cervical cancer (CC) cases and its precursors [11, 52]. HPV16 also causes most HPV-related cancers in other anogenital epithelia and the oropharynx. HPV18 is classified second in terms of etiological importance but accounts for a high proportion of adenocarcinomas [52].

M.A. Picconi (✉)
Laboratorio Nacional y Regional de Referencia de Papilomavirus, Instituto Nacional de Enfermedades Infecciosas-ANLIS Malbrán, Buenos Aires, Argentina
e-mail: mapicconi@anlis.gov.ar

L.L. Villa (✉)
Laboratorio de Inovação em Câncer, Faculdade de Medicina, Instituto do Câncer do Estado de São Paulo, Universidade de São Paulo, São Paulo, Brazil
e-mail: l.villa@hc.fm.usp.br

J.E. Ludert et al. (eds.), *Human Virology in Latin America*,
DOI 10.1007/978-3-319-54567-7_20

The varying carcinogenicity of HPV types partly relates to the expression of two early genes, the E6 and E7 oncogenes. Among other functions, E6 and E7 oncoproteins interfere with the functions of tumor suppressor proteins p53 and pRb, respectively. During the carcinogenic process, the HPV genome may integrate into the epithelial cell genome, and, during integration, parts of the HPV genome can be lost [76, 79]. However, the continued presence and expression of E6 and E7 gene regions are needed to sustain cancers and cancer cell lines.

Latin America (LA) has one of the highest incidence and mortality rates from CC in the world, with age-adjusted incidence rates ranging from 10 to 80 per 100,000 women/year [26, 32]. Overall, mortality rates are extremely high despite cytological screening in place in several countries. On the other hand, little is known about the rates of other HPV-associated tumors such as vulvar, vaginal, anal, penile, and oropharyngeal cancers. HPV DNA prevalence and type distribution are well known in many LA countries [15]. Moreover, data on the natural history of HPV infections and risk of disease development are available from large cohort studies and serve to propose new primary and secondary prevention modalities that include prophylactic HPV vaccination and HPV testing, respectively. The HPV vaccine was introduced in several LA national immunization programs, and multiple screening experiences using HPV testing were introduced in the region [45, 77]. Although promising, challenges to control HPV-related tumors are significant, mainly because as a comprehensive strategy it should include both components: vaccination and virological screening. Furthermore, information on HPV prevalence and type distribution in several LA countries is key both to measure the impact of HPV prophylactic vaccines and to establish appropriate post-vaccine epidemiological surveillance, with virological and disease endpoints.

2 HPV Natural History and Cervical Carcinogenicity

The cervix provides the best model of HPV and anogenital neoplasia natural history. The major stages in cervical carcinogenesis include infection of the cervical transformation zone metaplastic epithelium with one or more hr-HPV types, viral persistence, clonal progression of the persistently infected epithelium to cervical pre-cancer, and invasion.

Several epidemiological studies conducted in LA have contributed to establish these fundamental stages and to shed light on the factors that influence each of these transitions. Table 20.1 summarizes the main LA studies, pointing out their most relevant findings. The following findings should be highlighted: the cohort studies of the case-control studies conducted by IARC in Colombia, Paraguay, Brazil, and Peru; the cohort studies of Proyecto Guanacaste (Costa Rica) and Ludwig-McGill cohort (Brazil); and the HPV prevalence studies by IARC in Colombia, Mexico, Argentina, and Chile.

The great majority of sexually active women and men have been infected with HPV at least once in their lifetime [16]. HPV is the most common sexually

Table 20.1 Summary of major contributions of HPV epidemiological studies carried out in Latin American populations

Countries	Main contributions	References (first author, journal, year, vol:pages)
Colombia, Costa Rica, Mexico, and Panama (NCI case-control study)	First to show an association of HPV infection and CC [r1] Studied associations between CC and other risk factors (cervical screening, education, number of sexual partners, age at first sexual intercourse, male sexual behavior, and parity) [r2–r5]	r1 Reeves, *N Engl J Med* 1989;320:1437–1441 r2 Brinton, *Int J Cancer* 1989;44:199–203 r3 Herrero, *Bull Pan Am Health Organ* 1990;24:263–283 r4 Herrero, *Cancer* 1990;65:380–386 r5 Herrero, *Int J Epidemiol* 1992;21:1050–1056
Pooled analyses of these studies have established the current evidence on carcinogenic risks (IARC monographs), leading to the launch of HPV vaccine programs		
Colombia and Spain (IARC case-control studies)	First to show the higher sensitivity and specificity of PCR assays for HPV DNA detection, compared to other hybridization assays [r6] First to report a strong association between HPVs 16,18,31,33,35, and CIN3/CIS [r7] and of HPVs 16,18,31,33,35, and CC [r8] Evaluated the association of different risk factors and CIN3/CIS [r9] and CC [r10] and concluded that HPV infection could have a causal association with CC Comparison of risk factors' distribution in women of LAC and Europe [r11] Studied the husbands' risk factors, reported prevalence of penile HPV infection and its association with the risk of CC in their wives [r12, r13]	r6 Guerrero, *J Clin Microbiol* 1992;30:2951–2959 r7 Bosch, *Cancer Epidemiol Biomarkers Prev* 1993;2:415–422 r8 Muñoz, *Int J Cancer* 1992;52:743–749 r9 Muñoz, *Cancer Epidemiol Biomarkers Prev* 1993;2:423–431 r10 Bosch, *Int J Cancer* 1992;52:750–758 r11 de Sanjose, *Am J Public Health* 1996;86:1532–158 r12 Muñoz, *J Natl Cancer Inst* 1996;88:1068–1075 r13 Castellsague, *J Infect Dis* 1997;176:353–361
Paraguay (IARC case-control studies)	Strong association of HPV infection (types 16,18,31,33,45,58) and of other risk factors (education, lifetime number of sexual partners and screening history) with increased risk of CC	r14 Rolon, *Int J Cancer* 2000;85:486–491
Brazil (IARC case-control studies)	Strong association of HPV infection (types 16,18,31,33) and high parity with an increased risk of CC and history of previous cervical smears with reduced risk of CC [r15] A pooled analysis with the case-control studies in Colombia and Spain reported risk factors for HPV infection in middle-aged women and HPV distribution by age [r16]	r15 Eluf-Neto, *Br J Cancer* 1994;69:114–119 r16 Munoz, *Sex Transm Dis* 1996;23:504–510
Peru (IARC case-control studies)	HPV infection, long-term use of oral contraceptives, and smoking associated with increased risk of CC	r17 Santos, *Br J Cancer* 2001;85:966–971
Honduras	Prevalence of HPV types by histological grade (CIN1-CC); association of HPV16 and HPV18 infection and other high-risk types and risk of CC [r18] A sub-analysis reported the association between other cofactors and CC [r19]	r18 Ferrera, *Int J Cancer* 1999; 82:799–803 r19 Ferrera, *Int J Epidemiol* 2000;29:817–825

(continued)

Table 20.1 (continued)

Countries	Main contributions	References (first author, journal, year, vol:pages)
Costa Rica (the Guanacaste project, funded by the US NCI)	Risk factors associated with HSIL and CC and with the progression of HPV infection [r20-r22] First study to observe the second peak of HPV infection in women older than 65 in LA [r23] First to show that HPV testing (hc2) and liquid base cytology are more sensitive than conventional cytology in LA. Several screening strategies have been evaluated [r24] The incidence, persistence, and clearance of HPV infection in women aged 18–26 years [r25] Highest risk of CIN2+ among women younger than 30 years with HPV16 infections that persisted for at least 12 months; importance of not single-time HPV detection [r26] The rate of new infections declines with age, and new infections typically do not progress to CIN 2 or worse disease in older women [r27]	r20 Herrero R, *Rev. Panam Salud Publica* 1997;1:362–375 r21 Herrero R, *J Infect Dis* 2005;191:1796–1807 r22 Hildesheim A, *Br J Cancer* 2001;84:1219–1226 r23 Herrero R, *J Natl Cancer Inst* 2000;92:464–474 r24 Ferreccio C, *Cancer Epidemiol Biomarkers Prev* 2003;12:815–823 r25 Rodriguez AC, *Sex Transm Dis* 2007;34:494–502 r26 Rodriguez AC, *J Natl Cancer Inst* 2008;100:513–517 r27 Rodriguez AC, *J Natl Cancer Inst* 2010; 102:315–324
Brazil (the Ludwig-McGill cohort) Colombia (The Bogota cohort)	Strong associations of persistence of HPV16/18 infections with subsequent risk of cervical lesions [r28] Estimation of incidence and duration of infections by type and lesion sojourn time [r29, r30] Bimodal age distribution of hr-HPV types. Type-specific HPV prevalence [r31, r32] Characterization of the hazard rate of cumulative incidence of HPV infection [r29, r30] No difference in the likelihood of clearance by HPV type or woman's age, but lower clearance for HPV16 infection. Viral load inversely associated with clearance [r33]	r28 Trottier, Cancer *Epidemiol Biomarkers Prev* 2006;15:1274–1280 r29 Schlecht, *J Natl Cancer Inst* 2003;95:1336–1343 r30 Schlecht, *Am J Epidemiol* 2003;158:878–886 r31 Mendez, *J Infect Dis* 2005;192:1158–1165 r32 Muñoz, *J Infect Dis* 2004;190:2077–2087 r33 Muñoz N, *Br J Cancer* 2009; 100:1184–1190
Brazil and Argentina (The Latin American Screening Study - LAMS)	Comparisons of screening techniques; highest sensitivity of hc2 to detect either CIN2 or CIN3 and the overall increased sensitivity of different combinations of pair of tests with a corresponding loss of specificity [r34] Description of risk factors [r35, r36]	r34 Sarian, *J Med Screen* 2005;12:142–149 r35 Syrjanen, *Anticancer Res* 2005;25:3469–3480 r36 Gontijo, *Eur J Obstet Gynecol Reprod Biol* 2007;133:239–246

These studies have contributed to the description of age-specific and type-specific prevalence of HPV infection by country/region, giving the baseline information for the implementation of HPV vaccination programs in the world		
Colombia (IARC HPV prevalence survey)	Most common HPV types in negative cytology; bimodal age curve of HPV infection. Number of sexual partners and use of oral contraceptives as risk factors for HPV infection	r37 Molano, *Br J Cancer* 2002;87:324–33
Mexico (IARC HPV prevalence survey)	Most common HPV types in women with negative cytology. Main determinant of infection with high- and low-risk HPV types: number of sexual partners	r38 Lazcano-Ponce, *Int J Cancer* 2001;91:412–20
Argentina (IARC HPV prevalence survey)	HPV types in general population. Main factors for HPV infection: lifetime number of sexual partners and vaginal discharge. Decreasing trend of HPV infection with age	r39 Matos, *Sex Transm Dis* 2003;30:593–599
Chile (IARC HPV prevalence study)	HPV 16, 56, 31, 38, 39, 18, and 52 accounted for 75% of high-risk HPV infections. Main risk factor for abnormal cytology and HPV infection: lifetime number of sexual partners	r40 Ferreccio, *Cancer Epidemiol Biomarkers Prev* 2004;13:2271-6
Mexico and USA	Prevalence of HPV infection by both hc2 and PCR (low- and high-risk types) [r41] Risk factors for HPV infection by country. Mean viral load by the same risk factors, by different types of HPV infection, and by cytological lesions in the same population [r42]	r41 Giuliano, *Rev. Panam Salud Publica* 2001;9:172–81 r42 Flores, *Int J Cancer* 2006;118:1187–1193
Mexico (Morelos HPV study)	Sensitivity of conventional cytology and HPV testing (hc2) (both of self-collected and clinician-collected samples). Potential HPV testing role in primary screening. Risk factors in the general population	r43 Salmeron, *Cancer Causes Control* 2003;14:505–12
Mexico (Male factor studies)	Overall and type-specific prevalence of male genital HPV infection. Anal intercourse with males was positive associated with HPV acquisition, and being circumcised gave protection against persistent HPV infection [r44] HPV type-specific prevalence. Lifetime number of sexual partners was associated with HPV positivity. Condom use with both regular and sex worker partners and circumcision were protective against HPV prevalence [r45]	r44 Lajous, *Cancer Epidemiol Biomarkers Prev* 2005;14:1710–1716 r45 Vaccarella, *Int J Cancer* 2006;119:1934–1939

(continued)

Table 20.1 (continued)

Countries	Main contributions	References (first author, journal, year, vol:pages)
Nicaragua (Rivas screening project)	HPV type-specific prevalence by cytological grade and their association with CC [r46] Found better performance on VIA over cytology and difficulties of using VIA [r47] HPV type-specific prevalence by histological grade [r48]	r46 Claeys P, *Sex Transm Infect* 2002;78:204–207 r47 Claeys P, *Trop Med Int Health* 2003;8:704–709 r48 Hindryckx P, *Sex Transm Infect* 2006;82:334-336
Uruguay	HPV types 16, 18, and 45 have a very high prevalence in CC [r49] HPV infections increased from 16% in women with normal cytology to 96% in HSIL [r50] Type-specific HPV prevalence in samples from cervical screening [r51]	r49 Berois, *Int J Gynecol Cancer* 2013;23:527–532. r50 Ramas, *J Med Virol* 2013;85:845–851 r51 Berois, *J Med Virol* 2014;86:647–652
Peru (The TATI project)	A sub-analysis compared VIA, conventional cytology, LBC, and HPV testing (hc2). HPV testing had the highest sensitivity of all, and conventional cytology and VIA had low sensitivity but better specificity. Description of risk factors [r52] Effectiveness of cryotherapy treatment for CIN in this setting [r53]	r52 Almonte, *Int J Cancer* 2007;121:796–802 r53 Luciani, *Int J Gynaecol Obstet* 2008;101:172–177
Brazil, Mexico, and the USA (HPV Infection in Men: the HIM Study)	Incidence and clearance of oral HPV infection in men (r54) Broad HPV distribution in the genital region of men(r55) High genital prevalence of cutaneous HPVs on male genital skin (r56) Long-term persistence of oral HPV16 infection in men (r57) Data about the natural history of genital HPV among HIV-negative men having sex with men and men having sex with women (r58)	r54 Kreimer AR, *Lancet*, 2013;382:877–887 r55 Sichero L, *Virology*, 2013; 443:214–2170 r56 Sichero L, *BMC Infect Dis*, 2014; 9:14–16 r57 Pierce Campbell CM, *Cancer Prev Res (Phila)* 2015; 8:190–196 r58 Nyitray AG, *J Infect Dis*, 2015; 212:202–212
Argentina, Brazil, Chile, Colombia, Guatemala, Honduras, Mexico, Paraguay, Peru, and Venezuela (the ICO study)	HPV type-specific distribution in CC. HPV types 16, 18, 31, 33, 35, 45, 52, and 58 should be given priority when the cross-protective effects of current vaccines are assessed and for formulation of recommendations for the use of second-generation polyvalent HPV vaccines (r59) Stability on the HPV type distribution in a large series of CC over 70 years before vaccination (r60)	r59 de Sanjose, *Lancet Oncol* 2010; 11:1048–1056 r60 Alemany, *Int J Cancer* 2014; 135:88–95

Adapted from Almonte et al. [1]
CC invasive cervical cancer, *CIS* carcinoma in situ, *CIN* cervical intraepithelial neoplasia, *IARC* International Agency for Research on Cancer, *ICO* Instituto Catalán de Oncología, *NCI* National Cancer Institute (USA)

transmitted infection; thus, HPV prevalence peaks around the sexual debut age, when exposure is high. Infections become undetectable within 2 years in more than 90% of individuals. Approximately 60% of these infections will prompt type-specific seroconversion, and if cervical samples are collected during productive viral infection, they may be associated with mild cervical abnormalities [i.e., low-grade squamous intraepithelial lesions (LSILs) or cervical intraepithelial neoplasia 1 (CIN1)]. Most of them are "transient" infections (cleared by the immune system) and do not result in clinical complications. Genital warts are other benign and common clinical sequelae in low-risk cases of HPV infection [66]. On the other hand, the hr-HPV infections that "persist" are more likely to progress to true cervical cancer (CC) precursor lesions, that is, high-grade squamous intraepithelial lesions (HSIL) or CIN3; progression to cervical cancer may take several years if left untreated.

It is well known that cervical HPV infection is age dependent; an inverse relationship between age and HPV prevalence has been reported. HPV prevalence peaks below age 25 and declines with age. Using data from the Guanacaste cohort [39] and the TATI project (that only tested for 13 hr-HPVs) [2], the overall HPV prevalence was 26% in women younger than 25, dropping to 12% in those aged 35–44 and climbing again to 22% in those older than 54 (Table 20.1) [20]. This U-shaped age-specific curve of hr-HPV prevalence was previously shown by independent reports in Costa Rica [41], Mexico [49], Chile [34], Brazil (see r29, Table 20.1), and Colombia [57]. In Argentina, the curve peaked below age 25 and then dropped and plateaued around 30 to 35 years, reaching its minimum at 65 years of age or older [55]; this pattern resembles more those of Europe and North America.

CC risk is largely, and almost exclusively, defined by HPV natural history. Among HPV-infected women, the most important determinants of carcinogenic risk are persistence of infection and viral genotype, HPV16 being the most prominently carcinogenic [67].

Although hr-HPV DNA is detected in almost all CC cases, HPV infection alone is not sufficient to drive full carcinogenesis. A substantial part of the evidence of risk factors for HPV infection and progression to cervical cancer comes from studies conducted in LA. The lifetime number of male sexual partners and their sexual behavior are associated with an increased risk of HPV infection. High parity, long-term oral contraceptive use, and smoking are associated with an increased risk of HPV infection progression to CC; the role of chronic inflammation, especially from coinfection with *Chlamydia trachomatis*, and certain dietary deficiencies have also been reported [1, 29].

Immunity is obviously an important risk factor; an effective cell-mediated response to the early proteins is necessary for lesion regression. Host genetics and other influences on host immunity might affect the immune response to HPV infection; weak associations of HLA with risk of CIN3+ have been noted [31, 54]. Coinfection with HIV is important because HIV-induced immunosuppression impairs cell-mediated immune control of HPV infections [1].

3 HPV Prevalence and Type Distribution in Normal Cytology and Cervical Lesions in LA

The genotype distribution in normal cytology and LSIL reveals a wide spectrum of HPV types, both low- and high-risk types; as the severity of the cervical lesion increases, hr-HPVs become the most frequent types, being the only types in CC, with HPV16 and HPV18 accounting for about 70% of cases.

In one of the largest meta-analyses, including 48,171 women with normal cytology from studies in Trinidad and Tobago, Costa Rica, Honduras, Guatemala, Belize, Mexico, Argentina, Brazil, Chile, Colombia, Paraguay, and Peru, the prevalence of HPV (any type) was 16.1%. The vaccine-targeted HPV types (16 and/or 18) were identified in 4.3% of normal samples [15].

In LSIL, the most common viral types identified in samples from the LA region were HPV16 (26%), HPV33 (13%), HPV6 (11%), HPV58 (8%), and HPV31 (7%) [24].

In the regional meta-analyses including 2446 cases of HSIL and 5540 of CC, 46.5% of HSIL cases harbored HPV16 and 8.9% HPV18; in CC, 53.2% of cases harbored HPV16 and 13.2% HPV18, the next five most common types, in decreasing frequency, being HPV31, HPV58, HPV33, HPV45, and HPV52 [23].

The more recent worldwide meta-analysis of cross-sectional HPV-type distribution in HPV-positive women of all types of clinical status (from normal cytology to CC) included 35,895 samples from South and Central America studies, in which genotyping was performed by polymerase chain reaction (PCR)-based methods [36]. Overall HPV prevalence increased with growing severity of cervical disease from 24% in normal cytology (substantially higher than worldwide prevalence estimates) up to 90% in CC. HPV16 was the most frequently detected type in every grade. HPV16 positivity varied slightly across normal cytology (16.1%) and LSIL (25.1%), but increased substantially in HSIL (53.5%), to reach 59.5% in CC.

4 HPV Genetic Variability

Comparative nucleotide sequence analysis of these viruses has elucidated some features of their phylogenetic relationship and pathogenesis implications.

HPV genomes have been classified into *molecular variants* when they present more than 98% similarity to the prototype in the L1 gene sequence [27]. Nevertheless, more recently, the comparison of the complete nucleotide sequence of HPV16 isolates from different phylogenetic branches showed that 4% of the full genome may vary in the eight genes and that 9.9% of amino acid positions are variable [22].

The most extensive worldwide studies concerning HPV intratypic nucleotide heterogeneity by far have been conducted for HPV16 because of its global predominance, followed by HPV18 and HPV45, HPV6 and HPV11, HPV5 and HPV8, and, more recently, HPV58, HPV31, HPV33, HPV35, and HPV52 [17]. Table 20.2 presents a selection of the main studies on HPV variability performed in LA.

Table 20.2 Summary of selected HPV variants studies in Latin America

Country	Viral type and technique	Main findings	References (first author, journal, year, volume: pages)
Argentina	HPV16. Cross-sectional studies in aboriginal women PCR-based hybridization of L1 and E6 genes; sequencing of a LCR fragment	Ninety percent of the specimens from *Quechua* women had E variants. Only about 10% had non-E variants (AA, As, and NA-1); 87% of normal smears had EP, whereas non-EP were detected mainly in SIL and CC. A new variant was identified in the AA branch, with nucleotide substitutions adjacent to or within transcription factor binding sites Cervical samples from *Guarani* women with different clinical categories (normal cytology, LSIL, and HSIL) were identified: 51% EP, 32% E-350G, 9% Af1-a, 4% E-6862C, 3% Af2-a, and 1% AA-a Cervical samples from *Pilaga* women with normal cytology; 68.2% were E and 31.8% AA	Picconi MA, *J Med Virol* 2003; 69:546–552 Tonon SA, *Int J Infect Dis* 2007; 11:76–81 Deluca GD, *Medicina (B Aires)*, 2012; 72:461–466
Brazil (Ludwig-McGill cohort)	HPV16 and HPV18 Cohort study LCR sequencing and dot-blot hybridization of the E6 and L1 genes of HPV16 Sequencing of HPV18 LCR fragment	E variants were the most prevalent and diverse group. The same variant was detected in specimens of different visits. Non-E variants tended to persist more frequently than E variants. Women with non-E variants had higher risk for HSIL compared to those with E variants Persistent infections with HPV-18 associated with E variants; however, risk for simultaneous detection of HSIL and HPV DNA higher for non-E HPV16 variants. Same was observed with HSIL during follow-up, confirming the association between non-E variants and risk of CC	Villa LL, *J Gen Virol* 2000; 81:2959–2968 Sichero L, *Int J Cancer* 2007; 120:1763–1768
Costa Rica (Guanacaste cohort)	HPV16 Nested case-control study. Sequencing of HPV16 LCR HPV types: 31, 33, 35, 52, 58, 67, 73, 18, 39, 45, 70, 68, 51, 69, 53, 56, 66. Sequencing of a fragment of LCR/E6 genes	Non-E variants were seen in 5.8% of normal cytology, 14.3% of HSILs, and 43.7% of cancers. Women with non-E variants had a RR of 2.7 for HSIL and 11 for CC For HPV16, non-E variants were significantly more likely than E variants to cause persistence and CIN3+. HPV35 and HPV51 variant lineages also predicted CIN3+. A continued evolution of HPV types has led to even finer genetic discrimination linked to HPV natural history and cervical cancer risk	Hildesheim A, *J Natl Cancer Inst* 2001; 93:315–318 Schiffman M, *Cancer Res* 2010; 70:3159–3169
Honduras	HPV16. Cross-sectional study PCR–reverse hybridization assay for E6/E7 variants.	Most infections in all clinical groups (CIN I, II, and II and normal cytology) belong to the E6-E variants, suggesting that HPV-16 non-E variants do not represent an additional factor associated with increased occurrence of CC in this population. Mixed variants were detected mostly in normal cytology	Tabora N, *Int J Gynecol Cancer*, 2010; 20:323–328

(continued)

Table 20.2 (continued)

Country	Viral type and technique	Main findings	References (first author, journal, year, volume: pages)
Mexico	HPV16. Case-control study Sequencing of E6 and L1 regions HPV18. Site-directed mutagenesis and immunofluorescence analysis HPV16. Cross-sectional study Sequencing of E6 region	AA and E variants were found in 23.2% and 27.1% of cases and in 1.1% and 10% of controls. The frequency of AA variants was 21 times higher in CC than in controls, being the OR for CC associated with AA variants significantly higher than that of E variants Sequence variations in E1 and LCR affect ori function, particularly for Af and AA HPV18 variants Most of variants (82.12%) belonged to the E lineage, 17.58% to AA1, and 0.3% to Af2 sublineages, and eight new E6 variants were identified. AA-a variants showed the greatest risk of leading to the development of CC was AA-a, followed by E-A176/G350, AA-c, E-G350, and E-C188/G350	Berumen J, *J Natl Cancer Inst* 2001; 93:1325–1330 Amador-Molina A, *J Gen Virol* 2013; 94:393–402 Ortiz-Ortiz J, *Virol J* 2015; 22(12):29
Paraguay	HPV16 Cross-sectional study LCR sequencing	Most HPV-16 variants belonged to the E branch (82%) in all clinical groups (normal cytology, LSIL, HSIL, and CC). Two new E variants were characterized. Non-E variants, such as Af1 (1.5%) and AA (16.5%), were detected only among women with cervical lesions	Mendoza LP, *Int J Gynaecol Obstet* 2013; 122:44–47
Worldwide study[a]	HPV16 PCR-based hybridization of E6/L1	In CC, E variants were found in 76.4% of the samples, AA in 19.7%, Af1 in 2.2%, Af2 in 1.3%, and NA1 in 0.43%	Yamada, *J Virol* 1997; 71:2463–72
Worldwide IARC study[b]	HPV16 Complete sequencing of E6 /LCR	Two new sublineages within each of the lineages Af1 and Af2 were characterized. Improved classification system for HPV16 genomes based on phylogenetically distinguishing positions in E6 and the LCR, that distinguish nine HPV16 variant sublineages	Cornet I., *J Virol* 2012; 86:6855–6861
Worldwide IARC study[b]	HPV18 Complete sequencing of E6/LCR	No significant differences in the distribution of HPV18 variant lineages between CC and controls. Findings do not support the role of HPV18 (sub)lineages for discriminating cancer risk or explaining why HPV18 is more strongly linked with adenocarcinoma than squamous carcinoma	Chen AA, *J Virol*, 2015; 89:10680–10687
Worldwide study[c]	HPV 58 Sequencing of concatenated E6-E7-E2-E5-L1-LCR fragments	HPV-58 can be classified into four lineages that show some degree of ethnogeographic predilection in distribution. Lineage A was the most prevalent lineage across all regions	Chan PKS, *Int J Cancer* 2013; 132:2528–2536 Chan PKS, *J Infect Dis* 2011; 203:1565–1573

Phylogenetic branches: *E* European, *EP* European prototype, *AA* Asian-American, *As* Asian, *Af1* African 1, *Af2* African 2, *NA1* North American

Some authors grouped the variants in European (E) (including E branch) and non-European (non-E) (including AA, As, Af1, and Af2 branches)

CC cervical cancer, *LSIL* low-grade squamous intraepithelial lesions, *HSIL* high-grade squamous intraepithelial lesions

[a]Included samples from Argentina, Bolivia, Brazil, Chile, Colombia, Cuba, Panama, and Paraguay

[b]Included samples from Argentina Bolivia, Brazil, Chile, Colombia, Cuba, Panama, Paraguay, and Peru

[c]Included samples from Argentina, Brazil, Honduras, and Mexico

Investigations of HPV type diversity have identified different phylogenetic branches (variants); particularly for HPV16, there are six branches: European (E), Asian (As), African-1 (Af-1), African-2 (Af-2), Asian-American (AA), and North American (NA) [17]. Different HPV16 variants exhibit differences in their biological and biochemical properties.

The prevalence of HPV variants and their association with cervical cancer has been reported in three case-control studies [9, 42, 69] and five cross-sectional studies in LA women with different grades of cervical lesions [18, 47, 56, 62, 78]. Follow-up studies have reported the role of HPV variants in the persistence of infection and disease progression [69, 74]. Overall, studies conducted in Argentina, Brazil, Costa Rica, Honduras, Mexico, and Paraguay have shown a large diversity of variants, with a higher frequency of E variants compared to other phylogenetic branches (Table 20.2). Interestingly, a high prevalence (>80%) of E variants was also observed in indigenous groups from Argentina [28, 62, 71]. These studies also suggested that the colonization of the American continent by Europeans and Africans is reflected in the composition of its variants.

Studies carried out in Mexico, Costa Rica, and Brazil have shown that non-E HPV16 variants, mostly AA variants, are associated with a higher risk of viral persistence and/or HSIL and CC [9, 18, 42, 68, 69, 74]. These studies with a large number of samples provide enough study power to detect associations between low-prevalence variants and persistent infections or disease risk [1].

In vitro functional assays show that several HPV variants differ in their ability to induce p53 degradation, Bax degradation, activation of mitogen-activated protein kinase (MAPK) signaling, E2-related transcription, and immortalization activity. Specific studies with non-E variants have shown enhanced transcription and replication efficiency in HPV16 and HPV18 AA variants compared to E variants [17]. This information may explain the increased oncogenic potential reported for these variants and their contribution for the high incidence of CC.

HPV vaccines are based on virus-like particles (VLPs) composed of L1 protein, the viral capsid main component. So far, serological studies of different HPV16 variants have shown that the humoral immune response to HPV16 does not seem to discriminate between different molecular variants [60]. The cross-protection between variants was confirmed by the near 100% prophylactic efficacy of vaccines in multicenter studies [1].

Although the coevolution of human populations and HPV16 and HPV18 variants is well supported, the geographic association for variants of other types remains unresolved. Global studies of HPV variant lineages from worldwide populations are needed to better understand the relationship between HPV and the recent and past evolution and dispersion of their human hosts, as well as the genetic basis of the pathogenesis of specific HPVs, viral–host interactions, and host evolution, among other applications and scientific inquiries. Multicenter studies and/or meta-analyses will be useful to validate the nucleotide level of pathogenesis and provide insights into the molecular basis of HPV-associated disease [17].

5 Epidemiology of HPV-Related Neoplasias

About 1.1 million new cancer cases and 600,000 cancer deaths per year are estimated in Latin America and the Caribbean [32]. Estimates indicate a 72% increase in the incidence of cancer and 78% increase in mortality between 2012 and 2030 [26]. Cancer rates vary considerably within LA: although breast cancer remains the leading cause of death for women worldwide, CC is the main cause of death from cancer in Bolivia, Honduras, and Nicaragua. Also, cervical cancer incidence rates vary considerably in the region, ranging from 11.4 cases per 100,000 in Costa Rica to 47.7 cases per 100,000 in Bolivia [59].

Decades of Papanicolaou-based screening to detect precancerous cervical lesions have not had a major impact in reducing CC incidence and mortality rates, which are still high in the region (Fig. 20.1). Despite efforts to reorganize screening programs in a few countries of the region, only a slight reduction in cervical cancer mortality has been noted [59]. Among the difficulties to control the burden of cancer in the region are the uneven allocation of resources, variable infrastructure and ser-

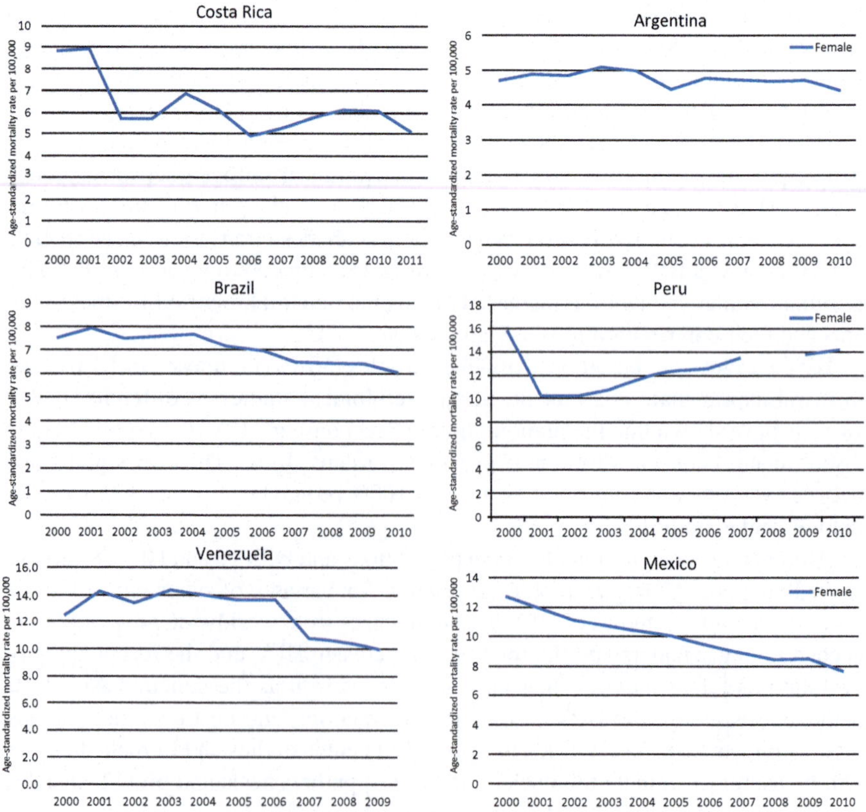

Fig. 20.1 Cervical cancer mortality rates for selected countries in Latin America. (Adapted from Pan American Health Organization [59])

vice availability, limited number of population-based cancer registries, and scarce distribution of public health posts, which is more evident in rural areas, distant from the large urban centers. These difficulties result in a scenario of disproportionate care provided to individuals affected by cancer.

The global burden of HPV infections and related diseases is significant [35]. HPV was associated with 83,195 new cases of cervical cancer and 35,673 associated CC deaths in LA in 2012 [14]. Most of the cases are associated with HPV16 and HPV18, followed by five additional hr types (HPVs 31, 33, 45, 52, 58), which together account for about 90% of CC cases worldwide.

Information concerning HPV-related tumors outside the uterine cervix in LA countries is scarce [14, 32]. A recent systematic review of the presence of HPV in noncervical sites suggests a high HPV prevalence and higher clearance rates than in the uterine cervix [70]. Anal cancer incidence rates vary from as low as 0.2 × 100,000/year to 1.4 × 100,000/year in the northeast of Brazil and some areas of Argentina [26]. Estimates for other LA are limited or nonexistent. Similarly to high-income countries, anal cancer incidence is increasing with time in both women and men. This neoplasia is associated with hr-HPV types, particularly HPV16. In fact, most HPV-positive neoplasias outside the cervix are related to HPV16 [70].

Vulvar and vaginal cancers are relatively rare tumors with incidence rates less than 1 × 100,000/year [59]. Information is very limited in LA. Regional data show that HPV16 is the most prevalent type and is found in 75% to 100% of the basaloid/warty vulvar cancers that are more common in young women. About two thirds of vaginal cancers are linked to HPV, in particular HPV16 [14].

In some LA countries, the incidence of penile cancer is significantly higher than in more developed parts of the world: the central region of Brazil and some areas in Colombia and Paraguay account for about 2.0 × 100,000/year as compared to other countries with incidence rates around 0.4 × 100,000/year [26, 73]. Studies performed in LA show HPV DNA positivity in 30% and 50% of penile cancers [10, 70].

In the head and neck anatomical sites, some cancers are linked to HPV, although in variable frequencies, being more HPV associated in the base of the tongue and tonsils [19]. The most common HPV type found is HPV16 worldwide and in series of cases from LA [19, 46, 64, 70]. Notwithstanding, there have been reports of lower HPV positivity in oropharyngeal cancers from LA countries as compared to other countries in the Northern Hemisphere [37, 53, 65]. Further studies are warranted to better understand the basis for such differences and the impact on cancer patient management.

6 Control of HPV Infections and Related Diseases

6.1 Primary Prevention: HPV Vaccines

Since 2006, two vaccines composed of HPV L1 proteins self-assembled into virus-like particles (VLP) have been approved in LA: one containing VLPs of HPV types 6, 11, 16, and 18 (Merck & Co.) and one composed of HPV 16 and 18 VLPs (GlaxoSmithKline). Large phase II and III clinical trials to assess prophylactic efficacy have been conducted in which both HPV infection and cervical disease endpoints were evaluated, particularly HSIL (CIN2 or CIN3) as well as vulvar and vaginal intraepithelial neoplasias (VIN and VaIN, respectively) and genital warts for the quadrivalent vaccine [75]. Very high efficacy rates were noted in different populations that included young women between 16 and 26 years and older (up to 55 years). The quadrivalent HPV vaccine has also proven to be efficacious in men to prevent genital and anal infection and disease caused by the types included in the vaccine [13, 40]. Importantly, several clinical trials of HPV prophylactic vaccines conducted in LA clearly demonstrated the safety, immunogenicity, and efficacy of such recombinant vaccines among Latin Americans [61]. Furthermore, data collected in these clinical trials concerning the incidence and prevalence of genital HPV-associated infection and disease have provided important insights on the burden of genital HPV in the region [40, 61]. Moreover, seminal demonstration studies and surveys have shown that the HPV vaccine acceptability is very high in the region [3, 6, 50].

Most LA countries have a well-developed public immunization infrastructure including adolescent vaccination, which has facilitated the introduction of national immunization programs in the region (Fig. 20.2). Organizations such as the United Nations Children's Emergency Fund (UNICEF), Global Alliance for Vaccines and Immunization (GAVI), and the PAHO revolving fund have enhanced HPV vaccine introduction in LA. After an initial phase, most countries are adopting the two-dose program supported by WHO [77] and are vaccinating girls aged 9 to 13 years at 0 and 6 months. Moreover, in several countries of the region, the HPV vaccine is offered to HIV-positive women up to 26 years of age.

Interestingly, a broad evaluation of the programs is driving a revision of the entire CC control strategies adopted by each country, which includes HPV vaccination for female adolescents and cytology/HPV testing for adult women [72]. Implementation of effective vaccine programs might seem straightforward and obvious in light of the vaccine efficacy and lack of serious adverse events to date; nonetheless, significant challenges remain. These problems include the cost of the program, covering two doses of vaccine and extending the vaccination to boys and other populations at risk including HIV-positive individuals. Equally important is to monitor the impact of this intervention that requires tools and strategies unavailable in many countries of the region. Introduction of cost-effective measures such as HPV vaccination only or vaccination supplemented with screening, with good coverage rates, will reduce HPV-related tumors in LA, as is happening in several countries of the world [13].

6.2 *Secondary Prevention: HPV Testing as Primary Screening*

For more than 50 years, cervical cytology [the Papanicolaou (Pap) smear) has been the standard of care for CC screening. Cytology-based mass screening programs have been successful in reducing incidence and mortality in developed countries (such as the U.S. and European countries). Unfortunately, most LA countries tried unsuccessfully to replicate these results, evidencing, however, after decades of efforts, high incidence and mortality rates, with little impact on the disease burden [58].

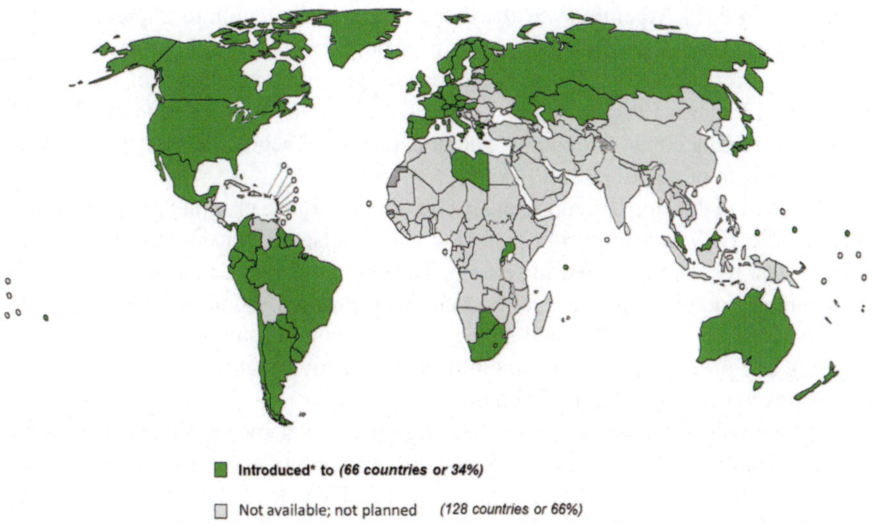

Fig. 20.2 Countries with prophylactic HPV vaccine in their immunization programs, 2016. Countries with partial introduction are not included. (Adapted from a WHO Immunization, Vaccines and Biologicals image 61, using data from the WHO Immunization, Vaccines and Biologicals database (accessed May 2016))

The limitations inherent in cervical cytology prompted the development of new screening technologies: tests to detect the presence of hr-HPV DNA, which should be clinically validated for this purpose. HPV testing offers numerous potential advantages compared to cytology-based screening, such as greater sensitivity, high negative predictive value (which allows to extend the screening intervals in HPV-negative women), and automation [21]. However, even among women over 30 years of age, the cancer to transient infection ratio is low, and HPV assays must overcome the intrinsic problem of low positive predictive value. This lower specificity of HPV testing requires an additional test ("triage") in women who are HPV DNA positive in the primary screening to identify those who are at risk of having a cervical cancer precursor and to reassure those who only have transient or low-risk infections. Triage includes visual inspection methods, cytology, and molecular biomarkers (high-risk HPV E6/E7 mRNA, high-risk HPV E6 proteins, p16, among others). Locally adapted algorithms employing primary screening with HPV testing are being developed in different settings [63]. The initial HPV tests were very expensive and unaffordable for several LA countries, but in recent years, more HPV tests became available and the prices have started to drop, making them more affordable.

During the past decade, there have been multiple experiences with HPV testing in LA, some as part of research studies and others to pilot the implementation of HPV tests in the public system and, more recently, the implementation of HPV testing as part of the public programs provided by the ministries of health [45]. Pilot studies that took place in Argentina [4, 5], Chile [33, 51], Colombia [58], El Salvador [25], Mexico [38, 48], and Nicaragua [7, 44] were highly efficacious to detect precancerous cervical lesions and good feasibility and acceptance of self-sampling.

Similarly, in 2011 Argentina was the first country in the region to implement HPV DNA testing for primary screening within its public system for all women aged 30 or older. In recent years, Mexico has expanded the implementation of HPV DNA testing to 17 sites across the country, applying its extensive knowledge in this field. El Salvador, Guatemala, Honduras, and Nicaragua are beginning to institutionalize HPV testing at population level [45].

The need to develop a comprehensive quality assurance program associated with the specific HPV test to be implemented should also be considered to guarantee reliable test results in real-world settings. Despite the fact that most tests have their own internal quality control, quality control procedures should be put in place to ensure proper transportation and storage of reagents and samples, correct sample labeling and processing, suitable monitoring of positivity rates, and other test characteristics to rule out contamination [45, 63].

LA is slowly shifting toward HPV testing for cervical cancer screening, with the endorsement of several regional experiences that have resulted in increased coverage and better detection of pre-cancer lesions using HPV tests. In line with this, the ESTAMPA study, recently launched in LA countries by the International Agency for Cancer Research, will contribute valuable information about the performance of emerging CC screening and triage techniques and the feasibility of different approaches to implement organized HPV-based screening programs in the region [43].

Finally, it is important to emphasize that the screening test is important, but it is only one component of many other aspects of population-based programs that should be implemented to effectively impact CC cancer incidence and mortality.

7 Conclusions and Perspectives

The prevalence and incidence of HPV-related infection and disease in LA underscore the importance of supporting CC prevention strategies in the region. CC is one of the leading killers among women in LA, a region where many countries have not been successful in implementing population-level cytology-based screening programs. Hence, a more comprehensive CC control approach is required, wherein primary and secondary prevention strategies are implemented with both high coverage and sustainability.

Although regional data seem to indicate a favorable trend in prevention, significant challenges still remain. In primary prevention these include the cost of the program, covering two doses of vaccine, and extending the vaccination to boys and other populations at risk, including HIV-positive individuals. Equally important is to monitor the impact of this intervention that requires tools and strategies unavailable in many countries of the region.

In secondary prevention, it is crucial to change the paradigm by implementing HPV testing as primary screening in the most appropriate way. Among several challenges for its implementation, it should take into account the need to update screening guidelines, strengthen treatment capacity, and develop a comprehensive quality assurance plan for HPV testing.

Finally, gaps still exist in the knowledge and the future lines of research, policy, and advocacy for noncervical HPV cancer prevention, mainly anal and oropharyngeal cancers and precancers; further studies are warranted to better understand their pathogenesis and the impact on cancer patient management. Public health commitment and research to implement HPV-based preventive strategies, together with stronger and common advocacy to counter barriers affecting the adoption of these strategies, are likely to yield major benefits in reducing the burden of HPV-associated diseases in LA.

Acknowledgments We are grateful to the contribution of hundreds of investigators, physicians, and students who diligently dedicated their work to generate information about HPV infections and related diseases in the Latin American region. As the amount of quality information available is vast, apologies are extended to the numerous authors whose work is not mentioned.

We wish to dedicate this chapter to the memories of Dr. Angélica R. Teyssié (1922–2015), a prestigious Argentine virologist, a pioneer in the research of viral oncogenesis, particularly HPV, and in the training human resources, and to Dr. Xavier Castellsagué (1959–2016), one of the most productive and influential epidemiologists in the field of HPV-related neoplasias.

References

1. Almonte M, Albero G, Molano M et al (2008) Risk factors for human papillomavirus exposure and co-factors for cervical cancer in Latin America and the Caribbean. Vaccine 26(suppl 11):L16–L36
2. Almonte M, Ferreccio C, Winkler JL et al (2007) Cervical screening by visual inspection, HPV testing, liquid-based and conventional cytology in Amazonian Peru. Int J Cancer 121:796–802
3. Arrossi S, Maceira V, Paolino M et al (2012) Acceptability and uptake of HPV vaccine in Argentina before its inclusion in the immunization program: a population-based survey. Vaccine 30:2467–2474
4. Arrossi S, Thouyaret L, Herrero R et al (2015) Effect of self-collection of HPV DNA offered by community health workers at home visits on uptake of screening for cervical cancer (the EMA study): a population-based cluster-randomised trial. Lancet Global Health 3(2):e85–e94
5. Arrossi S, Thouyaret L, Laudi R et al (2015) Implementation of HPV-testing for cervical cancer screening in programmatic contexts: the Jujuy demonstration project in Argentina. Int J Cancer 137:1709–1718
6. Baker ML, Figueiroa-Downing D, Chiang ED et al (2015) Paving pathways: Brazil's implementation of a national human papillomavirus immunization campaign. Rev Panam Salud Publica 38:163–166
7. Bansil P, Wittet S, Lim JL et al (2014) Acceptability of self-collection sampling for HPV-DNA testing in low-resource settings: a mixed methods approach. BMC Public Health 14:596
8. Bernard HU, Burk RD, Chen Z et al (2010) Classification of papillomaviruses (PVs) based on 189 PV types and proposal of taxonomic amendments. Virology 401:70–79
9. Berumen J, Ordonez RM, Lazcano E et al (2001) Asian-American variants of human papillomavirus 16 and risk for cervical cancer: a case-control study. J Natl Cancer Inst 93:1325–1330
10. Bezerra AL, Lopes A, Landman G et al (2001) Clinicopathologic features and human papillomavirus DNA prevalence of warty and squamous cell carcinoma of the penis. Am J Surg Pathol 25:673–678

11. Bosch FX, Burchell AN, Schiffman M et al (2008) Epidemiology and natural history of human papillomavirus infections and type-specific implications in cervical neoplasia. Vaccine 26(suppl 10):K1–K16
12. Bouvard V, Baan R, Straif K et al (2009) A review of human carcinogens. Part B: Biological agents. Lancet Oncol 10:321–322
13. Brotherton JM, Ogilvie GS (2015) Current status of human papillomavirus vaccination. Curr Opin Oncol 27:399–404
14. Bruni L, Barrionuevo-Rosas L, Albero G et al. (2015) Human papillomavirus and related diseases in Americas. Summary report. In: ICO Information Centre on HPV and Cancer (HPV Information Centre). Available via DIALOG. http://www.who.int/hpvcentre of subordinate document. Accessed 5 June 2016
15. Bruni L, Diaz M, Castellsagué X et al (2010) Cervical human papillomavirus prevalence in 5 continents: meta-analysis of 1 million women with normal cytological findings. J Infect Dis 202:1789–1799
16. Burchell AN, Winer RL, de Sanjosé S et al (2006) Chapter 6: epidemiology and transmission dynamics of genital HPV infection. Vaccine 24(suppl 3):S3/52–S3/61
17. Burk RD, Harari A, Chen Z (2013) Human papillomavirus genome variants. Virology 445:232–243
18. Calleja-Macias IE, Kalantari M, Huh J et al (2004) Genomic diversity of human papillomavirus-16, 18, 31 and 35 isolates in a Mexican population and relationship to European, African, and Native American variants. Virology 319:315–323
19. Castellsagué X, Alemany L, Quer M, ICO International HPV in Head and Neck Cancer Study Group et al (2016) HPV involvement in head and neck cancers: comprehensive assessment of biomarkers in 3680 patients. J Natl Cancer Inst 108(6):djv403
20. Castellsagué X, de Sanjosé S, Aguado T, Louie KS, Bruni L, Muñoz J, Diaz M, Irwin K, Gacic M, Beauvais O, Albero G, Ferrer E, Byrne S, Bosch FX (2007) HPV and cervical cancer in the world. 2007 Report. WHO/ICO information centre on HPV and cervical cancer (HPV Information Centre). Available at: www.who.int/hpvcentre
21. Castle PE, de Sanjosé S, Qiao YL et al (2012) Introduction of human papillomavirus DNA screening in the world: 15 years of experience. Vaccine 30(suppl 5):F117–F122
22. Chen Z, Terai M, Fu L et al (2005) Diversifying selection in human papillomavirus type 16 lineages based on complete genome analyses. J Virol 79:7014–7023
23. Ciapponi A, Bardach A, Glujovsky D et al (2011) Type-specific HPV prevalence in cervical cancer and high-grade lesions in Latin America and the Caribbean: systematic review and meta-analysis. PLoS One 6(10):e25493. doi:10.1371/journal.pone.0025493
24. Clifford GM, Rana RK, Franceschi S et al (2005) Human papillomavirus genotype distribution in low-grade cervical lesions: comparison by geographic region and with cervical cancer. Cancer Epidemiol Biomarkers Prev 14:1157–1164
25. Cremer ML, Maza M, Alfaro KM et al (2016) Introducing a high-risk HPV DNA test into a public sector screening program in El Salvador. J Low Genit Tract Dis 20:145–150
26. Curado MP, Souza DLB (2014) Cancer burden in Latin America and the Caribbean. Ann Global Health 80:370–377
27. de Villiers EM, Fauquet C, Broker TR et al (2004) Classification of papillomaviruses. Virology 324:17–27
28. Deluca GD, Basiletti J, González JV et al (2012) Human papilloma virus risk factors for infection and genotype distribution in aboriginal women from Northern Argentina. Medicina (B Aires) 72:461–466
29. Deluca GD, Basiletti J, Schelover E et al (2011) *Chlamydia trachomatis* as a probable cofactor in human papillomavirus infection in aboriginal women from northeastern Argentina. Braz J Infect Dis 15:567–572
30. Doorbar J (2007) Papillomavirus life cycle organization and biomarker selection. Dis Markers 23:297–313

31. Eiguchi K, Tatti S, Alonio LV et al (2008) Association of DRB1 and DQB1 HLA class II polymorphisms in high-grade and neoplastic cervical lesions of women from Argentina. J Low Genit Tract Dis 12:262–268
32. Ferlay J, Soerjomataram I, Ervik M et al. (2013) GLOBOCAN 2012 v1.2, Cancer incidence and mortality worldwide. In: IARC Cancer Base No. 11 [Internet]. Lyon: International Agency for Research on Cancer. Available from: http://globocan.iarc.fr. Accessed 15 June 2016
33. Ferreccio C, Barriga MI, Lagos M et al (2013) Screening trial of human papillomavirus for early detection of cervical cancer in Santiago, Chile. Int J Cancer 132:916–923. Erratum in: Int J Cancer 133:E1
34. Ferreccio C, Prado RB, Luzoro AV et al (2004) Population-based prevalence and age distribution of human papillomavirus among women in Santiago Chile. Cancer Epidemiol Biomarkers Prev 13:2271–2276
35. Forman D, de Martel C, Lacey CJ et al (2012) Global burden of human papillomavirus and related diseases. Vaccine 30(suppl 5):F12–F23
36. Guan P, Howell-Jones R, Li N et al (2012) Human papillomavirus types in 115,789 HPV-positive women: a meta-analysis from cervical infection to cancer. Int J Cancer 131:2349–2359
37. Heck JE, Berthiller J, Vaccarella S et al (2010) Sexual behaviours and the risk of head and neck cancers: a pooled analysis in the International Head and Neck Cancer Epidemiology (INHANCE) consortium. Int J Epidemiol 39:166–181
38. Hernandez-Marquez CI, Salinas-Urbina AA, Cruz-Valdez A et al (2014) Conocimientos sobre virus del papiloma humano (VPH) y aceptación de auto-toma vaginal en mujeres mexicana. Rev Salud Pública 16:697–708
39. Herrero R, Castle PE, Schiffman M et al (2005) Epidemiologic profile of type-specific human papillomavirus infection and cervical neoplasia in Guanacaste. J Infect Dis 191:1796–1807
40. Herrero R, González P, Markowitz LE (2015) Present status of human papillomavirus vaccine development and implementation. Lancet 16:e206–e216
41. Herrero R, Hildesheim A, Bratti C et al (2000) Population-based study of human papillomavirus infection and cervical neoplasia in rural Costa Rica. J Natl Cancer Inst 92:464–474
42. Hildesheim A, Schiffman M, Bromley C et al (2001) Human papillomavirus type 16 variants and risk of cervical cancer. J Natl Cancer Inst 93:315–318
43. International Agency for Research on Cancer. Latin American multicentre study of cervical cancer screening and triage with HPV testing (ESTAMPA). Available at: http://www.iarc.fr/en/research-groups/PRI/current-topics.php. Accessed 07 July 16
44. Jeronimo J, Bansil P, Lim J et al (2014) A multicountry evaluation of care HPV testing, visual inspection with acetic acid, and papanicolaou testing for the detection of cervical cancer. Int J Gynecol Cancer 24:576–585
45. Jeronimo J, Holme F, Slavkovsky R et al (2016) Implementation of HPV testing in Latin America. J Clin Virol 76(suppl 1):S69–S73
46. Kaminagakura E, Villa LL, Andreoli MA et al (2012) High-risk human papillomavirus in oral squamous cell carcinoma of young patients. Int J Cancer 130:1726–1732
47. Khouadri S, Villa LL, Gagnon S et al (2006) Human papillomavirus type 33 polymorphisms and high-grade squamous intraepithelial lesions of the uterine cervix. J Infect Dis 194:886–894
48. Lazcano-Ponce E, Lorincz AT, Torres L et al (2014) Specimen self-collection and HPV DNA screening in a pilot study of 100,242 women. Int J Cancer 135:109–116
49. Lazcano-Ponce E, Herrero R, Munoz N et al (2001) Epidemiology of HPV infection among Mexican women with normal cervical cytology. Int J Cancer 91:412–420
50. Lee FH, Paz-Soldan VA, Carcamo C et al (2010) Knowledge and attitudes of adult peruvian women vis-a-vis human papillomavirus (HPV), cervical cancer, and the HPV vaccine. J Low Genit Tract Dis 14:113–117
51. Léniz J, Barriga MI, Lagos M et al (2013) HPV vaginal self-sampling among women non-adherent to Papanicolaou screening in Chile. Salud Publica Mex 55:162–169

52. Li N, Franceschi S, Howell-Jones R et al (2011) Human papillomavirus type distribution in 30,848 invasive cervical cancers worldwide: variation by geographical region, histological type and year of publication. Int J Cancer 128:927–935
53. López RV, Levi JE, Eluf-Neto J et al (2014) Human papillomavirus (HPV) 16 and the prognosis of head and neck cancer in a geographical region with a low prevalence of HPV infection. Cancer Causes Control 25:461–471
54. Maciag PC, Schlecht NF, Souza PS et al (2002) Polymorphisms of the human leukocyte antigen DRB1 and DQB1 genes and the natural history of human papillomavirus infection. J Infect Dis 186:164–172
55. Matos E, Loria D, Amestoy GM et al (2003) Prevalence of human papillomavirus infection among women in Concordia, Argentina: a population-based study. Sex Transm Dis 30:593–599
56. Mendoza L, Picconi MA, Mirazo S et al (2013) Distribution of HPV-16 variants among isolates from Paraguayan women with different grades of cervical lesion. Int J Gynaecol Obstet 122:44–47
57. Molano M, Posso H, Weiderpass E et al (2002) Prevalence and determinants of HPV infection among Colombian women with normal cytology. Br J Cancer 87:324–333
58. Murillo R, Almonte M, Pereira A et al (2008) Cervical cancer screening programs in Latin America and the Caribbean. Vaccine Suppl 11:L37–L48
59. Pan American Health Organization (2013) Cancer in the Americas: country profiles 2013. Pan American Health Organization, Washington, DC
60. Pastrana DV, Vass WC, Lowy DR et al (2001) NHPV16 VLP vaccine induces human antibodies that neutralize divergent variants of HPV16. Virology 279:361–369
61. Perez G, Lazcano-Ponce E, Hernandez-Avila M et al (2008) Safety, immunogenicity, and efficacy of quadrivalent human papillomavirus (types 6, 11, 16, 18) L1 virus-like-particle vaccine in Latin American women. Int J Cancer 122:1311–1318
62. Picconi MA, Alonio LV, Sichero L et al (2003) Human papillomavirus type-16 variants in Quechua aboriginals from Argentina. J Med Virol 69:546–552
63. Picconi MA (2013) Human papillomavirus detection in cervical cancer prevention. Medicina (B Aires) 73:585–596
64. Quintero K, Giraldo GA, Uribe ML et al (2013) Human papillomavirus types in cases of squamous cell carcinoma of head and neck in Colombia. Braz J Otorhinolaryngol 79:375–381
65. Ribeiro KB, Levi JE, Pawlita M et al (2011) Low human papillomavirus prevalence in head and neck cancer: results from two large case-control studies in high-incidence regions. Int J Epidemiol 40:489–502
66. Rodríguez AC, Schiffman M, Herrero R et al (2008) Rapid clearance of human papillomavirus and implications for clinical focus on persistent infections. J Natl Cancer Inst 100:513–517
67. Schiffman M, Castle PE, Jeronimo J et al (2007) Human papillomavirus and cervical cancer. Lancet 370:890–907
68. Schiffman M, Rodriguez AC, Chen Z et al (2010) A population-based prospective study of carcinogenic human papillomavirus variant lineages, viral persistence, and cervical neoplasia. Cancer Res 70:3159–3169
69. Sichero L, Ferreira S, Trottier H et al (2007) High grade cervical lesions are caused preferentially by non-European variants of HPVs 16 and 18. Int J Cancer 120:1763–1768
70. Taylor S, Bunge E, Bakker M et al (2016) The incidence, clearance and persistence of non-cervical human papillomavirus infections: a systematic review of the literature. BMC Infect Dis 16:293–314
71. Tonon SA, Basiletti J, Badano I et al (2007) Human papillomavirus type 16 molecular variants in Guarani Indian women from Misiones, Argentina. Int J Infect Dis 11:76–81
72. Tota JE, Ramana-Kumar AV, El-Khatib Z et al (2014) The road ahead for cervical cancer prevention and control. Curr Oncol 21:e255–e264
73. Velazquez EF, Cubilla AL (2007) Penile squamous cell carcinoma: anatomic, pathologic and viral studies in Paraguay (1993–2007). Anal Quant Cytol Histol 29:185–198
74. Villa LL, Sichero L, Rahal P et al (2000) Molecular variants of human papillomavirus types 16 and 18 preferentially associated with cervical neoplasia. J Gen Virol 81(Pt 12):2959–2968

75. Villa LL (2011) HPV prophylactic vaccination: the first years and what to expect from now. Cancer Lett 305:106–112
76. Wentzensen N, Vinokurova S, Von KnebelDoeberitz M (2004) Systematic review of genomic integration sites of human papillomavirus genomes in epithelial dysplasia and invasive cancer of the female lower genital tract. Cancer Res 64:3878–3884
77. WHO (2014) Human papillomavirus vaccines: WHO position paper. Wkly Epidemiol Rep 89:465–492
78. Yamada T, Manos MM, Peto J et al (1997) Human papillomavirus type 16 sequence variation in cervical cancers: a worldwide perspective. J Virol 71:2463–2472
79. zur Hausen H (2002) Papillomaviruses and cancer: from basic studies to clinical application. Nat Rev Cancer 2:342–350

Part VI
Prevention and Treatment

Chapter 21
Rotavirus Vaccines: A Review of the Work, Progress, and Contributions Made in Latin America

Irene Pérez-Schael and Alexandre C. Linhares

1 Introduction

Rotavirus (RV) diarrhea is a public health problem, particularly in very poor countries. By the age of 5 years, almost every child around the world is infected by RV because this virus cannot be controlled by hygiene and sanitation conditions. In Latin America, RV causes 6,302 deaths and 229,656 hospitalizations every year [33].

Two World Health Organization (WHO) prequalified oral RV vaccines, Rotarix® and RotaTeq®, have been licensed worldwide. By now, more than 80 countries have introduced RV vaccines into their National Immunization Programs (NIPs), worldwide, of which 19 are in Latin America [103].

RV vaccine represents an example of changing paradigms in the process of development, licensure, and introduction into immunization programs. The first commercialized RV vaccine (Rotashield) was withdrawn from the market 9 months after its approval by the Food and Drug Administration (FDA) in the U.S. because of an association with intussusception. This severe and rare pathology is now recognized to be associated at a very low level to RV infection and new RV vaccines [84, 102]. This was terrible news for the scientific community, which had to develop a new vaccine promptly to recover lost time.

In consequence, by 2000, a global agenda was implemented with the participation of both public and private sectors, for the development of new RV vaccines. Clinical trials for vaccine evaluation were simultaneously performed in developed

I. Pérez-Schael, MSc (✉)
Instituto de Biomedicina, Enteric Disease Section, Universidad Central de Venezuela, Ministerio de Sanidad, Caracas, Venezuela
e-mail: ireneperezschael@gmail.com

A.C. Linhares, MD, PhD (✉)
Instituto Evandro Chagas, Virology Section, Health Surveillance Secretariat, Brazilian Ministry of Health, Belém, Pará, Brazil
e-mail: alexandrelinhares@iec.pa.gov.br

J.E. Ludert et al. (eds.), *Human Virology in Latin America*,
DOI 10.1007/978-3-319-54567-7_21

and developing countries, and a process to warrant the rapid introduction of RV vaccines into developing nations was considered.

In 2006, two new RV vaccines were ready for marketing. During this process, Latin America had a very significant role, as well as continuing to be critical in post-licensure studies [26, 92].

However, more than 94 million children in the world do not yet have access to RV vaccination [103], a situation that need to be improved to make the RV vaccine of imminent issue.

2 Rotavirus Vaccines: Latin America Participation

Although this section covers all RV vaccines, a special emphasis is given to the participation of Latin America (Table 21.1).

In 1979, the magnitude of rotavirus as a cause of diarrhea was established by WHO; hence, the development of a vaccine was justified and necessary. Because RV replicates in the intestine and being that local immunity is the main immune response for providing protection, particularly IgA, it was thought that the vaccine should be oral. Nevertheless, recently when more data have become available, serum secretory SIgA was suggested as a possible surrogate marker of vaccine protection [1, 38, 51].

2.1 Brief History of the First-Generation Rotavirus Vaccines

The first RV vaccine candidates were strains of animal origin, chosen as the base of vaccine development, because they embrace three important issues: (i) animal strains could be grown in cell culture, (ii) animal and human strains share antigens inducing immune response, and (iii) animal strains are usually naturally attenuated for humans and do not cause diarrhea in humans [57].

Three strains were available at that moment for being selected as vaccine candidates: RIT 4237 (bovine), WC3 (bovine), and MMU 186006 (RRV rhesus strain). This strategy was named the "Jennerian approach," in which a heterologous strain that mimics natural infection is used for immunization [57, 59]. These are live, oral, and attenuated RV vaccines.

Various studies in the early 1980s were fundamental to support this strategy. Foremost, a study showed protection induced by a bovine virus against a human virulent strain [57]. Moreover, a critical study was carried out by Dr. Ruth Bishop, discoverer of RV in Australia, where it was shown that RV neonatal infection induces protection during the first 3 years of life [13].

The first RV vaccine candidate proved in clinical trials was the bovine strain NCDV (RIT 4237) and it also was the first RV strain grown in cell culture. It was developed by Smith Kline Laboratories, in Rixensart, Belgium [31]. After being

Table 21.1 Efficacy of rotavirus (RV) vaccine against severe RV episodes

Vaccine	Country/region (study location)	Efficacy (95% CI)	Outcome	Reference
RRV/G3P5B [5]	Venezuela	85	Severe RVGE	[93]
	Finland	75 (−120–97)	Severe RVGE	[113]
RRV-TV/G1-G4	USA	82 (−9–97)	Very severe RVGE	[6]
	Arizona, USA	69 (29–88)	Severe RVGE	[109]
	Peru	24	Any RVGE	[63]
	Brazil	35	Any RVGE	[67]
	Venezuela	87 (43–97)	Severe RVGE	[94]
		70 (40–85)	RV hospitalization	
	Finland	91 (82–96)	Severe RVGE	[114]
	Ghana/South Africa	63 (27–82)	Any RVGE	[2]
WC3/G6P7 [5]	USA	82–100	Any RVGE	[20]
89–12/G1P1A[8]	USA	89 (65–95)	Any RVGE	[7]
		78 (14–94)	Severe RVGE	
RIX 4414/G1P1A [8]	Brazil, Mexico, Venezuela	86 (63–96)	Severe RVGE	[107]
		79 (48–92)	RV hospitalization	
		77 (18–96)	G9 RV hospitalization	
	Finland	85 (42–97)	Severe RVGE	[121]
	Singapore, Hong Kong, Taiwan	97 (88–97)	Severe RVGE	[97]
		95 (80–99)	Non-G1 severe RVGE	
Rotarix®/G1P1A [8]	Latin America, Finland	85 (72–92)	Severe RVGE	[104]
		85 (70–94)	RV hospitalization	
		87 (64–97)	Non-G1 RV (G3,G4,G9)	
		41 (−79–82)	G2 RV	
	Latin America (11 countries)	81 (71–87)	Severe RVGE	[70]
		80 (67–88)	Non-G1 (G2,G3,G4,G9)	
	Europe (6 countries)	90 (85–94)	Severe RVGE	[120]
		96 (84–100)	RV hospitalization	
		86 (24–99)	G2 RV	
	South Africa	72 (40–88)	Severe RVGE	[73]
		86 (55–97)	Non-G1 RV	
	Malawi/Africa	49 (11–72)	Severe RVGE	[73]
		50 (17–70)	Non-G1 RV	[73]
Rotavac®/G9P8 [11]	India/South Asia	55 (40–66)	Severe RVGE	[11]
RotaTeq®/G1-G4,P8	USA, Finland, Latin America	98 (88–100) 95 (91–97)	Severe RVGE RV hospitalization	[123]
	USA	95 (84–99)	RV hospitalization	[118]
	Europe	95 (91–97)	RV hospitalization	[118]

(continued)

Table 21.1 (continued)

Vaccine	Country/region (study location)	Efficacy (95% CI)	Outcome	Reference
	Latin America	90 (29–100)	RV hospitalization	[118]
	Finland	94 (91–96)	RV hospitalization	[119]
	Bangladesh/Asia	43 (10–64)	Severe RVGE	[132]
	Vietnam/Asia	64 (8–91)	Severe RVGE	[132]
	Ghana/Africa	56 (28–73)	Severe RVGE	[3]
	Kenya/Africa	64 (−6–90)	Severe RVGE	[3]
	Mali/Africa	18 (−23–45)	Severe RVGE	[3]

Rotarix® two doses, *RotaTeq*® three doses, *RV* rotavirus, *RVGE* rotavirus gastroenteritis

tested in pigs, the RIT 4237 vaccine was evaluated in humans. In Finland, the vaccine was safe and provided an efficacy of 23–100% against severe episodes [113, 115–117]. However, this vaccine did not succeed in Africa and Peru, nor in Native American infants in Arizona (0–75%) [49, 62, 108]. Thus, its development was discontinued.

Another bovine RV vaccine candidate, the WC3 strain, was developed by Fred Clark and Paul Offit at the Wistar Institute in Philadelphia, PA (USA). Similarly, initial results from WC3 studies were promising; however, its efficacy was inconsistent (0% in Africa–76% in the U.S.), and the process was discontinued [8, 18–20, 41].

The third vaccine of this group was the rhesus RRV strain developed by Albert Kapikian and collaborators at the National Institute of Allergy and Infectious Diseases, NIAID, National Institutes of Health (NIH), USA. The efficacy of this vaccine against severe diarrhea was variable in the U.S., Sweden, Finland, and Peru (0–80%) [59]. Nevertheless, a clinical trial phase II conducted in Venezuela showed that this vaccine was highly efficacious (85%) against severe gastroenteritis (GE) and induced fundamentally homotypic protection [93], as in Venezuela circulated the same serotype G3 of the RRV strain. In general, these monovalent vaccines showed poorest efficacy in developing countries in comparison with developed nations.

2.2 *Second-Generation Rotavirus Vaccines*

In view of the former disappointing results, the "modified Jennerian approach" was implemented, wherein single-gene substitution reassortants for human G1, G2, G3, and G4 serotypes were generated from RRV and WC3 RV strains to construct polyvalent vaccines. This methodology was created by Karen Midthun and collaborators at NIAID [76]. The reassortant viruses retained the attenuation of the animal strains but also express proteins responsible for inducing protective immune response in humans.

2.2.1 Polyvalent Reassortant Vaccines

RRV Reassortant Vaccines

From the RRV strain, three reassortants with specificity for G1, G2, and G4 were developed; the G3 serotype was covered by the RRV strain. Afterward, each individual reassortant in monovalent and later in bivalent, trivalent, and quadrivalent combinations was evaluated for safety, immunogenicity, and efficacy studies [59, 77]; this was a long process in which Finland, Venezuela, Brazil, and Peru had significant roles. Six important studies were carried out in Venezuela to decide on titer concentration and the number of doses for the quadrivalent vaccine (RRV-TV) for later evaluation in phase II and III trials [92].

Phase III clinical trials, administering three doses of RRV-TV vaccine, were conducted in the U.S., Finland, Venezuela, Peru, and Brazil. Efficacy was higher (82–91%) against severe diarrhea episodes in the U.S., Finland, and Venezuela (Table 21.1), but was lower in Native American infants in Arizona (69%), Peru (24%), and Brazil (35%) [6, 63, 67, 94, 109, 113, 122]. In Venezuela, in a large catchment study including 2480 infants, RRV-TV vaccine showed 88% protection against severe RV episodes and 70% reduction in hospital admissions by RV diarrhea [94].

On additional analysis of the Venezuelan clinical trials, the vaccine was efficacious despite socioeconomic conditions and seasonality, protection was sustained through the second year of life, and, moreover, vaccine strain transmission was observed in 13% of nonvaccinated infants [52, 91]. Thus, the vaccine could produce herd immunity, which is the induction of protection in the unvaccinated population. This effect has been demonstrated in various post-commercialization studies on impact RV vaccines in use [98].

The disappointing results of Peru and Brazil were probably, among other facts, associated with titer vaccine concentration but mostly with methodological issues, as the study design was focused not on efficacy determination for severe or moderate RV diarrhea but on efficacy against any RV diarrhea. This focus was confirmed by a reappraisal of data by pulling together both studies, in which efficacy was then comparable to the studies done in Venezuela and Finland [66, 69].

The Venezuelan pivotal catchment study was a great achievement because the rotavirus vaccine was successful for the first time in a developing country [94]. This study by Pérez-Schael and collaborators was commented on, in an editorial of the *New England Journal of Medicine*, as "the culmination of a long and highly creative process of research and development at the National Institutes of Health" [61].

In August 31, 1998, the U.S. Food and Drug Administration (FDA) granted a Biologic License for the RRV-TV vaccine under the trade name Rotashield® to Wyeth-Lederle Vaccines. Twenty-five years after the discovery of RV, Rotashield became the first RV vaccine licensed and marketed, following the recommendation of the Advisory Committee on Immunization Practices (ACIP-USA). However, 9 months later and after 1 million doses had been administered in the U.S., the Rotashield recommendation from ACIP was withdrawn, and the company removed the vaccine from market, because it was associated with intussusception, a rare but severe pathology that can cause intestinal obstruction if not treated in time.

Because RRV-TV vaccine preserved its license, later on it was acquired first by Biovir Laboratories and subsequently by International Medical Foundation (IMF), a nonprofit organization [58]. Under IMF auspices and with the collaboration of NIH, the vaccine was reformulated to obtain a product that does not need refrigeration. A phase II clinical trial with the RV vaccine RRV-TV carried out in Ghana, administering two doses in neonates, showed 63% efficacy against rotavirus episodes of any severity [2]. The advantage of this schedule, the first dose given at 0–4 weeks of age and the second dose at 4–8 weeks, is that vaccines are given at an age of less risk of intussusception and provide protection in early life.

The decision for the withdrawal of the Rotashield vaccine from the market has been under intense discussion including the magnitude of risk of intussusceptions associated with actual RV vaccines, the role of age at the time of vaccination, and the cost–benefit analysis of using the vaccine in countries with a high burden of rotavirus diseases [80, 102, 110, 126].

WC3 and UK Reassortant Vaccines

WC3 monovalent reassortants including VP7 specificities for G1–G4 and P [8], maintaining the bovine background, were developed and proved to be safe, immunogenic, and efficacious (69–100%) in clinical trials. Afterward, they were combined for constructing a quadrivalent and, subsequently, a pentavalent vaccine. Both vaccines were equally safe and efficacious [17, 20, 21, 50]. The RV pentavalent vaccine, manufactured by Merck in the U.S., is commercialized as RotaTeq.

On the other hand, human-bovine rotavirus (UK) reassortants were developed from the UK strain and were tested in monovalent and tetravalent combinations, containing serotypes G1, G2, G3, and G4, which were safe, immunogenic, and efficacious [60, 122]. Furthermore, it was planned to construct a hexavalent human-bovine rotavirus (UK) reassortant vaccine, including G8 and G9, to be administered in neonates [60]. Nonexclusive licenses of these vaccine strains were transferred by the NIH to vaccine manufacturers in Brazil, China, and India in 2005.

Subsequently, the Serum Institute of India Ltd developed a live attenuated RV pentavalent vaccine (BRV-PV) containing five RV human-bovine (UK) reassortant strains of serotypes G1, G2, G3, G4, and G9. After preclinical and clinical phase I and II studies with three doses of the vaccine in adults, toddlers, and infants, BRV-PV was found safe and immunogenic [131]. Results of two large phase III efficacy studies against severe RVGE, which are under way in India and Niger, will be available in 2017. These results will be used to achieve licensure in India and the prequalification from WHO.

2.2.2 Rotavirus Vaccines of Human Origin

There are two types of human RV vaccines: strains of neonatal origin (M37, RV3, and 116E) and a strain isolated from a child with RV diarrhea (89–12).

The neonatal strain 116E, isolated in India, is a natural and unique reassortant containing a bovine VP4 and a G9 human VP7. This live oral attenuated vaccine (116E) is the one in most advanced stages in this group and was investigated, developed, licensed, and marketed in India under the name Rotavac [42]; this is the first vaccine that entirely completed all process stages in a developing country as a result of team science work based in India [10]. Human 116E vaccine, administered in three doses, was safe and 55% efficacious in a phase III trial in India [11]. This efficacy, comparable to that of other vaccines when tested in low-income settings, was sustained through the first 2 years of life and provided protection against a wide variety of strains. The vaccine is manufactured by Bharat Biotech International with the financial support of the Indian government, Bill & Melinda Gates Foundation, and PATH, an example of changes in the RV vaccine paradigm.

The strain M37 (G1P2A [10]) was isolated in Venezuela by a scientific team at Instituto de Biomedicina, Universidad Central de Venezuela, and developed as vaccine candidate at NIAID. Strain M37 was safe and induced a good immune response against itself, but it was not efficacious in a phase II study in Finland [124]; thus, M37 vaccine development was abandoned. The human RV3 vaccine candidate was discovered by Ruth Bishop in Australia and is in phase I and II stages of development [4, 12].

Originally, the strain 89–12 was isolated from a child with diarrhea in the Cincinnati Children's Hospital Medical Center (U.S.), being developed as a candidate vaccine by David Bernstein and Richard Ward [7, 9]. Its development was based in the Mexican study, which provided the evidence for protection induced by natural RV infection against moderate or severe RVGE [112].

The 89–12 strain, which contains the most common human G serotype and P genotype, was initially attenuated by cell passages and, last, after it was acquired by GlaxoSmithKline (GSK), was cloned, further attenuated, and named RIX4414 [30]. This vaccine was tested in Finland, Singapore, Brazil, Mexico, and Venezuela [30, 107, 121]. Two doses of the vaccine were well tolerated and highly efficacious (85–97%) and did not interfere with the co-administered routine pediatric vaccines (Table 21.1). In Latin America, the vaccine reduced hospitalization by 79% and showed significant efficacy (86%) against severe RVGE and against non-G1 RV serotypes (77%) [107]. In addition, in Venezuela, it was found that the vaccine reduced significantly severe RV episodes regardless of nutritional status of vaccines [96]. This monovalent vaccine is manufactured in Belgium by GSK and commercialized as Rotarix.

2.2.3 Novel Strategy for Rotarix® and RotaTeq® Vaccine Development, Licensure, and Introduction into National Immunization Programs

After the withdrawal of Rotashield, there was an urgent need to accelerate the development of new RV vaccines, mainly for those countries where it was most needed. This situation led WHO and the scientific community to formulate some recommendations for the implementation of this novel strategy. Among those, the most important was to carry out a parallel testing of candidate vaccines in both developing and developed countries, given the differences in RV epidemiology and national regulatory norms [92, 95].

In view of the association of Rotashield with intussusception, it was necessary furthermore to conduct phase III clinical studies involving more than 60,000 infants for any vaccine to establish their safety against this infrequent pathology.

In consequence, Rotarix and RotaTeq were evaluated in phase III clinical trials. GSK undertook a large, multicenter, safety and efficacy trial in 63,225 infants from Finland and 11 Latin American countries (Argentina, Brazil, Chile, Colombia, the Dominican Republic, Honduras, Mexico, Nicaragua, Panamá, Peru, and Venezuela) [104]. Meanwhile, Merck carried out the study in 68,038 infants, mainly from the U.S. and Finland, but also including infants from Costa Rica, Guatemala, Jamaica, and Mexico [123]. Two doses of Rotarix were administered at 2–4 months of age, and three doses of RotaTeq were given at the ages of 2, 4, and 6 months.

Both vaccines were safe and not associated with an increased risk of intussusception. Rotarix provided high protection (85%) against severe RV diarrhea, reduced RV diarrheal hospitalization by 85%, and was highly efficacious (87%) against G3, G4, and G9 strains but less so against the G2 serotype (41%) [104]. Vaccine protection (79%) was sustained during the second year of life [70]. In Europe, Rotarix efficacy was 90% against severe episodes and 86% against G2 serotype [120]; the RotaTeq vaccine showed 98% of efficacy against severe RVGE and 95% against RV hospitalizations [123]. Reduction of hospitalization by this vaccine in the U.S., Europe, and Latin America was 95%, 95%, and 90%, respectively [118].

Both vaccines demonstrated good protection in America and Europe, but its efficacy was variable when tested in Africa and Asia. Rotarix was highly efficacious against severe RV episodes (97%) in Hong Kong, Singapore, and Taiwan, but provided less protection in South Africa (72%) and Malawi (49%) [73, 97]. Similarly, the efficacy of RotaTeq was 64% in Kenya and Vietnam, 56% in Ghana, 43% in Bangladesh, and 18% in Mali [3, 132].

Both vaccines presented an analogous efficacy gradient, being the lowest in poor socioeconomic settings and the highest in countries with higher socioeconomic conditions. Reduction of severe RV by region was similar for both vaccines: 90–95% in the U.S. and Europe, 81–90% in Latin America, 18–64% in Africa, and 43–64% in Asia, with the exception of Singapore, Hong Kong, and Taiwan, where efficacy was 97% (Table 21.1).

Generally, RV vaccines present variable levels of protection in distinct settings, as has been seen with other vaccines; differences that are probably associated with viral, environmental, and host factors among which low uptake caused by maternal antibodies, interference with other pathogens of the gastrointestinal tract or with other oral vaccines, microbiota, and nutritional deficiencies of infants may be involved [22, 114].

On the other hand, to understand better the reduced RV vaccine efficacy in low socioeconomic settings, it was compared with the protection conferred by natural RV infection with vaccination [71]. In this study, it was demonstrated that the reduced vaccine efficacy in low socioeconomic settings can be explained by intrinsic immunological and epidemiological factors of these populations, which need further evaluation. The authors suggested that modifying immunogenicity of vaccine or vaccination programs may improve the performance of RV vaccines in poor

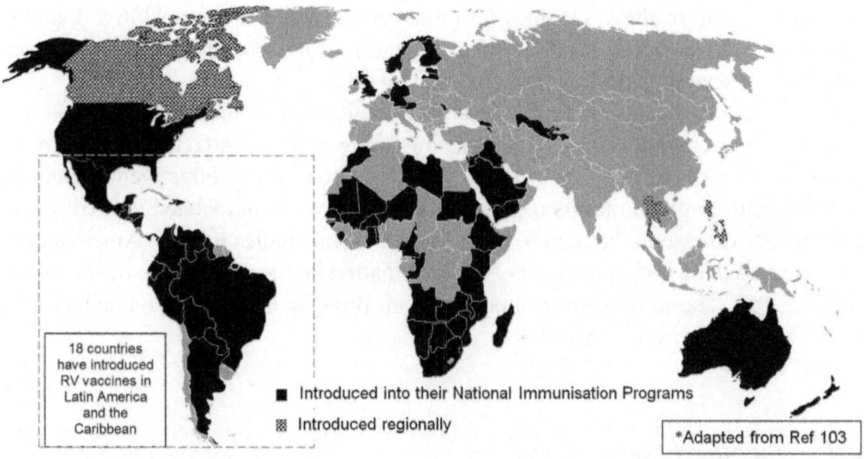

Fig. 21.1 Map shows the 80 countries that had introduced RV vaccine into their immunization programs by December 2015. (Adapted from [103])

socioeconomic countries. Nevertheless, it is necessary to have in mind that many vaccines would have low performance in some low socioeconomic settings, as occurred with polio vaccine.

For the first time in vaccine history, a vaccine was first licensed in another country distinct from the U.S. and was introduced at the same time in Latin America and the U.S., in 2006 [92, 95]. Rotarix and RotaTeq have been incorporated into NIPs from 80 countries, of which 18 are from Latin America Fig. 21.1) [103]. On the other hand, clinical trials conducted in Latin America facilitated licensure and universal introduction of RV vaccine in the region [95].

3 Other RV Vaccines

Because oral RV vaccine performance is diminished in low socioeconomic settings and has a low association with intussusception, the development of nonreplicating RV vaccines (inactivated/parenteral) has become an indispensable issue [54], and various nonreplicating RV vaccines that will be administered by intramuscular injection are in the development process [130].

4 Rotavirus Vaccine Post-licensure Studies

In 2009, WHO recommended that all countries should introduce Rotarix and RotaTeq vaccines into NIPs, particularly those countries where mortality rates among children aged less than 5 years were 10% or more [127]. As a result of the adoption of one of these vaccines into their public sectors, since 2006 a dramatic and progressive decline in RV infection rates, illnesses, and deaths has been observed in several countries throughout the world.

In the developing world, Latin American countries have pioneered the introduction of RV vaccination into NIPs, and this allowed the conduct of several post-licensure studies to assess its efficacy under real conditions (effectiveness) and its public health impact in terms of morbidity and mortality associated with all-cause GE and RV disease in the region. In addition, several studies in Latin America have also evaluated the potential effect of RV vaccination on the distribution of RV strains [68, 82, 95]. The most relevant findings from post-licensure studies conducted in Latin America are summarized next.

4.1 *Vaccine Effectiveness*

With the broad and progressive universal utilization of RV vaccines across Latin American countries, there was a need to evaluate effectiveness. In general, these studies have reassured the significant vaccine efficacy levels against severe RVGE that were achieved during the extensive phase III trials in Latin America. Overall, six case-control studies with Rotarix were conducted in Brazil (three studies), Bolivia (one study), Colombia (one study), and El Salvador (one study), whereas the effectiveness of RotaTeq was assessed in Nicaragua. The most relevant findings from these studies are detailed in Table 21.2. In addition to case-control studies, two cross-sectional studies and one cohort study were conducted in Brazil to assess vaccine effectiveness.

4.1.1 Case-Control Studies

In Bolivia, the first low-income country to adopt Rotarix into the public sector, vaccine effectiveness yielded rates as high as 80% [88]. Of note, the vaccine provided significant protection against either partially (G9P[8]) or fully (G2P[4]) heterotypic strains. In Brazil, at least three case-control studies were conducted, showing vaccine effectiveness rates as high as 85% depending on the characteristics of control groups used [23, 53, 56, 68]. A major finding from the Brazilian studies was that the RV vaccine provided significant protection against the G2P[4] strain, which was predominant all over the country and is known to be fully heterotypic as compared to that of the G1P[8] vaccine strain. Additional case-control studies have assessed the vaccine effectiveness in Colombia and El Salvador, yielding vaccine effectiveness rates as high as 79% and 76%, respectively [25, 28]. Of interest, predominant RV strains in Colombia and El Salvador were G2P[4] and G1P[8], respectively. Nicaragua was the first low-income country to introduce RotaTeq into its NIP. Overall, the effectiveness of three doses of the vaccine against RVGE requiring hospitalization or treatment with intravenous hydration yielded 46%; much higher

Table 21.2 Summary of results from post-licensure, case-control studies assessing RV vaccine effectiveness (full vaccination*) in Latin American countries

Vaccine	Country (year of introduction) study period	Main outcome(s) assessed	Main study characteristics	Predominant genotype (%)	Effectiveness (%) (95% CI)	Reference
Rotarix®*	Bolívia (2008), 2010–2011	RVGE hospitalization	Cases were children at least 8 weeks of age (born after June 1, 2008), hospitalized with confirmed RVGE	G9P[8] (36)	70 (56–79) using RV-negative GE controls	[88]
			Age-matched controls were either children hospitalized with RV-negative GE or admitted for other condition		80 (70–86) using non-GE controls	
Rotarix®	Brazil (2006), 2006–2008	RVGE hospitalization or requiring emergency visit	Cases were children aged <5 years with confirmed RVGE	G2P[4] (100)	85 (53–94) using RV-negative GE controls	[23]
			Age-matched controls were either children with severe RV-negative GE or children hospitalized with ARI		83 (50–93) using ARI controls	
Rotarix®	Brazil (2006), 2008–2009	RVGE hospitalization	Cases were children at least 12 weeks of age (born after Mach 2006) hospitalized with confirmed RVGE	G2P[4] (82)	40 (14–58) using hospital controls	[56]
			Hospital and neighborhood controls were age-matched children without GE		76 (58–85) using neighborhood controls	
Rotarix®	Brazil 2006), 2008–2011	RVGE hospitalization	Cases were children aged ≤2 years hospitalized with confirmed RVGE	G2P[4] (80)	76 (58–86)	[53]
			Age-matched controls were children hospitalized without GE			

(continued)

Table 21.2 (continued)

Vaccine	Country (year of introduction) study period	Main outcome(s) assessed	Main study characteristics	Predominant genotype (%)	Effectiveness (%) (95% CI)	Reference
Rotarix®	Colombia (2009), 2011–2013	RVGE requiring hospital ED treatment	Cases were children ≥8 weeks of age treated at the ED department with confirmed RVGE	G2P[4] (57)	79 (24–94)	[25]
			Controls were children treated at the ED with RV-negative GE			
Rotarix®	El Salvador (2006), 2007–2009	RVGE hospitalization	Cases were children aged <5 years with confirmed RVGE	G1P[8] (92)	76 (64–84)	[28]
			Neighborhood controls age matched to cases			
RotaTeq®**	Nicaragua (2006), 2007–2008	RVGE hospitalization or requiring intravenous hydration	Cases were children age-eligible to receive RV vaccine with confirmed RVGE, who were hospitalized or required intravenous hydration	G2P[4] (88)	46 (18–64)	[86]
			Age-matched neighborhood and hospital controls without GE were identified			

**Rotarix* two doses, ***RotaTeq* three doses, *RVGE* rotavirus gastroenteritis, ARI acute respiratory infection, *ED* emergency department protection (77%) was achieved if severe RVGE cases only were included in the analysis [86]. In this study, G2P[4] strains accounted for almost 90% of isolates.

In addition to the case-control studies, cross-sectional and cohort studies have also assessed effectiveness after RV vaccine introduction in Latin America, although in general small sample sizes used do not allow drawing firm conclusions [14, 45, 125]. Two cross-sectional studies in Brazil showed a remarkable decrease in both RV and all-cause GE-related hospitalization and treatment at emergency departments in the North-Eastern and Central-Western regions [14, 45].

In a longitudinal cohort study in North-Eastern Brazil, all-cause GE episodes were found to be more severe among unvaccinated children, as compared to vaccinated children ($P < 0.05$) [125].

4.2 Public Health Impact of Rotavirus Vaccine

To date, approximately 41% of the countries around the world have introduced RV vaccines into their NIPs. Following vaccine introduction, numerous observational studies conducted in Latin America have documented significant reductions in the frequency of hospitalizations related to GE of any cause, as well as in the amount of GE-associated mortality among children younger than 5 years of age (Fig. 21.2 [32]).

At least five ecological studies in Brazil have evaluated the changes in the proportions or trends of either all-cause or RV-related GE, as well as in GE-related childhood deaths before and after the introduction of Rotarix into the country NIP. do Carmo et al. [34] estimated the rates GE-related hospitalizations and deaths during the post-vaccination period, showing reductions as high as 25% and 22% in all causes of GE and GE-related deaths, respectively. In additional nationwide studies, Lanzieri et al. [64, 65] have reported decreasing trends as high as 48% and 39% in all-cause GE hospitalizations and childhood GE-related deaths, respectively. A similar, substantial nationwide reduction (as high as 36%) in the rates of childhood hospitalizations caused by GE of any cause was reported by Gurgel et al. [48]. Furthermore, recent country-level data demonstrated sustained significant decreases in GE-related hospital deaths and hospitalizations, yielding rates as high as 60% and 31%, respectively, among children less than 5 years of age [24]. The impact of RV vaccine in São Paulo, South-Eastern Brazil, translated into a reduction in RV-related hospitalizations in the range of 59% to 82% across age groups, with the highest decrease among infants aged 0–11 months [106].

At least two observational studies conducted in Mexico have reported major declines in GE-related deaths among children younger than 5 years of age, following rotavirus vaccine introduction in 2007. In an earlier study by Richardson et al. [101], a drastic reduction (29–41%) of child deaths was seen in 2008. In a subsequent analysis, a significant sustained decrease (43–55%) was observed in GE-related deaths during the 4 years after vaccine introduction [39]. Also, in Mexico a sharp decline in hospitalizations for GE was reported by Quintanar-Solares et al. [99], with rates ranging from 40% to 52% across age groups.

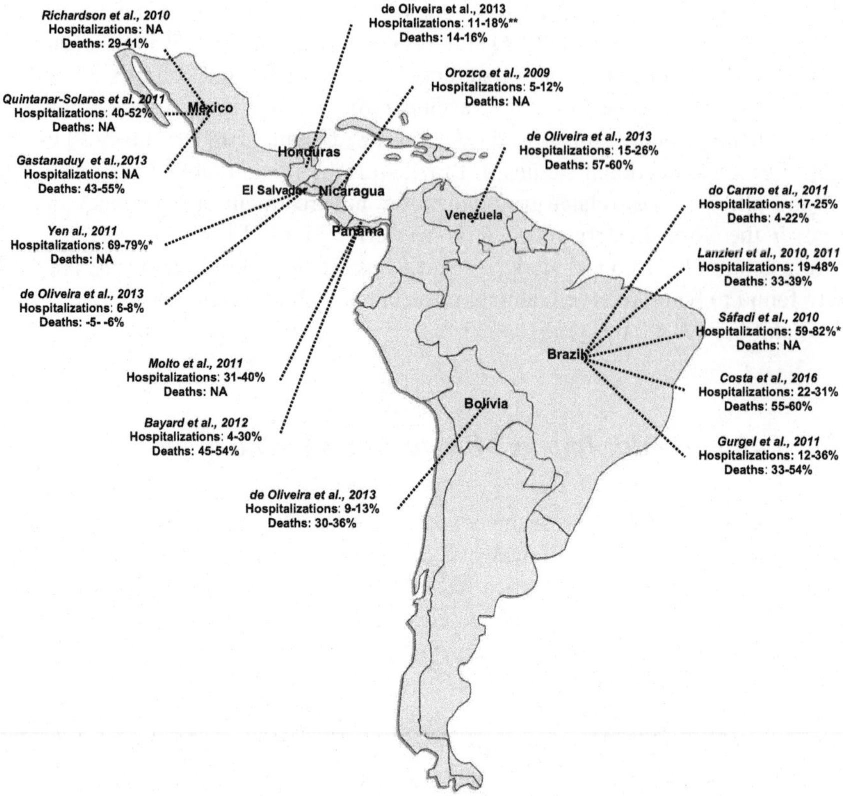

* RVGE; ** Ranges may include results obtained across different age groups (not specified in the figure) throughout the study period; NA: Not available

Fig. 21.2 Rotavirus vaccination impact on the reductions of acute gastroenteritis-associated hospitalizations and deaths in Latin America

Studies in Panama showed that GE-related hospitalizations decreased at rates as high as 30% [5] and 40% [78] across age groups. Furthermore, a substantial decline (up to 45%) in GE-related mortality was reported by Bayard et al. [5].

An apparent, relatively low impact of RV vaccination was reported in Nicaragua, translated into a reduction rate of only 12% in all-cause hospitalizations for GE among infants, probably resulting from the yet suboptimal vaccine coverage during the first year after introduction [83].

A recent multinational analysis was performed to assess the temporal trends in all-cause GE-related hospitalizations and deaths among children under 5 years of age, before and after introduction of RV vaccination in Bolivia, El Salvador, Honduras, and Venezuela [27]. Mortality reduction rates ranged from 14% (Honduras) to 60% (Venezuela) among children 0–1 year in age (Fig. 21.2).

Furthermore, the highest reduction rates in hospitalizations for all causes of GE were seen in Honduras, yielding 27% for children 0–1 year in age. An additional study conducted in El Salvador [129] showed that universal use of RV vaccination in the country NIP has reduced the prevalence of hospitalization from RV by as much as 79% among children younger than 1 year.

4.2.1 Strong Evidence for Herd Immunity

Similar to findings from studies in other regions worldwide, the remarkable public health impact of RV vaccine in Latin America appeared to not be restricted only to infants with less than 1 year of life, the age-eligible group for vaccination, but it was also evident among older, not age-eligible children [68, 82]. Indeed, such indirect benefits, known as the herd immunity effect, suggest that vaccination of a large proportion of the targeted population also translates into protection for those individuals who either were not vaccinated or had not developed immunity. This effect has been attributed to vaccination in reducing the rates of transmission among a diminished population of infants susceptible to RV infection [43]. In Brazil and El Salvador, for example, the rates of RV-related hospitalizations among children not vaccinated decreased by 41–81% and 24%, respectively [106, 129].

4.3 Effect on Circulating RV Strains

Following the implementation of universal RV vaccination in Latin America, a concern was raised as to whether the currently available vaccines would confer sufficient protection against strains not included in their composition. Moreover, it had been postulated that an evolutionary pressure exerted by the vaccines might potentially lead to strain replacement during the post-vaccination era, even resulting in the emergence of either unusual RV strains or strains that would escape vaccine protection [68, 74, 82, 89].

It has been extensively documented that circulating RV strain diversity and geographic distribution vary considerably throughout the world, and periodically emerging novel strains were identified in the pre-vaccine era, such as those bearing G9- and G12-type specificities [40]. Although more than 60 rotavirus strains are known to occur naturally in human beings, only 5 strains (G1P[8], G2P[4], G3P[8], and G9P[8]) are associated with 80–90% of childhood GE globally. Data from several clinical studies and post-licensure research have demonstrated that the two vaccines available today (Rotarix and RotaTeq) convincingly provide protection against a broad variety of circulating strains, including strains not incorporated in their composition [29, 56, 86, 87, 89], suggesting that immunity to RV is mostly heterotypic [1].

In certain countries in Latin America (e.g., Brazil), a marked increase in the prevalence of the fully heterotypic G2P[4] RV was particularly evident just following the implementation of the monovalent (G1P[8]) human rotavirus vaccine Rotarix. These findings raised the hypothesis that a serotype replacement was likely to be occurring as a result of vaccine-induced selective pressure [16, 47, 68, 82]. This high predominance of G2P[4] strains could also be seen in Nicaragua, where the use of the pentavalent (G1, G2, G3, G4, and P[8]) bovine-human reassortant vaccine RotaTeq was fully in place [85]. Notably, a sharp increase in the relative prevalence of G2P[4] was also concurrently observed in countries where RV vaccination had not even been implemented into the public sector, such as Argentina and Paraguay [36, 89].

Of interest, several further surveillance studies have shown a substantial decline in the occurrence of G2P[4] in more recent years, with an increase in prevalence of G1P[8] and several other non-G2 types [16, 35, 44, 46, 75, 79, 90]. Taking collectively with the already well-established good homotypic and heterotypic efficacy and effectiveness of available RV vaccines, these post-licensure surveillance data strongly strengthen the hypothesis of a natural secular strain fluctuation, rather than a potentially vaccine-driven effect on circulating RV strains. Nonetheless, the robustness of such an assumption will only be increased through the conduct of a continuous, long-term monitoring of circulating RV strains throughout Latin America countries and other settings.

5 Benefits Outweigh Potential Risks: The Issue of Intussusception

Several pre- and post-licensure studies conducted in every region of the world have convincingly demonstrated that both Rotarix and RotaTeq have strong safety records, albeit a slight increase in the risk of developing intussusception has been reported [15]. Nevertheless, it is currently well established that estimates for any potential risk are several times lower than the risk reported for RotaShield [81].

Intussusception represents an extremely rare condition in which the intestine folds on itself, occurring naturally (in the absence of vaccination) in infants between 2 and 9 months of age, at rates that vary considerably (9 to 328 per 100,000) from region to region around the world [55]. Similarly, in a prospective surveillance study conducted in Latin America before the introduction of universal RV vaccination, incidence rates of intussusception were found to vary widely across 11 countries, in the range between 3.8 and 105.3 per 100,000 children under 1 year of age [105].

At least two large post-marketing surveillance studies have assessed the risk of intussusception in real clinical practice in Latin America, both of which clearly show that benefits of vaccination far outweigh any potential risk of intussusception [85, 111].

In one study performed in parallel in Mexico and Brazil, a small transient increase in the occurrence of intussusception was detected within 7 days after the administration of the first vaccine dose among Mexican infants [incidence rate of 5.3 (95% CI, 3.0–9.3)], whereas in Brazil an even smaller proportion [2.6 (95% CI, 1.3–5.2)] was identified only after the second dose [85]. These data translate into the prevention by the vaccine of 11,551 RV-related hospitalizations and 663 deaths annually in Mexico, compared to an estimated 41 excess intussusceptions cases and 2 deaths attributable to the implementation of RV vaccination. In Brazil, the vaccine would prevent an estimated 69,572 hospitalizations and 640 deaths, compared to a predicted 55 excess intussusception cases and 3 deaths.

An additional large prospective, post-marketing active surveillance study for intussusception was conducted in Mexico, involving 66 hospitals and 1.5 million vaccinated infants [111]. The relative risk of intussusception during the 7-day period following the administration of the first RV vaccine dose [6.5 (95% CI, 4.2–10.1)] was comparable to that of previous studies in Mexico, with an estimated attributable risk of 3 or 4 additional cases per 100,000 vaccinated infants [85, 102].

In summary, taking all available data into account, studies in Latin America (and elsewhere) have shown a small increased risk of intussusception within the 7 days following administration of first vaccine dose and, to a lesser extent, second dose. Although it is recommended that monitoring for intussusception should continue in countries where RV vaccines are in use, it is currently well established that accumulating available evidence strongly supports the notion that benefits of vaccination far outweigh the risk [128].

6 Challenges and Recommendations for the Future

In contrast to the 18 Latin-American countries mentioned, 7 countries (Argentina, Chile, Costa Rica, Cuba, French Guiana, Suriname, and Uruguay) have not incorporated vaccination into their NIPs (Fig. 21.1). Taking into account the robust, aforementioned data available on the vaccine-derived health benefits in Latin America, decision makers and healthcare providers in these latter countries should urgently recognize that strong reasons exist to support introduction of RV vaccination into their NIPs [103]. In addition to this, it has been largely demonstrated that universal vaccination of infants is cost-effective for middle- and low-income settings in Latin America [100].

A number of general issues and recommendations remain to be addressed in the coming years toward an improved performance of RV vaccination, not only for Latin America but also for other regions throughout the world:

- In countries where RV vaccines have already been introduced, vaccination sustainability must be ensured through the support from national governments, funding agencies, and global health entities.

- Countries are strongly encouraged to gather continuously high-quality data on GE hospitalizations and GE-related deaths. There is also a need for continuous vaccine safety monitoring, particularly with respect to the occurrence of intussusception.
- A long-term monitoring of circulating RV strains is recommended following introduction.
- The adoption of strategies should address the issue of the yet great disparities in vaccination coverage rates across Latin-American countries. A delayed immunization, for instance, may have resulted in suboptimal coverage [37].
- It is recommended to build a broad scientific-based advocacy initiative for RV vaccination, stressing the overwhelming benefits over any risk of vaccination.

In the context of a proposed research agenda, the following specific aspects might also be put into perspective:

- Examine the potential effect of universal vaccination on the epidemiology, particularly in regard to seasonality and age distribution of RV.
- Assess the potential for vaccine-derived indirect benefits (herd immunity) among children not vaccinated.
- Undertake long-term monitoring for circulating RV strains to see whether changes in the ecology may result from universal vaccine use.
- Investigate whether factors such as oral polio vaccine, breastfeeding, and gut microbiome/intestinal enteropathy may interfere with vaccine effectiveness.
- Analyze the sustained effectiveness of RV vaccines to determine whether possible waning immunity occurs after the first year of life, particularly in regard to G2P[4] and a broad range of circulating RV strains.
- Improve the efficacy of vaccines by making adjustments to vaccine schedule, such as a birth dose, two versus three doses for Rotarix, timing and spacing of doses, and a booster dose at a later age.

A promising prospect for Latin America is the fact that Brazil is on its way toward fully developing the capacity to produce two RV vaccines through partnerships for technology transfer, as previously mentioned: first, a vaccine based on a UK bovine-human reassortant developed by the US National Institutes of Health and, second, a human-derived vaccine through an agreement with GSK [72, 95].

7 Conclusions

The RV vaccines currently licensed, monovalent Rotarix and pentavalent RotaTeq, have provided a long and broad protection against a range of RV strains in distinct settings. Additionally, these vaccines have been effective, and no increased risk of intussusception has been demonstrated in post-licensure studies up to the present. Research and clinical trials conducted in Latin America have in part made this reality possible. However, RV vaccines need further attention to broaden their use in NIPs in most countries.

References

1. Angel J, Steele AD, Franco MA (2014) Correlates of protection for rotavirus vaccines: possible alternative trial endpoints, opportunities, and challenges. Hum Vaccin Immunother 10:3659–3671
2. Armah GE, Kapikian AZ, Vesikari T et al (2013) Efficacy, immunogenicity, and safety of two doses of a tetravalent rotavirus vaccine RRV-TV in Ghana with the first dose administered during the neonatal period. J Infect Dis 208:423–431
3. Armah GE, Sow SO, Breiman RF et al (2010) Efficacy of pentavalent rotavirus vaccine against severe rotavirus gastroenteritis in infants in developing countries in sub-Saharan Africa: a randomised, double-blind, placebo-controlled trial. Lancet 376:606–614
4. Barnes GL, Lund JS, Mitchell SV et al (2002) Early phase II trial of human rotavirus vaccine candidate RV3. Vaccine 20:2950–2956
5. Bayard V, De Antonio R, Contreras R et al (2012) Impact of rotavirus vaccination on childhood gastroenteritis-related mortality and hospital discharges in Panama. Int J Infect Dis 16:e94–e98
6. Bernstein DI, Glass RI, Rodgers G et al (1995) Evaluation of rhesus rotavirus monovalent and tetravalent reassortant vaccines in US children. JAMA 273:1191–1196
7. Bernstein DI, Sack DA, Rothstein E et al (1999) Efficacy of live, attenuated, human rotavirus vaccine 89-12 in infants: a randomised placebo-controlled trial. Lancet 354:287–290
8. Bernstein DI, Smith VE, Sander DS et al (1990) Evaluation of WC3 rotavirus vaccine and correlates of protection in healthy infants. J Infect Dis 162:1055–1062
9. Bernstein DI, Ward RL (2006) Rotarix: development of a live attenuated monovalent human rotavirus vaccine. Pediatr Ann 35:38–43
10. Bhan MK, Glass RI, Ella KM et al (2014) Team science and the creation of a novel rotavirus vaccine in India: a new framework for vaccine development. Lancet 383:2180–2183
11. Bhandari N, Rongsen-Chandola T, Bavdekar A et al (2014) Efficacy of a monovalent human-bovine (116E) rotavirus vaccine in Indian infants: a randomised, double-blind, placebo-controlled trial. Lancet 383:2136–2143
12. Bines JE, Danchin M, Jackson P et al (2015) Safety and immunogenicity of RV3-BB human neonatal rotavirus vaccine administered at birth or in infancy: a randomised, double-blind, placebo-controlled trial. Lancet Infect Dis 15:1389–1397
13. Bishop RF, Barnes GL, Cipriani E et al (1983) Clinical immunity after neonatal rotavirus infection. A prospective longitudinal study in young children. N Engl J Med 309:72–76
14. Borges AMT, Dias e Souza M, Fiaccadori FS et al (2011) Monitoring the circulation of rotavirus among children after the introduction of the Rotarix™ vaccine in Goiânia, Brazil. Mem Inst Oswaldo Cruz 106:499–501
15. Buttery JP, Standish J, Bines JE (2014) Intussusception and rotavirus vaccines: consensus on benefits outweighing recognized risk. Pediatr Infect Dis J 33:772–773
16. Carvalho-Costa FA, Volotão Ede M, de Assis RM et al (2011) Laboratory-based rotavirus surveillance during the introduction of a vaccination program, Brazil. Pediatr Infect Dis J 30:S35–S41
17. Clark HF, Bernstein DI, Dennehy PH et al (2004) Safety, efficacy, and immunogenicity of a live, quadrivalent human-bovine reassortant rotavirus vaccine in healthy infants. J Pediatr 144:184–190
18. Clark HF, Borian FE, Bell LM et al (1988) Protective effect of WC3 vaccine against rotavirus diarrhea in infants during a predominantly serotype 1 rotavirus season. J Infect Dis 158:570–587
19. Clark HF, Furukawa T, Bell LM et al (1986) Immune response of infants and children to low-passage bovine rotavirus (strain WC3). Am J Dis Child 140:350–356
20. Clark HF, Offit PA, Ellis RW et al (1996) The development of multivalent bovine rotavirus (strain WC3) reassortant vaccine for infants. J Infect Dis 174:S73–S80

21. Clark HF, Offit PA, Plotkin SA et al (2006) The new pentavalent rotavirus vaccine composed of bovine (strain WC3): human rotavirus reassortants. Pediatr Infect Dis J25:577–583
22. Clarke E, Desselberger U (2015) Correlates of protection against human rotavirus disease and the factors influencing protection in low-income settings. Mucosal Immunol 8:1–17
23. Correia JB, Patel MM, Nakagomi O et al (2010) Effectiveness of monovalent rotavirus vaccine (Rotarix) against severe diarrhea caused by serotypically unrelated G2P[4] strains in Brazil. J Infect Dis 201:363–369
24. Costa I, Linhares AC, Cunha MH et al (2016) Sustained decrease in gastroenteritis-related death and hospitalizations in children aged <5 years after the introduction of rotavirus vaccination: a time trend analysis in Brazil (2001–2010). Pediatr Infect Dis J 35:e180–e190
25. Costes-Cantillo K, Paternima-Caicedo A, Coronell-Rodriguez W et al (2014) Effectiveness of the monovalent rotavirus vaccine in Colombia: a case-control study. Vaccine 32:3035–3040
26. de Oliveira LH, Camacho LA, Coutinho ES et al (2015) Rotavirus vaccine effectiveness in Latin American and Caribbean countries: a systematic review and meta-analysis. Vaccine 33:A248–A254
27. de Oliveira LH, Giglio N, Ciapponi A et al (2013) Temporal trends in diarrhea-related hospitalizations and deaths in children under age 5 before and after the introduction of the rotavirus vaccine in four Latin American countries. Vaccine 31:C99–C108
28. de Palma O, Cruz L, Ramos H et al (2010) Effectiveness of rotavirus vaccination against childhood diarrhoea in El Salvador: case-control study. BMJ 340:c2825
29. De Vos B, Han HH, Bouckenooghe A et al (2009) Live attenuated human rotavirus vaccine, RIX4414, provides protection in infants against rotavirus strains with and without shared G and P genotypes: integrated analysis of randomized controlled trials. Pediatr Infect Dis 28:261–266
30. De Vos B, Vesikari T, Linhares AC et al (2004) A rotavirus vaccine for prophylaxis of infants against rotavirus gastroenteritis. Pediatr Infect Dis J 23:S179–S182
31. Delem A, Lobmann M, Zygraich N (1984) A bovine rotavirus developed as a candidate vaccine for use in humans. J Biol Stand 12:443–445
32. Desai R, de Oliveira LH, Parashar UD et al (2011) Reduction in morbidity and mortality from childhood diarrhoeal disease after species A rotavirus vaccine introduction in Latin America: a review. Mem Inst Oswaldo Cruz 106:907–911
33. Desai R, Parashar UD, Lopman B et al (2012) Potential intussusception risk versus health benefitis from rotavirus vaccination in Latin America. Clin Infect Dis 54:1397–1405
34. do Carmo GM, Yen C, Siqueira AA et al (2011) Decline in diarrhea mortality and admissions after routine childhood immunization in Brazil: a time-series analysis. PLoS Med 8:e1001024
35. Dulgheroff AC, Figueiredo EF, Moreira LP et al (2013) Distribution of rotavirus genotypes after vaccine introduction in Triângulo Mineiro region of Brazil: 4-year follow-up study. J Clin Virol 55:67–71
36. Esteban LE, Rota LP, Gentsch JR et al (2010) Molecular epidemiology of group A rotavirus in Buenos Aires, Argentina 2004–2007: reemergence of G2P[4] and emergence of G9P[8] strains. J Med Virol 82:1083–1093
37. Flannery B, Samad S, de Moraes JC et al (2013) Uptake of oral rotavirus vaccine and timeliness of routine immunization in Brazil's National Immunization Program. Vaccine 31:1523–1528
38. Franco MA, Greenberg HB (2013) Rotavirus. Microbiol Spectr 1:1–10. doi:10.1128/microbiolspec. AID-001-2013
39. Gastañaduy PA, Sanchez-Uribe E, Esparza-Aguilar M et al (2013) Effect of rotavirus vaccine on diarrhea mortality in different socioeconomic regions of Mexico. Pediatrics 131:e1115–e1120
40. Gentsch JR, Laird AR, Bielfelt B et al (2005) Serotype diversity and reassortment between human and animal rotavirus strains: implications for rotavirus vaccine programs. J Infect Dis 192:S146–S159

41. Georges-Courbot MC, Monges J, Siopathis MR et al (1991) Evaluation of the efficacy of a low-passage bovine rotavirus (strain WC3) vaccine in children in Central Africa. Res Virol 142:405–411
42. Glass RI, Bhan MK, Ray P et al (2005) Development of candidate rotavirus vaccines derived from neonatal strains in India. J Infect Dis 192:S30–S35
43. Glass RI, Parashar UD, Patel M (2014) Rotavirus vaccines: successes and challenges. J Infect 68:S9–S18
44. Guerra SFS, Linhares AC, Mascarenhas JDP et al (2015) Rotavirus strain surveillance for three years following the introduction of rotavirus vaccine into Belém, Brazil. J Med Virol 87:1303–1310
45. Gurgel RQ, Bohland AK, Vieira SC et al (2009) Incidence of rotavirus and all-cause diarrhea in northeast Brazil following the introduction of a national immunization program. Gastroenterology 137:1970–1975
46. Gurgel RQ, Alvarez Ade J, Rodrigues A et al (2014) Incidence of rotavirus and circulating genotypes in Northeast Brazil during 7 years of national rotavirus vaccination. PLoS One 9:e110217
47. Gurgel RQ, Correia JB, Cuevas LE (2008) Effect of rotavirus vaccination on circulating virus strains. Lancet 371:301–302
48. Gurgel RQ, Ilozue C, Correia JB et al (2011) Impact of rotavirus vaccination on diarrhoea mortality and hospital admissions in Brazil. Trop Med Int Health 16:1180–1184
49. Hanlon P, Hanlon L, Marsh V et al (1987) Trial of an attenuated bovine rotavirus vaccine (RIT 4237) in Gambian infants. Lancet 1:1342–1345
50. Heaton PM, Ciarlet M (2007) Vaccines: the pentavalent rotavirus vaccine: discovery to licensure and beyond. Clin Infect Dis 45(12):1618–1624
51. Herrera D, Vásquez C, Corthésy B et al (2013) Rotavirus specific plasma secretory immunoglobulin in children with acute gastroenteritis and children vaccinated with an attenuated human rotavirus vaccine. Hum Vaccin Immunother 9:2409–2417
52. Hoshino Y, Wagner M, Yan XY et al (2003) Horizontal transmission of rhesus monkey rotavirus-based quadrivalent vaccine during a phase 3 clinical trial in Caracas, Venezuela. J Infect Dis 187:791–800
53. Ichihara MYT, Rodrigues LC, Teles Santos CAS et al (2014) Effectiveness of rotavirus vaccine against hospitalized rotavirus diarrhea: a case-control study. Vaccine 32:2740–2747
54. Jiang B, Gentsch JR, Glass RI (2008) Inactivated rotavirus vaccines: a priority for accelerated vaccine development. Vaccine 26:6754–6758
55. Jiang J, Jiang B, Parashar U et al (2013) Childhood intussusception: a literature review. PLoS One 8:e68482
56. Justino MC, Linhares AC, Lanzieri TM (2011) Effectiveness of the monovalent G1P[8] human rotavirus vaccine against hospitalization for severe G2P[4] rotavirus gastroenteritis in Belém, Brazil. Pediatr Infect Dis J 30:396–401
57. Kapikian AZ (1994) Jennerian and modified Jennerian approach to vaccination against totavirus diahrrea in infants and young children: an introduction. In: Kapikian AZ (ed) Viral infections of the gastrointestinal tract, 2nd edn. Dekker, New York, pp 409–415
58. Kapikian AZ (2011) History of rotavirus vaccines. Part I: RotaShield. In: Plotkin S (ed) History of vaccine development. Springer, New York/Dordrecht/Heidelberg/London, pp 285–314
59. Kapikian AZ, Hoshino Y, Chanock RM et al (1996) Efficacy of a quadrivalent rhesus rotavirus-based human rotavirus vaccine aimed at preventing severe rotavirus diarrhea in infants and young children. J Infect Dis 174:S65–S72
60. Kapikian AZ, Simonsen L, Vesikari T et al (2005) A hexavalent human rotavirus-bovine rotavirus (UK) reassortant vaccine designed for use in developing countries and delivered in a schedule with the potential to eliminate the risk of intussusception. J Infect Dis 192:S22–S29
61. Keusch GT, Cash RA (1997) A vaccine against rotavirus – when is too much too much? N Engl J Med 337:1228–1229

62. Lanata CF, Black RE, del Aguila R et al (1989) Protection of Peruvian children against rotavirus diarrhea of specific serotypes by one, two, or three doses of the RIT 4237 attenuated bovine rotavirus vaccine. J Infect Dis 159:452–459
63. Lanata CF, Midthun K, Black RE et al (1996) Safety, immunogenicity, and protective efficacy of one and three doses of the tetravalent rhesus rotavirus vaccine in infants in Lima, Peru. J Infect Dis 174:268–275
64. Lanzieri TM, Costa I, Shafi FA et al (2010) Trends in hospitalizations from all-cause gastroenteritis in children younger than 5 years of age in Brazil before and after human rotavirus vaccine introduction. Pediatr Infect Dis 29:673–675
65. Lanzieri TM, Linhares AC, Costa I et al (2011) Impact of rotavirus vaccinations on childhood deaths from diarrhea in Brazil. Int J Infect Dis 15:206–210
66. Linhares AC, Bresee JS (2000) Rotavirus vaccines and vaccination in Latin America. Rev Panam Salud Publica 8:305–331
67. Linhares AC, Gabbay YB, Mascarenhas JD et al (1996) Immunogenicity, safety and efficacy of tetravalent rhesus-human, reassortant rotavirus vaccine in Belém, Brazil. Bull World Health Organ 74:491–500
68. Linhares AC, Justino MCA (2014) Rotavirus vaccination in Brazil: effectiveness and health impact seven years post-introduction. Expert Rev Vaccines 13:43–57
69. Linhares AC, Lanata CF, Hausdorff WP et al (1999) Reappraisal of the Peruvian and Brazilian lower titer tetravalent rhesus-human reassortant rotavirus vaccine efficacy trials: analysis by severity of diarrhea. Pediatr Infect Dis J18:1001–1006
70. Linhares AC, Velázquez FR, Pérez-Schael I et al (2008) Efficacy and safety of an oral live attenuated human rotavirus vaccine against rotavirus gastroenteritis during the first 2 years of life in Latin American infants: a randomised, double-blind, placebo-controlled phase III study. Lancet 371:1181–1189
71. Lopman BA, Pitzer VE, Sarkar R et al (2012) Understanding reduced rotavirus vaccine efficacy in low socio-economic settings. PLoS One 7:e41720
72. Luna EJ, Frazatti-Gallina NM, Timenetsky MC et al (2013) A phase I clinical trial of a new 5-valent rotavirus vaccine. Vaccine 31:1100–1105
73. Madhi SA, Cunliffe NA, Steele D et al (2010) Effect of human rotavirus vaccine on severe diarrhea in African infants. N Engl J Med 362:289–298
74. Matthijnssens J, Nakagomi O, Kirkwood CD et al (2012) Group A rotavirus universal mass vaccination: how and to what extent will selective pressure influence prevalence of rotavirus genotypes? Expert Rev Vaccines 11:1347–1354
75. Matthijnssens J, Van Ranst M (2012) Genotype constellation and evolution of group A rotaviruses infecting humans. Curr Opin Virol 2:426–433
76. Midthun K, Greenberg HB, Hoshino Y et al (1985) Reassortant rotaviruses as potential live rotavirus vaccine candidates. J Virol 53:949–954
77. Midthun K, Kapikian AZ (1996) Rotavirus vaccines: an overview. Clin Microbiol Rev 9:423–434
78. Molto Y, Cortes JE, de Oliveira LH et al (2011) Reduction of diarrhea-associated hospitalizations among children aged <5 years in Panama following the introduction of rotavirus vaccine. Pediatr Infect Dis J 30:S16–S20
79. Munford V, Gilio AE, de Souza EC (2009) Rotavirus gastroenteritis in children in 4 regions in Brazil: a hospital-based surveillance study. J Infect Dis 200:S106–S113
80. Murphy BR, Morens DM, Simonsen L et al (2003) Reappraisal of the association of intussusception with the licensed live rotavirus vaccine challenges initial conclusions. J Infect Dis 187:1301–1308
81. Murphy TV, Gargiullo PM, Massoudi MS et al (2001) Intussusception among infants given an oral rotavirus vaccine. N Engl J Med 344:564–572
82. O'Ryan M, Lucero Y, Linhares AC (2011) Rotarix®: vaccine performance 6 years postlicensure. Expert Rev Vaccines 10:1645–1659
83. Orozco M, Vasquez J, Pedreira C et al (2009) Uptake of rotavirus vaccine and national trends of acute gastroenteritis among children in Nicaragua. J Infect Dis 200:S125–S130

84. Parashar UD, Cortese MM, Payne DC et al (2015) Value of post-licensure data on benefits and risks of vaccination to inform vaccine policy: the example of rotavirus vaccines. Vaccine 33:D55–D59
85. Patel MM, Lopes-Collada VR, Bulhões MM et al (2011) Intussusception risk and health benefits of rotavirus vaccination in Mexico and Brazil. N Engl J Med 364:2283–2292
86. Patel M, Pedreira C, de Oliveira LG et al (2009) Association between pentavalent rotavirus vaccine and severe rotavirus diarrhea among children in Nicaragua. JAMA 301:2243–2251
87. Patel M, Pedreira C, de Oliveira LH et al (2016) Effectiveness of pentavalent rotavirus vaccine against a diverse range of circulating strains in Nicaragua. Clin Infect Dis 62:S127–S132
88. Patel MM, Patzi M, Pastor D et al (2013) Effectiveness of monovalent rotavirus vaccine in Bolivia: case-control study. BMJ 346:f3726
89. Patel MM, Steele D, Gentsch JR et al (2011) Real-world impact of rotavirus vaccination. Pediatr Infect Dis J 30:S1–S5
90. Peláez-Carvajal D, Cotes-Cantillo K, Caicedo-Paternina A et al (2014) Characterization of rotavirus genotypes before and after the introduction of a monovalent rotavirus vaccine in Colombia. J Med Virol 86:1083–1086
91. Pérez ME, Glass R, Alvarez G et al (2000) Rhesus rotavirus-based quadrivalent vaccine is efficacious despite age, socioeconomic conditions and seasonality in Venezuela. Vaccine 19:976–981
92. Pérez Schael I (2012) Vacuna de Rotavirus. Una agenda global para su desarrollo y aplicación universal. Editorial Médica Panamericana, Colombia
93. Pérez-Schael I, Garcia D, Gonzalez M et al (1990) Prospective study of diarrheal diseases in Venezuelan children to evaluate the efficacy of rhesus rotavirus vaccine. J Med Virol 30:219–229
94. Pérez-Schael I, Guntiñas MJ, Pérez M et al (1997) Efficacy of the rhesus rotavirus-based quadrivalent vaccine in infants and young children in Venezuela. N Engl J Med 337:1181–1187
95. Pérez-Schael I, O'Ryan M, Saez-Llorens X et al (2012) The Latin American experience. Trials Vaccinol 1:10–20
96. Pérez-Schael I, Salinas B, Tomat M et al (2007) Efficacy of the human rotavirus vaccine RIX4414 in malnourished children. J Infect Dis 196:537–540
97. Phua KB, Lim FS, Lau YL et al (2012) Rotavirus vaccine RIX4414 efficacy sustained during the third year of life: a randomized clinical trial in an Asian population. Vaccine 30:4552–4557
98. Pollard SL, Malpica-Llanos T, Friberg IK et al (2015) Estimating the herd immunity effect of rotavirus vaccine. Vaccine 33:3795–3800
99. Quintanar-Solares M, Yen C, Richardson V et al (2011) Impact of rotavirus vaccination on diarrhea-related hospitalizations among children <5 years of age in Mexico. Pediatr Infect Dis J 30:S11–S15
100. Rheingans RD, Constenla D, Antil L et al (2007) Potential cost-effectiveness of vaccination for rotavirus gastroenteritis in eight Latin American and Caribbean countries. Rev Panam Salud Publica 21:205–216
101. Richardson V, Hernandez-Pichardo J, Quintanar-Solares M et al (2010) Effect of rotavirus vaccination on death from childhood diarrhea in Mexico. N Engl J Med 362:299–305
102. Rosillon D, Buyse H, Friedland LR et al (2015) Risk of intussusception after rotavirus vaccination: meta-analysis of postlicensure studies. Pediatr Infect Dis J 34:763–768
103. ROTA Council (2016). Rotavirus: common, severe, devastating, preventable. Available from: http://tinyurl.com/rota4all. Accessed 21 May 2016
104. Ruiz-Palacios GM, Pérez-Schael I, Velázquez FR et al (2006) Safety and efficacy of an attenuated vaccine against severe rotavirus gastroenteritis. N Engl J Med 354:11–22
105. Sáez-Llorens X, Velázquez FR, Lopez P et al (2013) A multi-country study of intussusception in children under 2 years of age in Latin America: analysis of prospective surveillance data. BMC Gastroenterol 13:95
106. Safadi MA, Berezin EN, Munford V et al (2010) Hospital-based surveillance to evaluate the impact of rotavirus vaccination in São Paulo, Brazil. Pediatr Infect Dis J 29:1019–1022

107. Salinas B, Pérez Schael I, Linhares AC et al (2005) Evaluation of safety, immunogenicity and efficacy of an attenuated rotavirus vaccine, RIX4414: a randomized, placebo-controlled trial in Latin American infants. Pediatr Infect Dis J 24:807–816
108. Santosham M, Letson GW, Wolff M et al (1991) A field study of the safety and efficacy of two candidate rotavirus vaccines in a Native American population. J Infect Dis 163:483–487
109. Santosham M, Moulton LH, Reid R et al (1997) Efficacy and safety of high-dose rhesus-human reassortant rotavirus vaccine in Native American populations. J Pediatr 131:632–638
110. Simonsen L, Viboud C, Elixhauser A et al (2005) More on RotaShield and intussusception: the role of age at the time of vaccination. J Infect Dis 192:S36–S43
111. Velázquez FR, Colindres RE, Grajales C et al (2012) Postmarketing surveillance of intussusception following mass introduction of the attenuated human rotavirus vaccine in Mexico. Pediatr Infect Dis J 31:736–744
112. Velázquez FR, Matson DO, Calva JJ et al (1996) Rotavirus infections in infants as protection against subsequent infections. N Engl J Med 335:1022–1028
113. Vesikari T (1993) Clinical trials of live oral rotavirus vaccines: the Finnish experience. Vaccine 11:255–261
114. Vesikari T (1997) Rotavirus vaccines against diarrhoeal disease. Lancet 350:1538–1541
115. Vesikari T, Isolauri E, Delem A et al (1983) Immunogenicity and safety of live oral attenuated bovine rotavirus vaccine strain RIT 4237 in adults and young children. Lancet 2:807–811
116. Vesikari T, Isolauri E, Delem A et al (1985) Clinical efficacy of the RIT 4237 live attenuated bovine rotavirus vaccine in infants vaccinated before a rotavirus epidemic. J Pediatr 107:189–194
117. Vesikari T, Isolauri E, D'Hondt E et al (1984) Protection of infants against rotavirus diarrhoea by RIT 4237 attenuated bovine rotavirus strain vaccine. Lancet 1:977–981
118. Vesikari T, Itzler R, Matson DO et al (2007) Efficacy of a pentavalent rotavirus vaccine in reducing rotavirus-associated health care utilization across three regions (11 countries). Int J Infect Dis 11:S29–S35
119. Vesikari T, Karnoven A, Ferrante SA et al (2010) Efficacy of the pentavalent rotavirus vaccine, RotaTeq®, in Finnish infants up to 3 years of age: the Finnish Extension Study. Eur J Pediatr 169:1379–1386
120. Vesikari T, Karvonen A, Prymula R et al (2007) Efficacy of human rotavirus vaccine against rotavirus gastroenteritis during the first 2 years of life in European infants: randomised, double-blind controlled study. Lancet 370:1757–1763
121. Vesikari T, Karvonen A, Puustinen L et al (2004) Efficacy of RIX4414 live attenuated human rotavirus vaccine in Finnish infants. Pediatr Infect Dis J 23:937–943
122. Vesikari T, Karvonen AV, Majuri J et al (2006) Safety, efficacy, and immunogenicity of 2 doses of bovine-human (UK) and rhesus-rhesus-human rotavirus reassortant tetravalent vaccines in Finnish children. J Infect Dis 194:370–376
123. Vesikari T, Matson DO, Dennehy P et al (2006) Safety and efficacy of a pentavalent human-bovine (WC3) reassortant rotavirus vaccine. N Engl J Med 354:23–33
124. Vesikari T, Ruuska T, Koivu HP et al (1991) Evaluation of the M37 human rotavirus vaccine in 2- to 6-month-old infants. Pediatr Infect Dis J 10:912–917
125. Vieira SCF, Gurgel RQ, Kirby A et al (2011) Acute diarrhoea in a community cohort of children who received an oral rotavirus vaccine in Northeast Brazil. Mem Inst Oswaldo Cruz 106:330–334
126. Weijer C (2000) The future of research into rotavirus vaccine. Benefits of vaccine may overweight risk for children in developing countries. BMJ 321:525–526
127. WHO (2009) Global advisory committee on vaccine safety report. Wkly Epidemiol Rec 84:213–236
128. WHO (2014) Update on intussusception following rotavirus vaccine administration: World Health Organization 2014. Available from: http:www.who.int/vaccine_safety/committee/topics/rotavirus/rotarix_and_rotateq/dec_2013/en/. Accessed 21 May 2016

129. Yen C, Armero Guardado JÁ, Alberto P et al (2011) Declines in rotavirus hospitalization and health care visits for childhooh diarrhea following rotavirus vaccination in El Salvador. Pediatr Infect Dis J 30:S6–S10
130. Yen C, Tate JE, Hyde TB et al (2014) Rotavirus vaccines: current status and future considerations. Hum Vaccin Immunother 10:1436–1448
131. Zade JK, Kulkarni PS, Desai SA et al (2014) Bovine rotavirus pentavalent vaccine development in India. Vaccine 32S:A124–A128
132. Zaman K, Dang DA, Victor JC et al (2010) Efficacy of pentavalent rotavirus vaccine against severe rotavirus gastroenteritis in infants in developing countries in Asia: a randomised, double-blind, placebo-controlled trial. Lancet 376:615–623

Chapter 22
Progress for Antiviral Development in Latin America

Viviana Castilla, Claudia S. Sepúlveda, Cybele C. García, and Elsa B. Damonte

1 Introduction

The search of novel effective antiviral agents for treatment of viral infections is a constant challenge for human health. Even when successful drugs have been developed and employed to combat well-known human pathogens, many current therapies face the difficulty of a high rate of genetic change exhibited by viruses, enabling the selection of drug-resistant mutants. This shift adds to the increasing periodic emergence of new viral pathogens or the reemergence of old ones lacking a reliable drug therapy. Therefore, the finding of new antiviral approaches is a continuously demanding effort.

In Latin America, antiviral research has been mainly focused on herpes simplex virus (HSV) as a global model system of antiviral studies because resistance to acyclovir, the currently standard antiherpetic agent, is still a therapeutic challenge, particularly for immunocompromised patients requiring prolonged treatment [110]. Viruses causing neglected diseases that represent a serious threat for public health in the region, such as dengue virus (DENV) [72]. Junín virus (JUNV), the agent of

V. Castilla
Departamento de Química Biológica, Facultad de Ciencias Exactas y Naturales, Universidad de Buenos Aires, Buenos Aires, Argentina

C.S. Sepúlveda
IQUIBICEN-CONICET, Ciudad Universitaria, Pabellón 2, Piso 4, 1428 Buenos Aires, Argentina

C.C. García • E.B. Damonte (✉)
Departamento de Química Biológica, Facultad de Ciencias Exactas y Naturales, Universidad de Buenos Aires, Buenos Aires, Argentina

IQUIBICEN-CONICET, Ciudad Universitaria, Pabellón 2, Piso 4, 1428 Buenos Aires, Argentina
e-mail: edamonte@qb.fcen.uda.ar

J.E. Ludert et al. (eds.), *Human Virology in Latin America*,
DOI 10.1007/978-3-319-54567-7_22

Argentine hemorrhagic fever [64], and Mayaro virus (MAYV), an emerging arbovirus causing outbreaks of a debilitating febrile illness in many regions of South America [59], among other ones, were also tested in several studies intending to solve local health problems.

Much investigation has been performed with crude extracts and pure compounds derived from the very abundant American natural sources (such as plants, sea algae, marine invertebrates). These products present interesting characteristics such as high chemical diversity, mild undesirable side effects, and, very important for our region, low cost. The collaboration of organic chemical and virological teams has also allowed the finding of synthetic compounds with promising antiviral properties. To complement traditional virus-targeted inhibitors, agents that act via host factors required for the virus life cycle have attracted increasing interest in recent years based on the possibility of obtaining a wide antiviral spectrum because different viruses may share a host factor.

This chapter summarizes the main contributions in different experimental approaches for antiviral development from scientists in Latin America in the last decade and the challenges and perspectives for the near future.

2 Natural Antiviral Compounds

2.1 Plant Products

Plants have been widely used to alleviate diverse human diseases, and they constitute an important source of novel antiviral molecules. Based on their ethnomedicinal use, numerous plants in South America have been selected to investigate the ability of derived extracts or essential oils to affect virus infectivity. Essential oils from different plant species displayed virucidal action against HSV-1 and HSV-2, yellow fever virus (YFV), DENV, and JUNV [60, 66, 92]. On the other hand, extracts derived from different plant families—*Asteraceae* [141, 150], *Euphorbiaceae* [17], *Fabaceae* [74], *Phyllanthaceae* [9, 58], *Aquifoliaceae* [81], *Meliaceae* [54], and *Verbenaceae* [122]—exhibited antiviral activity against different viruses including HSV-1, HSV-2, rotavirus (RV), astrovirus, poliovirus (PV), and influenza A virus (IAV).

An extract of *Achyrocline satureioides* showed anti-HSV-1 activity mainly related to the presence of flavonoid aglycones [18]. Moreover, the extract included in nanoemulsions proved to be an efficient delivery system for anti-HSV-1 topical application, improving the retention and accessibility of the active flavonoids in an animal model [10]. Other studies showed that soybean (*Glycine max*) isoflavonoids as well as C-glycosylflavonoids from *Cecropia glaziovii* exhibited antiviral action against HSV-1 and HSV-2 [10, 29]. Interestingly, antiviral activity against MAYV was described for flavonoids obtained from extracts of *Cassia australis* and *Bauhinia longifolia* [51, 137].

A partially purified extract (MA) from the leaves of *Melia azedarach* L. exhibited a potent antiviral effect against several RNA and DNA viruses. MA impaired DNA synthesis and virus assembly in HSV-1-infected cultures, whereas in vivo studies demonstrated that this active principle prevented the development of herpetic stromal keratitis (HSK) in mice and exhibited protective effect in a mouse model of genital HSV-2 infection [7, 104]. The tetranortriterpenoid 1-cinnamoyl-3,11-dihydroxymeliacarpin (CDM) was then identified as the molecule responsible for the broad spectrum of MA action. CDM blocked intracellular transport of viral HSV-1 glycoproteins and altered cytokine production in infected conjunctival and corneal cells by blocking NF-κB pathway activation [14, 23] and in HSV-stimulated macrophages by an NF-κB independent pathway [105].

Remarkably, diterpenes isolated from *Euphorbia laurifolia* and *E. lacteal* were able to induce human immunodeficiency virus (HIV) reactivation in an in vitro latency system, suggesting that these molecules would be useful, in combination with highly active antiretroviral therapy, to eradicate the pool of latently HIV-infected CD4+ T cells [13].

On the other hand, nordihydroguaiaretic acid (NDGA), the main metabolite of the creosote bush (*Larrea tridentata*), inhibited DENV infection by reduction of viral genome replication and inhibition of virion assembly by its ability to reduce the levels of cell lipid droplets, structures where virus assembly takes place [132].

Table 22.1 summarizes the antiviral spectrum of main plant-derived compounds studied by Latin American researchers and includes information about other molecules of plant origin such as polysaccharides that were chemically sulfated [49, 98], alkaloids [39, 76], cardenolides [16], and the antioxidant compound gallic acid [67].

2.2 *Marine Products*

Marine organisms also represent a formidable source of natural products with biological activities. In particular, in past decades, great interest has been placed in the search of antiviral compounds from marine macroalgae or seaweeds. Latin America has an important and diverse group of seaweed species, and the main algal-derived compounds studied in the region comprise sulfated polysaccharides (SP) and terpenoids.

SP can be obtained from the three main classes of seaweed: fucans and alginates from brown algae; agaroids, galactans, and carrageenans from red seaweeds; and heteropolysaccharides and ulvans from green seaweeds. The collaborative work of chemists and virologists from Argentina, Brazil, and Chile allowed the finding of potent antiviral SP from various American-located red, brown, and green seaweeds, as summarized in Table 22.1. Both purified SP and crude extracts were similarly able to block HSV and DENV replication in cell culture at concentrations as low as about 0.1–1 μg/ml without cell toxicity. Sulfated galactans and carrageenans (CGN), derived from red seaweeds, were the most effective in vitro viral inhibitors. Additionally, a significant protective effect of unformulated CGN against HSV-2

Table 22.1 Studies on antiviral activity of natural compounds

Compound	Source	Virus	References
Plants			
Flavonoid aglycones	*Achyrocline satureioides*	HSV-1	[18, 19]
Isoflavonoids	*Glycine max*	HSV-1 HSV-2	[10]
C-Glycosylflavonoids	*Cecropia glaziovii*	HSV-1 HSV-2	[129]
Flavonoids	*Cassia australis*	MAYV	[51, 137]
Tetranortriterpenoid	*Melia azedarach* L.	HSV-1 HSV-2	[14, 23, 105]
Diterpenes	*Euphorbia laurifolia*	HIV	[13]
Galactomannans (chemically sulfated)	*Mimosa scabrella* *Leucaena leucocephala*	YFV DENV-1	[98]
Seed polysaccharide (chemically sulfated)	*Adenanthera pavonina*	HSV-1 PV-1	[49]
Emetine (alkaloid)	*Psychotria ipecacuanha*	HIV-1	[39]
Alkaloid	*Maytenus ilicifolia*	HCV	[76]
Cardenolide	*Digitalis lanata*	HSV-1 HSV-2	[16]
Gallic acid	Various land and aquatic plants	HCV	[67]
Seaweeds			
Sulfated agarans	*Acanthophora spicifera*	HSV-1 HSV-2	[53]
Sulfated galactans	*Bostrychia montagnei,* *Schizymenia binderi,* *Cryptonemia seminervis*	HSV-1 HSV-2 HMPV	[52, 87] [91]
Sulfated hybrid D-L galactan	*Gymnogongrus torulosus,* *Cryptonemia crenulata*	HSV-1 HSV-2 DENV-2 DENV-3	[115, 138]
λ-, μ/ν, κ/ι-, κ-,ι-carrageenan	*Gigartina skottsbergii* *Hypnea musciformis*	HSV-1 HSV-2	[29, 42]
λ-, ι-carrageenan	*Gigartina* spp., *Eucheuma spinosa*	DENV-2 DENV-3 DENV-4	[139]
Hybrid carrageenan	*Gymnogongrus griffithsiae,* *Meristiella gelidium,* *Callophyllis variegata,* *Stenogramme interrupta*	HSV-1 HSV-2 DENV-2 DENV-3	[26, 119, 138, 142]
Fucoidans	*Leathesia difformis* *Adenocystis utricularis*	HSV-1 HSV-2 HIV-1	[57, 112, 147]
Sulfated fucan	*Laminaria abyssalis*	HTLV-1	[120]
Sulfated heterorhamnan	*Gayralia oxysperma*	HSV-1 HSV-2	[34]

(continued)

Table 22.1 (continued)

Compound	Source	Virus	References
Dolabellane diterpenes	*Dictyota pfaffii, D. menstrualis*	HSV-1 HIV-1	[1, 40, 93, 101, 103, 149]
Merodíterpenes	*Stypopodium zonale*	HSV-1 HMPV	[90, 130]
Glycolipid SQDG	*Osmundaria obtusiloba, Sargassum vulgare*	HSV-1 HSV-2	[50, 111]
Sponges			
Alkaloid	*Aaptos aaptos*	HSV-1	[133]
Halistanol fraction	*Petromica citrina*	HSV-1	[70]
Miscellaneous			
Sulfated glucomannan	*Agaricus brasiliensis*	HSV-1 HSV-2	[27, 28]
Aureonitol	*Chaetomium coarctatum*	IAV IBV	[123]
Protein-based metabolites	*Lactobacillus casei Bifidobacterium adolescentis*	RV	[97]
S-layer protein	*Lactobacillus acidophilus*	JUNV	[84]
Subtilosin	*Bacillus amyloliquefaciens*	HSV-1 HSV-2	[116, 146]
Phospholipases A_2	*Bothrops leucurus* *Crotalus durissus terrificus*	DENV Wide spectrum	[38] [96]
Protein	*Lonomia obliqua*	Wide spectrum	[31, 69]
Lactoferrin	Bovine secretions	MAYV	[33]

DENV dengue virus, *HCV* hepatitis C virus, *HIV* human immunodeficiency virus, *HMPV* human metapneumovirus, *HSV* herpes simplex virus, *HTLV* human T-cell lymphotropic virus, *IAV* influenza A virus, *IBV* influenza B virus, *JUNV* Junín virus, *MAYV* Mayaro virus, *RV* rotavirus, *YFV* yellow fever virus

infection was demonstrated in a murine genital model [30], a property that may be improved by an adequate formulation or a chemical modification to meliorate the pharmacokinetics.

Mechanistic studies demonstrated that the main target of CGN is virion adsorption to the host cell for HSV infection [29] together with a post-adsorption event blocking the viral nucleocapsid penetration into the cytoplasm for DENV [139]. These findings are consistent with the fact that SP structure resembles heparan sulfate chains present in proteoglycans of mammalian cells acting as the initial cell receptor for HSV [128] and DENV [36]. The specific mode of action of CGN was confirmed by the isolation of HSV- and DENV-resistant variants that showed phenotypic, genomic, and virulence alterations in comparison to the original virus [86, 140].

With DENV in particular, the antiviral studies with SP also provided new information about the mode of entry of DENV into the host cell. In fact, a variable level of anti-DENV effectiveness was reported according to the compound, the virus serotype, the host cell, and the source of virus [5, 138, 139]. This differential susceptibility led to pioneer studies that demonstrated for the first time the existence of alternative endocytic and cell trafficking pathways for the infective entry of DENV-1, DENV-2, and DENV-3 into mammalian cells [3, 4, 107], a phenomenon that must be considered for the development and evaluation of safe entry-targeted antiviral agents effective against all DENV serotypes.

Interestingly, effective antiviral activity in seaweed-derived SP was also reported against other human pathogenic viruses, such as HIV-1, human T-lymphotropic virus (HTLV), and human metapneumovirus (HMPV) (Table 22.1).

Brown seaweeds are one of the richest sources of diterpenes, a group of molecules widely used in medicine. Several dollabelane diterpenes isolated from the Brazilian brown algae *Dictyota pfaffii* and *D. menstrualis* showed antiviral activity against HSV-1 and HIV-1 (Table 22.1), by decreasing some early HSV-1 proteins [1] and affecting proviral DNA in primary HIV-1-infected cells [103], respectively. Enzymatic studies demonstrated that the diterpene dolabelladienetriol (THD) isolated from *D. pfaffii* was a typical noncompetitive inhibitor of HIV-1 reverse transcriptase (RT), blocking synthesis and integration of provirus. It was additive with AZT, synergistic with protease inhibitors, and did not exhibit crossed resistance with clinical available nonnucleoside RT inhibitors [40]. Molecular docking studies confirmed the interactions of THD with important residues in RT [93]. More recently, two new dolabelladienols were isolated with even more potent anti-HIV-1 activity than THD [101], indicating the excellent perspectives of these molecules as antiretroviral agents. Meroditerpenes, a class of polycyclic diterpenes obtained from *Stypopodium zonale*, were also strong inhibitors of HSV-1 and HMPV [90, 130].

Finally, glycolipids are a less studied class of antiviral seaweed secondary metabolites inspiring increasing interest in past years. Recent investigation has found promising anti-HSV activity in sulfoquinovosyldiacylglycerols (SQDG) isolated from the Brazilian seaweeds *Osmundaria obtusiloba* and *Sargassum vulgare* [50, 111].

Other marine organisms investigated for antiviral agents include sponges from the Brazilian coast. The alkaloid 4-methylaaptamine from the sponge *Aaptos aaptos* impaired HSV-1 penetration and immediate early protein synthesis [133], whereas the halistanol-enriched fraction from extract of *Petromica citrina* affected adsorption and penetration [70].

2.3 Other Miscellaneous Natural Products

Microorganisms are also an interesting source of antiviral agents. An SP obtained by sulfation of the glucomannan extracted from the mycelia of *Agaricus brasiliensis*, a basidiomycete fungus native to the Atlantic forest in Brazil, was a selective inhibitor of HSV-1 and HSV-2 and showed synergistic effect with acyclovir [27]. The antiherpetic efficacy of this SP was assessed in murine models of HSV as an oral agent and as a topical microbicide [28]. Aureonitol, a tetrahydrofuran derivative isolated from the fungus *Chaetomium coarctatum*, was an inhibitor of IAV and IBV by impairment of virus adsorption. Molecular modeling studies showed that this compound docked in the sialic acid-binding site of surface hemagglutinin (HA) [123].

In the search of treatment alternatives for viral diseases, the use of probiotic bacteria that provide a beneficial effect against multiple pathological agents has also been tested. Clinical trials have shown a significant reduction in duration of acute RV diarrhea when oral doses of *Saccharomyces boulardii* preparations were given to hospitalized children in Cochabamba, Bolivia [68], or Goiás, Brazil [41]. A recent in vitro study reported that protein-based metabolites of *Lactobacillus casei* and *Bifidobacterium adolescentis* were able to interfere with the intracellular amount of NSP4, a known RV enterotoxin, and Ca^{2+} liberation from the cells, suggesting a novel mechanism exerted by probiotics to reduce the impact of RV infection by preventing electrolyte loss [97]. The surface (S)-layer protein of *Lactobacillus acidophilus* was inhibitor of JUNV [84]. The inhibition was caused by the interaction between S-layer and DC-SIGN, a cell-surface adhesion factor that enhances entry of several viruses, indicating the potential wide-spectrum activity for this microbial protein. Other microbial peptides were also inhibitors of HSV-1 and HSV-2 by affecting late events of the viral cycle [116, 146].

Among natural products, venoms are complex mixtures that can provide clues for designing therapeutic molecules. This strategy has been used for the isolation of antiviral proteins from snakes and insects. Anti-DENV activity was reported for phospholipase A_2 isolated from *Bothrops leucurus* [38] and *Crotalus durissus terrificus* [96] venoms by virucidal effect, probably the result of glycerophospholipid cleavage and disruption of the lipid envelope. In addition, phospholipase A_2 inactivated other enveloped viruses, highlighting its potential as a natural leader for developing broad-spectrum antivirals. On the other hand, an antiviral protein effective against several human viruses was identified in the hemolymph of *Lonomia obliqua* caterpillars [69]. The protein did not display virucidal activity but probably acts by an intracellular mechanism that affects the innate antiviral immune response. The caterpillar *Lonomia obliqua* gained prominence in biotechnology in Brazil because of the various active properties identified in its venom and hemolymph, and a recombinant version of this protein was very effective to reduce HSV and rubella virus replication [31].

Bovine lactoferrin, a natural iron-binding glycoprotein found in various mucosal secretions important in the primary defense against diverse microorganisms, was another natural product with reported antiviral activity against MAYV by impairment of virus entry [33].

3 Synthetic Compounds

Several Latin American researcher groups have studied the antiviral activity of a great variety of synthetic molecules. Here we comment on the main findings of the past 10 years.

3.1 *Steroids*

Various studies have demonstrated the ability of steroids to inhibit in vitro and in vivo viral multiplication. Earlier reports described the antiviral activity of synthetic derivatives of brassinosteroids, plant steroids involved in growth and development, against arenaviruses, measles virus (MV), and HSV [35]. Synthetic analogues of the natural brassinosteroid 24(*S*) ethylbrassinone exerted a potent in vitro antiviral activity against arenaviruses, and studies about the mode of action of the most active derivative revealed that it would affect JUNV RNA synthesis [35]. This compound was also an inhibitor of HSV-1 replication (TK+ and TK− strains) that mainly prevented late viral protein [151]. Other antiherpetic brassinosteroid derivatives were able to ameliorate the signs of HSK in mice. Because no inhibition of virus titers in animal eyes was observed, the derivatives may play a role in immune-mediated stromal inflammation rather than an antiviral action [94]. According with these observations, two of the anti-HSV derivatives exhibited in vitro immunomodulatory activity modifying cytokine production in HSV-1-infected corneal, conjunctival, and nervous cells [95, 106]. Another set of synthetic sterol analogues with a diamide side chain also exhibited anti-HSV-1 activity [46].

Dehydroepiandrosterone (DHEA), one of the most abundant circulating steroid hormones in humans, displays a great variety of biological properties [35]; however, prolonged treatment with DHEA may cause masculinization in women, whereas epiandrosterone (EA), a DHEA metabolite, lacks androgenic activity. DHEA, EA, and their synthetic derivatives exhibited antiviral activity against JUNV, adenovirus type 5 (ADV5), and HSV-1 by preventing viral protein synthesis. However, it should be noted that natural compounds were more selective than the synthetic analogues assayed so far [2, 121, 145].

On the other hand, azasteroids derived from pregnenolone, an important endogenous steroid in mammals, showed antiviral activity against HSV-1, impairing the maturation of viral glycoproteins [47].

3.2 *Acridone Derivatives*

N-Substituted acridones exhibited selective antiviral activity against different arenaviruses and the four DENV serotypes. Mechanistic analysis of the most effective derivatives indicated that they caused a strong inhibition of viral RNA synthesis. The addition of exogenous guanosine partially reversed inhibition, suggesting that the reduction of the GTP pool contributed to acridone antiviral activity [88, 126].

On the other hand, structure–activity relationship studies on 1-hydroxyacridone derivatives with proved antiherpetic activity allowed the design and synthesis of new naphthyridine derivatives with promissory anti-HSV activity [12].

3.3 *Quinolinic Acid Derivatives*

An oxoquinolinic acid derivative with antiherpetic activity inhibited viral adsorption by impairing viral binding to HSV entry mediator (HVEM) receptor [134]. As this compound is a novel oxoquinolinic ribonucleoside, and taking into account the inhibitory effect of quinolone acyclonucleoside derivatives on HSV DNA polymerase [80], the inhibitory effect of the derivative on post-adsorption steps was analyzed. The results obtained indicated that the compound inhibited HSV DNA polymerase activity in a noncompetitive way [135]. In addition, this derivative was also active against HIV-1, and kinetic studies revealed that it inhibited HIV-1 RT in a dose-dependent manner and was additive with other RT inhibitors [136]. Interestingly, it has been demonstrated that new synthetic 1,2,3-triazolyl-4-oxoquinolines inhibited the multiplication of IAV and IBV, including those strains resistant to oseltamivir, the most used anti-influenza drug [20].

3.4 *Zinc-Finger Active Compounds*

Zinc-binding proteins with cysteine-rich Zn-finger motifs represent a potential target for antiviral chemotherapy. For example, the arenavirus Z protein, involved in the regulation of viral RNA synthesis and viral budding, possesses a Cys3HisCys4 RING-finger motif that coordinates two zinc ions. Evaluation of the antiviral action of a group of antiretroviral Zn-finger active compounds revealed that disulfides, as well as azoic and hydrazide derivatives, exhibited a broad range of antiviral activity against arenaviruses [125]. One of the active aromatic disulfides also displayed virucidal activity, and inactivated JUNV particles were unable to uncoat and perform viral RNA replication. Z protein appeared to be the inactivation target because treatment of viral particles with the compound induced Z-protein unfolding and oligomerization [61, 62]. In addition, this active compound affected the interaction of Z protein with host cell factors [63]. Another potent virucidal aromatic disulfide impaired JUNV RNA replication. Moreover, a virus-like particle (VLP) assay demonstrated that even though this compound caused the formation of high molecular weight multimers of Z protein, this alteration did not affect the ability of Z to drive budding of VLPs [127].

3.5 *Naphthoquinone Derivatives*

Nonstructural DENV proteins with enzymatic activities are promising targets to the development of anti-DENV compounds. In particular, the nonstructural protein 3 (NS3) has essential activities for viral replication such as protease, ATPase, and helicase activities. The evaluation of the anti-DENV activity of a library of synthetic naphthoquinones revealed that 1,4-pyran naphthoquinones are potent inhibitors of DENV-2 replication in cells and impact the in vitro NS3 ATPase activity [43]. Moreover, other naphthoquinone derivatives, aminomethylnaphthoquinones, were able to inhibit in vitro HSV-1 multiplication [109].

3.6 Acetylsalicylic Acid

Because salicylates, nonsteroidal antiinflammatory drugs that inhibit cellular cyclooxygenase (COX) activity, hinder the replication of several flaviviruses, the antiviral action of acetylsalicylic acid (ASA) on hepatitis C virus (HCV) replication was explored using an HCV subgenomic replicon system. This study showed that the reduction of viral RNA and protein levels after ASA treatment were related to the inhibitory effect on COX-2 expression [148]. The activation of inducible nitric oxide synthase (iNOS) by core and NS3 viral proteins has a major role in cell damage during chronic HCV infection. Remarkably, another study demonstrated that the modulation by iNOS partially mediated the antiviral activity of ASA [117].

The antiviral activity of porphyrins [12], terpenes [24], and 1α,25-dihydroxyvitamin D_3 [114] has also been studied. Major findings are summarized in Table 22.2.

4 Host–Target Antivirals

Licensed antiviral drugs in current use usually target viral proteins and are virus specific. However, a number of processes for virus multiplication within the infected cell that involve cellular pathways and enzymes also have proven to be attractive for chemotherapeutic intervention against several unrelated viruses. Besides the chance of a wide antiviral spectrum, this strategy offers the advantage of a significantly higher barrier to the emergence of drug resistance. Figure 22.1 illustrates host factors involved in virus multiplication analyzed as potential antiviral targets.

Several cellular proteins that promote viral infection have been evaluated as antiviral candidates. Such is the case of the heterogeneous nuclear ribonucleoproteins (hnRNPs) that participate in splicing, trafficking, translation, and turnover of mRNAs and participate in the life cycle of several cytoplasmic RNA viruses. Studies have confirmed the interaction between hnRNP A1 protein and the JUNV nucleoprotein N during acute and persistent infection [82]. Moreover, JUNV and DENV-2 infections induced the hnRNP K cytoplasmic translocation to improve viral multiplication. It is worth noting that the knockdown of these cellular proteins significantly reduced virus extracellular production [22]. Furthermore, molecular docking studies revealed that DENV NS1 protein interacts with glyceraldehyde-3-phosphate dehydrogenase (GAPDH), promoting its relocalization to the perinuclear region and its glycolytic activity [8]. Altogether, these interesting results show some perspectives to consider these proviral proteins as possible antiviral targets.

Table 22.2 Antiviral spectrum and mode of action of synthetic compounds

Compound	Antiviral spectrum	Mode of action	References
Brassinosteroid derivatives	HSV-1 HSV-2 JUNV PICV TCRV MV	Inhibition of viral protein and RNA synthesis. In vitro and in vivo immunomodulatory effect	[35, 94, 95, 106, 151]
Sterol analogues with a diamide side chain	HSV-1	Unknown	[46]
Dehydroepiandrosterone derivatives	JUNV PICV TCRV HSV-1 ADV	Inhibition of viral protein synthesis	[2, 121, 145]
Azasteroids derived from pregnenolone	HSV-1	Altered viral glycoprotein processing	[47]
N-Substituted acridones	JUNV TCRV LCMV DENV 1–4	Inhibition of viral RNA synthesis	[88, 126]
1-Hydroxyacridone derivatives	HSV-1	Unknown	[15]
Oxoquinolinic acid derivatives	HSV-1 HSV-2 HIV-1 IAV IBV	Inhibition of HSV-1 adsorption and HSV DNA polymerase Inhibition of HIV reverse transcriptase	[20, 80, 134–136]
Zinc-finger active compounds	JUNV PICV TCRV LCMV	Virucidal. Z-protein oligomerization and inhibition of viral uncoat and RNA synthesis	[61–63, 125]
1,4-Pyran naphthoquinones	DENV	Inhibition of ATPase activity of viral NS3 protein	[43]
Aminomethylnaphthoquinones	HSV-1	Unknown	[109]
Acetylsalicylic acid	HCV	Inhibition of cell enzyme expression (COX-2 and iNOS)	[117, 149]
Porphyrins	DENV YFV	Virucidal	[12]
1a,25-Dihydroxy-vitamin D_3	DENV	Unknown. In vitro immunomodulatory effect	[114]
Diterpenes	HSV-1	Unknown. In vitro immunomodulatory effect	[24]

ADV adenovirus, *DENV* dengue virus, *HCV* hepatitis C virus, *HIV* human immunodeficiency virus, *HSV* herpes simplex virus, *IAV* influenza A virus, *IBV* influenza B virus, *JUNV* Junín virus, *LCMV* lymphocytic choriomeningitis virus, *MV* measles virus, *PICV* Pichindé virus, *TCRV* Tacaribe virus, *YFV* yellow fever virus

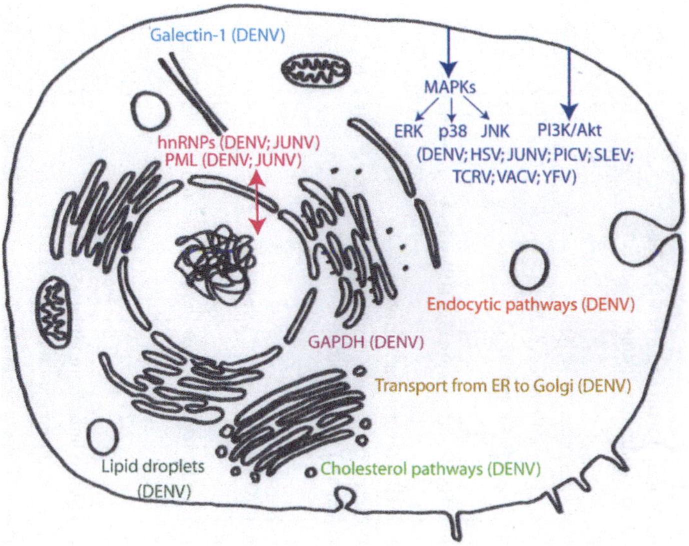

Fig. 22.1 Host factors involved in virus multiplication analyzed as potential antiviral targets. *DENV* dengue virus, *HSV* herpes simplex virus, *JUNV* Junín virus, *PICV* Pichindé virus, *SLEV* Saint Louis encephalitis virus, *TCRV* Tacaribe virus, *VACV* vaccinia virus, *YFV* yellow fever virus, *ER* endoplasmic reticulum, *ERK* extracellular signal-regulated kinases, *GAPDH* glyceraldehyde-3-phosphate dehydrogenase, *hnRNPs* heterogeneous nuclear ribonucleoproteins, *JNK* c-Jun N-terminal kinases, *MAPKs* mitogen-activated protein kinases, *PI3K/Akt* phosphatidylinositol-3-kinase/protein kinase B, *PML* promyelocytic leukemia protein

The antimalarial agent chloroquine, an amine acidotropic drug known to affect intracellular endocytic pathways by increasing endosomal pH, significantly reduced virus production and pro-inflammatory cytokine expression in DENV-2-infected cells [55]. Interestingly, the treatment of infected *Aotus* monkeys reduced the levels of viremia, pro-inflammatory cytokines tumor necrosis factor (TNF)-α and interferon (IFN)-γ, and systemic liver aspartate aminotransferase, suggesting that chloroquine effectively interferes with DENV-2 replication in this monkey model [56]. More importantly, it was detected that DENV-infected patients treated with chloroquine had an improvement in their quality of life, but there were no alterations in the length of the disease or degree and days of fever [21].

Other groups have focused their studies on the cell membranes and their components. Evaluation of the subcellular localization of the DENV capsid protein C showed that the cytoplasmic C protein accumulates around endoplasmic reticulum (ER)-derived structures called lipid droplets (LDs), which increase in number in DENV-infected cells, suggesting a link between lipid metabolism and viral replication [124]. The pharmacological manipulation of the amount of LDs in the cell to control DENV replication was postulated [75]. Also, cholesterol and lipid rafts present on the plasma membrane are important during DENV infection. Statins or other drugs that affect cholesterol biosynthesis can modulate DENV infection. The treatment of differentiated U937 cells with nystatin, filipin, or β-methyl cyclodextrin significantly reduced the antibody-dependent enhancement (ADE) of DENV-4 infection in vitro without any effect on viability [113]. Lovastatin (LOV) affected several cellular mechanisms, resulting in a moderate inhibition of DENV productive infection. When LOV was added before virus inoculation, virus entry was altered by a reduction in membrane cholesterol, and the addition of LOV after infection affected the transport from the ER to the Golgi apparatus and thus prevented the budding of viral progeny [85].

Another approach to be considered is the modulation of survival signaling pathways as a critical event in the viral replication cycle. Pharmacological and genetic inhibition of PI3K/Akt resulted in a significant decline of vaccinia virus (VACV) yields as well as in the apoptosis of infected cells, because viral receptor-mediated signals transmitted via this pathway are required for the expression of viral anti-apoptotic genes [131]. In line with these findings, the inhibition of the PI3K/Akt signaling pathway impaired JUNV protein synthesis and expression, leading to virus yield reduction without blocking the onset of the persistent stage of infection [77]. Furthermore, cell pretreatment with SP600125, a pharmacological inhibitor of c-JUN N-terminal kinase (JNK1/2), prevented poxvirus replication by blockade of virus-stimulated JNK phosphorylation and through another unknown mechanism [102]. Additionally, inhibition of JNK and p38 MAPK pathways in DENV-infected macrophages resulted in a significant reduction in viral yield and protein synthesis [37].

The Raf/MEK/ERK signaling pathway was also subject of study as an antiviral target. Pharmacological inhibition of MEK1/2 through U0126 cell treatment or pathway silencing by small interfering RNAs (siRNAs) blocked YFV- and JUNV-stimulated ERK1/2 phosphorylation [6, 118] and also inhibited the replication of DENV, Saint Louis encephalitis virus (SLEV), Tacaribe virus (TCRV), and Pichindé virus (PICV) [6, 118]. Nevertheless, DHEA, a known Raf/MEK/ERK activator, showed anti-HSV properties [145] that were not dependent in its ability to modulate ERK phosphorylation.

Another potential antiviral strategy is the manipulation of viral restriction factors, proteins constitutively expressed in the cells that viruses must block to enhance further replication. Galectin 1 (Gal-1) is a widely expressed mammalian lectin with several functions in cell–pathogen interactions and immunoregulatory mechanisms. DENV-1 infection inhibited the expression of Gal-1, and the overexpression of human recombinant Gal-1 (hrGal-1) inhibited DENV-1 production in different human cell lines [143]. Furthermore, the promyelocytic leukemia (PML) nuclear bodies (NBs) are discrete nuclear foci that contain several cellular proteins involved in intrinsic antiviral responses against several viruses. The JUNV and DENV-2 yields observed in PML-silenced cell cultures were significantly higher. However, the PML overexpression induced partial resistance to JUNV multiplication, but significantly reduced DENV-2 production [65, 78].

5 Novel Antiviral Strategies

Other experimental approaches aimed at unconventional targets or strategies for antiviral chemotherapy and represent new points for the development of novel effective therapies.

Many RNA and DNA viral pathogens, such as JUNV [11], YFV [100], HCV [32], HTLV-1 [73], and HSV-1 [45], have shown in vitro susceptibility to silencing with specific siRNAs. Only YFV and HSV-1 siRNAs were in vivo evaluated: two constructions protected the adult Balb/c mice against YFV challenge [100], whereas one against HSV UL-39 region reduced the herpetic encephalitis signs [45].

The use of bioinformatics tools, molecular modeling programs, and high-performance computing has been leading the process of design and in silico search of therapeutically useful molecules, an approach also applied by Latin American investigators. One of the strategies at the forefront of drug discovery is the virtual screening of databases. Molecular dynamics (MD) simulations were performed to explore the DENV-3 NS2B/NS3 protein binding-site flexibility, achieving identification of conformations of potential importance for virtual screening studies [48, 108]. Computer-based methods also allowed the design of peptides derived from IAV HA with high affinity for the target protein that were effective against several virus strains in cell culture [144]. Docking studies provided evidence that the active peptides bound to the HA suggesting that they might impair its fusion activity [79]. Besides HCV NS3 serine protease [44], thymidylate kinase from variola virus (VARV) [71], HSV-1 protease [89], and respiratory syncytial virus (RSV) proteins [149] were targets for the search of new inhibitory molecules using algorithms and computational methods.

On the other hand, plants provide a convenient system for peptide production, being cost-effective and easier to scale up to an industrial level than other platforms. RhoA, a small host GTPase located on the cell membrane that controls multiple cell functions, bound RSV-F protein mediating the virus-induced syncytium formation. A RhoA-derived peptide fused to carrier molecules from plants showed enhanced anti-RSV activity compared to the RhoA peptide alone [99]. Although additional evaluation in animal models is required, this study is relevant because carrier molecules would be important to facilitate in vivo peptide delivery and bioavailability. Another recent example of engineering production of antiviral peptides in plants is the HRA2pl peptide expressed in *Nicotiana tabacum* plants that inhibited the binding of HMPV to HEp-2 cells at the fusion stage [83].

The inhibition of viral proteases is a well-established way for preventing viral infection. By using the structural information on DENV NS2B/NS3 protein, a virtual screening was performed to find low molecular weight molecules that could be potentially employed in the treatment of DENV infection. This approach allowed the identification of three compounds that inhibited DENV protease activity and significantly reduced virus production [25]. This study constitutes one example of the use of bioinformatics methods, molecular modeling programs, and high-performance computing in the search for new specific therapeutic molecules.

6 Conclusions and Future Perspectives

As illustrated in this chapter, efforts in the search for new antiviral candidates in Latin America have increased significantly in the past decade. Given the continued emergence of new pathogenic agents, accessible antiviral therapy remains an urgent need, and, consequently, antiviral research has gained predominance in the Latin American scientific community. The extensive investigations carried out have led to the discovery of diverse products derived from natural sources or programmed synthesis with promising antiviral activity against several human pathogens relevant to regional public health.

However, there will be still a significant delay before drugs manufactured at low cost, as required in our countries, may be available for patient treatment. Particularly, most of the developed inhibitors await in vivo experimentation. To overcome the current limitations for in vivo validation of promising in vitro inhibitors, first in animal models and later in clinical trials, is a hard challenge to exploit the real potential that these agents may possess to be developed as chemotherapeutic drugs.

Noticeably, new structural insight into viral components and their interactions with cellular molecules during infection together with the recent progress in virtual screening and docking techniques in the region have opened a new avenue for structure-based design of viral inhibitors, a tendency with increasing perspectives to change the strategy for antiviral development in the near future.

Acknowledgments Research in the authors' laboratory was supported by Agencia Nacional de Promoción Científica y Tecnológica, Consejo Nacional de Investigaciones Científicas y Técnicas (CONICET), and Universidad de Buenos Aires (UBA), Argentina. C.S.S., C.C.G., and E.B.D. are members of Research Career from CONICET.

Conflict of Interest The authors declare that they have no conflict of interest.

References

1. Abrantes JL, Barbosa J, Cavalcanti D et al (2010) The effects of the diterpenes isolated from the Brazilian brown algae *Dictyota pfaffii* and *Dictyota menstrualis* against the herpes simplex type-1 replicative cycle. Planta Med 76:339–344
2. Acosta EG, Bruttomesso AC, Bisceglia J et al (2008) Dehydroepiandrosterone, epiandrosterone and synthetic derivatives inhibit Junín virus replication *in vitro*. Virus Res 135(2):203–212
3. Acosta EG, Castilla V, Damonte EB (2009) Alternative infectious entry pathways for dengue virus serotypes into mammalian cells. Cell Microbiol 11:1533–1549
4. Acosta EG, Castilla V, Damonte EB (2012) Differential requirements in endocytic trafficking for dengue virus penetration. PLoS One 7:e44835
5. Acosta EG, Piccini LE, Talarico LB et al (2014) Changes in antiviral susceptibility to entry inhibitors and endocytic uptake of dengue-2 virus serially passaged in Vero or C6/36 cells. Virus Res 184:39–43
6. Albarnaz JD, De Oliveira LC, Torres AA et al (2014) MEK/ERK activation plays a decisive role in yellow fever virus replication: implication as an antiviral therapeutic target. Antiviral Res 111:82–92

7. Alché LE, Barquero AA, Sanjuan NA, Coto CE (2002) An antiviral principle present in a purified fraction from *Melia azedarach* L. leaf aqueous extract restrains herpes simplex virus type 1 propagation. Phytother Res 16(4):348–352
8. Allonso D, Andrade IS, Conde JN et al (2015) Dengue virus NS1 protein modulates cellular energy metabolism by increasing glyceraldehyde-3-phosphate dehydrogenase activity. J Virol 89(23):11871–11883
9. Alvarez AL, del Barrio G, Kourí V et al (2009) *In vitro* anti-herpetic activity of an aqueous extract from the plant *Phyllanthus orbicularis*. Phytomedicine 16(10):960–966
10. Argenta DF, Silva IT, Bassani VL et al (2015) Antiherpes evaluation of soybean isoflavonoids. Arch Virol 160(9):2335–2342
11. Artuso MC, Ellenberg PC, Scolaro LA et al (2009) Inhibition of Junín virus replication by small interfering RNAs. Antiviral Res 84:31–37
12. Assunção-Miranda I, Cruz-Oliveira C, Neris RL et al (2016) Inactivation of dengue and yellow fever viruses by heme, cobalt-protoporphyrin IX and tin-protoporphyrin IX. J Appl Microbiol 120(3):790–804
13. Avila L, Perez M, Sanchez-Duffhues G et al (2010) Effects of diterpenes from latex of *Euphorbia lactea* and *Euphorbia laurifolia* on human immunodeficiency virus type 1 reactivation. Phytochemistry 71(2-3):243–248
14. Barquero AA, Michelini FM, Alché LE (2006) 1-Cinnamoyl-3,11-dihydroxymeliacarpin is a natural bioactive compound with antiviral and nuclear factor-kappaB modulating properties. Biochem Biophys Res Commun 344(3):955–962
15. Bernardino AM, Azevedo AR, Pinheiro LC et al (2012) Synthesis and anti-HSV-1 evaluation of new 3*H*-benzo[*b*]pyrazolo[3,4-*h*]-1,6-naphthyridines and 3*H*-pyrido[2,3-*b*] pyrazolo[3,4-*h*]-1,6-naphthyridines. Org Med Chem Lett 2(1):3
16. Bertol JW, Rigotto C, de Pádua RM et al (2011) Antiherpes activity of glucoevatromonoside, a cardenolide isolated from a Brazilian cultivar of *Digitalis lanata*. Antiviral Res 92(1):73–80
17. Betancur-Galvis LA, Morales GE, Forero JE, Roldan J (2002) Cytotoxic and antiviral activities of Colombian medicinal plant extracts of the Euphorbia genus. Mem Inst Oswaldo Cruz 97(4):541–546
18. Bettega JM, Teixeira H, Bassani VL et al (2004) Evaluation of the antiherpetic activity of standardized extracts of *Achyrocline satureioides*. Phytother Res 18(10):819–823
19. Bidone J, Argenta DF, Kratz J et al (2015) Antiherpes activity and skin/mucosa distribution of flavonoids from *Achyrocline satureioides* extract incorporated into topical nanoemulsions. Biomed Res Int 2015:238010
20. Boechat Fda C, Sacramento CQ, Cunha AC et al (2015) 1,2,3-Triazolyl-4-oxoquinolines: a feasible beginning for promising chemical structures to inhibit oseltamivir-resistant influenza A and B viruses. Bioorg Med Chem 23(24):7777–7784
21. Borges MC, Castro LA, Fonseca BA (2013) Chloroquine use improves dengue-related symptoms. Mem Inst Oswaldo Cruz 108(5):596–599
22. Brunetti JE, Scolaro LA, Castilla V (2015) The heterogeneous nuclear ribonucleoprotein K (hnRNP K) is a host factor required for dengue virus and Junín virus multiplication. Virus Res 203:84–91
23. Bueno CA, Alché LE, Barquero AA (2010) 1-Cinnamoyl-3,11-dihydroxymeliacarpin delays glycoprotein transport restraining virus multiplication without cytotoxicity. Biochem Biophys Res Commun 393(1):32–37
24. Bueno CA, Michelini FM, Pertino MW et al (2015) Natural and semisynthetic diterpenoids with antiviral and immunomodulatory activities block the ERK signaling pathway. Med Microbiol Immunol 204(5):575–584
25. Cabarcas-Montalvo M, Maldonado-Rojas W, Montes-Grajales D et al (2016) Discovery of antiviral molecules for dengue: *in silico* search and biological evaluation. Eur J Med Chem 110:87–97
26. Cáceres PJ, Carlucci MJ, Damonte EB et al (2000) Carrageenans from chilean samples of *Stenogramme interrupta* (Phyllophoraceae): structural analysis and biological activity. Phytochemistry 53:81–86

27. Cardozo FT, Camelini CM, Mascarello A et al (2011) Antiherpetic activity of a sulfated polysaccharide from *Agaricus brasiliensis* mycelia. Antiviral Res 92:108–114
28. Cardozo FT, Larsen IV, Carballo EV et al (2013) *In vivo* anti-herpes simplex virus activity of a sulfated derivative of *Agaricus brasiliensis* mycelial polysaccharide. Antimicrob Agents Chemother 57:2541–2549
29. Carlucci MJ, Ciancia M, Matulewicz MC et al (1999) Antiherpetic activity and mode of action of natural carrageenans of diverse structural types. Antiviral Res 43:93–102
30. Carlucci MJ, Scolaro LA, Noseda MD et al (2004) Protective effect of a natural carrageenan on genital herpes simplex virus infection in mice. Antiviral Res 64:137–141
31. Carmo AC, Yamasaki LH, Figueiredo CA et al (2015) Discovery of a new antiviral protein isolated from *Lonomia obliqua* analysed by bioinformatics and real-time approaches. Cytotechnology 67:1011–1022
32. Carneiro B, Braga AC, Batista MN et al (2015) Evaluation of canonical siRNA and Dicer substrate RNA for inhibition of hepatitis C virus genome replication: a comparative study. PLoS One 10(2):e0117742
33. Carvalho CA, Sousa IP Jr, Silva JL et al (2014) Inhibition of Mayaro virus infection by bovine lactoferrin. Virology 452-453:297–302
34. Cassolato JEF, Noseda MD, Pujol CA et al (2008) Chemical structure and antiviral activity of the sulfated heterorhamnan isolated from the green seaweed *Gayralia oxysperma*. Carbohydr Res 343:3085–3095
35. Castilla V, Ramírez J, Coto CE (2010) Plant and animal steroids a new hope to search for antiviral agents. Curr Med Chem 17(18):1858–1873
36. Castilla V, Piccini LE, Damonte EB (2015) Dengue virus entry and trafficking: perspectives as antiviral target for prevention and therapy. Future Virol 10:625–645
37. Ceballos-Olvera I, Chávez-Salinas S, Medina F et al (2010) JNK phosphorylation, induced during dengue virus infection, is important for viral infection and requires the presence of cholesterol. Virology 396:30–36
38. Cecilio AB, Caldas S, Oliveira RA et al (2013) Molecular characterization of Lys49 and Asp49 phospholipases A2 from snake venom and their antiviral activities against dengue virus. Toxins (Basel) 5:1780–1798
39. Chaves Valadão AL, Abreu CM, Dias JZ et al (2015) Natural plant alkaloid (emetine) inhibits HIV-1 replication by interfering with reverse transcriptase activity. Molecules 20(6):11474–11489
40. Cirne-Santos CC, Souza TM, Teixeira VL et al (2008) The dolabellane diterpene dolabelladienetriol is a typical noncompetitive inhibitor of HIV-1 reverse transcriptase enzyme. Antiviral Res 77:64–71
41. Corrêa NB, Penna FJ, Lima FM et al (2011) Treatment of acute diarrhea with *Saccharomyces boulardii* in infants. J Pediatr Gastroenterol Nutr 53:497–501
42. Cosenza VA, Navarro DA, Pujol CA et al (2015) Partial and total C-6 oxidation of gelling carrageenans. Modulation of the antiviral activity with the anionic character. Carbohydr Polym 128:199–206
43. da Costa EC, Amorim R, da Silva FC et al (2013) Synthetic 1,4-pyran naphthoquinones are potent inhibitors of dengue virus replication. PLoS One 8(12):e82504
44. da Cunha EF, Matos KS, Ramalho TC (2013) QSAR and docking studies of HCV NS3 serine protease inhibitors. Med Chem 9(6):774–805
45. da Silva AS, Raposo JV, Pereira TC et al (2016) Effects of RNA interference therapy against herpes simplex virus type 1 encephalitis. Antivir Ther 21(3):225–235
46. Dávola ME, Alonso F, Cabrera GM et al (2012) Sterol analogues with diamide side chains interfere with the intracellular localization of viral glycoproteins. Biochem Biophys Res Commun 427(1):107–112
47. Dávola ME, Mazaira GI, Galigniana MD et al (2015) Synthetic pregnenolone derivatives as antiviral agents against acyclovir-resistant isolates of herpes simplex virus type 1. Antiviral Res 122:55–63

48. de Almeida H, Bastos IM, Ribeiro BM (2013) New binding site conformations of the dengue virus NS3 protease accessed by molecular dynamics simulation. PLoS One 8(8):e72402
49. de Godoi AM, Faccin-Galhardi LC, Lopes N et al (2015) Characterization and antiherpetic activity of native and chemically sulfated polysaccharide from *Adenanthera pavonina*. Curr Pharm Biotechnol 16(11):1024–1031
50. de Souza LM, Sassaki GL, Romanos MT et al (2012) Structural characterization and anti-HSV-1 and HSV-2 activity of glycolipids from the marine algae *Osmundaria obtusiloba* isolated from Southeastern Brazilian coast. Mar Drugs 10:918–931
51. dos Santos AE, Kuster RM, Yamamoto KA et al (2014) Quercetin and quercetin 3-*O*-glycosides from *Bauhinia longifolia* (Bong.) Steud. show anti-Mayaro virus activity. Parasit Vectors 7:130
52. Duarte MER, Noseda DG, Noseda MD et al (2001) Inhibitory effect of sulfated galactans from the marine alga *Bostrychia montagnei* on herpes simplex virus replication *in vitro*. Phytomedicine 8:53–58
53. Duarte MER, Cauduro JP, Noseda MD et al (2004) The structure of the agaran sulfate from *Acantophora spicifera* (Rhodomelaceae, Ceramiales) and its antiviral activity. Relation between structure and antiviral activity in agarans. Carbohydr Res 339:335–347
54. Espada SF, Faccin-Galhardi LC, Rincao VP et al (2015) Antiviral activity of *Trichilia catigua* bark extracts for herpesvirus and poliovirus. Curr Pharm Biotechnol 16(8):724–732
55. Farias KJ, Machado PR, de Almeida RF Jr et al (2014) Chloroquine interferes with dengue-2 virus replication in U937 cells. Microbiol Immunol 58(6):318–326
56. Farias KJ, Machado PR, Muniz JA et al (2015) Antiviral activity of chloroquine against dengue virus type 2 replication in *Aotus* monkeys. Viral Immunol 28(3):161–169
57. Feldman SC, Reynaldi S, Stortz CA et al (1999) Antiviral properties of fucoidan fractions from *Leathessia difformis*. Phytomedicine 6:335–340
58. Fernández Romero JA, Del Barrio AG, Romeu Alvarez B et al (2003) *In vitro* antiviral activity of *Phyllanthus orbicularis* extracts against herpes simplex virus type 1. Phytother Res 17(8):980–982
59. Figueiredo ML, Figueiredo MT (2014) Emerging alphaviruses in the Americas: Chikungunya and Mayaro. Rev Soc Bras Med Trop 47:677–683
60. García CC, Talarico LB, Almeida NV et al (2003) Virucidal activity of essential oils from aromatic plants of San Luis, Argentina. Phytother Res 17(9):1073–1075
61. García CC, Djavani M, Topisirovic I et al (2006) Arenavirus Z protein as an antiviral target: virus inactivation and protein oligomerization by zinc finger-reactive compounds. J Gen Virol 87(Pt 5):1217–1228
62. García C, Ellenberg P, Artuso M et al (2009) Characterization of Junín virus particles inactivated by a zinc finger-reactive compound. Virus Res 143(1):106–113
63. García CC, Topisirovic I, Djavani M et al (2010) An antiviral disulfide compound blocks interaction between arenavirus Z protein and cellular promyelocytic leukemia protein. Biochem Biophys Res Commun 393(4):625–630
64. García CC, Sepúlveda CS, Damonte EB (2011) Novel therapeutic targets for arenavirus hemorrhagic fevers. Future Virol 6:27–44
65. Giovannoni F, Damonte EB, García CC (2015) Cellular promyelocytic leukemia p protein is an important dengue virus restriction factor. PLoS One 10(5):e0125690
66. Gómez LA, Stashenko E, Ocazionez RE (2013) Comparative study on *in vitro* activities of citral, limonene and essential oils from *Lippia citriodora* and *L. alba* on yellow fever virus. Nat Prod Commun 8(2):249–252
67. Govea-Salas M, Rivas-Estilla AM, Rodríguez-Herrera R et al (2016) Gallic acid decreases hepatitis C virus expression through its antioxidant capacity. Exp Ther Med 11(2):619–624
68. Grandy G, Medina M, Soria R et al (2010) Probiotics in the treatment of acute rotavirus diarrhoea. A randomized, double-blind, controlled trial using two different probiotic preparations in Bolivian children. BMC Infect Dis 10:253
69. Greco KN, Mendonça RM, Moraes RH et al (2009) Antiviral activity of the hemolymph of *Lonomia obliqua* (Lepidoptera: Saturniidae). Antiviral Res 84:84–90

70. Guimarães TR, Quiroz CG, Borges CR et al (2013) Anti HSV-1 activity of halistanol sulfate and halistanol sulfate C isolated from Brazilian marine sponge *Petromica citrina* (Demospongiae). Mar Drugs 11:4176–4192
71. Guimarães AP, de Souza FR, Oliveira AA et al (2014) Design of inhibitors of thymidylate kinase from Variola virus as new selective drugs against smallpox. Eur J Med Chem 91:72–90
72. Guzmán MG, Harris E (2015) Dengue. Lancet 385:453–465
73. Haddad R, Kashima S, Rodrigues ES et al (2011) Silencing of HTLV-1 gag and env genes by small interfering RNAs in HEK 293 cells. J Virol Methods 173(1):92–98
74. Herrero Uribe L, Chaves Olarte E, Tamayo Castillo G (2004) *In vitro* antiviral activity of *Chamaecrista nictitans* (Fabaceae) against herpes simplex virus: biological characterization of mechanisms of action. Rev Biol Trop 52(3):807–816
75. Iglesias NG, Mondotte JA, Byk LA et al (2015) Dengue virus uses a non-canonical function of the host GBF1-Arf-COPI system for capsid protein accumulation on lipid droplets. Traffic 16:962–977
76. Jardim AC, Igloi Z, Shimizu JF et al (2015) Natural compounds isolated from Brazilian plants are potent inhibitors of hepatitis C virus replication *in vitro*. Antiviral Res 115:39–47
77. Linero FN, Scolaro LA (2009) Participation of the phosphatidylinositol 3-kinase/Akt pathway in Junín virus replication *in vitro*. Virus Res 145:166–170
78. Linero FN, Sepúlveda CS, Giovannoni F et al (2012) Host cell factors as antiviral targets in arenavirus infection. Viruses 4:1569–1591
79. López-Martínez R, Ramírez-Salinas GL, Correa-Basurto J et al (2013) Inhibition of influenza A virus infection *in vitro* by peptides designed *in silico*. PLoS One 8(10):e76876
80. Lucero BD, Gomes CR, Frugulhetti IC et al (2006) Synthesis and anti-HSV-1 activity of quinolonic acyclovir analogues. Bioorg Med Chem Lett 16(4):1010–1013
81. Lückemeyer DD, Müller VD, Moritz MI et al (2012) Effects of *Ilex paraguariensis* A. St. Hil. (yerba mate) on herpes simplex virus types 1 and 2 replication. Phytother Res 26(4):535–540
82. Maeto CA, Knott ME, Linero FN et al (2011) Differential effect of acute and persistent Junín virus infections on the nucleo-cytoplasmic trafficking and expression of heterogeneous nuclear ribonucleoproteins type A and B. J Gen Virol 92:2181–2190
83. Márquez-Escobar VA, Tirado-Mendoza R, Noyola DE et al (2015) HRA2pl peptide: a fusion inhibitor for human metapneumovirus produced in tobacco plants by transient transformation. Planta (Berl) 242(1):69–76
84. Martínez MG, Prado Acosta M, Candurra NA, Ruzal SM (2012) S-layer proteins of *Lactobacillus acidophilus* inhibits JUNV infection. Biochem Biophys Res Commun 422:590–595
85. Martínez-Gutierrez M, Castellanos JE, Gallego-Gómez JC (2011) Statins reduce dengue virus production via decreased virion assembly. Intervirol 54:202–216
86. Mateu CG, Recalde MP, Artuso MC et al (2011) Emergence of herpes simplex virus-1 syncytial variants with altered virulence for mice after selection with a natural carrageenan. Sex Transm Dis 38:555–561
87. Matsuhiro B, Conte AF, Damonte EB et al (2005) Structural analysis and antiviral activity of a sulfated galactan from the red seaweed *Schizymenia binderi* (Gigartinales, Rhodophyta). Carbohydr Res 340:2392–2402
88. Mazzucco MB, Talarico LE, Vatansever S et al (2015) Antiviral activity of an *N*-allyl acridone against dengue virus. J Biomed Sci 22:29
89. Mello JF, Botelho N, Souza AM et al (2015) Computational studies of benzoxazinone derivatives as antiviral agents against herpesvirus type 1 protease. Molecules 20(6):10689–10704
90. Mendes GS, Soares AR, Sigiliano L et al (2011) *In vitro* anti-HMPV activity of meroditerpenoids from marine alga *Stypopodium zonale* (Dictyotales). Molecules 16:8437–8450
91. Mendes GS, Duarte ME, Colodi FG et al (2014) Structure and anti-metapneumovirus activity of sulfated galactans from the red seaweed *Cryptonemia seminervis*. Carbohydr Polym 101:313–323
92. Meneses R, Ocazionez RE, Martínez JR, Stashenko EE (2009) Inhibitory effect of essential oils obtained from plants grown in Colombia on yellow fever virus replication *in vitro*. Ann Clin Microbiol Antimicrob 8:8

93. Miceli LA, Teixeira VL, Castro HC et al (2013) Molecular docking studies of marine diterpenes as inhibitors of wild-type and mutants HIV-1 reverse transcriptase. Mar Drugs 11:4127–4143
94. Michelini FM, Berra A, Alché LE (2008) The *in vitro* immunomodulatory activity of a synthetic brassinosteroid analogue would account for the improvement of herpetic stromal keratitis in mice. J Steroid Biochem Mol Biol 108(1-2):164–170
95. Michelini FM, Zorrilla P, Robello C et al (2013) Immunomodulatory activity of an anti-HSV-1 synthetic stigmastane analog. Bioorg Med Chem 21(2):560–568
96. Muller VD, Soares RO, dos Santos NN Jr et al (2014) Phospholipase A2 isolated from the venom of *Crotalus durissus terrificus* inactivates dengue virus and other enveloped viruses by disrupting the viral envelope. PLoS One 9:e112351
97. Olaya Galán NN, Ulloa Rubiano JC, Velez Reyes FA et al (2016) *In vitro* antiviral activity of *Lactobacillus casei* and *Bifidobacterium adolescentis* against rotavirus infection monitored by NSP4 protein production. J Appl Microbiol 120:1041–1051
98. Ono L, Wollinger W, Rocco I et al (2003) *In vitro* and *in vivo* antiviral properties of sulfated galactomannans against yellow fever virus (BeH111 strain) and dengue 1 virus (Hawaii strain). Antiviral Res 60(3):201–208
99. Ortega-Berlanga B, Musiychuk K, Shoji Y et al (2016) Engineering and expression of a RhoA peptide against respiratory syncytial virus infection in plants. Planta (Berl) 243(2):451–458
100. Pacca CC, Severino AS, Mondini A et al (2009) RNA interference inhibits yellow fever virus replication *in vitro* and *in vivo*. Virus Genes 38(2):224–231
101. Pardo-Vargas A, de Barcelos OI, Stephens PR et al (2014) Dolabelladienols A-C, new diterpenes isolated from Brazilian brown alga *Dictyota pfaffii*. Mar Drugs 12:4247–4259
102. Pereira AC, Soares-Martins JA, Leite FG et al (2012) SP600125 inhibits orthopoxviruses replication in a JNK1/2-independent manner: Implication as a potential antipoxviral. Antiviral Res 93(1):69–77
103. Pereira HS, Leão-Ferreira LR, Moussatché N et al (2004) Antiviral activity of diterpenes isolated from the Brazilian marine alga *Dictyota menstrualis* against human immunodeficiency virus type 1 (HIV-1). Antiviral Res 64:69–76
104. Petrera E, Coto CE (2009) Therapeutic effect of meliacine, an antiviral derived from *Melia azedarach* L., in mice genital herpetic infection. Phytother Res 23(12):1771–1777
105. Petrera E, Coto CE (2014a) Effect of the potent antiviral 1-cinnamoyl-3,11-dihydroxymeliacarpin on cytokine production by murine macrophages stimulated with HSV-2. Phytother Res 28(1):104–109
106. Petrera E, Níttolo AG, Alché LE (2014b) Antiviral action of synthetic stigmasterol derivatives on herpes simplex virus replication in nervous cells *in vitro*. Biomed Res Int 2014:947560
107. Piccini LE, Castilla V, Damonte EB (2015) Dengue-3 virus entry into Vero cells: role of clathrin-mediated endocytosis in the outcome of infection. PLoS One 10:e1140824
108. Piccirillo E, Merget B, Sotriffer CA et al (2016) Conformational flexibility of DENV NS2B/NS3pro: from the inhibitor effect to the serotype influence. J Comput Aided Mol Des 30:251
109. Pires de Mello CP, Sardoux NS, Terra L et al (2016) Aminomethyl naphthoquinones and HSV-1: in vitro and in silico evaluations of potential antivirals. Antivir Ther. 21(6):507–515
110. Piret J, Boivin G (2011) Resistance of herpes simplex viruses to nucleoside analogues: mechanisms, prevalence and management. Antimicrob Agents Chemother 55:459–472
111. Plouguerné E, de Souza LM, Sassaki GL et al (2013) Antiviral sulfoquinovosyldiacylglycerols (SQDGs) from the Brazilian brown seaweed *Sargassum vulgare*. Mar Drugs 11:4628–4640
112. Ponce NMA, Pujol CA, Damonte EB et al (2003) Fucoidans from the brown seaweed *Adenocystis utricularis*: extraction methods, antiviral activity and structural studies. Carbohydr Res 338:153–165
113. Puerta-Guardo H, Mosso C, Medina F et al (2010) Antibody-dependent enhancement of dengue virus infection in U937 cells requires cholesterol-rich membrane microdomains. J Gen Virol 91:394–403

114. Puerta-Guardo H, Medina F, De la Cruz Hernández SI et al (2012) The 1α,25-dihydroxy-vitamin D3 reduces dengue virus infection in human myelomonocyte (U937) and hepatic (Huh-7) cell lines and cytokine production in the infected monocytes. Antiviral Res 94(1):57–61
115. Pujol CA, Estévez JM, Carlucci MJ et al (2002) Novel DL-galactan hybrids from the red seaweed *Gymnogongrus torulosus* are potent inhibitors of herpes simplex virus and dengue virus. Antivir Chem Chemother 13:83–89
116. Quintana VM, Torres NI, Wachsman MB et al (2014) Antiherpes simplex virus type 2 activity of the antimicrobial peptide subtilosin. J Appl Microbiol 117:1253–1259
117. Ríos-Ibarra C, Lozano-Sepúlveda S, Muñoz-Espinosa L et al (2014) Downregulation of inducible nitric oxide synthase (iNOS) expression is implicated in the antiviral activity of acetylsalicylic acid in HCV-expressing cells. Arch Virol 159(12):3321–3328
118. Rodríguez ME, Brunetti JE, Wachsman MB et al (2014) Raf/MEK/ERK pathway activation is required for Junín virus replication. J Gen Virol 95:799–805
119. Rodriguez MC, Merino EC, Pujol CA et al (2005) Galactans from cystocarpic plants of the red seaweed *Callophyllis variegata* (Kallymeniaceae, Gigartinales). Carbohydr Res 340:2742–2751
120. Romanos MT, Andrada-Serpa MJ, Mourão PA et al (2002) A sulphated fucan from the *Laminaria abyssalis* inhibits the human T cell lymphotropic virus type 1-induced syncytium formation in HeLa cells. Antivir Chem Chemother 13:219–221
121. Romanutti C, Bruttomesso AC, Castilla V et al (2010) Anti-adenovirus activity of epiandrosterone and dehydroepiandrosterone derivatives. Chemotherapy 56(2):158–165
122. Ruffa MJ, Wagner ML, Suriano M et al (2004) Inhibitory effect of medicinal herbs against RNA and DNA viruses. Antivir Chem Chemother 15(3):153–159
123. Sacramento CQ, Marttorelli A, Fintelman-Rodrigues N et al (2015) Aureonitol, a fungi-derived tetrahydrofuran, inhibits influenza replication by targeting its surface glycoprotein hemagglutinin. PLoS One 10:e139236
124. Samsa MM, Mondotte JA, Iglesias NG et al (2009) Dengue virus capsid protein usurps lipid droplets for viral particle formation. PLoS Pathog 5(10):e1000632
125. Sepúlveda CS, García CC, Damonte EB (2010) Inhibition of arenavirus infection by thiuram and aromatic disulfides. Antiviral Res 87(3):329–337
126. Sepúlveda CS, García CC, Fascio ML et al (2012) Inhibition of Junín virus RNA synthesis by an antiviral acridone derivative. Antiviral Res 93(1):16–22
127. Sepúlveda CS, García CC, Levingston Macleod JM et al (2013) Targeting of arenavirus RNA synthesis by a carboxamide-derivatized aromatic disulfide with virucidal activity. PLoS One 8(11):e81251
128. Shukla D, Spear PG (2001) Herpesviruses and heparan sulfate: an intimate relationship in aid of viral entry. J Clin Invest 108:503–510
129. Silva IT, Costa GM, Stoco PH et al (2010) *In vitro* antiherpes effects of a C-glycosylflavonoid-enriched fraction of *Cecropia glaziovii* Sneth. Lett Appl Microbiol 51(2):143–148
130. Soares AR, Abrantes JL, Lopes Souza TM et al (2007) *In vitro* antiviral effect of meroditerpenes isolated from the Brazilian seaweed *Stypopodium zonale* (Dictyotales). Planta Med 73:1221–1224
131. Soares JA, Leite FG, Andrade LG et al (2009) Activation of the PI3K/Akt pathway early during vaccinia and cowpox virus infections is required for both host survival and viral replication. J Virol 83(13):6883–6899
132. Soto-Acosta R, Bautista-Carbajal P, Syed GH et al (2014) Nordihydroguaiaretic acid (NDGA) inhibits replication and viral morphogenesis of dengue virus. Antiviral Res 109:132–140
133. Souza TM, Abrantes JL, de A Epifanio R et al (2007a) The alkaloid 4-methylaaptamine isolated from the sponge *Aaptos aaptos* impairs herpes simplex virus type 1 penetration and immediate-early protein synthesis. Planta Med 73:200–205
134. Souza TM, De Souza MC, Ferreira VF et al (2007b) The chloroxoquinolinic derivative 6-chloro-1,4-dihydro-4-oxo-1-(beta-D-ribofuranosyl) quinoline-3-carboxylic acid inhibits HSV-1 adsorption by impairing its adsorption on HVEM. Arch Virol 152(7):1417–1424

135. Souza TM, De Souza MC, Ferreira VF et al (2008) Inhibition of HSV-1 replication and HSV DNA polymerase by the chloroxoquinolinic ribonucleoside 6-chloro-1,4-dihydro-4-oxo-1-(beta-D-ribofuranosyl) quinoline-3-carboxylic acid and its aglycone. Antiviral Res 77(1):20–27
136. Souza TM, Rodrigues DQ, Ferreira VF et al (2009) Characterization of HIV-1 enzyme reverse transcriptase inhibition by the compound 6-chloro-1,4-dihydro-4-oxo-1-(beta-D-ribofuranosyl) quinoline-3-carboxylic acid through kinetic and *in silico* studies. Curr HIV Res 7(3):327–335
137. Spindola KC, Simas NK, Salles TS et al (2014) Anti-Mayaro virus activity of *Cassia australis* extracts (Fabaceae, Leguminosae). Parasit Vectors 7:537
138. Talarico LB, Pujol CA, Zibetti RG et al (2005) The antiviral activity of sulfated polysaccharides against dengue virus is dependent on virus serotype and host cell. Antiviral Res 66:103–110
139. Talarico LB, Damonte EB (2007) Interference in dengue virus adsorption and uncoating by carrageenans. Virology 363:473–485
140. Talarico LB, Damonte EB (2016) Characterization of *in vitro* dengue virus resistance to carrageenan. J Med Virol 88:1120–1129
141. Téllez MA, Téllez AN, Vélez F, Ulloa J (2015) *In vitro* antiviral activity against rotavirus and astrovirus infection exerted by substances obtained from *Achyrocline bogotensis* (Kunth) DC. (Compositae). BMC Complement Altern Med 15(1):428
142. Tischer PC, Talarico LB, Noseda MD et al (2006) Chemical structure and antiviral activity of carrageenans from *Meristiella gelidium* against herpes simplex and dengue virus. Carbohydr Polym 63:459–465
143. Toledo KA, Fermino ML, Andrade Cde C et al (2014) Galectin-1 exerts inhibitory effects during DENV-1 infection. PLoS One 9(11):e112474
144. Torres J, López R, Cetina A et al (2012b) Prediction of antiviral peptides derived from viral fusion proteins potentially active against herpes simplex and influenza A viruses. Bioinformation 8(18):870–874
145. Torres NI, Castilla V, Bruttomesso AC et al (2012a) *In vitro* antiviral activity of dehydroepiandrosterone, 17 synthetic analogs and ERK modulators against herpes simplex virus type 1. Antiviral Res 95:37–48
146. Torres NI, Noll KS, Xu S et al (2013) Safety, formulation, and *in vitro* antiviral activity of the antimicrobial peptide subtilosin against herpes simplex virus type 1. Probiotics Antimicrob Proteins 5:26–35
147. Trinchero J, Ponce NM, Córdoba OL et al (2009) Antiretroviral activity of fucoidans extracted from the brown seaweed *Adenocystis utricularis*. Phytother Res 23:707–712
148. Trujillo-Murillo K, Rincón-Sánchez AR, Martínez-Rodríguez H et al (2008) Acetylsalicylic acid inhibits hepatitis C virus RNA and protein expression through cyclooxygenase 2 signaling pathways. Hepatology 47(5):1462–1472
149. Vargas JE, Puga R, de Faria PJ et al (2015) A network flow approach to predict protein targets and flavonoid backbones to treat respiratory syncytial virus infection. Biomed Res Int 2015:301635
150. Visintini Jaime MF, Redko F, Muschietti LV et al (2013) *In vitro* antiviral activity of plant extracts from *Asteraceae* medicinal plants. Virol J 10:245
151. Wachsman MB, Castilla V, Talarico LB et al (2004) Antiherpetic mode of action of (22*S*,23*S*)-3β-bromo-5α,22,23-trihydroxystigmastan-6-one *in vitro*. Int J Antimicrob Agents 23(5):524–526

Index

J.E. Ludert et al. (eds.), *Human Virology in Latin America*,
DOI 10.1007/978-3-319-54567-7

B